Concepts and Applied Principles of Nitroxides

Concepts and Applied Principles of Nitroxides

Edited by **Saul Rowen**

NY RESEARCH PRESS
New York

Published by NY Research Press,
23 West, 55th Street, Suite 816,
New York, NY 10019, USA
www.nyresearchpress.com

Concepts and Applied Principles of Nitroxides
Edited by Saul Rowen

International Standard Book Number: 978-1-63238-095-1 (Hardback)

Printed in China.

Contents

Preface

Every book is a source of knowledge and this one is no exception. The idea that led to the conceptualization of this book was the fact that the world is advancing rapidly; which makes it crucial to document the progress in every field. I am aware that a lot of data is already available, yet, there is a lot more to learn. Hence, I accepted the responsibility of editing this book and contributing my knowledge to the community.

This book presents an elaborative account of information on nitroxide (aminoxyl) radicals, which have come to emerge as one of the most studied and developing topics of modern chemical physics. It has found critical uses in biophysics, molecular biology, polymer sciences and medical advancements. This book has been compiled with research works by experts renowned in the sphere of spin label/probe technique. The authors have presented an in-depth study of the latest trends and functions of nitroxide radicals. This book is an important source of academic literature for chemists, physical chemists, biophysicists, biologists, physicians and other masters in diverse disciplines. It will also be helpful in using nitroxide applications for promoting the progress in natural sciences.

While editing this book, I had multiple visions for it. Then I finally narrowed down to make every chapter a sole standing text explaining a particular topic, so that they can be used independently. However, the umbrella subject sinews them into a common theme. This makes the book a unique platform of knowledge.

I would like to give the major credit of this book to the experts from every corner of the world, who took the time to share their expertise with us. Also, I owe the completion of this book to the never-ending support of my family, who supported me throughout the project.

Editor

Section 1

Theory

History of the Use of Nitroxides (Aminoxyl Radicals) in Biochemistry: Past, Present and Future of Spin Label and Probe Method

Lawrence J. Berliner

Additional information is available at the end of the chapter

1. Introduction

The perspective of this chapter is very much historical. The author was fortunate enough to have begun his graduate studies at the very inception of the technique of spin labeling. Mind you, the topic of this book is nitroxides a.k.a. aminoxyl radicals which in fact preceded the spin labeling method and its inception. Hence the chapter will cover a history of the synthetic developments with nitroxides, the history of the development of spin labels, and the use of nitroxides and will provide an overview to the future of its applications. The intent is to cover the very beginning and then discuss some of the key areas (always dominated by synthetic organic chemistry) that allowed this technique to blossom more and more. Needless to say, while the definition of spin labeling is the incorporation of a stable, free radical into a macromolecular system of choice, we have yet to find anything (aside from the trityl radicals) that will fulfill this purpose. And the trityl radicals, which may be covered briefly in this chapter, give no dynamic or structural information whatsoever as they yield a single narrow line spectrum. The author feels confident to discuss these areas since he was deeply involved in the synthetic organic chemistry of these compounds, frequently repeating the syntheses of the basic starting compounds that were either commercially unavailable or quite expensive at the time. This includes the synthesis of phorone/triacetonamine from ammonia, acetone and calcium, in a 1898 synthesis[1], which was the basis for the nitroxide TEMPONE(2,2,6,6-tetramethylpiperidinone 1-oxyl).

Perhaps one of the earliest papers describing 'nitroxides' was from the American Cyanamid Company laboratories about the reaction of t-nitrobutane with metallic sodium[2]. They found a g value of 2.0065 and a single line linewidth of 8.5G (probably because they observed the compound in neat form where exchange and dipolar broadening were predominant). A follow-up publication produced a plethora of compounds derived from phenyl derivatives. They were able to measure a hyperfine coupling constant for DTBN of 15.25G [3]. We should recall that the earliest example of these radicals was the famous Fremy's salt, used to calibrate EPR machines to this day. This long-lived free radical, shown below, was discovered in 1845 by Edmond Frémy [4].

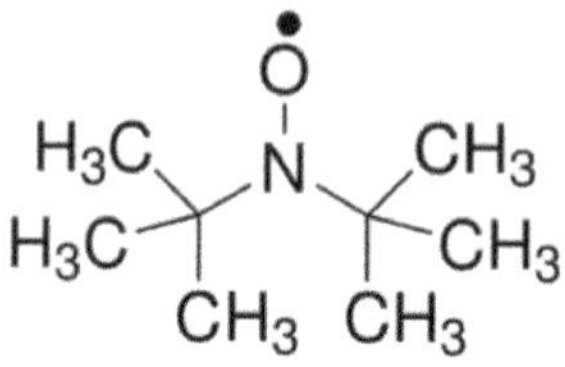

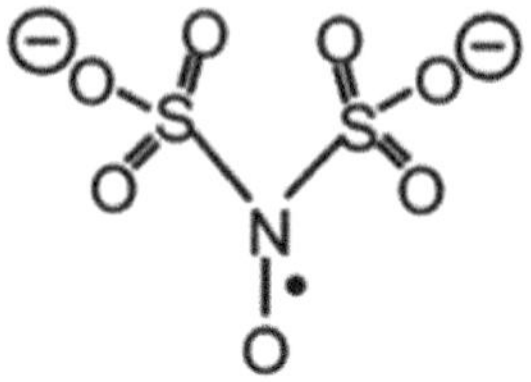

Figure 1. (left) di-t- butylnitroxide[2];(right) Fremy's salt [1]

In the 1960s, stable paramagnetic compounds were developed extensively in the USSR Academy of Sciences, Institute of Chemical Physics, that contained aminoxyl (or iminoxyl or nitroxyl or nitroxide) 'reporter' groups. Until these compounds became commercially available, one was obligated to prepare them homemade, but their syntheses were fairly straightforward (starting with the either phorone or triacetoneamine) [1]. The Russian group was led by organic chemists M.B. Neiman and E.G. Rozantsev and the group expanded these syntheses into a broad range of compounds, some of which could be applied as protein modification reagents [5-6].

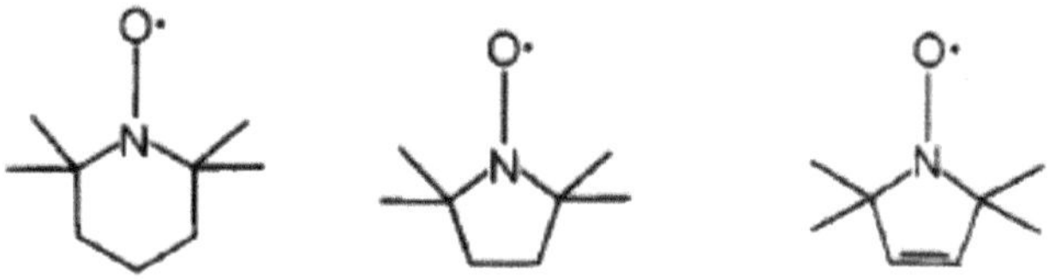

Figure 2. Piperidine, pyrrolidine and pyrroline nitroxides.

Let us not overlook the tremendous advantages of nitroxides that contribute to their versatility in the study of (biological) macromolecules i.e., they are very stable in most solvents over a wide range of pH values. The paramagnetic N-O bond moiety is quite tolerant to various synthetic conditions, specifically those in the tetramethyl flanked piperidine, pyrroline or pyrrolidine rings. Freezing, thawing, distilling or boiling usually impart no adverse affects on their stability, ie, paramagnetism is retained. Since EPR does not require optical transparency, and is not sensitive to magnetic susceptibility effects (which plagues NMR), one can work in opaque solutions, solids or mixtures. And the EPR sensitivity is 600- 700 times higher per spin compared with a proton in NMR. Thus, with very narrow linewidth spectra, one could detect nitroxide spectra in solution down to nanomolar levels with high sensitivity cavities. The EPR spectral lineshape reflects nitroxide tumbling motion, hence one can distinguish freely tumbling, 'unattached' or unreacted label in a sample with other bound species. The only real drawback of spin labels is their susceptibility to reduction to the corresponding diamagnetic hydroxylamine in the presence of organic or biological reducing agents, which will be addressed under in-vivo studies. Yet, where some synthetic recipes may utilize e.g., $NaBH_4$ which reduces the N-O moiety, the radical is easily regenerated in mild H_2O_2 or exposure to O_2.

Figure 3. First meeting of L. J. Berliner with E. G. Rozantsev, USSR Institute of Chemical Physics, 1979

2. Horrible, terrible nomenclature: IUPAC versus 'common usage'

Spin labels are commonly called 'nitroxides,' also the title of this book. In addition the terms iminoxyl or nitroxyl have been used as well as the occasional use of the term aminoxyl. Yet IUPAC RNRI Rule RC-81.2.4.D defines compounds with the structure $R_2NO·$ as 'radicals derived from hydroxylamines by removal of the hydrogen atom from the hydroxy group, and they are in many cases isolable.' While Chemical Abstracts Service uses nitroxide as the parent name for $H_2N–O·$, e.g., $(ClCH_2)\ _2N–O·$ or bis(chloromethyl) nitroxide, the IUPAC name is bis(chloromethyl) aminoxyl. It is correct to state that nitroxide should not be used as a name of a class of compounds that are specifically and correctly (a la IUPAC) aminoxyl radicals. As for the use of iminooxy or iminoxyl radicals, this has been used incorrectly for alkylideneaminoxyl radicals (also called iminoxyl radicals, $R_2C=N–O·$. Its use is strongly discouraged. In the Sigma/Aldrich catalog, the spin probe TEMPO is listed as 2,2,6,6-tetramethylpiperidine 1-oxyl.

Hence, the most inappropriate term for these radicals, nitroxides, has been used most widely and, as of 2009, has been cited about 115,000 times, nitroxyl about 29,500 times, iminoxyl (initiated by E.G. Rozantsev and coworkers) about 4,150 times. Aminoxyl, the most correct, has been cited 3,910 times. Obviously, the term nitroxyl is way out of line, pushed only bnefly by the late Andre Rassat, but is not relevant to this class of radicals (although I have two colleagues who continue to propogate this misuse!) I recall a friendly conversation with my long time colleague, Jim Hyde, who emphasized that if it becomes common usage, it's here so stay and to just give up on the issue. However when we academics teach organic chemistry to our young students, we try to imbue them with the correct terminology. Furthermore, standard states and nomenclature were designed so that scientists in the world can understand one another. It is clear that the correct nomenclature that the spin-label community should be using is aminoxyl radicals. It would be great if, from this point in our history moving forward, we might correct this error in the future and abide by the IUPAC rules.

3. Early applications to studying subtle aspects of protein/enzyme structure

The spin label method is a reporter group technique, a concept in the 1960s [7], as depicted in Fig. 4.

Figure 4. Schematic representation of an enzyme-substrate complex in native protein (top left), protein containing reporter group (solid black area) adjacent to substrate binding area (top right), and reporter group distant from substrate binding area (bottom right). From [7] with permission

The revolutionary developments in organic synthesis of nitroxide spin labels helped us overcome a major challenge for biochemical studies, where the plan was to attempt to fit the spin label to the biological system in as subtle a manner as possible, that is to "fool" the system into thinking it was binding to, or interacting with, a real, natural substrate or cofactor. One of the early attempts in the McConnell laboratory was the synthesis, a nitrophenylester of 1-carboxyl-2,2,5,5-tetramethylpyrollidine, depicted in Figure 5 (left), so that one could take advantage of the esterase activity of the enzyme α–chymotrypsin and virtually 'hook' the enzyme during its action on the compound. The approach was to isolate the active acyl-enzyme intermediate, covalently attached at serine 195, which is where the intermediate in the enzyme catalyzed hydrolysis resides. Indeed, the spin labeled acyl-enzyme intermediate reflected a tightly bound (possibly rigid, uniquely oriented) spin label at the active site [8]. However, it became much more difficult when one wanted details as to how the enzyme handled this spin labeled substrate analog and business area to do single crystal studies of the spin labeled chymotrypsin order to derive information about the precise orientation of the label. This could be gleaned from knowledge of the anisotropic hyperfine constants and anisotropic g-factors. One could determine orientation of the N-O bond and work backwards to find the orientation of the nitroxide five-membered ring on the protein. With knowledge of the x-ray structure of α-chymotrypsin and its reactive intermediate structures already known, the process was straightforward, Bauer and Berliner were able to obtain individual binding orientation of the R-and S-enantiomers of this particular house substrate panel and from that understand why the more slowly released enantiomer substrate acyl group was mis-oriented at the active site making hydrolysis by an activated water molecule quite difficult [9].

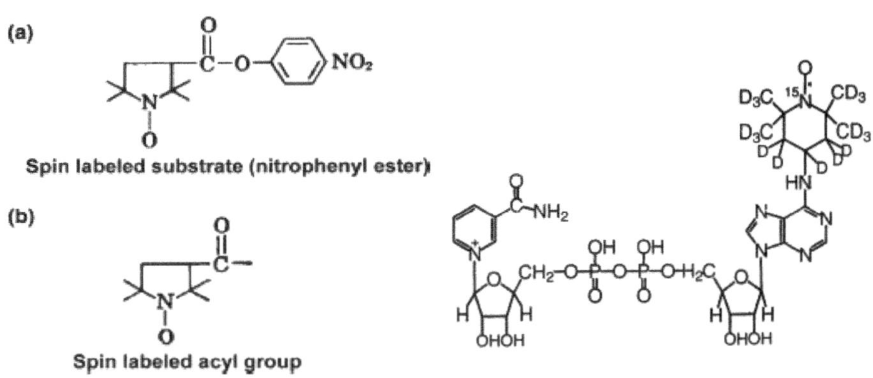

Figure 5. (left). Spin-labeled substrate (R,S)-2,2,5,5-tetramethyl-3-carboxy-pyrrolidine-p-nitrophenyl ester (a). The acyl-nitroxide group that is covalently linked to Ser 195 of α-chymotrypsin (b). It was later found that the "specificity" for a spin-labeled acyl-chymotrypsin was the S- enantiomer, although both enantiometers can be isolated as acyl enzymes. (right) Chemical structure of SL-NAD$^+$,nicotinamide N^6-([^{15}N,^{2}H$_{17}$]2,2,6,6-tetramethylpiperidine-4-yl-1-oxyl)adenine nucleotide [10]

The real sophistication came in studies of enzymes that bound nucleotide analogues or, in fact, DNA and nucleotide complexes. In some beautiful work Trommer and colleagues synthesized NAD analog, SL-NAD$^{+,}$ where the nitroxide ring was fused onto the nicotinamide, the structure, shown above in Figure 5b (right) [10]. The enzyme bound very tightly to this NAD analog and its precise orientation could be determined. What was interesting was that in the example glyceraldehyde phosphate dehydrogenase, a tetrameric enzyme that binds one NAD per subunit in each tetramer, the distance between two NAD spin labeled analogs could be determined from the electron-electron dipole interaction. This was the first example of distance measurements involving two spin labels within a protein structure and, due to the fortuitous situation of a perfectly, rigidly bound spin label, distances could be determined precisely [10]. This study still remains the gold standard of distance measurements by electron-electron dipolar interactions.

4. Lipid spin probes (oxazolidinyl or doxyl, proxyl)

The development of spin labeling and spin probes expanded to lipids and membranes. In order to probe these biological structures, one needs a label that mimics or looks like a lipid and can be incorporated into a phospholipid membrane structure. As late as the late 1960s, one could only prepare an ester of a fatty acid with one of piperidine or pyrrolidine nitroxides, but one could not incorporate a probe somewhere in the middle of the lipid chain in order to probe various depths of a membrane. It was not until John Keana demonstrated that one can incorporate an oxazolidine ring at specifically placed ketone (keto) groups in a lipid, resulting in a rigid five membered ring fused to the lipid chain that was easily oxidized to the radical nitroxide (doxyl) [11]. This virtually led to a revolution in our ability to probe membrane structure and dynamics with structural and dynamic accuracy. Several of these compounds are shown below. It took a while before these were commercially available, however, the synthesis was reasonably straightforward and scientists in the area were willing to share their compounds with one another. This was a clear departure from the relatively straightforward chemistry of the piperidine, pyrroline and pyrrolidine based aminoxyl radicals that had been developed by the Russian groups up to that point. The synthesis is relatively straightforward: take a lipid of interest, which can be purchased as a halo-derivative or occasionally as the desired keto derivative. Then the oxazolidine ring is formed at this position on the chain, then oxidized to the radical. Spin labeled lipid probes became available with aminoxyl radical group at the 5-, 12-, or 16 position in the lipid chain, and later at other positions. The resulting biochemistry, i.e., to incorporate these lipid nitroxides at either the 2- or 3- position of a phospholipids, was fairly straightforward as the fatty acid interchange or ester interchange chemistry was already well known. The synthetic schemed and some example probes are shown below, with a phospholipid analog in Figure 6.

Some years later, the problem of the oxazolidine ring being essentially reversible, (i.e. hydrolyzable), a newer development involved the incorporation directly of a five-membered ring (proxyl) into the structure of a lipid molecule at a strategically placed double bond. The chemistry again was somewhat sophisticated but straightforward; the synthetic route (Figure 7) leads to a side-chain-substituted 2,2,5,5-tetramethylpyrrolidine-N-oxyl (proxyl) nitroxide lipid spin probes from a commercially available nitrone is treated with an organometallic reagent, which after Cu^{2+}-air oxidation gives a new intermediate nitrone, followed by a second selected organometallic reagent, which after Cu^{2+}-air oxidation yields the proxyl spin probe [11]. The advantage of proxyl chemistry over oxazolidine chemistry in order to make lipid spin probes was that one could tailor the orientation of the N-O group with respect to the lipid axis. This became important since the hyperfine coupling constant along the z-axis of the label (i.e. directly above and perpendicular to the N-O plane) yielded a large splitting, upwards of 32G, that allowed a quite accurate estimate of the orientation, order parameter and dynamics of this portion of the lipid spin probe within the membrane.

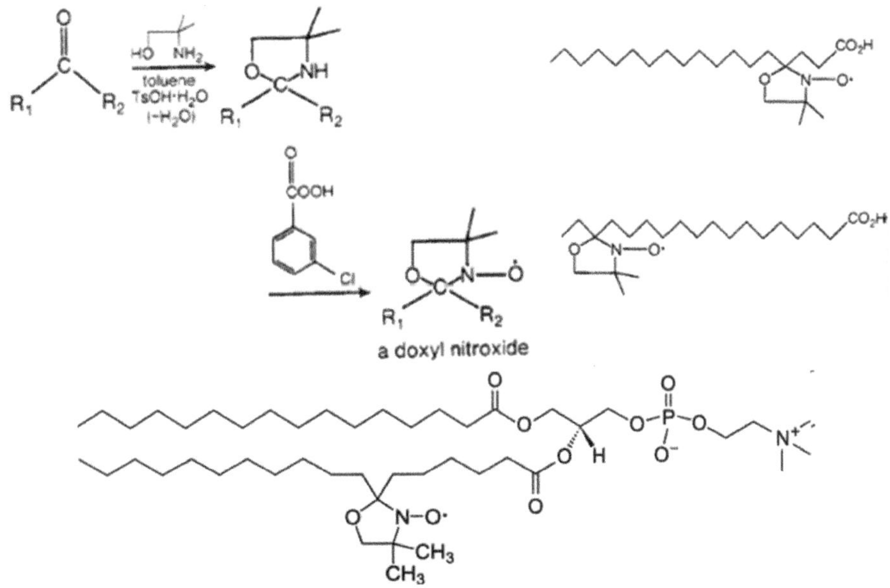

Figure 6. 16:0-7 Doxyl PC 1-palmitoyl-2-stearoyl-(7-doxyl)-sn-glycero-3-phosphocholine.

Even more rigid lipid probes were possible with the advent of racemic azethoxyl lipid probes nitroxides (called minimum steric perturbation spin labels). In the azethoxyl the nitrogen atom is actually embedded in the hydrocarbon chain. Cis-trans isomerism is possible and modeling suggests that the trans isomer should resemble a saturated lipid, whereas the cis isomer introduces a bend in the chain which approximates that observed with a cis carbon-carbon double bond.

The general synthetic route to the azethoxyl nitroxide spin labels is similar to that of the proxyl nitroxides, except that a different nitrone is used in the beginning (Figure8), where in this specific example, the trans isomer predominates [11].

Synthetic development was also carried out by several chemists in Ljubljana, Slovenia, as well as other synthetic organic labs, all of which were principally in Eastern Europe. In the U.S.A., the plight of an organic chemist attempting to obtain tenure in an academic department required the synthesis of complex natural products for the development of new synthetic reactions. Frequently the synthetic procedures for preparing these aminoxyl radicals, spin labels or spin probes were albeit modern but not new and novel; the organic chemist simply adapted the new, clever synthetic procedures to obtain the required label. It wasn't until the late 1990s, or perhaps the new millennium, where chemistry departments accepted applied chemistry as a valid academic area of new ideas and novel techniques. Certainly, it was the synthetic organic chemist who solved this problem and, for that matter, most biophysical studies involving probes depend on clever synthetic abilities. EPR had a great advantage in membrane and cell studies and cell membranes since the technique did not require optical transparency, did not have the magnetic susceptibility problems encountered in NMR, and required a fairly low level of spin label doping of the biological system in order to obtain a strong, highly sensitive spectrum. Indeed, it is fair to say that EPR added a tremendous amount of knowledge to our understanding of lipid, membrane and related polymeric systems, which was a great complement to that learned from NMR, solid-state NMR and microscopic methods The real leadership in the implications of these problems started, again, in the McConnell lab at Stanford University and with people like Joe Seelig, Wayne Hubbell and others who followed. Nobel laureate Roger Kornberg was also a graduate student in this laboratory, and his work also was involved in studies of lipids and membranes through the use of spin labels and spin probes [12].

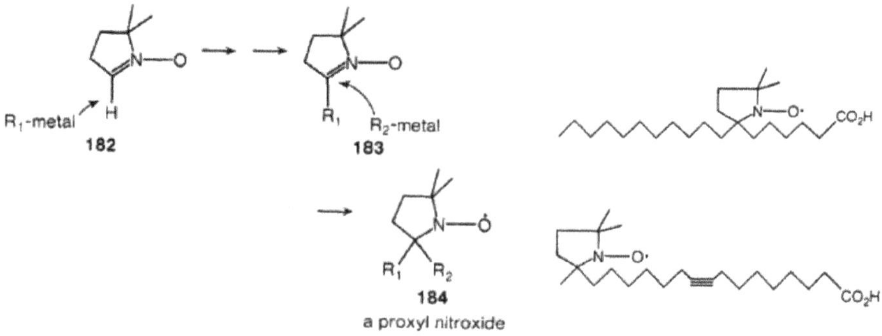

Figure 7. Synthesis of a proxyl nitroxide

5. Nucleic acid analogues

The Bobst laboratory at the University of Cincinnati synthesized some very novel nucleotide analogues where the label was covalently tethered to various purine and pyrimidine rings in such a manner that the tether did not distort the DNA structure and was rigid enough to not create ambiguities in an interpretation of the backbone or sidechain motion of the polynucleotide where the label was incorporated [13]. A series of these novel, unique structures are shown in Figure 9. This work was then copied and extended by other groups, particularly the Seattle group (University of Washington) that also designed nucleotide analogues for probing DNA [14]. In all cases the syntheses were truly challenging, could only be carried out by very proficient organic chemists, and support the view of this author that synthetic organic chemistry is the rate-limiting step in many of these biophysical probe experiments.

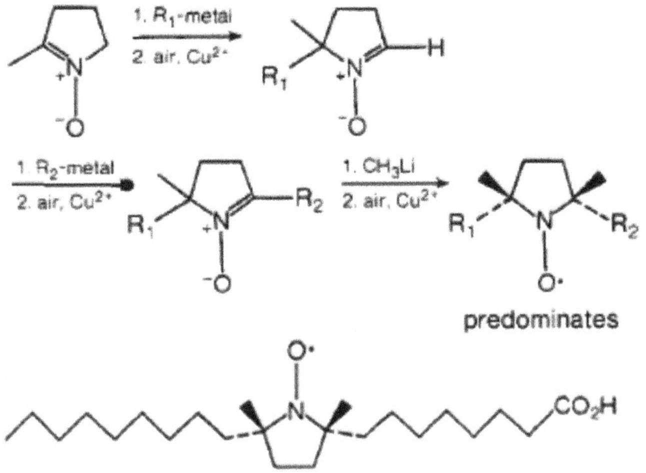

Figure 8. Synthesis of an azethoxyl nitroxide

6. Specificity in protein labeling: Thiol groups

The ideal goal with spin labeling is a universal method to label any tailored site with high specificity. Let's face it; spin labeling of proteins is a protein chemical modification methodology. That aside, it is the chemistry of the functional groups utilized in order to label a protein. If one examines the 20 common amino acids, one finds that the advantageous modification chemistry is both quite limited, ambiguous, and is very much dependent on pKa values where charged sidechains are targeted. This leaves us only with the cysteine thiol as the best candidate for any sort of specific modification. If one looks at the standard array of protein modification functional groups, at least what existed in the 1960s, 1970s and 1980s, we were limited to the maleimde, the alpha halo-acetamide groups and a few disulfide-based reagents, all of which had limitations, particularly the former two. Reagents such as iodoacetamide or N-ethylmaleimide (NEM) will react with thiol groups, amino groups (alpha- amino groups, lysine) and occasionally with hydroxyl groups of a nucleophilic serine or threonine or a tyrosyl side chain. Secondly, the preponderance of these sidechains is usually multifold in proteins, while thiol groups are usually small in number, or occasionally nonexistent. In a quest for highly specific reversible thiol reagents, Berliner and Hideg capitalized on the chemistry pioneered by George Kenyon with methylthiomethane sulfonate, a reagent that undergoes disulfide interchange with a cysteine eliminating the methylsulfonate leaving group [15]. This was definitely advantageous over dithiol reagents, where one loses half of the label in the exercise and it also created other problems involving thiol interchange that could eventually negate of the advantages. Hence the label shown in Figure 10 (bottom), affectionately known as MTSSL or MTSL, was synthesized and was shown to be highly reactive, uniquely specific for cysteine thiol groups and could be easily released with a small concentration of mercaptoethanol or dithiothreitol, allowing one to recover the protein and also allowing for a second labeling stoichiometry quantitation based on the released label [16]. Berliner and Hideg showed eloquently how this works with the reactive protease papain, which contains a cysteine SH at the active site analogous to the serine OH in chymotrypsin [16]. Initially this label wasn't used much by other research groups, but the advent of molecular biology and the power of site-specific mutation triggered a revolution in this area, pioneered by Wayne Hubbell. The technique, named site directed spin labeling, has really been the method of choice since the 1990s and has created a renaissance in spin labeling [17].

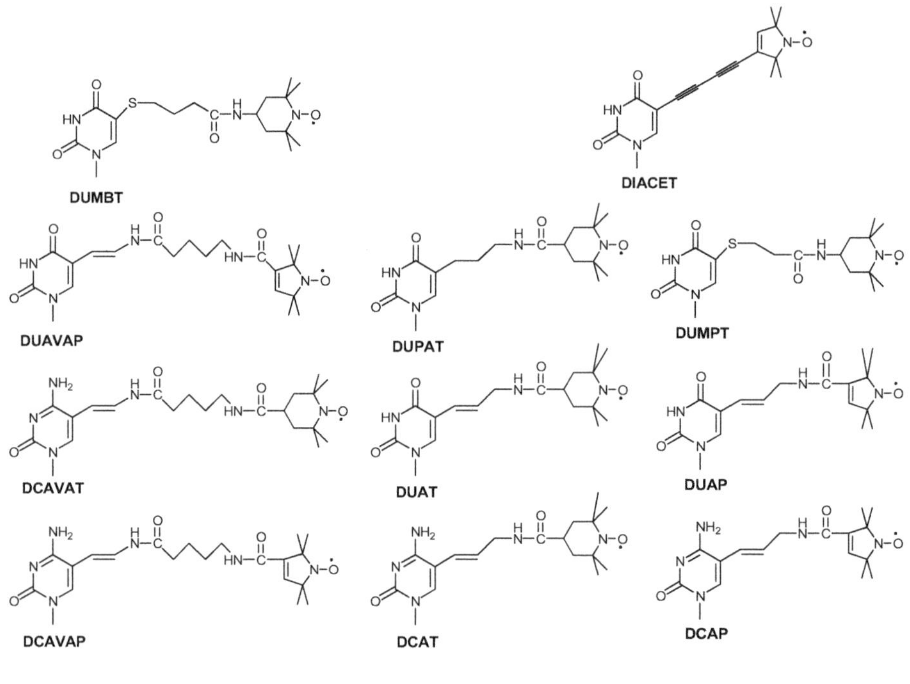

Figure 9. Representative spin labeled pyrimidine bases that can be incorporated into nucleic acid structures. Adapted from [13] with permission.

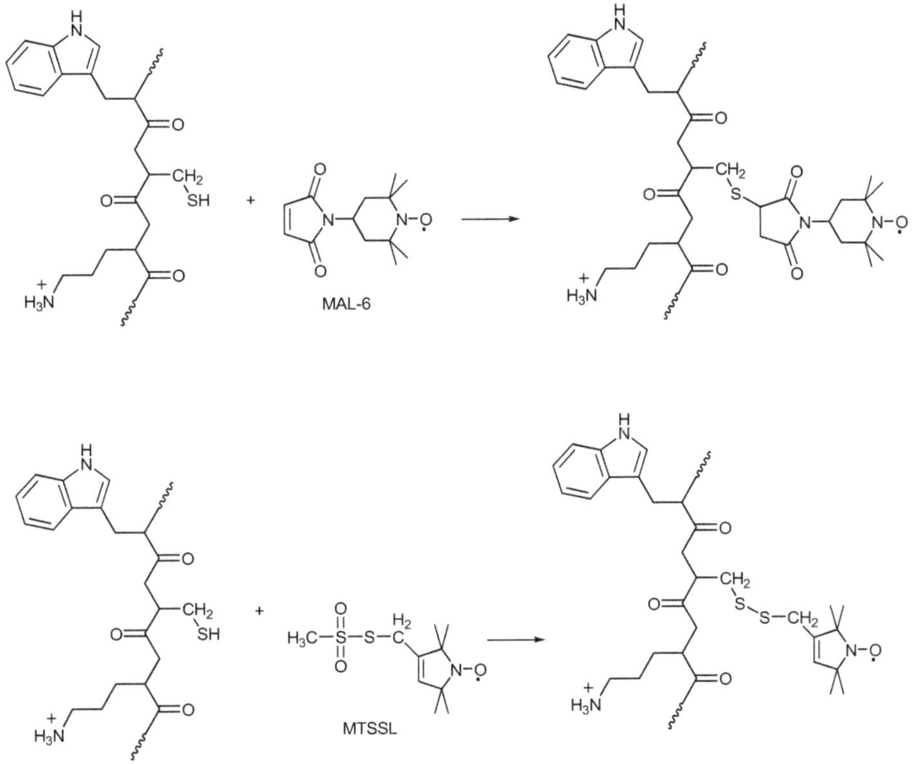

Figure 10. Comparison of thiol labeling reagents. The top reaction utilizes the NEM analog, which reacts irreversibly with the protein thiol group, but can also partially react with other nucleophiles. On the other hand, MTSSL reacts specifically with SH groups and can be reversed in the presence of another thiol reagent, such as mercaptoethanol or DTT. Adapted from [18] with permission

Wayne Hubbell's important contribution was to realize that one could incorporate thiol groups into protein sequences with ease, almost at choice. If there was an example where the disulfide bridge or a few free thiol group caused a major perturbation in the structure or the folding of the protein, it was usually pretty obvious by some functional or conformational (e.g. CD, ORD) analysis. Hubbell attacked the most pressing problems in protein science, which is membrane proteins, which are neither soluble nor amenable to x-ray crystallography or NMR. He started first in collaboration with Nobel Laureate H. G. Khorana on bacteriorhodopsin, a protein whose structure and function had eluded us up to this point, particularly with respect to the light induced conformational changes that occur [19]. The technique, in concert with the well known molecular oxygen Heisenberg exchange relaxation (broadening) of the N-O, allowed assessment of secondary structure characteristics, particularly that of bundled helical structures, which are typical of membrane spanning proteins. If one mutates every residue in a helical protein to a Cys, in each case labels the protein, and then assesses accessibility under increased oxygen, a periodicity of about 3 -4 would be expected in the oscillation of oxygen exposure (since O_2 has a higher solubility in the interior vs the solvent environment). For β-sheet structures, the periodicity would be 2 since every other residue is exposed to the solvent and vice-versa. The two figures below depict the theoretical behavior for a β-sheet and α-helical domain, respectively. Further confirmation occurs when using aqueous paramagnetic reagents such as chromium oxalate or potassium ferricyanide, which selectively relax (broaden) spin labels on the exterior of the protein pointing into the solvent [17]. Hence these accessibility parameters could be quantified and used as sensitive probes of secondary and super-secondary structure. Theoretical plots are shown below in Figure 11. In a study on lac permease, an SDSL 'scan' was done and the results are shown in Figure 12 below [21].

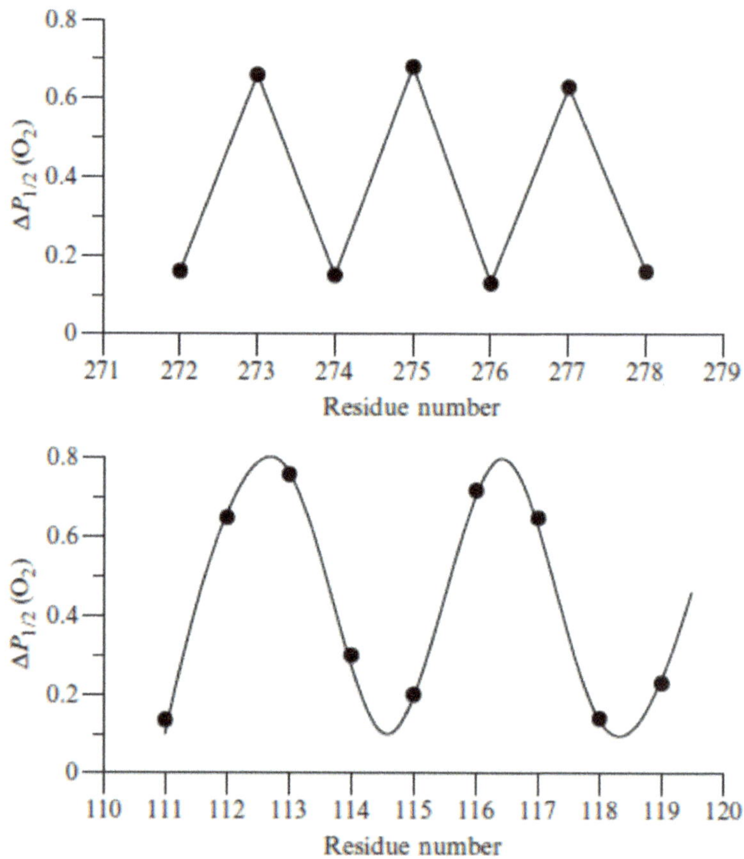

Figure 11. Idealized accessibility data plots indicating β-strand and α-helical secondary structure. From [20] with permission

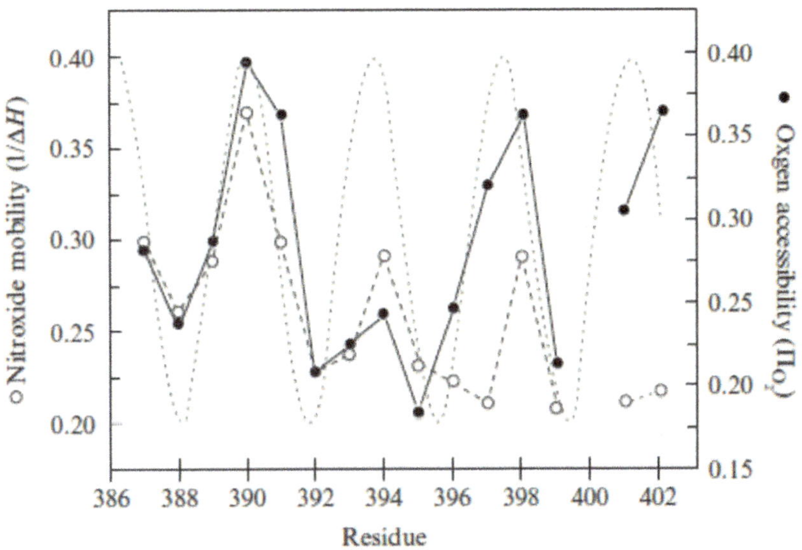

Figure 12. $\Pi(O_2)$ (solid line) and $1/\Delta H$ (broken line) versus sequence position for the nitroxide-labeled single-Cys residues at positions 387-402 in lac permease. The dotted curve is that for a function of period 3.6, and comparison with the $\Pi(O_2)$ and $1/\Delta H$ functions confirms that the data are consistent with an α-helical structure. Ada[ted from [21] with permission.

The site directed spin labeling (SDSL) method blossomed by the early to mid-1990s, with dedicated sessions at meetings on the use of SDSL and EPR in protein structure. Of course, the MTSL label had some disadvantages: it still had some conformational flexibility, it could perturb the protein structure and lastly, in order to obtain an unambiguous assessment of protein structure and function, the use of additional spin labels would be desirable. Hence the Hideg lab propagated several more labels and analogues [22]. Recent work has involved distance measurements within proteins, i.e., the incorporation of two cysteines at selected positions in protein with the idea of mapping the structure by distance triangulation. This is a major effort since one obtains only the distance between the electrons on the two labels, respectively, and each spin label must be correlated back to the protein backbone with inferences from amino acid side chain structure and the aspects of motion of the label in multiple orientations. Consequently, one pair of incorporated cysteines yields just one distance. Figure 13 shows the dilemma of attaining very accurate distance measurements from a double labeling experiment. Nonetheless, the technique has still been valuable and people have developed sophisticated motional simulations in order to localize the label in the protein structure. One looks at motion around a cone and workers have attempted to come up with distances within 15 to 20% accuracy. Needless to say, the x-ray crystallographers have a major advantage (assuming that the protein can be crystallized) but the NMR spectroscopists have a bigger advantage because they can incorporate one single, relatively rigid, spin label and obtain more than 100 distances from electron-proton paramagnetic relaxation enhancements over distance ranges of 5 to 15 Å. In fact, the EPR double label method was quite limited on distances, about that of the electron-proton distance limits. Remembering that we only obtain one distance for each labelled pair, it wasn't until Jack Freed and coworkers incorporated DHQC methodology enabling one to assess distances upwards of 50 to 60 Å [23].

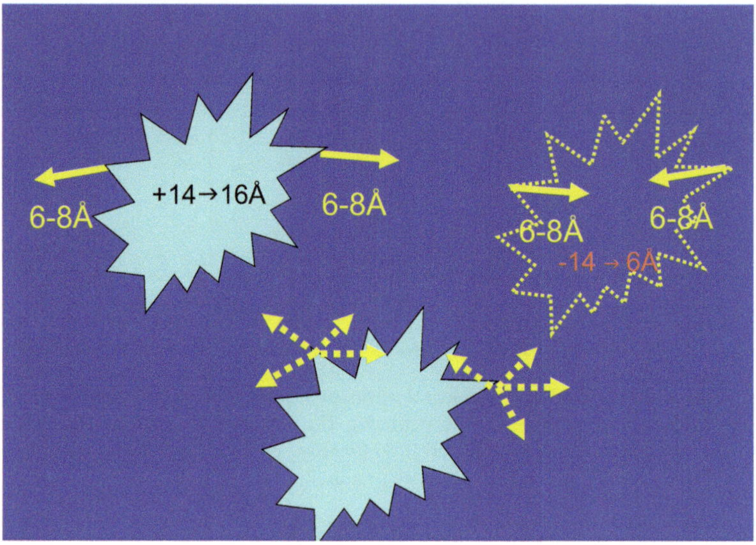

Figure 13. Uncertainty about spin label motion and orientation(s). The spin label ring can orient in several directions, depending on the flexibility of the tether. Ideally, one would like a rigid unique orientation, but the likelihood is not high.

Some possible approaches are to anchor the label in two points on the protein. The label shown below in Figure 14 would require two mutated cysteines at proper spacing in order to meet that requirement.

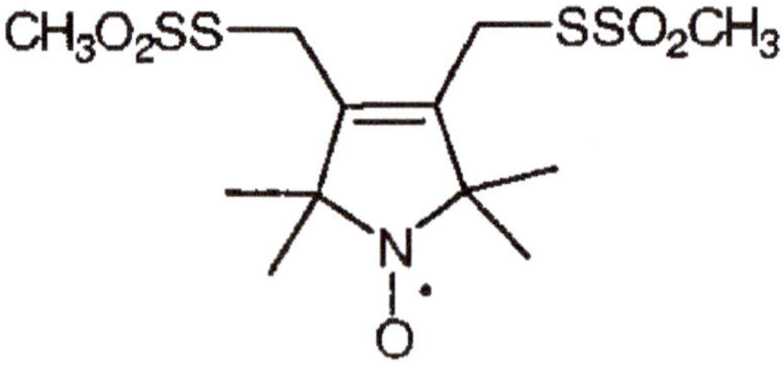

Figure 14. Rigid two site attachment spin label. bis-MTSSL [pyrrolinyl] 2,5-Dihydro-3,4-bis(methanesulfonylthiomethyl)-2,2,5,5-tetramethyl-1H-pyrrol-1-yloxyl radical

7. Nitrones and spin traps: The adducts form nitroxides

These compounds are actually a class of chemical functional groups that had been known quite early since one of the synthetic methods of producing of nitroxides is by a controlled, specific oxidation of a nitrone compound. However, these have found tremendous use in the characterization of free radicals in solution, particularly in the biological field where a plethora of potential radicals are possible. In fact, the reaction of a nitrone with carbon or oxygen-based radicals yields a nitroxide adduct with the spectrum that is characteristic of the chemistry of that particular initial radical (with some caveats that are discussed below). The use of radical-addition reactions to detect short-lived radicals was first proposed by E. G. Janzen in 1965 [24]. The early pioneers in this field studied two classes/types of spin traps which were commercially available at the time and are still on the market today: DMPO, 5,5-dimethyl-1-pyrroline-N-oxide, and PBN, alpha-phenyl N-tertiary-butyl nitrone. These and other second generation spin traps are shown in Figure 15 below.

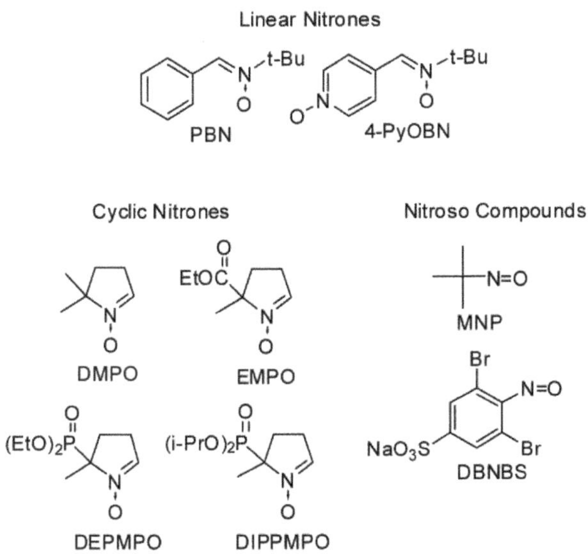

Figure 15. Structures of various spin trap types including second generation nitrones. Structure abbreviations: PBN α-phenyl N-tert-butyl nitrone, 4PyOBN; DMPO 5,5-dimethylpyrroline N-oxide, EMPO 5-(ethoxycarbonyl)-5-methyl-1-pyrroline N-oxide, DEPMPO 5-(diethoxyphosphoryl)-5-methyl-1-pyrroline-N-oxide, DIPPMPO 5-diisopropoxy-phosphoryl-5-methyl-1-pyrroline-N-oxide, AMPO 5-carbamoyl-5-methyl-1-pyrroline N-oxide, MNP 2-methyl-2-nitroso propane, DBNBS 3,5-dibromo-4-nitrosobenzenesulfonate, CPCOMPO 7-oxa-1-azaspiro[4.4]non-1-en-6-one 1-oxide

DMPO could cross the cell membrane and was in fact sensitive to radicals that were present in both the aqueous media as well as the lipid medium. Some derivatives of DMPO with selected substitutions at the 1- or 4- position allow one to affect its partitioning properties between aqueous and lipid environments. Janzen successfully showed that DMPO could trap the important reactive oxygen species: superoxide and hydroxyl radical [25]. He clearly demonstrated the differences in their EPR spectra, allowing characterization of these radicals in vivo. However it wasn't until a few years later that Rosen and colleagues found that the superoxide radical adduct of DMPO could decompose to the hydroxyl adduct by a mechanism which, to date, is still not totally clear. Hence one has to take special care, e.g., including SOD in an experiment in order to include or exclude superoxide [26].

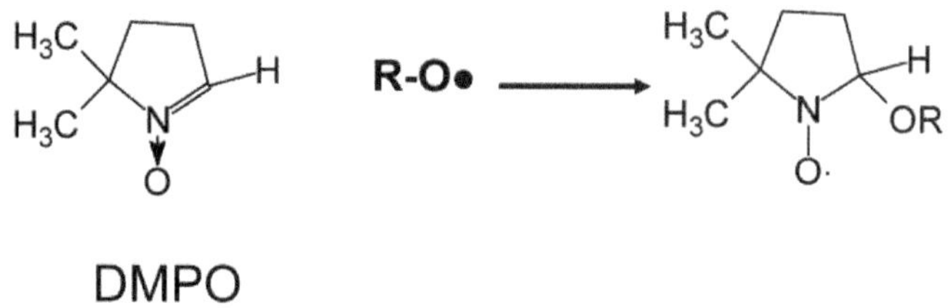

Figure 16. Reaction of DMPO with an oxyradical.

The PBN spin trap was, by virtue of its non-polar lipophilic behavior, quite valuable for trapping lipid radicals and those in a membrane milieu. If PBN reacted with oxygen radicals, such as hydroxyl radical, it decomposed without any radical adduct remaining. On the other hand, the PBN-lipid adducts, if partitioned into the membrane, were stable for long periods of time and one could e.g., isolate lipid-radical adducts of PBN in erythrocytes, then extract and concentrating them later. In fact, some lipid biologically produced lipid radicals are stable in lipid media and erythrocytes could be 'post labeled' with PBN after some oxidative stress event [26].

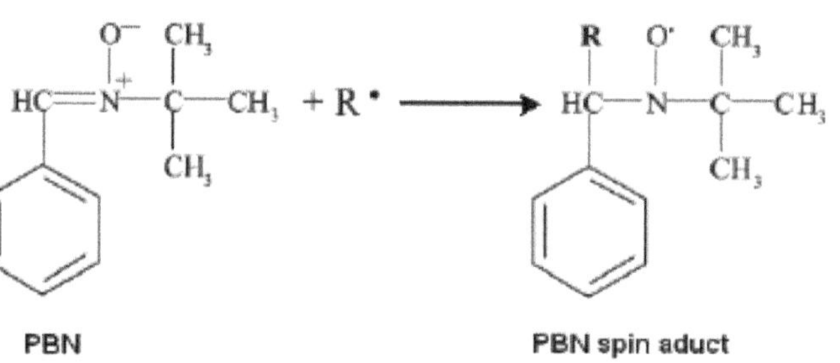

Figure 17. Reaction of PBN with a carbon based radical.

These two spin traps remained with us for almost 20 years and, along with the particular annoying side reaction noted with DMPO and superoxide, the DMPO adducts tended to be fairly unstable. In the late 1980s, Tordo's group prepared a DMPO analog that contained a phosphoester-type group in one of the positions of the flanking methyl groups [28]. One of these compounds, DEPMPO, was quite successful in that the half life of the radical adduct was much longer than that known for DMPO. It is important to point out at this juncture that the reaction kinetics for all of these radical traps were quite poor, involving the necessity of having 50 – 100 mM concentrations of spin trap in the solutions. Further development of other analogs by Tordo's group and also some quality control by the commercial laboratory that were selling DEPMPO was quite an advance for the community.

However the real, major effort in our understanding and designing nitrone spin traps, based on the DMPO skeleton, was a result of an intense effort at Ohio State University by Villamena, who did synthetic, kinetic and computational design studies of these traps as well as their aminoxyl radical adducts. Hence Villamena studied both their reactivity and the stability of the adducts [29]. Overall, spin traps are powerful reagents, albeit more limited for in vivo studies due to their low sensitivity and kinetics and the concentration limits of reactive radicals in vivo. The hope was to accumulate radicals in vivo up to levels where the trapped adducts exceeded the normal in vivo level. Suffice it to say we are 'part way' there. But we still suffer from the breakdown of the radical adducts and have not yet attained optimal kinetics. Some future concepts for applications of these type of compounds would be to prepare 'spin trap labels' that could be incorporated at specifically targeted organ sites in vivo which would then would convert to the radical adduct at the time and place of radical generation. There have also been some efforts to attach fluorescent moieties onto these labels, whereby the fluorescence is quenched upon formation of the nitroxide radical adduct [30-31]. This area has great promise and some examples are diagrammed below under future developments (Section 8).

8. In-vivo EPR using aminoxyl radicals: History and fate

The last frontier of applications of aminoxyl (a.k.a. nitroxide) radicals is in their applications to in-vivo studies. Since the aminoxyl (nitroxide) radicals were the first, and for a long time the only radicals that were stable and detectable in aqueous solution, such as cell and other components, it was straightforward and logical to try to examine the fate and behavior of these radicals in living systems. Early on, attempts were made to mix development of spin labeling that one attempted to mix aminoxyl radicals and living systems. In fact, in an undocumented experiment in the McConnell laboratory the toxicity of a nitroxide was tested on a goldfish. A beaker-full of the radical, t-butylnitroxide, was emptied into a Bell jar containing a goldfish. While the concentration was not accurately estimated, it was certainly in the tens to hundreds of millimolar; needless to say the fish lived, and as we learned later these labels are actually life-sustaining compounds). But someone accidentally left the hot water slowly dripping into the Bell jar and eventually the fish expired. One postdoc in the lab actually monitored his urine for the ingestion of these compounds as several of them are volatile and there was an extensive amount of synthesis and gas chromatography ongoing them lab with these volatile compounds. Needless to say, no adverse effects were found.

So what happens if you mix a nitroxide with e.g., a cell or tissue suspension? If it is the six-membered ring species, i.e., the piperidinyl nitroxides, and easily those that can cross the cell membrane into the cytoplasm, they are rapidly reduced, i.e., 'neutralized,' within a few minutes, since a plethora of intracellular biological reducing agents ready to take on their antioxidant role and convert the nitroxide to its corresponding hydroxylamine. For example,

TEMPONE, or for that matter TEMPOL, are rapidly converted to the hydroxylamine with an immediate loss of the paramagnetism. This occurs within a few minutes. The five-membered ring species, however, have much longer half-life, i.e. of the order of 15 to 30 min., allowing one to study some aspects of the metabolism and perhaps the ability to image this paramagnetic material in a living species. The first experiments were done by the Brasch group where they were evaluating nitroxides as MRI contrast agents[33-34]. This was followed by a plethora of studies on animals, tissue samples, blood samples, etc. where we obtained a wealth of pharmacokinetic data (although no totally clear understanding of the mechanism and detailed rate constants) [35-36]

Suffice it to say, imaging by EPR methods is challenging, if not hopelessly low resolution, since most nitroxide labels have linewidths of, at best, 0.3-0.5G for a compound that is deuterium and N-15 enriched. It has only been with the trityl radicals mentioned early where any hope of imaging was possible. However, if one takes advantage of the power and high resolution of magnetic resonance imaging (MRI) and the fact that the contrast agents in this methodology are paramagnet, then organic radicals have a place. Therefore nitroxide spin label/spin probe analogs have been tested as MRI contras agents and have met with some success. One must overcome the problem of biological reduction, also a problem with the radical adducts of nitrone spin traps since the cellular/tissue milieu contain many reducing agents such as NADH, ascorbic acid, and mitochondrial reduction sources [33]. The real quest here is to produce a well protected, aminoxyl radical that is highly resistant to biological reduction yet can be incorporate into the tissue system of choice. A few examples have been reported to date, particularly where the tetramethyl groups that flank the N-O group are replaced by long aliphatic chains such as lipids or tertiary butyl chains or cages.

9. Conclusions/prognosis/summary/future developments

This author has frequently concluded, about once per decade, that spin labeling has met its limits and should go into the category of the 'on the shelf' routine technique given all of its limitations. However, we have found one or two cases of a renaissance in the use of nitroxides, particularly the inception of the SDSL technique using MTSL labels which have given it a major rebirth. The future should involve marrying various techniques that can utilize paramagnetic materials, some of which that have already been mentioned earlier: NMR, fluorescence, dynamic nuclear polarization (DNP) and other technologies yet to be developed or discovered. Some examples are shown below.

Optical probes, e.g., absorb in the visible or are fluorescent, when coupled to a paramagnetic moiety, experience shifts or lifetime relaxation from a nearby free electron. One example below (Figure 18), is a nitroxide fluorophore, which exhibits significantly quenched fluorescence emission. Some applications are cartooned in Figures 19 and 20.

This nitrone spin trap depicted in Fig. 21 is one of several synthesized and tested by David Becker [32] at Florida International University. Upon reaction with reactive oxygen species, the absorption properties of the nitrone shifts are depicted below.

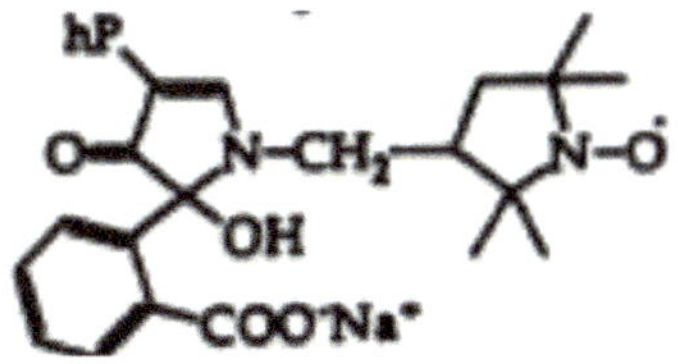

Figure 18. ((2-Carboxy)phenyl)-5-hydroxy-1-((2,2,5,5-tetramethyl-1-oxypyrrolidin-3-yl)methyl)-3-phenyl-2-pyrrolin-4-one sodium salt [30]

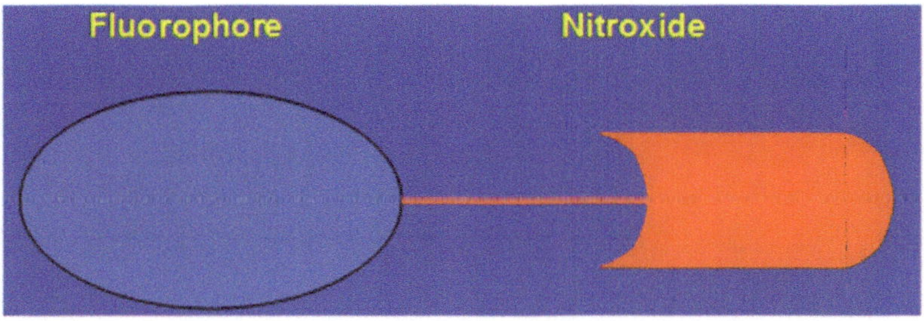

Figure 19. Schematic of the spin label in Figure 18 , where the red nitroxide depicts the paramagnetic N-O group, while the weak fluorescence reflects quenching by the paramagnet.

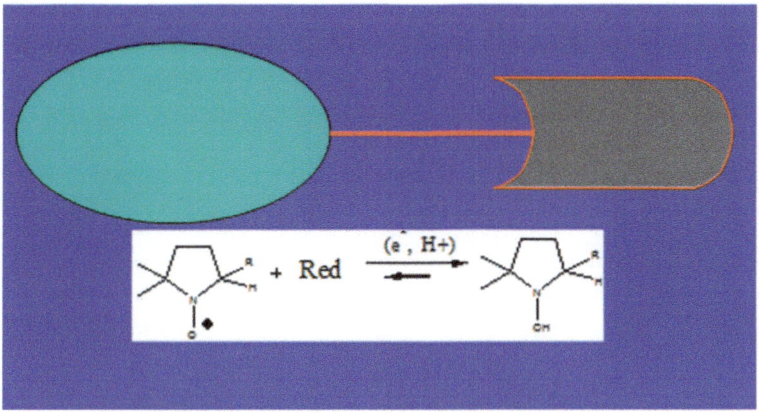

Figure 20. Upon reduction of the spin label the corresponding hydroxylamine, e.g., in a biological system by NADH or ascorbic acid, the fluorescence emission is strong and the EPR spectrum from the spin label has disappeared.

One can take advantage of NMR/MRI by spin labeling a cell surface with multiple nitroxide labels. The highly labeled surface now acts as an excellent paramagnetic relaxation enhancement site for exchanging water molecules, enhancing contrast in MRI [33-34] and being uniquely sensitive to changes in conformation, permeability and flexibility of the cell membrane surface as depicted below.

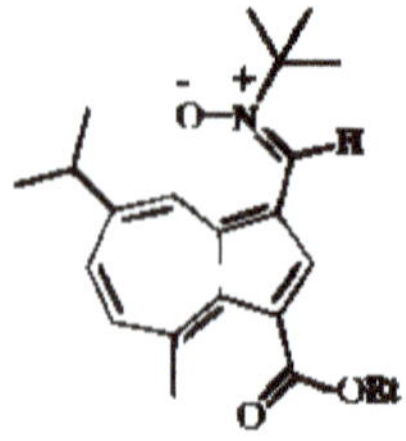

Figure 21. A colorimetric nitrone spin trap.

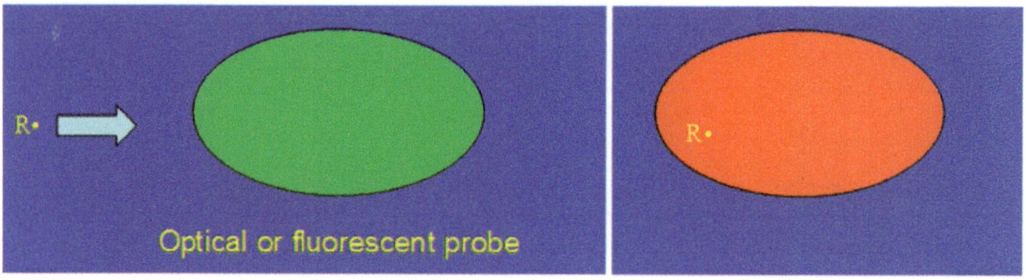

Figure 22. Schematic of the reaction of a colorimetric or fluorescent nitrone spin trap with a radical

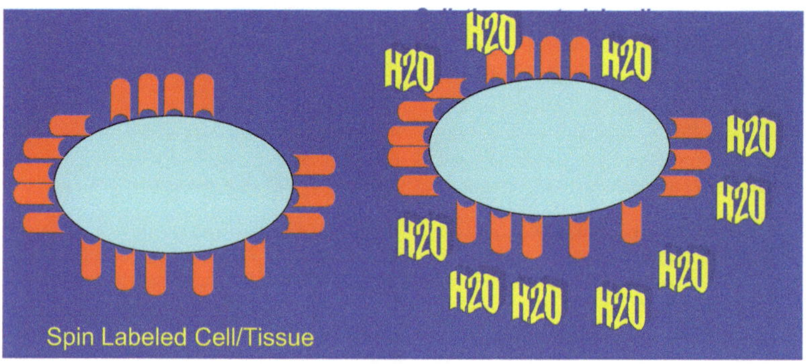

Figure 23. Schematic of a proton relaxation enhancement (PRE) spin labeled cell. Multiple paramagnetic labels are affixed to the cell surface by specific binding or covalent attachment. This results in a significantly enhanced PRE, which is detectable in MRI.

Hence future developments in several of these areas should show great promise for the future.

Author details

Lawrence J. Berliner
Dept. of Chemistry and Biochemistry, University of Denver, USA

10. References

[1] Pauly, H. (1898). Über die Einwirkung von Brom auf Triacetonamin der deutschen chemischen Gesellschaft, 31: 668–674.

[2] Hoffmann AK, Henderson AT (1961) A New Stable Free Radical: Di-T-Butyl-Nitroxide, J. am. chem. soc 83:4671–4672.

[3] Hoffmann AK, Feldman AM, Gelblum E (1964) Reactions of Organoalkali Compounds with Nitro Compounds: A New Synthesis of Nitroxides, J. am. chem. soc 86:646-7650.

[4] Fremy E (1845) Nitrosodisulfonate de Potassium Ann. chim. phys 15:408.

[5] Rosantzev EG, Neiman MB (1964) Organic Radical Reactions Involving No Free Valence, Tetrahedron 20:131–137

[6] Rozantsev EG (1970) Free Nitroxyl Radicals, New York:Plenum Press.

[7] Burr M, Koshland DE Jr (1964) Use of Reporter Groups In Structure–Function Studies of Proteins, C:1017–1024.

[8] Berliner L J, McConnell, HM (1966) A spin labeled substrate for α chymotrypsin, Proc. nat. acad. sci., U.S. 55:708-712

[9] Bauer RS, Berliner LJ (1979) Spin label investigations of chymotrypsin active site structure in single crystals, J. mol. biol. 128:1-19.

[10] Beth AH, Robinson BH, Cobb CE, Dalton LR, Trommer WE, Birktoft JJ, Park JH, (1984) Interactions and spatial arrangement of spin-labeled bound to glyceraldehyde-3-phosphate dehydrogenase, J. biol.chem. 259:9717-9728.

[11] Keana, JFW (1978) Newer Aspects of the Synthesis and Chemistry of Nitroxide Spin Labels, Chemical reviews 78:37-64

[12] Kornberg RD, McConnell, HM (1971) Lateral Diffusion of Phospholipids in a Vesicle Membrane. Proc. nat. acad. sci. U.S. 68:2564-2568.

[13] Bobst, A (1979) Application of Spin Labeling to Nucleic Acids, In: Berliner LJ, editor, Spin Labeling II: Theory and Applications Academic Press: New York, (1979), pp 291-345.

[14] Shelke, S. A. and Sigurdsson, S. T. (2012) Site-Directed Spin Labelling of Nucleic Acids. Eur. j. org. chem.. doi: 10.1002/ejoc.201101434

[15] Smith D, Maggio E., Kenyon G (1975) Simple Alkanethiol Groups for Temporary Blocking of Sulfhydryl Groups of Enzymes, Biochem 14:766-761.

[16] Berliner LJ, Grunwald J, Hankovszky HO, Hideg K (1982) A Novel Reversible Thiol Specific Spin Label: Papain Active Site Labeling and Inhibition, Anal. biochemistry 119:450-455.

[17] Hubbell WL, Cafiso DS, Altenbach C (2000) Identifying Conformational Changes With Site-Directed Spin Labeling, Nature structural biology 7:735-739.

[18] Feix J B, Klug C S (1998). Site-directed Spin Labeling of Membrane Proteins and Peptide-Membrane Interactions, , In: Berliner L J, Reuben J, editors Spin Labeling: The Next Millennium, Biological Magnetic Resonance14, Plenum Press: New York, pp. 251-281.

[19] Farrens DL, Altenbach C, Yang K, Hubbell WL,Khorana HG (1996) Requirement of Rigid-Body Motion of Transmembrane Helices for Light Activation of Rhodopsin, Science 274: 768-770.

[20] Klug C S, Feix J B (2008). Methods and Applications of Site-Directed Spin Labeling EPR Spectroscopy, Methods in cell biology, 84:617-658

[21] Voss J, He MM, Hubbell WL, Kaback HR (1996) Site-Directed Spin Labeling Demonstrates That Transmembrane Domain XII in the Lactose Permease of Escherichia coli Is an α-Helix, Biochemistry *35:*12915-12918.
[22] Fawzi NL, Fleissner MR, Anthis NJ, Ka′lai T, Hideg K, Hubbell WL, Clore GM (2011) A Rigid Disulfide-Linked Nitroxide Side Chain Simplifies the Quantitative Analysis of PRE Data, J biomol NMR 51:105–114.
[23] Borbat PP, Costa-Filho AJ, Earle KA, Moscicki JK, Freed JH (2001) Electron Spin Resonance in Studies of Membranes and Proteins, Science 291:266 - 269
[24] Janzen EG, Blackburn B J (1968) Detection and Identification of Short-Lived Free Radicals by an Electron Spin Resonance Trapping Technique, J. am. chem. soc. 90:4481-4490
[25] Sang H, Janzen EG, Poyer JL, McKay PB (1997) The Structure of Free Radical Metabolites Detected by EPR Spin Trapping and Mass Spectroscopy from Halocarbons in Rat Liver Microsomes, Free radical biology & medicine 22:843–852.
[26] Britigan BE, Rosen GM, Chai Y, Cohen MS (1986) Stimulated Human Neutrophils Limit Iron-Catalyzed Hydroxyl Radical Formation as Detected by Spin-Trapping Techniques J. biol. chem. 261:4426–4431.
[27] Mergner GW, Weglicki WB, Kramer JH (1991) Postischemic Free Radical Production in the Venous Blood of the Regionally Ischemic Swine Heart. Effect of Desferoxamlne. Circulation 84:2079-2090.
[28] Roubaud V, Sankakarpandi S, Kuppusamy P, Tordo P, Zweier J (1997) Quantitative Measurement of Superoxide Generation Using the Spin Trap 5-(Diethoxyphosphoryl)-5-Methyl-1-Pyrroline-N-Oxide. Anal. biochem. 247:404–411;
[29] Villamena FA (2009) Superoxide Radical Anion Adduct of 5,5-Dimethyl-1-pyrroline N-Oxide (DMPO). 5. Thermodynamics and Kinetics of Unimolecular Decomposition, J. phys. chem A. 113: 6398-6403.
[30] Pou S, Bhan A, Bhadti VS, Wu SY, Hosmane RS, Rosen GM (1995) The Use of Fluorophore-Containing Spin Traps as Potential Probes to Localize Free Radicals in Cells With Fluorescence Imaging Methods, FASEB j 9:1085-1090.
[31] Ka′lai T, Hideg E, Imre V, Hideg K (1998) Double (Fluorescent and Spin) Sensors for Detection Of Reactive Oxygen Species in the Thylakoid Membrane Free radical biology & medicine 24:649–652.
[32] Becker DA (1996) Highly Sensitive Colorimetric Detection and Facile Isolation of Diamagnetic Free Radical Adducts of Novel Chromotropic Nitrone Spin Trapping Agents Readily Derived from Guaiazulene, J. am. chem. soc. 118, 905-906.
[33] Couet WR, Eriksson UG, Tozer TN, Tuck LD, Wesbey GE, Nitecki D, Brasch RC (1984) Pharmacokinetics and Metabolic Fate of Two Nitroxides Potentially Useful as Contrast Agents for Magnetic Resonance Imaging, Pharmaceutical research 1:203-209.
[34] Wikstrom MG, White DL, Moseley ME, Dupon JW, Brasch RC (1989) Ascorbate-Induced Cancellation of Nitroxide Contrast Media Enhancement of MR Images. Invest radiol 24:692-696.
[35] Swartz HM, Sentjurc M, Morse PD (1986) Cellular Metabolism of Water-Soluble Nitroxides: Effect on Rate of Reduction of Cell/Nitroxide Ratio, Oxygen Concentrations and Permeability of Nitroxides. Biochim biophy acta 888:82-90.
[36] Kocherginsky N, Swartz HM (1995) Nitroxide Spin Labels: Reactions In Biology And Chemistry, Boca Raton CRC Press

Chapter 2

ESR Spectroscopy of Nitroxides: Kinetics and Dynamics of Exchange Reactions

Günter Grampp and Kenneth Rasmussen

Additional information is available at the end of the chapter

1. Introduction

Stable nitroxide radicals have proved to be helpful in solving many problems in chemistry, biochemistry, biophysics, material science etc. as model compounds. To understand in detail the great variety of different chemical reactions, a good knowledge of the underlying "simple" exchange reactions related to nitroxide radicals is necessary. This chapter focuses on electron-self and spin exchange reactions of various nitroxide radicals in different solvents. The corresponding rate constants, the activation parameters, like activation enthalpies and volumes are obtained from temperature- and pressure dependent ESR-line broadening effects. A short introduction to dynamic ESR-spectroscopy is given.

1.1. From ESR spectrum to exchange rate constant

There are several chemical reactions giving rise to dynamic line shape effects, such as spin-exchange, also known as Heisenberg exchange, electron self-exchange and proton or counter ion transfer. A theoretical illustration of how an ESR spectrum may be affected by chemical reactions is presented in figure 1. Here, the interconvertion between two states, a and b, can influence the spectrum dramatically as the exchange rate increases.

Apart from the two extreme limits of infinitely slow or fast exchange, usually three regions are considered: slow, intermediate and fast. In the slow region the two lines broaden as if the life, or relaxation, times of the spins had been reduced due to quenching. When the exchange enters the intermediate region, the interconvertion rate leads to an averaging of the lines and correspondingly to line shifts. At one point, the exchange so fast that the spectrum appears as a single line due to overlap of the two original ones. This marks the beginning of the fast region, where a further increase in the exchange rate leads to narrowing of the single line.

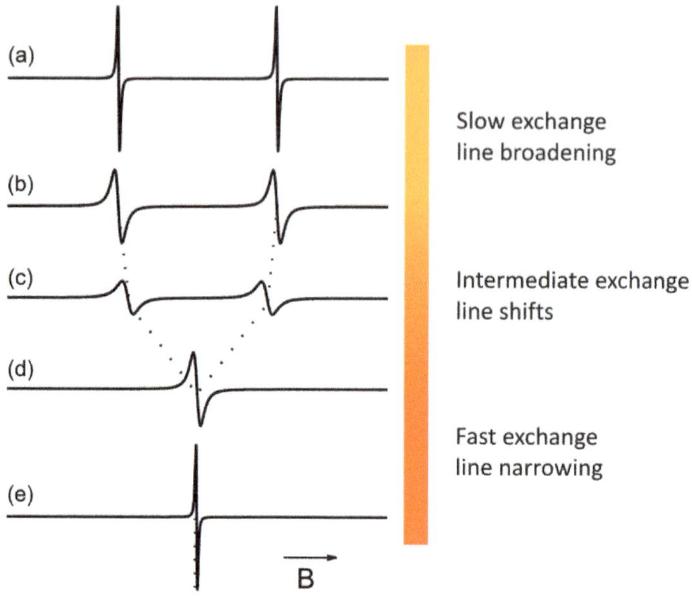

Figure 1. Theoretical illustration of the first derivative ESR spectra arising from the interconversion between forms *a* and *b*. (a) Slow limit. (b) Moderately slow region. (c) Intermediate region showing line shifts. (d) Moderately fast region. (e) Fast limit.

A chemical reaction corresponding to this illustration could, for example, be an exchange reaction like the one shown in equation (1), where the forms *a* and *b* are interpreted as being molecules having different nuclear spin configurations.

$$A_{(a)} + A^{\bullet -}_{(b)} \xrightarrow{k_{obs}} A^{\bullet -}_{(a)} + A_{(b)} \tag{1}$$

Transferring the unpaired electron from one molecule to the other thus corresponds to the interchange between the two forms. Note, that reactions between molecules of the same form, e.g. *a*, also take place. If this happens no ESR line broadening is expected, but nevertheless the electron transfer takes place and these 'uneventful' reactions must be taken into account. The resulting line width in the case of slow exchange then becomes:

$$\Delta B = \Delta B^{o} + \frac{1 - p_j}{|\gamma_e \tau|} \tag{2}$$

with p_j being the probability that the reactants have the same nuclear spin configuration, here *j*.

In general, an ESR spectrum may show more than just a few lines (states), of which some are originating from transitions involving degenerate energy levels. This can give rise to substantially different values of p_j and it is important to recognize that when a self-exchange reaction occurs, one expects the ESR lines to have different widths. Strictly speaking, equation (2) is therefore only valid for a single state and its corresponding ESR line.

Looking at equation (1) from the point of view of chemical kinetics it is seen that although the self-exchange reaction is bimolecular, one may express it using a first order rate law since the concentration of diamagnetic species is constant. Recalling the relationship between life times and first order rate constants one obtains

$$\tau = \frac{1}{k} = \frac{1}{k_{obs}[A]} \tag{3}$$

which in combination with equation (2) and taking the ESR signal in its first derivative Lorentzian form results in the following relationship between line broadening and rate constant:

$$k_{obs} = \frac{\sqrt{3}\pi|\gamma_e|\left(\Delta B_{pp} - \Delta B_{pp}^{o}\right)}{\left(1-p_j\right)\left[A\right]} \tag{4}$$

where ΔB_{pp}^{o} and ΔB_{pp} are the peak-to-peak widths of the first derivative line in the absence and presence of the self-exchange reaction.

Similarly, an expression of k_{obs} may be derived for the fast exchange region, where the broadening of the singe ESR line obeys

$$\left(\Delta B_{pp} - \Delta B_{pp}^{o}\right) = \frac{4\pi|\gamma_e|\nabla_2}{3k_{obs}\left[A\right]} \tag{5}$$

Here ∇_2 is the second moment of the ESR spectrum, that is

$$\nabla_2 = \sum_j p_j\left(\overline{B} - B_j\right)^2 \tag{6}$$

with $\overline{B}$ being the center field of the spectrum and B_j the resonant magnetic field strength of the j'th ESR line.

Note that the fact that an ESR spectrum belonging to an exchange reaction appears as a single line does not always imply that the fast limit is reached. Such systems may still be in the intermediate region and equation (5) does not hold. In an effort to decide if a given system is indeed exhibiting fast exchange, the following relation may be used [1].

$$Z = \frac{\sqrt{3}}{2}\cdot\frac{\left(\Delta B_{pp} - \Delta B_{pp}^{o}\right)}{\sqrt{\nabla_2}} \leq 0,2 \tag{7}$$

That is, if $Z \leq 0.2$ the system may safely be considered to be in the fast exchange region.

In the intermediate exchange region, the pseudo-first order rate constant, $k = \tau^{-1} = k_{obs}[A]$, can been extracted from the experimental spectra by means of density matrix simulations of the dynamically broadened line shapes [2]. Assuming a weakly coupled spin system undergoing degenerate electron exchange in the high temperature limit and a relaxation superoperator diagonal in the subspace of induced transitions, *i.e.* neglecting cross relaxations, the evolution of the relevant terms of the spin density operator in the rotating frame, σ, is governed by:

$$\frac{\partial \sigma_{\beta\{v\},\alpha\{v\}}}{\partial t} = \left[i\left(\omega_{\alpha\{v\},\beta\{v\}} - \omega\right) - R_{\beta\{v\},\alpha\{v\}} - \frac{1}{\tau}\right] \sigma_{\beta\{v\},\alpha\{v\}} + \frac{2}{N\tau} \sum_{v' \neq v} \sigma_{\beta\{v'\},\alpha\{v'\}} + \frac{i\omega_{1,e}\omega_e}{2k_B TN}, \tag{8}$$

where $\{v\}$ denotes the set of quantum numbers describing the nuclear spin states and N the dimensionality of the Hilbert space [3]. $R_{\beta\{v\}\alpha\{v\}}$ is the element of the relaxation matrix corresponding to the induced transition and $\omega_{\alpha\{v\},\beta\{v\}} = \omega_{\alpha\{v\}} - \omega_{\beta\{v\}} = (E_{\alpha\{v\}} - E_{\beta\{v\}})/\hbar$. The absorption signal is obtained from the expectation value of the out of phase magnetization assuming steady state conditions, *i.e.*

$$Y' = Tr\left(\hat{S}_y \hat{\sigma}\right) \propto \mathrm{Im} \frac{-i\sum_k \omega_k n_k F_k}{1 + \frac{2}{N\tau}\sum_k n_k F_k} \tag{9}$$

where the sum now runs over all completely equivalent spin packets of degeneracy n_k and resonance frequency ω_k. N equals the sum over all n_k and τ denotes the average lifetime of the nuclear spin configurations, *i.e.* $\tau^{-1} = k = k_{obs}$ [A]. Furthermore, F_k is defined by $F_k^{-1} = i(\omega_k - \omega) - T_{2,k}^{-1} - \tau^{-1}$, with $T_{2,k}$ denoting the transversal relaxation time for spin packet k and ω the microwave irradiation frequency, respectively.

Equation (9) has been implemented in Matlab and Fortran 95. Local (trusted region Newton, Simplex) as well as global (Lipschitz, adaptive simulated annealing) optimization algorithms have been employed to fit the experimental spectra. Field modulation effects (100 kHz) have been accounted for by using pseudo-modulation as introduced by Hyde [4].

1.2. Experimental considerations

To obtain the best possible results from ESR line broadening investigations, a certain amount of experimental diligence is required. It may seem superfluous to mention this here, but as it is said: 'God is in the detail'. Albeit information on this subject is ample in literature [5, 6], some key points shall be reiterated.

Any spectrometer setting that might influence the line shape, e.g. modulation amplitude or microwave power, should remain unchanged throughout a series of measurements. Temperature control can be very advantageous as a means of reducing the experimental error. The reason why can be illustrated in the following example: Assuming an Arrhenius-like behaviour with an activation energy of 15 kJ mol^{-1}, a deviation of the temperature by 1 K around 298K corresponds to a change in rate constant of approximately 2%.

Other things that are often overlooked are the position of the sample in the ESR resonator as well as the so-called filling factor. These are particularly important when working with lossy solvents, as the effective microwave power felt by the paramagnetic species, often expressed by the Q factor, is strongly dependent on the amount of solvent used [7].

Finally, a few remarks on the design of the experiment: In the case of self-exchange, the concentration of the radical is kept constant while varying that of the diamagnetic compound. It is advisable to keep the concentration of the radical low in order to avoid effects from spin exchange, especially when interested in temperature or pressure effects. Often radical concentrations of around 10^{-4} mol l^{-1} are used, whereas the concentration of the diamagnetic compound preferably spans several orders of magnitude. Paramagnetic exchange is treated analogously.

1.3. Experimental techniques in pressure dependent ESR-measurement

Several commercially available temperature-control units exist for magnetic resonance spectroscopic measurements, ranging from liquid helium to very high (1100°C) temperatures.

In contrast units for pressure dependent experiments must be constructed by the experimentalist themselves.[1] Fortunately, several excellent books are published on high-pressure techniques covering a great variety of experimental measurements and different spectroscopies [8, 9]. To our knowledge no special review concerning ESR-spectroscopy under high pressure exists, but several articles describing high-pressure cells for NMR- and ESR-spectroscopy. Many experimental details developed for high-pressure NMR-cells can be adapted for ESR-spectroscopy often in simpler versions [10-13]. Several publications deal with high-pressure magnetic resonance cells for solid state investigations [14-21]. Even high-pressure cells for ENDOR-studies [22], for K-band (23 GHz) ESR-cavities [23] and for helix ESR-resonators [24, 25] are reported.

Since this article focuses on liquid samples, only details of pressure-dependent papers will be considered here. A short overview on different investigations is given.

A detailed description for a high-pressure ESR-cell up to 900 MPa using quartz glass capillaries is reported by Yamamoto et al. [26]. An apparatus for studying ESR of fluids in a flow cell under high pressure and high temperature is described by Livingston and Zeldes [27]. Measurements up to 12.4 MPa and 500°C are used to study polymerization and catalysis [28]. Even a combination of high-pressure and rapid mixing stopped-flow techniques is realized [29]. A simple system for measurements in water up to 60 MPa is described by Cannistraro [30].

---

[1] The only commercial high-pressure ESR-cell was offered by: High Research Center, Unipress Equipment Division, Warszawa, Poland.

Important Russian contributions to high pressure ESR-cells for solid state investigations appeared in the early 1970s [31, 32].

An early contribution published by Maki et al. describes the temperature- and pressure dependence on the spin exchange kinetics and the change in ESR-line widths of the di-tert-butyl nitroxide (DTBN) radical in various solvents [33, 34]. Also the nitrogen ESR-hyperfine splitting, a^N, of DTBN was measured in solution as a function of the pressure [35]. Freed et al. looked at the pressure dependence of ordering and spin relaxation in liquid crystals [36].

Solvated electrons and their reaction behavior under high pressure is published by Schindewolf et al. together with experimental details of the high-pressure cell [37]. The line width of vanadyl acetylacetonate was investigated in a number of non-hydrogen bonded solvent together with its anisotropic interactions [38]. Biochemical spin label studies in solution up to 300 MPa are reported together with a detailed cell construction [39]. Well studied kinetic phenomena like the cation migration within p-quinone radical anions have been studied up to 63.7 MPa. Activation volumes for different cations are reported [40]. Even CIDEP-experiments were performed to get information on the pressure dependence of the spin-lattice relaxation time [41].

Several papers deal with investigations on various nitroxides. The pressure of the spin exchange rate constants of 2,2,6,6-tetramethyl-4-oxo-1-piperidinoxyl (4-oxo-TEMPO) was measured in different solvents, both unpoar and polar. Up to 58.8 MPa the rate constants vary between 10^9 – 10^{10} $M^{-1}s^{-1}$ [42]. Experimental and calculated activation volumes are compared for these reactions [43]. These investigations were extended to DTBN and 2,2,6,6-tetramethyl-1-piperidinoxyl radical (TEMPO) [44] and finally to 4-Hydroxy-TEMPO, 4-Amino-TEMPO, Carboxy-PROXYL, Carbamoyl-PROXYL, 5-DOXYL and 10-DOXYL. From the rate constants and the activation volumes obtained the authors were able to calculate the corresponding exchange integral J for each reaction [45]. The pressure dependence of the inclusion equilibrium of diphenylmethyl t-butyl nitroxide and DTBN with β- and γ-cyclodextrins were studied in detail [46,47]. The effect of pressure and the solvent dependence of the intramolecular spin-exchange of biradicals with two nitroxide fragments linked by a long flexible chain were obtained from ESR.-lifetimes studies of the radical fragments inside and outside the cage. The nearly cyclic conformation in the cage is reported as the favorable one in solution [48]. A short review in Japanese appeared on the spin exchange kinetics of nitroxides [49]. Applications of the spin-label technique at high pressures is reported up to the high pressure of 700 MPa [50]. Recently Hubbell et al. [51] published the design of a high-pressure cell using polytetrafluoroethylene (PTFE) coating fused silica capillary tubes up to 400 MPa. Bundles of these capillaries are placed inside an ESR-cavity. This cell is adapted from an NMR-cell designed by Yonker et al. [52].

Another recent high pressure setup was used to investigate electron self-exchange reactions [53]. The experimental arrangement used by the authors is illustrated in figure 2. A hand-driven high-pressure pump is connected with the medium exchanger, separating sample solution and pressure medium, using ethylene glycol as pressure medium. The sample-side of the system can evacuated for sample changing and cleaning.

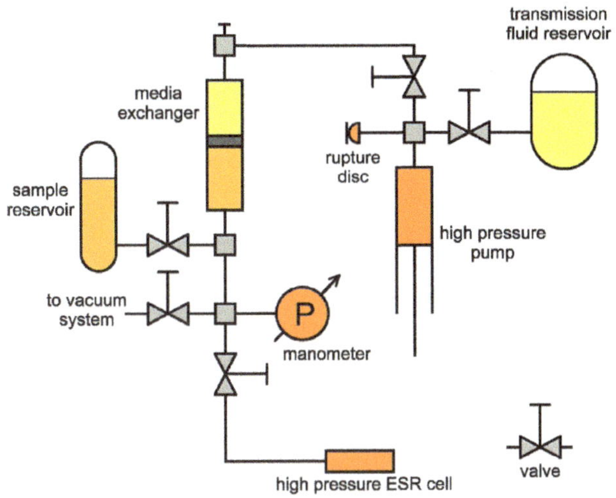

Figure 2. Diagram of a typical arrangement for experiments at elevated pressures [53].

2. Electron self-exchange reactions

The ESR line broadening method was first used by Ward and Weissman [54] in 1957 when they investigated the self-exchange reaction between naphthalene and its radical anion. In the following decade, Weissman continued his work, adding information on further systems as well as refining the theoretical treatment of the experiment [55]. Other important contributors were Miller and Adams [56], who investigated several systems using electrochemically generated radicals. These authors were among the first to use this technique to make a comparison between homogeneous and heterogeneous electron transfer according to the Marcus Theory [57]. Since then, numerous other authors have made their contributions, covering many aspects of the self-exchange reaction, including temperature dependencies and effects from counter ions and solvent.

2.1. Theory

The theoretical aspects of electron transfer in solution originate in the 1950s with the pioneering work of Marcus [58-60], who offered a description of the energetics of both the self-exchange and the general reaction. Since then the theory has greatly evolved [61-63], as shall be discussed in the theoretical part of this chapter, where especially solvent (dynamic) effects are of interest.

In general, self-exchange denotes the one-electron transfer between the partners of a redox couple, which results in no net chemical reaction. The mechanism is shown in figure 3, using a radical anion, $A^{\bullet -}$, and its neutral parent compound, A.

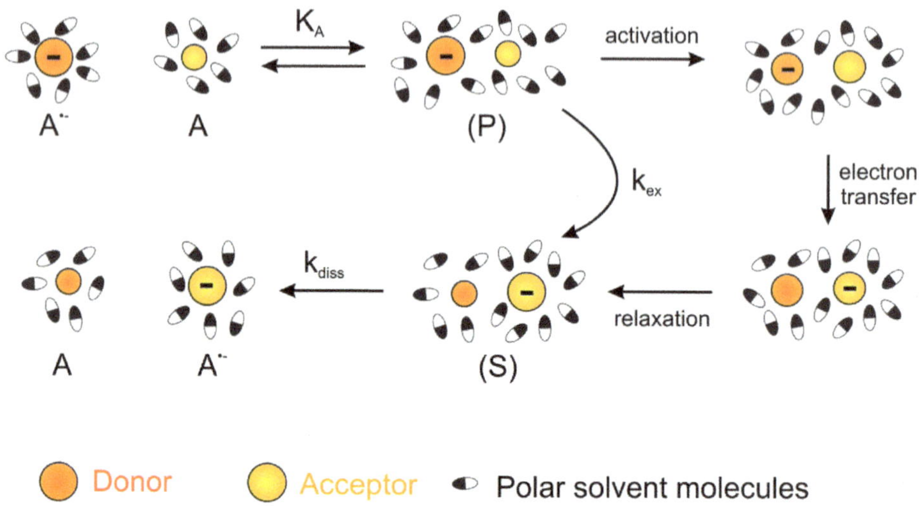

Figure 3. The reaction mechanism of an electron transfer reaction.

The bimolecular electron transfer rate constant, $k_{et} = K_A k_{ex}$, is expressed as

$$k_{et} = K_A \kappa_{el} \nu_n \exp\left(\frac{-\Delta G^*}{RT}\right) \tag{10}$$

Here, $K_A = 4\pi N_A \sigma^2 \delta\sigma$ is the formation constant of the precursor complex, which in turn is expressed by Avogadro's number, N_A, the reaction distance σ and the reaction zone thickness $\delta\sigma$. Further in equation (10), κ_{el} is the electronic transmission coefficient, υ_n the nuclear collision frequency and ΔG^* energy of activation according to Marcus.

Figure 4 depicts the energy surfaces of precursor and successor complexes, simplified in a two-dimensional representation of energy as a parabolic function of reaction coordinate. From this it is possible to express the activation energy as $\Delta G^* = \lambda/4 - V_{PS}$, using the reorganization energy, λ, and the resonance splitting energy, V_{PS}.

Strictly speaking, the Marcus equation only applies to the precursor and successor complexes. In most cases, however, no information is available about these and therefore it is desirable to express the Marcus equation in terms of the reactants, A and $A^{\bullet -}$. In the case where both reactants have nonzero charges, zA and zD, a work term, W, accounting for the Coulombic work needed to bring the reactants together must be included,

$$W = \frac{z_A z_D e_0^2}{4\pi\varepsilon_0 \varepsilon_S \sigma} \tag{11}$$

with e_0 representing the electronic charge, ε_S the dielectric constant of the medium, σ the reaction distance.

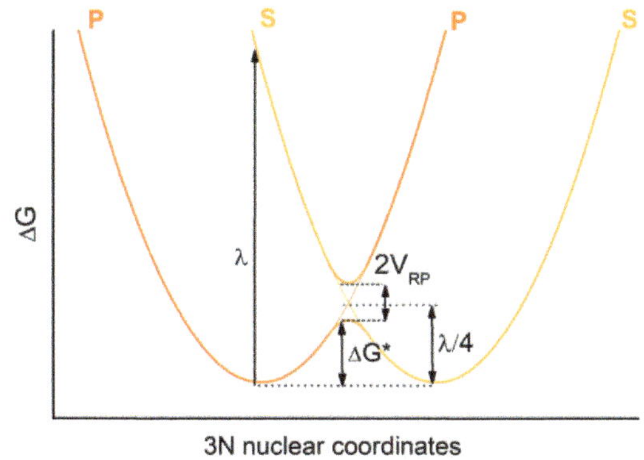

Figure 4. Energy diagram of the self-exchange reaction

Resulting, the expression of the Marcus equation takes the following form:

$$\Delta G^{*} = W + \frac{\lambda}{4} - V_{PS} \tag{12}$$

The reorganization energy, λ, is an energetic measure of the changes the reactants and their surroundings undergo as the reaction proceeds.

Commonly, the reorganization energy is split into two contributions, the inner and outer reorganization energies, λ_i and λ_o , respectively.

$$\lambda = \lambda_i + \lambda_o \tag{13}$$

λ_i, takes into account the changes in bond lengths and/or angles in the reactants undergoing electron transfer. In the classic high temperature limit, where all vibrational frequencies fulfil $\nu_n << k_BT/h$ (not confusing ν_n with the collision frequency), λ_i is given by:

$$\lambda_i^{\infty} = \sum_j \frac{f_R^j \cdot f_P^j}{f_R^j + f_P^j} \left(\Delta q_j\right)^2 \tag{14}$$

Here, $f_{R,P}$ is the force constant of bond j, either in the reactants, R, or the products P, and Δq_j expresses the change in length of bond j. Unfortunately this approach has certain drawbacks when organic redox couples are considered, as it becomes increasingly difficult to find all important vibrations contributing to the reaction coordinate. The situation worsens further as the high temperature approximation is not always valid, meaning that corrections for quantum-mechanical tunnelling should be included. This was pointed out by Holstein [64], and used by several other authors like Bixon and Jortner [61, 65], and Sutin [66] yielding an equation of the following type:

$$\lambda_i(T) = \lambda_i^{\infty} \frac{4k_BT}{h\overline{\nu}} \tanh\left(\frac{h\overline{\nu}}{4k_BT}\right) \tag{15}$$

where $\bar{\nu}$ =5·10^{13} s^{-1} is used as the mean vibration frequency for organic redox couples.

A method of calculating λ_i^∞ , without explicitly knowing the changes in bond-lengths and force constants, has been published by Nelsen *et al.* [67,68], needing the energies of four distinct points on the energy surfaces of products and reactants. By calculating the energies of the reactants in their most stable geometrical configuration, as well as in the most stable configuration of the products, together with analogous calculations for the products, λ_i^∞ may be obtained.

The outer reorganization energy, λ_o, which is sometimes also referred to as the solvent reorganization energy, describes the change in polarization of the solvent molecule surrounding the activated complex. Marcus showed that the outer reorganization energy may be expressed as in equation (16), using a simple dielectric continuum model and assuming that the reactants are spherical.

$$\lambda_o = \frac{e_0^2 N_A}{4\pi \varepsilon_0} g(r,\sigma) \cdot \gamma \tag{16}$$

Here, apart from the permittivity of vacuum, ε_0, new parameters are the function g(r,σ) which depends on the geometry of the transition state, and γ, the so-called Pekar factor, which depends on the dielectric properties of the solvent.

In its simplest form, assuming $r_A = r_{A\bullet-} = r$, using spherical models of the reactants, g(r, σ) is given by:

$$g(r,\sigma) = \frac{1}{r} - \frac{1}{\sigma} \tag{17}$$

Often the reaction distance, σ, is assumed to be that of close contact, here being twice the radius of the reactants: σ=2r. For organic molecules, this approach is often too simple, as the geometry can no longer be assumed to be spherical. If, in stead, an ellipsoidal model is used, r is replaced by the mean elliptic radius, $\bar{r}$:

$$\bar{r} \approx \frac{\left(a^2 - c^2\right)^{1/2}}{F(\varphi,\alpha)} \tag{18}$$

with the elliptic semi axes fulfilling a > b > c and F(ϕ,α) being elliptic integrals of the first kind, using $\varphi = \arcsin\left[\left(a^2 - c^2\right)^{1/2} / a\right]$ and $\alpha = \arcsin\left[\left(a^2 - b^2\right) / \left(a^2 - c^2\right)\right]^{1/2}$ [69]. A good approximation has been found to be: $\bar{r} = \frac{1}{3}(a + b + c)$.

Another approach has been suggested by Compton and co-workers [70, 71] who, instead of calculated radii, use experimentally determined hydrodynamic radii.

As intermolecular orientation becomes more difficult to describe, the reaction distance may also be modified using the ellipsoid model [72, 73], replacing σ, with the apparent reaction distance, σ′:

$$\frac{1}{\sigma'}=\frac{1}{\sigma}\left[1+\frac{2c^2-a^2-b^2}{3\sigma^2}+\frac{abc}{\sigma^3}+\frac{4\left(8c^4+3\left(a^4+b^4\right)-8c^2\left(a^2+b^2\right)+2a^2b^2\right)}{15\sigma^4}\right] \tag{19}$$

The second contribution to the reorganization energy, the solvent dependent Pekar factor, γ, in turn, is given by

$$\gamma=\frac{1}{\varepsilon_\infty}-\frac{1}{\varepsilon_S} \tag{20}$$

where ε_S and ε_∞ are the static and optical dielectric constants, respectively. Unfortunately, reliable experimental values of ε_∞, can be very hard to come by, so it has become common practice to use the refractive index, n_D, in an approximation: $\varepsilon_\infty = n_D^2$.

Apart from the obvious solvent dependence of the reorganization energy, the rate constant of an electron transfer reaction may also be influenced by the solvent in terms of the nuclear frequency factor, ν_n.

Originally, Marcus expressed ν_n using a simple gas-phase collision model [74], yielding equation (21),

$$\nu_n=\sigma^2 N_A\left(\frac{16\pi RT}{M}\right)^{1/2} \tag{21}$$

with the only new parameter being the molar mass, M.

The solvent dependence is introduced by taking the formation of the precursor complex into account, expressing the total frequency factor as combination of those of the solvent, ν_o and the reactants, ν_i, weighted by the respective reorganization energies.

$$\nu_n=\left(\frac{\nu_i^2\lambda_i+\nu_o^2\lambda_o}{\lambda}\right)^{1/2} \tag{22}$$

Equation (22) is valid only for rapid dielectric relaxation in the solvent. For slow dielectric relaxation, the longitudinal relaxation time of the solvent, τ_L, must be included as the solvent relaxation now happens on the same time scale as the reaction. This observation was first made by Kramers [75], who spoke of a solvent dynamic effect or solvent friction, and the theory was later expanded by authors like Zusman [76, 77], Jortner and Rips [61, 78, 79], Marcus and Sumi [63] as well as several others.

The longitudinal relaxation time may be determined from the usual dielectric parameters of the solvent [80],

$$\tau_{\mathrm{L}} = \frac{\varepsilon_{\infty}}{\varepsilon_{\mathrm{S}}}\tau_{\mathrm{D}} \tag{23}$$

with, τ_D being the Debye relaxation time, which itself may be expressed using the molar volume, V_M, and the viscosity, η, of the solvent:

$$\tau_D = \frac{3V_M\eta}{RT} \tag{24}$$

Assuming that $\lambda_i << \lambda_o$, the nuclear frequency factor then becomes:

$$\nu_n = \frac{1}{\tau_L}\left(\frac{\lambda_o}{4\pi RT}\right)^{1/2} \tag{25}$$

So far it has been supposed that the electron transfer reaction was adiabatic, assuming the transmission factor, κ_{el}, to be unity. This applies to all reactions studied in this work, but for completeness, the expression valid in the case of diabatic electron transfer [69, 76, 81, 82] is given here:

$$\kappa_{el}\nu_n = \frac{2\pi V^2}{\hbar N_A\left(4\pi\lambda_o RT\right)^{1/2}} \tag{26}$$

It is now possible to discern the nature of the electron transfer by the use of simple mathematical treatment, calculating the experimental value, $(\kappa_{el}\nu_n)_{obs}$ as suggested by Weaver [83, 84],

$$\left(\kappa_{el}\nu_n\right)_{obs} = k_{et,obs}K_A^{-1}\exp\left(\frac{\Delta G^*_{calc}}{RT}\right) \tag{27}$$

where $k_{et,obs}$ is the experimentally determined rate constant and ΔG^*_{calc} the activation energy calculated using equation (12). Double logarithmic plots determine the type of electron transfer observed. In the general Marcus case, a constant value of ln $(\kappa_{el}\nu_n)_{obs}$ is expected, whereas the diabatic and adiabatic cases both produce straight lines with slopes of unity when plotted against ln $(\gamma^{-1/2})$ or ln $(\gamma^{1/2}\tau_L^{-1})$, respectively.

It is well-known that, k_{ex} depends on temperature and pressure. The temperature dependence is seen directly in equation (28), but also in the dependencies of the activation energy, via γ, and that of the nuclear frequency factor.

$$k_{ex}\left(p,T\right) = \kappa_{el}\nu_n\left(p,T\right)\exp\left(\frac{-\Delta G^*\left(p,T\right)}{RT}\right) \tag{28}$$

The temperature dependence of ν_n is often neglected in the general Marcus case and that exhibiting diabatic solvent dynamics. However, when dealing with adiabatic solvent dynamics, the temperature dependence must be included, due to the fact that τ_L may vary strongly with temperature, according on the solvent in question. In some cases even, this variation may be of more than an order of magnitude, as in the case of propylene carbonate, where τ_L is 42.6 ps at 228 K and 2.6 ps at 308 K [85].

The pressure dependence is usually expressed using the activation volume, $\Delta^{\ddagger}V$:

$$\Delta^{\ddagger}V = -RT\left(\frac{\partial \ln k_{ex}}{\partial P}\right)_T \tag{29}$$

In terms of transition state theory, the activation volume may be expressed using the volume of the transition state, $^{\ddagger}V$ and the molar volumes of the reactants, V_A and $V_{A\bullet-}$, cf. equation (30),

$$\Delta^{\ddagger}V = {}^{\ddagger}V - \left(V_A + V_{A\bullet-}\right) \tag{30}$$

and thus represents the changes in volume involved in the electron transfer process. This is also seen in figure 5, where the volume profile of a reaction with a positive volume of activation is shown.

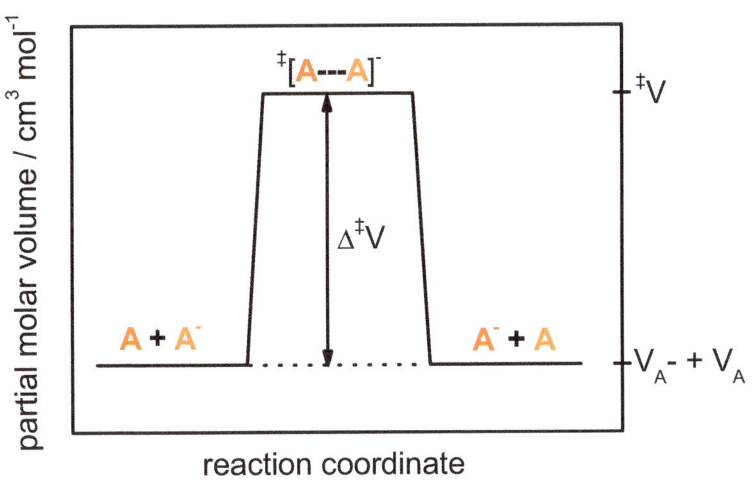

Figure 5. Volume profile of an electron transfer reaction

Generally, one may also define a reaction volume, Δ_rV, as the difference in the volumes of products and reactants, but for self-exchange reactions, this becomes zero.

As in the case of ΔG^*, one may seek to describe the activation volume by separating the various contributions. This has been discussed in great detail by Swaddle [10, 86-88], who derived expressions of the contributions from inner and outer reorganization, $\Delta^{\ddagger}V_i$ and $\Delta^{\ddagger}V_o$, Coulombic work, $\Delta^{\ddagger}V_{Coulomb}$, Debye-Hückel effects, $\Delta^{\ddagger}V_{DH}$, solvent dynamics, $\Delta^{\ddagger}V_{SD}$, and finally a contribution from the non-Coulombic part of the precursor complex formation constant, $\Delta^{\ddagger}V_{PREC}$. Summing up the different parts, the volume of activation of the self-exchange reaction is calculated as expressed in equation (31).

$$\Delta^{\ddagger}V_{calc} = \Delta^{\ddagger}V_i + \Delta^{\ddagger}V_o + \Delta^{\ddagger}V_{Coulomb} + \Delta^{\ddagger}V_{DH} + \Delta^{\ddagger}V_{SD} + \Delta^{\ddagger}V_{PREC} \tag{31}$$

Each of these contributions is treated separately by differentiation of the respective energetic expressions.

The contribution to the activation volume, arising from inner reorganization is usually neglected, as the increase in volume of the acceptor almost cancels the decrease in volume of the donor. Stranks [89] showed that in most cases, a value of 0.6 cm^3 mol^{-1} may be taken for $\Delta^{\ddagger}V_i$, which is usually close to the experimental uncertainty of $\Delta^{\ddagger}V$. The contribution from $\Delta^{\ddagger}V_{PREC}$ is also expected to be small since both σ and $\delta\sigma$ are practically pressure independent [88].

Taking the Marcus expression for the outer reorganization energy, λ_o, and assuming that only the Pekar factor, γ, is pressure dependent, and additionally taking into account compression of the solvent, $\Delta^{\ddagger}V_o$, must be given as shown in equation (32), remembering that λ_o only contributes with a quarter of its value to the activation energy.

$$\Delta^{\ddagger}V_o = \frac{1}{4}\left(\frac{\partial \lambda_o}{\partial P}\right)_T = \frac{e_0^2 N_A}{16\pi\varepsilon_0}\left[\left(\frac{1}{r}-\frac{1}{\sigma}\right)\left(\frac{\partial \gamma}{\partial P}\right)_T - \gamma\frac{\beta}{3\sigma}\right] \tag{32}$$

The Pekar factor is approximated as $\gamma = n^{-2} - \varepsilon_S^{-1}$ using the refractive index, n, and static dielectric constant, ε_S, of the solvent and the isothermal compressibility of the solvent, β, is defined via its density, ρ, as $\beta = \rho^{-1}\left(\frac{\partial \rho}{\partial P}\right)_T$.

By simple differentiation of the work term given in equation (11), the Coulombic contribution to the activation volume is found.

$$\Delta^{\ddagger}V_{Coulomb} = \frac{z_A z_D e_0^2 N_A}{4\pi\varepsilon_0\sigma}\left(\frac{\partial \varepsilon_s^{-1}}{\partial P}\right)_T \tag{33}$$

The expression for $\Delta^{\ddagger}V_{DH}$ is derived from the extended Debye-Hückel theory, and takes the following, rather cumbersome, form:

$$\Delta^{\ddagger}V_{DH} = \frac{RTz_A z_D C I^{1/2}}{\left(1+BaI^{1/2}\right)^2}\left[\left(\frac{\partial \ln\varepsilon}{\partial P}\right)_T\left(3+2BaI^{1/2}\right)-\beta\right] \tag{34}$$

with a, B; and C being the Debye-Hückel constants and I the ionic strength.

These two contributions are of opposite sign and usually, for values of I between 0.1 and 0.5, tend to cancel each other quite conveniently [86, 87]. In the case where neutral molecules are involved, as it is in the experimental part of this work, both $\Delta^{\ddagger}V_{Coulomb}$ and $\Delta^{\ddagger}V_{DH}$ are zero.

Recalling that the solvent dynamic effect is manifested in the pre-exponential term of the Arrhenius equation, the following definition offers itself readily:

$$\Delta^{\ddagger}V_{SD} = RT\left(\frac{\partial \ln\left(\kappa_{el}\nu_n\right)}{\partial P}\right)_T \tag{35}$$

This may then be further elaborated in the cases of diabatic and adiabatic solvent dynamics, as shown shortly. Note that even in the general Marcus case, with $\kappa_{el}\nu_n$ depending on σ, a slight pressure dependence is expected, but as argued earlier, this may be neglected. However, due to the logarithmic form of equation (35), it is now considerably simpler to describe the influence of σ, as all constants in equation (21) are conveniently eliminated. Assuming that σ compresses approximately as the solvent, i.e. $\sigma \propto \rho^{-1/3}$, the contribution to the activation volume should be obtainable with relative ease.

In the diabatic case given in equation (36), $\Delta^{\ddagger}V_{SD}$ becomes:

$$\Delta^{\ddagger}V_{SD} = RT\left(\frac{\partial}{\partial P}\left(\ln\frac{2\pi V^2}{\hbar N_A\left(4\pi\lambda_o RT\right)^{1/2}}\right)\right)_T = -\frac{RT}{2}\left(\frac{\partial \ln\gamma}{\partial P}\right)_T \tag{36}$$

with all pressure independent quantities eliminated as above. In cases of strong diabaticity, where a noticeable dependence of k_{ex} on σ is expected, one must include the pressure dependence of V_{PS}, which follows the principle discussed in the Marcus case, and shall not be further elaborated here.

For electron transfer exhibiting adiabatic solvent dynamic behaviour, the situation becomes slightly more complicated, as now also the longitudinal relaxation time contributes to $\Delta^{\ddagger}V_{SD}$.

$$\Delta^{\ddagger}V_{SD} = RT\left(\frac{\partial}{\partial P}\left(\ln\frac{1}{\tau_L}\left(\frac{\lambda_o}{4\pi RT}\right)^{1/2}\right)\right)_T = RT\left[\frac{1}{2}\left(\frac{\partial \ln\gamma}{\partial P}\right)_T - \left(\frac{\partial \ln\tau_L}{\partial P}\right)_T\right] \tag{37}$$

Returning to electron self-exchange reactions, where the activation volume consists of the five parts, which have just been introduced, by far the greatest contribution is expected from $\Delta^{\ddagger}V_o$. As already mentioned, $\Delta^{\ddagger}V_{Coulomb}$ and $\Delta^{\ddagger}V_{DH}$ tend to cancel each other, or even vanish when an uncharged molecule is reacting, and $\Delta^{\ddagger}V_i$ is generally negligible. Consequently, one may write:

$$\Delta^{\ddagger}V = \Delta^{\ddagger}V_o + \Delta^{\ddagger}V_{SD} \tag{38}$$

In other words, the activation volume may be interpreted as owing to a term described by the energy barrier crossed during the reaction, $\Delta^{\ddagger}V_o$, and a term linked to the actual crossing of the barrier, $\Delta^{\ddagger}V_{SD}$. Even here, the contribution from the pre-exponential factor is unimportant in the general Marcus case and in that of diabatic solvent dynamics, leaving only the adiabatic case with an appreciable contribution.

Typical values of $\Delta^{\ddagger}V$ are found to be in the range of -30 to +30 $cm^3\ mol^{-1}$ [10], corresponding to a retardation or acceleration of the reaction of about a factor of 4 at 100 MPa as compared to ambient pressure (0,1 MPa).

2.2. Experimental

As mentioned above, much work has been done using ESR line broadening experiments to study self-exchange reactions; however, only in surprisingly few cases nitroxides have been involved. Nitroxides are readily oxidized [90-92] to form the diamagnetic compound needed in the self exchange with the neutral radical. In most cases, the ESR spectra remain within the slow exchange region due to the relatively large nitrogen hyperfine splitting constant. The reported results have dealt with the self-exchange of various nitroxide couples in acetonitrile [93] as well as of that of TEMPO with its oxidized counterpart in several different solvents [94]. For the latter couple, the temperature dependence of the exchange reaction has also been investigated [95].

Some typical spectra are shown in figure 6 together with the accompanying plot to determine k_{obs}. Subsequently, the electron transfer rate constant, k_{et}, can be obtained applying a correction for diffusion as shown in equation (39).

$$\frac{1}{k_{obs}} = \frac{1}{k_{et}} + \frac{2}{k_d} \tag{39}$$

The factor of two seen on the last term of equation (39), as opposed to that of one which is seen in more general reactions, is due to the nature of the self-exchange reaction.

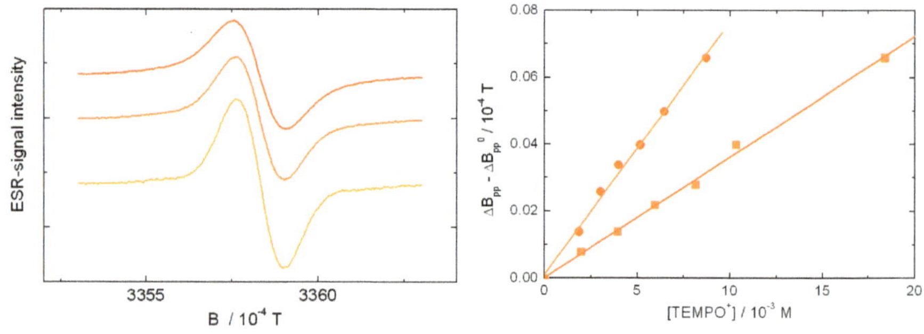

Figure 6. Left: Central lines of the ESR spectra of TEMPO in acetonitrile. Concentrations of $TEMPO^+\ Cl^-$ were 0 mM (bottom) 8 mM (middle) and 18 mM (top). Right: Concentration dependence of the change in peak-to-peak line width in water (squares) and propylene carbonate (circles).

The experimental results of references [94] and [95] serve as a good example of how ESR line broadening experiments can be utilized to study solvent effects on electron transfer.

Table 1 presents the electron transfer rate determined as given above, together with relevant solvent properties and the quantities needed to investigate the presence of a solvent effect.

Solvent	k_{et} / 10^8 l mol^{-1} s^{-1}	k_{diff} / 10^8 l mol^{-1} s^{-1}	γ	τ_L / ps	ΔG^*_{calc} / kJ mol^{-1}	$(\kappa_{el}\nu_n)_{obs}$ / 10^{14}s^{-1}
Acetonitrile [94]	2.6±0.2	190	0.527 [96]	0.21 [96]	35	20
Benzonitrile [94]	2.2±0.1	52.4	0.389 [85]	5.75 [85]	27,2	0.67
Deuterium Oxide [94]	1.4±0.1	51.9	0.554 [97]	0.593 [97]	36.5	20
Methanol [95]	0.50±0.05	109	0.537 [98]	9.36 [98]	35.5	4.7
Propylene Carbonate [94]	2.1±0.2	23.5	0.480 [85]	3.27 [85]	32.3	5.2
Tetrahydrofuran [94]	5.0±0.3	114	0.37 [69]	1.7 [69]	26	1.1
Water [94]	0.82±0.07	64.8	0.550 [99]	0.812 [99]	36.3	10
Water [95]	0.72±0.06	64.8	0.550 [99]	0,812 [99]	36.3	8.5

Numbers in brackets correspond to the relevant reference.

Table 1. Self-exchange rate constants TEMPO/TEMPO$^+$ in pure solvents at 293K

The values of ΔG^*_{calc} listed in table 1 were determined using a λ_i of 33 kJ mol^{-1} and λ_o of 191 $\times\gamma$ kJ mol^{-1}. Both values were calculated [95] as shown above, and have been confirmed experimentally by other authors [100].

Values of $(\kappa_{el}\nu_n)_{obs}$ were obtained using equation (27), and as it is clearly not solvent independent, the general Marcus type of electron transfer is immediately ruled out and the existence of a solvent dynamic effect is successfully demonstrated. The nature of this is shown in figure 7 where the appropriate double logarithmic plots are given.

From figure 7 it can clearly be seen that the presence of a diabatic solvent effect can be excluded, since a line with a slope of unity would be needed to confirm the corresponding assumption. The line shown in the right hand plot of the figure has been fixed to this slope and thereby shows that an adiabatic solvent effect is likely in the case of self-exchange in the TEMPO/TEMPO$^+$ system.

Since the literature values of the dielectric properties of the solvents are often subject to large uncertainties error bars have been introduced in each diagram to illustrate the range of values corresponding to errors of 10% in each γ and τ_L. In the case of the former an error of such magnitude is highly unexpected, but for the latter, 10% might even be considered conservative in some cases.

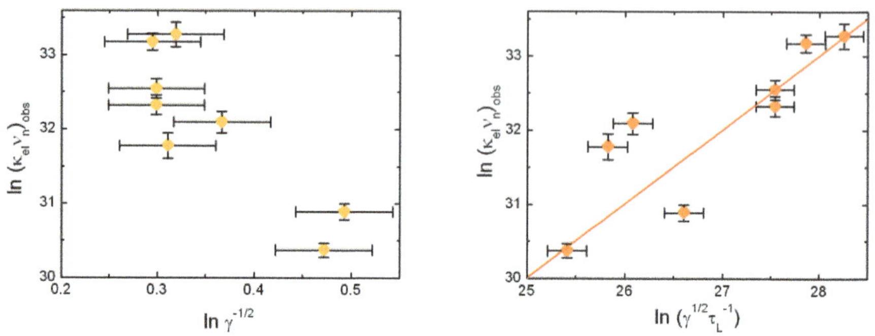

Figure 7. Diagrams assuming diabatic (left) and adiabatic (right) solvent dynamic effects. X-axis error bars correspond to a 10% error on each γ and τ_L. The line has a fixed slope of one.

In fact, regarding the longitudinal relaxation time needed in the calculations concerning the adiabatic case a few remarks must be made. In many solvents and electrolyte solutions, high precision experiments have indicated the existence of several relaxation pathways in the solvent. Usually it is assumed that only the slowest of the relaxation processes will influence the electron transfer reaction [76, 101] and the τ_L to be used in investigations of solvent dynamic effects, is thus the largest of the relaxation times:

$$\tau_{iL} = \frac{\varepsilon_{i\infty}}{\varepsilon_{iS}}\tau_{iD} \quad i = 1, 2, \ldots \tag{40}$$

As an example, take methanol [98] which has two relaxations with parameters at 293K: $\varepsilon_S = \varepsilon_{1S} = 32.5$, $\varepsilon_{1\infty} = \varepsilon_{2S} = 5.91$, $\varepsilon_{2\infty} = \varepsilon_\infty = 4.9$ and $\tau_{D1} = 51.49$ ps, $\tau_{D2} = 7.09$ ps, leading to $\tau_{L1} = 9.36$ ps and $\tau_{L2} = 5.88$ ps. For comparison, both older and newer works [102-105] report only a single relaxation process for methanol, resulting in a τ_L which ranges from 4.58 ps to 10.6 ps. The reason that these latter experiments have only detected one relaxation time is, that they have been conducted using a relatively narrow frequency range and thus a long extrapolation for determination of ε_∞.

Having established that the solvent dynamic effect is adiabatic, a plot of ln $(k_{et}\tau_L\gamma^{-1/2})$ vs. γ allow the calculation of g(r,σ′) since,

$$\ln\left(k_{et}\tau_L\gamma^{-1/2}\right) = const - \frac{\lambda_o}{4RT} = const - \frac{e_0^2 g(r,\sigma')}{16\pi\varepsilon_0 k_B T}\gamma \tag{41}$$

From the slope of this plot, while taking $r = \bar{r} = 3.11$ Å, the reaction distance becomes σ′ = 5.0 Å, which is in reasonably good agreement with the value of σ′ = 5.4 Å found in the theoretical calculations [94]. The experimental value of σ′ corresponds to $K_A = 0.21$ l mol^{-1}.

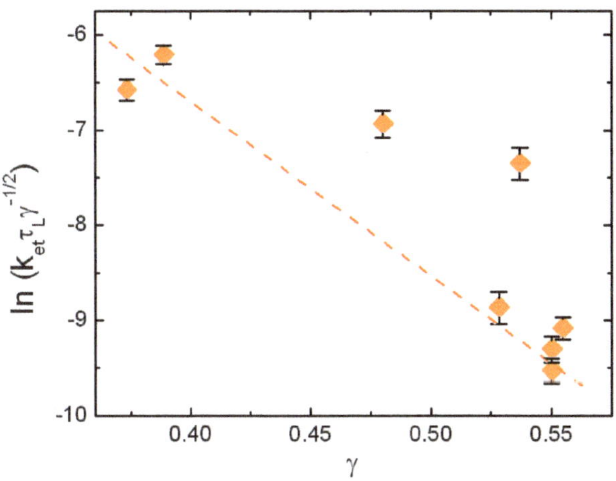

Figure 8. 'Traditional' diagram for illustration of the adiabatic solvent dynamic effect. The slope of the dotted line corresponds to a reaction distance of 5.0 Å.

The listed rate constants were used to determine the activation energy of the reaction in each of the solvents, which, due to the adiabatic solvent dynamics, was done by plotting ln $(k_{et}\tau_L\gamma^{-1/2}T^{1/2})$ vs. T^{-1} as can be seen in table 2. For comparison, the values obtained from common Arrhenius plots (ln k_{et} vs T^{-1}) are also shown. The table also contains the longitudinal relaxation energies of the solvent as well as the quantity ΔH^{*}_{calc}, which represents a theoretical prediction of the activation enthalpy, taking into account the temperature dependency of the reorganization energy [95].

The concerns regarding the physical properties of solvents become even more pronounced when dealing with temperature dependencies. Once again, the longitudinal relaxation time provides the greatest problem as the relaxation energy, H_L, varies tremendously in literature. Using the example of methanol again, H_L could only be obtained from data using a single relaxation time, whereas for THF no less than four values are reported, ranging from -8.3 kJ mol^{-1} [106] to 11.1 kJ mol^{-1} [107].

Returning to the experimental activation energies and enthalpies, it is seen that the relation $E_a = \Delta H^* + H_L$ is fulfilled within the experimental error, as expected due to the Arrhenius-like temperature dependence of the involved quantities, with any deviations being mainly due to the slight temperature dependence of γ. The results illustrate the importance of knowing whether or not solvent dynamics are present in an electron transfer system before determining the activation parameters. The activation energies from the traditional Arrhenius plots do not correspond to Marcus' ΔG^* due to the involvement of τ_L. To make matters worse, neither does the energy obtained from the plots of ln $(k_{et}\tau_L\gamma^{-1/2}T^{1/2})$ vs. T^{-1}, since both λ_i and λ_o are temperature dependent. For this reason the values in the table have been labeled as enthalpies in order to signify that they correspond to the temperature independent part of ΔG^*. Taking the temperature dependence of ΔG^* into account, it is possible to reach a theoretical estimate of this enthalpy, ΔH^*_{calc} [95]. As can be seen, for acetonitrile, benzonitrile, tetrahydrofuran and water, good agreement is found between experiment and theory. This is not the case for the three remaining solvents, where large discrepancies are found. Except for propylene carbonate, the main reason for the poor agreement is likely to be the quality of the dielectric data as already discussed.

Solvent	Energy / kJ mol^{-1}			
	E_a a)	ΔH^* b)	H_L c)	ΔH^*_{calc} d)
Acetonitrile [94]	18.2±0.7	18.1±0.7	1.0 [96]	19
Benzonitrile [94]	23.9±0.7	14.1±0.4	11.0 [85]	15
Deuterium Oxide [94]	15±2	-4.3±0.5	19.0 [97]	25
Methanol [95]	21±2	8.8±0.5	12.8 [102]	31
Propylene Carbonate [94]	13±1	-7±1	20.9 [85]	18
Tetrahydrofuran [94]	6.7±0.8	16.6±0.4	-8.3 [106]	19
Water [94]	6.5±0.2	22.7±0.5	-14.9 [99]	24
Water [95]	6.8±0.1	22.9±0.4	-14.9 [99]	24

a) Activation energy determined from a traditional Arrhenius plot (ln k_{et} vs. T^{-1})

b) Activation energy determined assuming adiabatic solvent dynamics (ln $k_{et}\tau_L\gamma^{-1/2}T^{1/2}$ vs. T^{-1})

c) Longitudinal relaxation energy obtained from the empirical relationship $\ln \tau_L = \frac{H_L}{RT} + \ln b$.

d) From ref [95].

Numbers in brackets correspond to the relevant reference.

Table 2. Characteristic energies of the of the TEMPO/TEMPO$^+$ self-exchange in pure solvents at 293K

3. Spin exchange reactions

Spin exchange and its effect on the acquired ESR spectra is well understood, experimentally as well as theoretically [108-115]. A large amount work has been done using nitroxide radicals, e.g. by Plachy and Kivelson who studied the spin exchange between di-tertiary-butyl nitroxide radicals in n-pentane and propane [116]. Another interesting work is one of Bales, where the spin exchange between various nitroxide radicals was used to study the structure of micelles as well as chemical reactions taking place within them [117].

Other authors have studied intramolecular spin exchange in nitroxide biradicals [118-122] or the so-called paramagnetic exchange between a nitroxide radical and a paramagnetic line broadening agent [123-126]. Several reports have been given on temperature and pressure dependent spin exchange reactions [33, 42-45].

Due to the wealth of information available on spin exchange of nitroxide radicals, here, focus shall be on the pressure dependence of the spin exchange reaction, as this arguably has received the least attention in the past.

3.1. Theory of pressure dependent spin exchange

With respect to the pressure dependence of the spin exchange reaction, an approach similar to the one shown for self-exchange can be made. Thus, the activation volume takes the form shown below, containing contributions from diffusion, Coulombic work and Debye-Hückel effects.

$$\Delta^{\ddagger}V_{calc} = \Delta^{\ddagger}V_{diff} + \Delta^{\ddagger}V_{Coulomb} + \Delta^{\ddagger}V_{DH} - RT\beta \tag{42}$$

$\Delta^{\ddagger}V_{Columb}$ and $\Delta^{\ddagger}V_{DH}$ can be calculated as shown above and $\Delta^{\ddagger}V_{diff}$ is obtained analogously to equation (29):

$$\Delta^{\ddagger}V_{diff} = -RT\left(\frac{\partial \ln k_{diff}}{\partial P}\right)_T \tag{43}$$

The expression of k_{diff} varies with the used theoretical description of diffusion, but often a simple Smoluchowski approach, as above, is sufficient. The last term in eq. 5 takes into account the compression of the solvent due to elevated pressures and typically has a value of 1 – 3 cm^3/mol.

3.2. Experimental results

Spin exchange rate constants, $k_{obs} = k_{se}$, obtained in some pressure dependent studies are shown in table 3 together with the corresponding activation volumes for spin exchange and diffusion.

From the pressure dependent rate constants, volumes of activation were determined for each system according to equation (29).

The differences between the values of $\Delta^{\ddagger}V_{exp}$ and $\Delta^{\ddagger}V_{diff}$ vary tremendously with the solvent and radical used. Several explanations for this have been offered. For example, it was noted in [43] that for non-polar solvents, the agreement between the volumes is very good, whereas it fails for protic polar solvents. The explanation for this is suggested to lie in the formation of the encounter complex prior to the spin exchange. If the solvent assists in this, e.g. through hydrogen bonds, a noticeable reduction in the experimental activation volume is to be expected.

The theoretical description of k_{se} has been thoroughly dealt with in literature [108, 109]. relating it to the diffusional rate constant, k_{diff}, by

$$k_{se} = pf_S k_{diff} \tag{44}$$

where p is the probability of spin exchange upon collision of the radicals and f_s is a steric factor that takes into account the anisotropy of the spin exchange, i.e., its dependence on the orientation of the colliding particles.

Compound	Solvent	k_{se} / 10^9 l mol^{-1} s^{-1}	$\Delta^‡V_{exp}$ / cm^3 mol^{-1}	$\Delta^‡V_{diff}$ / cm^3 mol^{-1}
TEMPO	Acetone [44]	7.8	10.7	15.8
	Acetonitrile [95]	3.9±0.3	9.4±0.7	8,1
	Methanol [95]	2.8±0.2	8.5±0.9	8.8
4-hydroxy-TEMPO	Acetone [45]	6.42	8.1	15.8
	n-Hexane [45]	10.7	13.3	23.2
4-oxo-TEMPO	Acetone [44]	8.1	8.7	15.8
	n-Hexane [44]	11.5	12.1	23.2
	Methanol [42]	5.5±0.2	6.7±1.1	8.8
	Water [42]	1.9±0.1	-7.6±0.4	-2.5
	Benzene [43]	5.3±0.3	18.4±1.0	19.1
	Toluene [42]	5.27±0.09	13.7±0.9	14.2
	o-Xylene [43]	4.2±0.2	15.3±1.2	15.5
4-amino-TEMPO	Acetone [45]	5.82	12.6	15.8
	n-Hexane [45]	8.94	11.5	23.2
DTBN	Acetone [44]	9.4	7.4	15.8

Numbers in brackets correspond to the relevant reference.

Table 3. Rate constants and activation volumes of selected spin exchange reactions.

The probability factor can be determined as shown in equation (45), using the exchange integral, J, and the collision time τ_1 as well as a factor p_{max} which depends on the value of the spin and the spin-lattice relaxation time of the radical [108].

$$p = p_{max} \frac{J^2 \tau_1^2}{1 + J^2 \tau_1^2} \tag{45}$$

The pressure dependence of the probability factor has been shown to differ from one solvent to another [44]. For many solvents, p is practically pressure independent while for others, such as acetone or hexane, it varies greatly with pressure. This is believed to be connected to the strength of the exchange. Systems with a pressure independent p, are believed to exhibit strong exchange, i.e. p≈1, whereas those showing a pressure dependent p are in the weak exchange region. This illustrates the large discrepancies between $\Delta^‡V_{exp}$ and $\Delta^‡V_{diff}$ in those solvents.

For reactions where the spin exchange is realised by the use of a paramagnetic broadening agent, the situation is similar. Some data obtained from this kind of experiments are shown in table 4.

Compound	Solvent	k_{se} /10^9 l mol^{-1} s^{-1}	$\Delta^‡V_{exp}$ / cm^3 mol^{-1}	$\Delta^‡V_{diff}$ / cm^3 mol^{-1}
	Tetrahedral complexes			
4-oxo-TEMPO /[CoCl2(solv)2]	1-Propanol [127]	1.42	15.1	16.4
	1-Butanol [127]	0.95	24.4	20.7
	1-Pentanol [127]	0.75	4.4	9.5
4-oxo-TEMPO /[CoBr2(solv)2]	1-Propanol [127]	1.75	16.6	16.4
10-DOXYL /[CoCl2(solv)2]	1-Propanol [127]	4.22	1.3	16.4
	Octahedral complexes			
4-oxo-TEMPO /[Co(acac)2(C5H5N)2]	1-Propanol [127]	3.06	-0.2	16.4
4-oxo-TEMPO /[Co(acac)2(C5H5N)2]	Chloroform [127]	1.88	-0.7	9.9
4-oxo-TEMPO /[Ni(acac)2(C5H5N)2]	Chloroform [127]	1.15	-0.2	9.9
DTBN /[Co(acac)2(C5H5N)2]	Chloroform [127]	5.03	-1.2	9.9
DTBN /[Ni(acac)2(C5H5N)2]	Chloroform [127]	2.46	0.3	9.9
TEMPO /Fe(acac)3	Methanol [128]	4.4±0.4	12.2	8.8
	Acetone [128]	7.7±0.7	13.6	15.8
	Chloroform [128]	3.2±0.3	11.8	9.9

Numbers in brackets correspond to the relevant reference.

Table 4. Rate constants and activation volumes of selected paramagnetic exchange reactions.

When the paramagnetic exchange takes place between a nitroxide and a tetragonal transition metal complex, only slight differences between $\Delta^‡V_{exp}$ and $\Delta^‡V_{diff}$ are observed, thus indicating strong exchange. An exception is the reaction between 10-DOXYL and $[CoCl_2(C_3H_7OH)_2]$, where the steric hindrance around the >N-O$^•$ group leads to weak exchange.

However, changing the structure of the complex to orthogonal, a change in the effectiveness of the paramagnetic exchange is observed. The steric hindrance is now increased and many complexes show weak exchange.

4. Conclusion

ESR is a suitable tool for investigating the kinetics and thermodynamics of exchange reactions. Such reactions are often used to model elementary processes like electron transfer and diffusion. Electron Self- and spin exchange rate constants can be measured quite accurately and from their temperature and pressure dependencies, the corresponding activation energies and volumes can be determined. The obtained experimental data may be interpreted within the framework of the appropriate theories, e.g. Marcus theory in the case of electron transfer.

Due to their great stability and the relative simplicity of their ESR spectra, nitroxides are prime candidates in ESR line broadening experiments.

Author details

Günter Grampp and Kenneth Rasmussen
Institute of Physical and Theoretical Chemistry, Graz University of Technology, Graz, Austria

Acknowledgement

The authors would like to thank Dr. A. Kokorin for helpful discussion through many years. The collaboration of Dr. T. Hussain, Dr. D. Kattnig, Dr. B. Mladenova-Kattnig and Dr. S. Landgraf is also greatly appreciated.

Financial supports are acknowledged from different foundations: The German Science Foundation (DFG), the Volkswagen-Foundation (Hanover, Germany), the Austrian Science Foundation (FWF), the WTZ-programs from the Austrian Exchange Service (OeAD) and the Asea-Uninet Network Program for travel grants.

5. References

[1] Grampp G (1998) Intermolecular Electron-Self Exchange Kinetics Measured by Electron Paramagnetic Resonance-Linebroadening Effects: Useful Rate Constants for the Application of Marcus Theory. Spectrochim. Acta A. 54A: 2349-2358.
[2] G. Grampp and D. Kattnig, to be published.
[3] Freed J H (1967) Theory of Saturation and Double Resonance Effects in Electron Spin Resonance Spectra. II. Exchange *vs.* Dipolar Mechanisms. J. Phys. Chem. 71: 38-51.
[4] Hyde J S, Pasenkiewicz-Gierula M, Jesmanowicz A and Antholine W E (1990) Pseudo Field Modulation in EPR Spectroscopy. Appl. Magn. Reson. 1: 483-496.
[5] Weil J A, Bolton J R (2007) Electron Paramagnetic Resonance, Elementary Theory and Practical Applications. New York: Wiley-Interscience. 664 p.
[6] Poole C P (1996) Electron Spin Resonance, A Comprehensive Treatise on Experimental Techniques. New York: Dover Publications. 780 p.

[7] Dalal P D, Eaton S S, Eaton R E (1981) The Effects of Lossy Solvents on Quantitative EPR Studies. J. Magn. Reson. 44: 415-428

[8] Eremets M. (Ed.), (1996) High Pressure Experimental Methods. New York: Oxford University Press. 390 p.

[9] Holzapfel, W. B., Isaacs N.S.(Eds.) (1997) High-Pressure Techniques in Chemistry and Physics - The Practical Approach in Chemistry. New York: Oxford University Press. 388p.

[10] Eldik van R., Klärner F.-G. (Eds.)(2002) High Pressure Chemistry. Weinheim: Wiley-VCH. 458 p.

[11] Gaarz U, Lüdemann H.-D (1976) Ber. Bunsenges. Phys. Chem. 80: 607-614.

[12] Yamada H (1974) Pressure-resisting Glass Cell for High Pressure, High Resolution NMR Measurements. Rev. Sci. Instrum. 45: 640-642.

[13] Zahl A, Igel P, Weller M van Eldik R (2004) A High-pressure MNR Probehead for Measurements at 400 MHz. Rev. Sci. Instrum. 75: 3152-3157.

[14] Walch W M Jr. and Bloembergen M (1957) Paramagnetic Resonance of Nickel Fluosilicate under High Hydrostatic Pressure. Phys. Rev. 107: 904-905.

[15] Jaworski M, Checinski K, Bujnowski W, Porowski S (1978) High-pressure EPR Cavity. Rev. Sci. Instrum. 49: 383-384.

[16] Stankowski J, Galezewski A, Krupski M, Waplak S, Gierszal H (1976) Microwave Resonators for EPR studies ar Hgh Hydrostatic Pressure. Rev. Sci. Instrum. 47: 128-129.

[17] Sakai N, Piler H J (1985) Electron paramagnetic Resonance at High Pressure using a Diamond Anvil Cell. Rev. Sci. Instrum. 56: 726-731.

[18] Cecv P, Srinivasan R (1978) Microwave Cavity for EPR at High Pressure. Rev. Sci. Instrum. 49: 1282-1284.

[19] Bromberg S E, Chan I Y (1992) Enhanced Sensitivity for High-pressure EPR using Dielectric resonators. Rev. Sci. Instrum. 63: 3670-3673.

[20] Rupp W L Jr., Peercy P S, and Walsh W M Jr. (1977) Moderate Pressure, Low-Temperature Microwave Spectrometer. Rev. Sci. Instrum. 48: 877-878.

[21] Wu R K, Dutton T F, Doyle W T (1979) 20-Kilobar Pressure System. Rev. Sci. Instrum. 50: 586-589.

[22] Doyle W T, Dutton T F, Wolbarst A B (1972) High Prerssure ENDOR Cavity. Rev. Sci. Instrum. 43: 1668-1670.

[23] Wu R K, Dutton T F, Doyle W T (1979) K-band High-pressure Cell for ENDOR Studies. Rev. Sci. Instrum. 50: 590-594.

[24] Plachy W Z, Schaafsma T J, Wu R K, Dutton T F, Doyle W T (1969) Microwave Helices in Electron Spin Resonance Studies at High Pressure or Very Stable Temperatures. Rev. Sci. Instrum. 40: 1590-1594.

[25] Yoshioka H, Kazama S, Mitani T, Horigome T (1985) Construction of a High-Pressure Electron Spin Resonance Spectrometer Using a Helix Resonator. Anal Chem. 57: 2517-2519.

[26] Sueishi Y, Yamamoto S, Nishimura N (1993) Pressure-resisting glass cell for high-pressure ESR Measurements. Meas. Sci. Technol. 4: 1171-1172.

[27] Livingston R, Zeldes H (1981) Apparatus to Study the Electron Spin Resonance of Fluids under High Pressure Flowing at High Temperature. Rev. Sci. Instrum. 52: 1352-1357.

[28] Grandy D W, Petrakis L (1980) A High-Pressure, High-Temperature Electron Paramagnetic Resonance Cavity. J. Magn. Reson., 41: 367-373.

[29] Sienkiewicz A, Jaworski M, Garaj S, Forro L, . Scholes C P (2002) Application of Electron Spin resonance in Biophysics: From rapid Mixing Stopped-Flow to High-Hydrostatic Pressure ESR.Defect and Diffusion Forum, 208-209: 1-18.

[30] Cannistraro S (1984) Simple System for High-Pressure EPR Measurements on Aqueous Samples. Rev. Sci. Instrum., 55: 996-997.

[31] Alaeva T E, Vereshagin L F, Gvozdev V V, Timofeev U A, Shanditsev V A, Yakovlev E H (1972)A cell for the investigation of electron resonance spectra under quasi-hydrostatic pressures up to 100 kbar. Instr. Exp. Techn. 5: 206-208

[32] Filippov A I, Yablokov U V (1971) A cell to EPR spectrometer for the study of substances under hydrostatic pressures. Instr. Exp. Techn. 6: 161-164

[33] Edelstein N, Kwok A, Maki A H (1964) Effects of Hydrostatic Pressure and Temperature on Spin Exchange between Free Radicals in Solution. J. Chem. Phys. 41: 3473-3478.

[34] Edelstein N, Kwok A, Maki A H (1964) Effects of Hydrostatic Pressure on Linewidths of Free Radicals in Solution. I. Anisotropic Region. J. Chem. Phys., 41:179-183 (1964).

[35] Griller D (1978) High Pressure Electron Paramagnetic Resonance Experiments. Effect of Solvent Contraction on Hyperfine Splitting. J. Am. Chem. Soc. 100 :5240-5241.

[36] Hwang J W, Rao K V S, Freed H J (1976) An electron Spin Resonance Study of the Pressure Dependence of Ordering and Spin Relaxation in a Liquid Crystalline Solvent. J. Phys. Chem. 80: 1490-1501.

[37] Böddeker W K, Lang G, Schindewolf D U (1969) ESR Measurem,ents under Pressure: Effect of Pressure on Chemical Equilibria involving Solvated Electrons. Angew. Chem. Internat Edit. 8:138-139.

[38] Hwang J, Kivelson D, Plachy W (1973)ESR Linewitdths in Solution. VI. Variation with Pressure and Study of Functional Dependence of Anisotropic Interaction Parameter κ. J. Chem. Phys. 58: 1753-1765.

[39] Yoshioka H, Kazama S, Mitani T, Horigome T (1985) Construction of a High-Pressure Electron Spin Resonance Spectrometer Using a Helix Resonator. Anal. Chem., 57: 2517-2519.

[40] M. Kasahara, Sueishi Y, Yamamoto S (2001) Pressure Effects on Cation Migration in 2,5-Di-tert-Butyl-1,4-Benzoquinone Radical Anion. Int. J. Chem. Kinet. 33: 397-401.

[41] Sueishi Y, Kuwata K (1989) Pressure Effects on the Electron Spin Relaxation of Several Radicals in Solution. Chem. Phys. Lett., 160: 640-643.

[42] Sueishi Y, Nishimura N, Hirata K, Kuwata K (1988) ESR Studies of Solvent and Pressure Effects on Spin Exchange of Nitroxide Radicals in Solution. Bull. Chem. Soc. Jpn., 61: 4253-4257.

[43] Sueishi Y, Nishimura N, Hirata K, Kuwata K (1990) ESR Studies at high Pressure II. Spin Exchange of 2,2,6,6-Tetramethyl-4-oxo-1-piperidininyoxyl in Protic and Aprotic Solvents. Bull. Chem. Soc. Jpn.63, 252-254.

[44] Sueishi Y, Kuzukawa M, Yamamoto S, Nishimura N (1991) ESR Studies at high Pressures.III. Spin-Exchange Reactions of Some Nitroxide Radicals in Acetone and Hexane. Bull. Chem. Soc. Jpn., 64: 2188-2191.

[45] Sueishi Y, Kuzukawa M, Yamamoto S, Nishimura N (1992) ESR Studies at high Pressure. IV. Spin-exchange Reactions in various Types of Nitroxides in Some Low-Viscosity Solvents. Bull. Chem. Soc. Jpn. 65: 3118-3121.

[46] Sueishi Y, Kasahara M, Kotake Y (2000) Differential Pressure Effect on Bimodal Inclusion Complex of β-Cyclodextrin with a Nitroxide Probe as Studies with Electron Paramagnetic Resonance. Chem. Lett. 792.

[47] Sueishi Y, Nishimura N, Hirata K, Kuwata K (1991) An ESR Study of Pressure Effects on the Inclusion-Complex Formation ov cyclodextrins with Di-tert-butyl Nitroxide. J. Phys. Chem. 95: 5359-5361.

[48] Sueishi Y, Noritake K, Liu Y J (1999), An EPR Study on Intramolecular Spin Exchange and Rotational Mobility of Nitroxides Linked with a Long Chain. Chemical Research in Chinese Universities, 15: 174-181.

[49] Sueishi Y, Nishimura N (1993) High Pressure Electron spin resonance of radicals in Solution , Koatsuryoku no Kagaku to Gijutsu, 2: 139-145.

[50] Goldhammer E, Paul J, Wenzel H R (1992) Applications of Spin-Label Techniques at High Pressure. In: Zhdanov R I, editor. Bioact. Spin Labels .Berlin: Springer. pp.611-629.

[51] McCoy J, Hubbell L W (2011) High-pressure EPR reveals Conformational Equilibria and Volumetric Properties of Spin-Labeled Proteins. PNAS 108: 1331-1336.

[52] Pfund D M, Zemanian T S, Linehan J C, Fulton J L, Yonker C R (1994) Fluid Structure in Supercritical Xenon by Nuclear magnetic Resonance Spectroscopy and Small Angle Y-Ray Scattering. J.Phys.Chem. 98:11846-11857.

[53] Rasmussen K, Hussain T, Landgraf S, Grampp G (2012) High Pressure ESR Studies of Electron Self-Exchange Reactions of Organic Radicals in Solution. J. Phys. Chem. 116: 193-198.

[54] Ward R L, Weissman S I (1957) Electron Spin Resonance Study of the Electron Exchange between Naphthalene Negative Ion and Naphthalene. J. Am. Chem. Soc. 79: 2086-2090.

[55] Zandstra P J, Weissman S I (1961) Note on Measurements of Rates of Electron Transfer Processes by Broadening of ESR Lines. *J. Chem. Phys.* 35: 757

[56] Layloff T, Miller T, Adams R N, Fah H, Horsfield A, Proctor W (1965) Homogeneous Electron Exchange Reactions of Aromatic Molecules. Nature 205: 382-383.

[57] Malachesky P A, Miller T A, Layloff T, Adams R N (1965) Electron-exchange reactions of aromatic molecules. *Exchange Reactions Proc. Symp., Upton, N.Y.*: 157-170.

[58] Marcus R A (1956) On the Theory of Oxidation-Reduction Reactions Involving Electron Transfer. I. J. Chem. Phys. 24: 966-978.

[59] Marcus R A (1957) On the Theory of Oxidation-Reduction Reactions Involving Electron Transfer. II. Applications to Data on the Rates of Isotopic Exchange Reactions. J. Chem. Phys. 26: 867-871.

[60] Marcus R A (1957) On the Theory of Oxidation-Reduction Reactions Involving Electron Transfer. III. Applications to Data on the Rates of Organic Redox Reactions. J. Chem. Phys. 26: 872-877.

[61] Bixon M, Jortner J (1993) Solvent Relaxation Dynamics and Electron Transfer. Chem. Phys. 176: 467-481.

[62] Hush N S (1961) Adiabatic theory of outer sphere electron-transfer reactions in solution. Trans. Faraday Soc. 57: 557-580.

[63] Sumi H, Marcus R A (1986) Dynamical Effects in Electron Transfer Reactions. J. Chem. Phys. 84: 4894-4914.

[64] Holstein T (1978) Quantal Occurrence-Probability Treatment of Small-Polaron Hopping. Philos. Mag. B 37: 49-62.

[65] Jortner J, Bixon M (1988) Intramolecular Vibrational Excitations Accompanying Solvent-Controlled Electron Transfer Reactions. J. Chem. Phys. 88: 167-170.

[66] Sutin N (1983) Theory of Electron Transfer Reactions: Insights and Hindsights. Prog. Inorg. Chem. 30: 441-498.

[67] Nelsen S F, Newton M D (2000) Estimation of Electron Transfer Distances from AM1 Calculations. J. Phys. Chem. A. 104: 10023-10031.

[68] Nelsen S F, Blackstock S C, Kim Y (1987) Estimation of Inner Shell Marcus Terms for Amino Nitrogen Compounds by Molecular Orbital Calculations. J. Am. Chem. Soc. 109: 677-682.

[69] Grampp G, Jaenicke W (1991) Kinetics of Diabatic and Adiabatic Electron Exchange in Organic Systems Comparison of Theory and Experiment. Ber. Bunsenges. Phys. Chem. 95: 904-927.

[70] Clegg A D, Rees N V, Klymenko O V, Coles B A, Compton R G (2004) Marcus Theory of Outer-Sphere Heterogeneous Electron Transfer Reactions: Dependence of the Standard Electrochemical Rate Constant on the Hydrodynamic Radius from High Precision Measurements of the Oxidation of Anthracene and Its Derivatives in Nonaqueous Solvents Using the High-Speed Channel Electrode. J. Am. Chem. Soc. 126: 6185-6192.

[71] Rees N V, Clegg A D, Klymenko O V, Coles B A, Compton R G (2004) Marcus Theory for Outer-Sphere Heterogeneous Electron Transfer: Predicting Electron-Transfer Rates for Quinones. J. Phys. Chem. B. 108: 13047-13051.

[72] German E D, Kuznetsov A M (1981) Outer Sphere Energy of Reorganization in Charge Transfer Processes. Electrochim. Acta 26: 1595-1608.

[73] German E D, Kharkats I Yu (1995) Calculation of the polar media reorganization energy for the model of two dielectric spheres. Chem. Phys. Lett. 246: 427-430.

[74] Marcus R A (1965) On the Theory of Electron-Transfer Reactions. VI. Unified Treatment for Homogeneous and Electrode Reactions. J. Chem. Phys. 43: 679-701.

[75] Kramers H A (1940) Brownian Motion in a Field of Force and the Diffusion Model of Chemical Reactions. Physica 7: 284-304.

[76] Zusman L D (1988) The Theory of Electron Transfer Reactions in Solvents with two Characteristic Relaxation Times. Chem. Phys. 119: 51-61.
[77] Zusman L D (1994) Dynamical Solvent Effects in Electron Transfer Reactions. Z. Phys. Chem. 186: 1-29.
[78] Rips I, Jortner J (1987) Dynamic Solvent Effects on Outer-Sphere Electron Transfer. J. Chem. Phys. 87: 2090-2104.
[79] Rips I, Jortner J (1987) Outer Sphere Electron Transfer in Polar Solvents. Activationless and Inverted Regimes. J. Chem. Phys. 87: 6513-6519.
[80] Fröhlich H (1958) Theory of Dielectrics: Dielectric Constant and Dielectric Loss. Oxford: Claredon Press. 192 p.
[81] Zusman L. D (1980) Outer-Sphere Electron Transfer in Polar Solvents. Chem. Phys. 49: 295-304.
[82] Zusman L D (1983) The Theory of Transitions between Electronic States. Application to Radiationless Transitions in Polar Solvents. Chem. Phys. 80: 29-43.
[83] Weaver M J (1992) Dynamical Solvent Effects on Activated Electron-Transfer Reactions: Principles, Pitfalls, and Progress. Chem. Rev. (Washington, DC, United States). 92: 463-480.
[84] Weaver M J, McManis G E III (1990) Dynamical Solvent Effects on Electron-Transfer Processes: Recent Progress and Perspectives. Acc. Chem. Res. 23: 294-300.
[85] Barthel J, Buchner R, Hölzl C G, Münsterer M (2000) Dynamics of Benzonitrile, Propylene Carbonate and Butylene Carbonate: the Influence of Molecular Shape and Flexibility on the Dielectric Relaxation Behaviour of Dipolar Aprotic Liquids. Z. Phys. Chem. 214: 1213-1231.
[86] Swaddle T W (1990) "Pressure-Testing" Marcus-Hush Theories of Outer-Sphere Electron-Transfer Kinetics. Inorg. Chem. 29: 5017-5025.
[87] Swaddle T W (1996) Reflections on the Outer-Sphere Mechanism of Electron Transfer. Can. J. Chem. 74: 631-638.
[88] Zahl A, van Eldik R, Matsumoto M, Swaddle T W (2003) Self-Exchange Reaction Kinetics of Metallocenes Revisited: Insights from the Decamethylferricenium – Decamethylferrocene Reaction at Variable Pressure. Inorg. Chem. 42: 3718-3722.
[89] Stranks D R (1974) The Elucidation of Inorganic Reaction Mechanisms by High Pressure Studies. Pure Appl. Chem. 38: 303-323.
[90] Zhdanov R I, Golubev V A, Gida, V M (1971) Interaction of iminoxyl radicals with antimonypentachloride. Proc. Acad. Sci. USSR 196: 856-857.
[91] Miyazawa T, Endo T, Shiihashi S (1985) Selective oxidation of alcohols by oxoammonium salt ($R_2N = O^+X^-$). J. Org. Chem. 50: 1332-1334
[92] Liu Y C, Ding Y B, Liu Z L (1990) Preparation of single crystal and molecular structure of phenothiazine radical cation hexachloroantimonates. Acta Chimica Sinica 48: 1199-1203
[93] Wu L, Guo X, Wang J, Guo Q, Liu Z, Liu Y (1999) Kinetic Studies on the Single Electron Transfer Reaction between 2,2,6,6-Tetramethylpiperidine Oxoammonium Ions and Phenothiazines: the Application of Marcus Theory. Sci. China Ser. B 42: 138-144.

[94] Grampp G, Landgraf S, Rasmussen K, Strauss S (2002) Dimerization of Organic Free Radicals in Solution.: 1. Temperature Dependent Measurements. Spectrochim. Acta A 58A: 1219-1226.

[95] Rasmussen K, (2006) High-Pressure ESR Spectroscopy Applied to the Kinetics of Electron Self-Exchange Reactions in Solution. Dissertation, Graz University of Technology

[96] Mansingh K, Mansingh A (1964) Dielectric relaxation and viscosity of liquids. Ind. J. Pure Appl. Phys. 2: 176-178.

[97] Owen B B, Miller R C, Milner C E, Cogan H L (1961) The Dielectric Constant of Water as a Function of Temperature and Pressure. J. Phys. Chem. 65: 2065-2070.

[98] Barthel J, Buchner R, Munsterer M (1996) Electrolyte Data Collection. Part 2a Dielectric Properties of Nonaqueous Electrolyte Solutions. Frankfurt/Main: Dechema. 387 p.

[99] Buchner R, Barthel J, Stauber J (1999) The Dielectric Relaxation of Water between 0°C and 35°C. Chem. Phys. Lett. 306: 57-63.

[100] Finklea H O, Madhiri N (2008) Reorganization energies of $TEMPO^{\bullet}/TEMPO^{+}$ in water. J. Electroanal. Chem. 621: 129-133.

[101] Fawcett W R (1992) Time Dependence of the Relaxation Parameters in non-Debye Solvents. Chem. Phys. Lett. 199: 153-160.

[102] Shirke R M, Chaudhari A, More N M, Patil P B (2001) Temperature Dependent Dielectric Relaxation Study of Ethyl Acetate — Alcohol Mixtures Using Time Domain Technique. J. Mol. Liq. 94: 27-36.

[103] Shirke R M, Chaudhari A, More N M, Patil P B (2000) Dielectric Measurements on Methyl Acetate + Alcohol Mixtures at (288, 298, 308, and 318) K Using the Time Domain Technique. J. Chem. Eng. Data 45: 917-919.

[104] Simon J D (1988) Time-Resolved Studies of Solvation in Polar Media. Acc. Chem. Res. 21: 128-134.

[105] Bertolini D, Cassettari M, Salvetti G (1983) The Dielectric Properties of Alcohols–Water Solutions. I. The Alcohol Rich Region. J. Chem. Phys. 78: 365-372.

[106] Chaudhari A, Khirade P, Singh R, Helambe S N, Narain N K, Mehrotra S C (1999) Temperature dependent dielectric relaxation study of Tetrahydrofuran in Methanol and Ethanol at microwave frequency using Time Domain Technique. *J. Mol. Liq.* 82: 245-253.

[107] Carvajal C, Tolle K J, Smid J, Szwarc M (1965) Studies of Solvation Phenomena of Ions and Ion Pairs in Dimethoxyethane and Tetrahydrofuran. J. Am. Chem. Soc. 87: 5548-5553.

[108] Molin Yu N, Salikhov K M, Zamaraev K I (1980) Spin Exchange Principles and Applications in Chemistry and Biology. Berlin: Springer-Verlag. 242 p.

[109] Nayeem A, Rananavare S B, Sastry V S S, Freed J H (1989) Heisenberg Spin Exchange and Molecular Diffusion in Liquid Crystals. J. Chem. Phys. 91: 6887-6905.

[110] Berner B, Kivelson D (1979) The Electron Spin Resonance Line Width Method for Measuring Diffusion. A Critique. J. Phys. Chem. 83: 1406-1412.

[111] Bales B L, Stenland C (1995) Statistical Distributions and Collision Rates of Additive Molecules in Compartmentalized Liquids Studied by EPR. 2. Sodium Dodecyl Sulfate Micelles, 5-Doxylstearic Acid Ester, and Copper(II). J. Phys. Chem. 99: 15163-15171.
[112] Salikhov K M (2010) Contributions of Exchange and Dipole-Dipole Interactions to the Shape of EPR Spectra of Free Radicals in Diluted Solutions. Appl. Magn. Reson. 38: 237-265.
[113] Kivelson D J (1960) Theory of ESR Linewidths of Free Radicals. J. Chem. Phys. 33: 1094-1106.
[114] Currin J D (1962) Theory of Exchange Relaxation of Hyperfine Structure in Electron Spin Resonance. Phys. Rev. 126: 1995-2001.
[115] Johnson Jr, C S (1967) Theory of Line Widths and Shifts in Electron Spin Resonance Arising From Spin Exchange Interactions. Mol. Phys. 12: 25-31
[116] Plachy W, Kivelson D (1967) Spin Exchange in Solutions of di-Tertiary-Butyl Nitroxide. J. Chem. Phys. 47: 3312-3318
[117] Bales B L, Ranganathan R, Griffiths P C (2001) Characterization of Mixed Micelles of SDS and a Sugar-Based Nonionic Surfactant as a Variable Reaction Medium. J. Phys. Chem. B 105: 7465–7473
[118] Parmon V N, Kokorin A I, Zhidomirov G M (1977) Conformational Structure of Nitroxide Biradicals Use of Biradicals as Spin Probes. J. Struct. Chem. 18: 104-147.
[119] Parmon V N, Kokorin A I, Zhidomirov G M (1980) Stable Biradicals. Moscow: Nauka 240 p.
[120] Rassat A (1971) Application of Electron Spin Resonance to Conformational Analysis. Pure. Appl. Chem. 25: 623-634.
[121] Shapiro A B, Baimagambetov K, Goldfield M G, Rozantsev E G (1972) Long-chain iminoxyl biradicals. Zh. Org. Khim. 8: 2263-2269.
[122] Parmon V N, Kokorin A I, Zhidomirov G M, Zamaraev K I (1975) On the Mechanism of Spin Exchange in Long-Chain Nitroxide Biradicals. Molec. Phys. 30: 695-701.
[123] Salikhov K M, Doctorov A B, Molin Yu N, Zamaraev K I (1971) Exchange Broadening of ESR Lines for Solutions of Free Radicals and Transition Metal Complexes. J. Magn. Reson. 5: 189-205.
[124] Sueishi Y, Inoue H, Oka T, Tsukube H, Yamamoto S (1998) EPR Studies at High-Pressures. V. Spin-Exchange Reactions of Nitroxide Radicals with Cobalt and Nickel Complexes. Bull. Chem. Soc. Jpn. 71: 817-823.
[125] Sueishi Y, Hori A (2001) EPR Studies at High-Pressure. VI. Spin-Exchange Reaction of Nitroxide Radical with Copper(II) and Iron(III) Complexes. Bull. Chem. Soc. Jpn. 74: 2431-2432.
[126] Kokorin A I, Pridantsev A A (1997) Diffusion of Molecules and Ions in Solutions of Complexing Polymers. Zh. Fiz. Khim. (Russ. J. Phys. Chem. A) 71(12): 1963-1969.
[127] Sueishi Y, Inoue H, Oka T, Tsukube H, Yamamoto S (1998) EPR Studies at High-Pressures. V. Spin-Exchange Reactions of Nitroxide Radicals with Cobalt and Nickel Complexes. Bull. Chem. Soc. Jpn. 71: 817-823.

[128] Hussain T, (2009) Pressure and Temperature Dependence of Spin- and Electron-Exchange Reactions Measured by ESR Spectroscopy. Dissertation, Graz University of Technology

Chapter 3

Forty Years of the d_1/d Parameter

Alexander I. Kokorin

Additional information is available at the end of the chapter

1. Introduction

Spin labeling, as a part of electron paramagnetic resonance (EPR), became one of the most sensitive and adequate methods for investigation the structure, properties of different biological systems, their dynamics, and mechanisms of various processes after opening a new class of chemical reactions of stable nitroxide radicals, in which the unpaired electron remained untouched and retained its paramagnetic properties (Neiman et al., 1962, Rozantsev, 1964, 1970, Rozantzev & Neiman, 1964). Just at once, spin labels, attached to biological (proteins, oligopeptides, polysaccharides, nucleic acids) or synthetic macromolecules, and probes, incorporated into biological or artificial membranes, polymers, solid materials and solutions, have been applied for investigation structural and functional properties of such complex and supra-molecular systems. Harden M. McConnell, O. Hayes Griffith, Gertz I. Likhtenstein, Anatoly L. Buchachenko, Lawrence J. Berliner, Alexander Kalmanson, Geoffrey R. Luckhurst, Jack H. Freed, Andrey N. Kuznetsov, and some others, were first pioneers in this area. Much detailed, the historical aspects of spin label technique are described in chapter written by L. J. Berliner. Great advances have been achieved in practical applications of numerous amount of mono- bi- and poly-radicals synthesized in groups headed by Eduard Rozantsev, John Keana, Andre Rassat, Kalman Hideg, George Sosnovsky, Leonid Volodarsky, Igor Grigor'ev, and their pupils.

First books concerning new method, which are actual up to now, were published approximately in ten years later. Among them, the most cited, are written or edited by Buchachenko & Wasserman, 1973, Likhtenshtein, 1974, Berliner, 1976 & 1979, and Kuznetsov, 1976. Among recent publications, I would like to press attention on several once performing further development and success of this method: Likhtenshtein, 2008, Moebius & Savitsky, 2009, Brustolon & Giamello, 2009, Hemminga & Berliner, 2007, Bender & Berliner, 2006, Schlick, 2006, Webb, 2006. Indeed, high-field (high frequencies) EPR spectroscopy, pulse technique, various double- and multy- resonance methods, time-resolved EPR, etc., enlarged the area of magnetic resonance applications pretty much.

One of the most important possibilities giving by EPR methods is connected with distance measurements in chemical, biological and nanostructured systems and materials (Parmon et al., 1980, Berliner et al., 2001, Eaton & Eaton, 2004, Steinhoff, 2004, Webb, 2006, Tsvetkov et al., 2008, Moebius & Savitsky 2009), which allows determine distances, two- and three-dimensional distribution of paramagnetic centers, and their mutional orientation in the case of not too long distances.

Stable nitroxide spin probes and labels contain paramagnetic >N–O group with the unpaired electron, surrounding usually with four methyl groups in the appropriate piperidine (R_6, R_6''), pirrolidine (R_5) or imidazoline (R_5^N, R_5^{NO}) rings, which have different "tails" with functional residues in fourth or third positions of the ring, by which probes can be attached to macromolecules, surfaces, etc., becoming spin labels. These radicals are shown in Fig. 1.

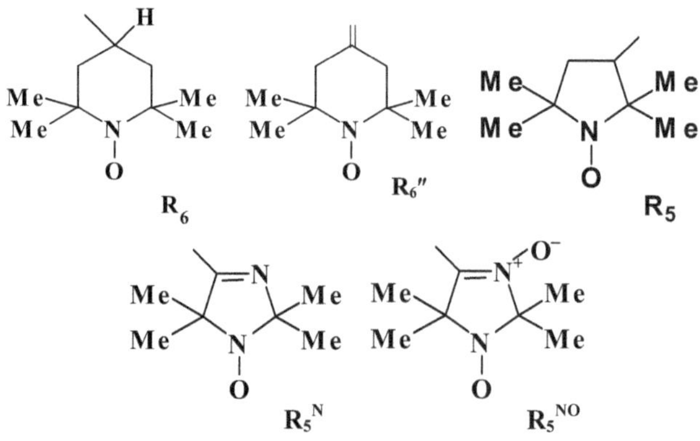

Figure 1. Structures of paramagnetic fragments of nitroxide radicals.

Among spectroscopic methods for determination distances between spin probes, which are discussed in the next section, the simplest one, was developed forty years ago based on empirical parameter d_1/d (Fig. 2) characterizing the shape of the X-band EPR spectrum of the nitroxide radical solution frozen at 77 K (Kokorin et al., 1972). d_1/d is measured with high precision as the ratio of the summed amplitudes of two lateral (low and high field) lines, recorded as the first derivative of the absorption EPR spectrum, to the amplitude of the central component (Fig. 2). It was shown that d_1/d is straightly connected with the efficiency of dipole-dipole coupling between radical paramagnetic >N–O groups, the type of spatial distribution of radicals, as well as with polarity of the surrounding media and temperature of the sample. A methodic procedure of measuring distances is described in detail in Section 3, allows characterize quantitatively the spatial organization of nitroxide biradicals, proteins, nucleic acids, frozen two-component solutions, synthetic polymers, and nanostructured materials. Unfortunately, the most part of the d_1/d features and the majority of the resulta obtained have been published in original only in Russian, though translated into English, scientific journals, and are not known well to practical scientists and students.

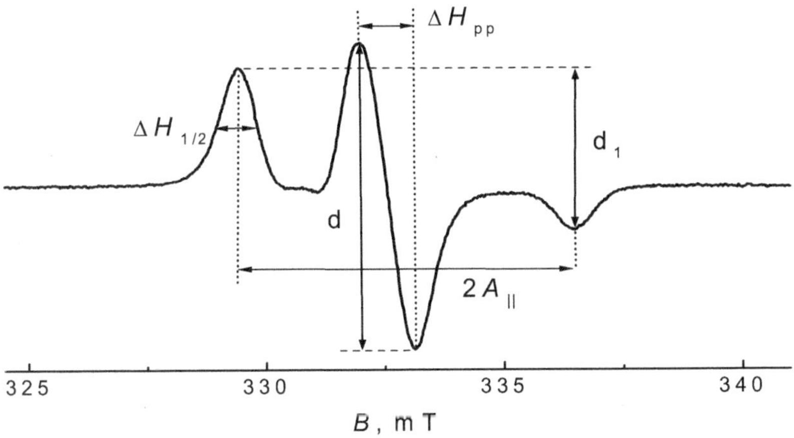

Figure 2. The EPR spectrum of TEMPOL radical at 77 K and c = 0.001 M

Therefore, the main goal of this chapter is describing the present state of this method and the analysis of regularities and peculiarities of its application to various objects and systems under EPR investigation including molecules, macromolecules and supramolecular systems. Theoretical aspects, experimental details, obtained results and conclusions will be reported in the following sections.

2. Methods of the local concentration and the distance measurement from EPR spectra

It is well known from the basic theory of EPR spectroscopy that the value of magnetic dipole-dipole interaction between electron spins of paramagnetic centers depends strongly on the distance between them (Abragam, 1961, Blumenfeld et al., 1962, Altshuler & Kozirev, 1964), and this information is very important for understanding the structure and properties of such systems. There are several independent methods for measuring distances, which have been analyzed in detail in many publications, for example, in the following books: Likhtenshtein, 2008, Moebius & Savitsky, 2009, Berliner et al., 2001, Schweiger & Jeschke, 2001, Eaton & Eaton, 2004, Weil & Bolton, 2007, Tsvetkov et al., 2008, Parmon et al., 1980, Lebedev & Muromtsev, 1971.

The most complete analysis of different approaches and techniques allow measuring distances with high accuracy was presented by Berliner et al., 2001 and Eaton & Eaton, 2004. Below, for better understanding, we shortly report about some of them widely used. Complete information about these approaches one can read in works cited above. It should be mentioned that methods based on double electron-electron resonance, pulsed EPR and spin echo measurements, high frequency and high field EPR are usually used in the case of pairwise interaction between two spins distributed in an immobilized sample (Berliner et al., 2001). As a rule, a distance between two interacting spins is much less than between different pairs, for example, in the case of nitroxide biradicals in diluted solid solutions or spin labeled proteins.

2.1. Dipolar interaction measured by EPR

Fig. 3 shows typical changes in EPR spectra of R_6OH dissolved in 1-butanol at three different concentrations. Broadening of CW X-band EPR spectra of nitroxide radicals in frozen solutions, due to dipolar interaction, results in changes of the whole spectrum shape which can be characterized with changes in widths of the EPR spectrum lines and in their relative intensities.

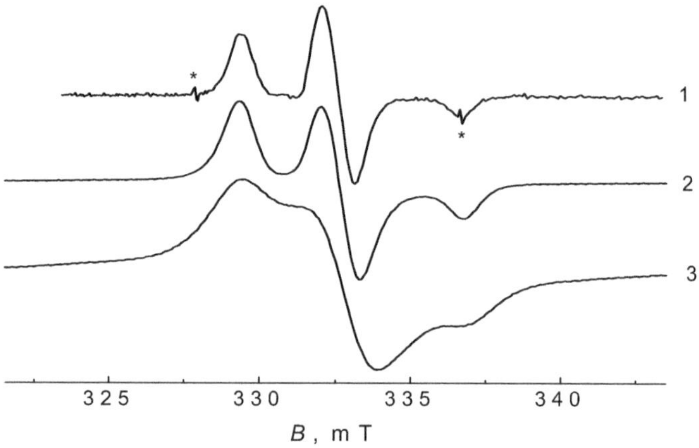

Figure 3. Experimental EPR spectra of R_6OH dissolved in 1-butanol at 0.001 (1), 0.1 (2), and 0.4 mol/l (3) at 77 K. Asterisks * show the 3rd and 4th lines of Mn^{2+} ions in MgO matrix

The line width of the separate line and the shape of whole EPR spectra depend on several various factors such as the distance between paramagnetics, type of their spatial distribution, temperature, polarity, viscosity and organization of the solvent, longitudinal relaxation time T_1, etc. On practice, there are two most common types of distribution: random and pairwise. A quantitative structural characteristic for the first one is the local concertration, C_{loc}. The mean distance among interacting spins, <r>, can be calculated from C_{loc} in an assumption of the distribution type, for the simplest example: $<r> = (C_{loc})^{-1/3}$ for cubic regular lattice. For distribution of spins in pairs, for instance, in stable nitroxide biradicals, if they are dissolved at low concentration in solvents glazed under freezing in liquid nitrogen at 77 K, the biradical structure can be completely determined by the certain distance r between two interacting unpaired electrons localized at >N–O bonds, and angles of their mutual spatial orientation (Parmon et al., 1977a, 1980). At high temperatures, if radicals are dissolved in low-viscous liquids with fast rotational and translational mobility, the dipolar coupling is averaged decreasing up to zero.

Dipole-dipole interaction between paramagnetic centers is manifested in dipolar splitting of EPR lines in the case of two coupled spins (biradicals) or in dipolar broadening of EPR lines in the case of interacting of several spins at random distribution. It was shown for the dipolar broadening δH that (Abragam, 1961, Lebedev & Muromtsev, 1971):

$$\delta H = \Delta H - \Delta H_0 = A \bullet C \quad (1)$$

Here ΔH is the width of the homogeneous individual EPR spectrum line, ΔH_0 is a linewidth at the absence of the dipolar coupling, C is concentration, and A is a coefficient, which depends on the character of the spatial distribution of the paramagnetic centres in the sample, the shape of the individual line, and the longitudinal relaxation time T_1. Values of A for different cases of radical distributions (regular, random, in pairs, etc.) are published in (Lebedev & Muromtsev, 1971). For example, the theoretical value $A_{theor} = 5.8 \bullet 10^{-20}$ G/cm^3 = 34.8 G•l/mol for the width at a half-height $\Delta H_{1/2}$ for the Gaussian line shape (Grinberg et al., 1969). Eq.(1) is valid at not too high concentrations, such as $(4/3)\pi r_0^3 C \ll 1$, where r_0 is the characteristic size of the paramagnetic particle (a distance of the closest approach). Values of r_0 calculated from the experiment were equal to 6.2 ± 0.5 for $R_6OCOC_6H_5$ in toluene, 5.8 ± 0.6, and 6.0 ± 0.4 for R_6OH in ethanol and 50% H_2O-glycerol mixture (Kokorin, 1974).

If the experimental A_{exp} value measured in a wide concentration interval, coincide or close to the A_{theor} value, it is the objective confirmation that paramagnetic species (radicals) are distributed in the whole volume of the sample. In the case of rather often real situation when paramagnetic species are localized in a certain part of the sample only (spin probes in emulsions, in lipid vesicles, spin labels attached to polymer macromolecules, etc.), the experimentally measured value of the parameter $A_{exp} = \delta H/C_{loc}$ characterize the magnitude of the dipolar broadening and can be significantly greater the value of A_{theor}. At the same time, if it is known or can be assumed that for paramagnetic centres used, the type of spatial distribution in the areas of their localization remains the same (random, regular, etc.), the real value of A_{exp} in these areas does not change and has to be equal to A_{theor}. Therefore, one can calculate C_{loc} values by the equation analogues to Eq. (1) (Kokorin, 1992):

$$C_{loc} = \delta H/A \quad (2)$$

Eq. (2) is valid for nitroxide radicals and paramagnetic metal ions with long $T_1 > 10^{-10}$ s (Kokorin, 1992).

It has been experimentally shown that in accordance with Eq. 1, the dipolar broadening of the low-field "parallel" line $\delta H_{1/2}$ and of the central complex line δH_{PP} (see Fig. 2) depends on the radical concentration linearly but with different values of ΔH_0 and A for $\Delta H_{1/2}$ and ΔH_{PP} (Fig. 4).

From these experimental dependences and Eq. 1, one can calculate: A_{PP} = 24.9 ± 0.5 G•l/mol, $\Delta H(0)_{PP}$ = 10.2 ± 0.07 G for all solvents studied; $A_{1/2}$ = 40.6 ± 0.7 G•l/mol, $\Delta H(0)_{1/2}$ is equal to 6.0 ± 0.07, 7.6 ± 0.08, 8.6 ± 0.1 G for R_6OH in frozen at 77 K toluene, ethanol, and 50% glycerol water mixture correspondingly. Very important contribution concerning correct

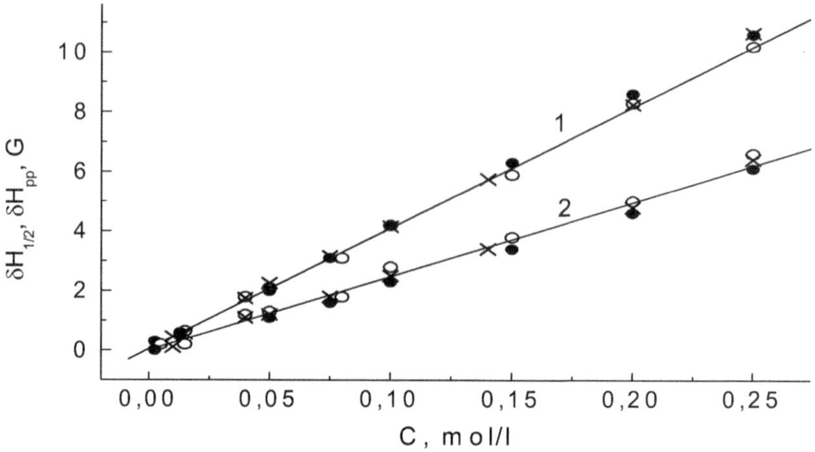

Figure 4. $\delta H_{1/2}$ (1) and δH_{pp} (2) as a function of concentration at 77 K: R_6OH in glycerol:H_2O = 1:1 mixture (○) and in ethanol (×), $R_6OCOC_6H_5$ in toluene (•)

determination of different impacts (dipole-dipole interaction, spin exchange coupling, electron spin relaxation, etc.) to the line broadening of EPR spectra was recently done by Salikhov, 2010.

2.2. Second central moment measured by EPR spectra

Any EPR spectrum can be characterized by its second central moment, which can be determined as:

$$M_2 = \int(H - H_0)^2F(H)dH/\int F(H)dH, \tag{3}$$

where H_0 is the value of the spectrum centre, F(H) gives the shape of the spectrum absorption line as a function of the magnetic field H, and $\int F(H)dH$ is the normalization condition of the EPR spectrum. The value of H_0 one can find from the equation:

$$\int(H - H_0)F(H)dH = 0 \tag{4}$$

The classical theory of spin-spin interaction connected the value of M_2 with the distance between interacting spins S (Van Vleck, 1948, Pryce & Stevens, 1950):

$$M_2 = (3/4)g^2\beta^2S(S + 1)\bullet\sum[(1 - 3\cos^2\theta_{j,k})^2/r_{j,k}^6] \tag{5}$$

Here $r_{j,k}$ is the distance between spins j and k, $\theta_{j,k}$ is the angle between the line connecting these spins and the external magnetic field H, g is a g-factor, and β is Bohr magneton. Later, it was shown by Lebedev, 1969, that the second central moment of the absorption EPR spectrum in the case of the random distribution of paramagnetic centres in a solid matrix depends linearly on their concentration C:

$$M_2 = (2\pi/15)\xi^2 g^2 \beta^2 C_0 C \quad (6)$$

Here $\xi = 3/2$ in the case of the equivalent spins, and $C_0 = r_0^{-3}$ means the characteristic volume occupied by a paramagnetic molecule in the matrix. Hence, linear dependence between M_2 and radical concentration C in solid solutions is evidence of a random distribution of spins in the matrix. Dipole-dipole interaction contributes to the broadening of the spectrum and its second moment M_2:

$$M_2 = M_2(0) + B \bullet C, \quad (7)$$

where $M_2(0)$ is the M_2 value at the absence of dipolar broadening, and a coefficient B is a characteristic of a certain solid matrix, for instance, in magnetically diluted frozen solutions. This dependence is illustrated well in Fig. 5:

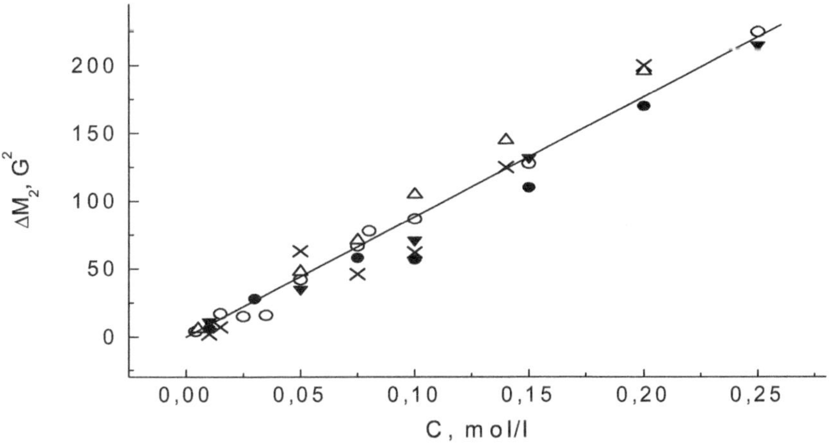

Figure 5. ΔM_2 as a function of concentration at 77 K: R_6OH in glycerol:H_2O = 1:1 mixture (○) and in ethanol (×), $R_6OCOC_6H_5$ in toluene (•), $R_5^NCHCOCH_2I$ in ethanol (Δ), $R_5^NCH_2Br$ in ethanol (▼)

As in the case of dipolar broadening (section 2.1) the local spin concentrations can also be measured from the spectral second central moment M_2 by Eq. (8) in the agreement with Fig. 5:

$$\Delta M_2 = M_2 - M_2(0) = B \bullet C_{loc} \quad (8)$$

The following values were calculated: B = 910 ± 30 $G^2 \bullet$l/mol, and $M_2(0)$ is equal to 340 ± 10, 390 ± 12 G^2 for R_6OH in frozen at 77 K ethanol and 50% glycerol-water mixture correspondingly; 305 ± 6 G^2 for $R_6OCOC_6H_5$ in toluene, 250 ± 10 G^2 for $R_5^NCH_2Br$, and 270 ± 12 G^2 for $R_5^NCHCOCH_2I$ (in ethanol both). $M_2(0)$ is a characteristic of the solvent and the radical structure, while ΔM_2 depends on the magnitude of the dipolar interaction.

Several other useful relations between various spectroscopic parameters are presented in section 3.

2.3. Measurements based on the relaxation time and saturation effects

A method for estimation distances between nitroxide radicals or between spins of the spin label and paramagnetic metal center, based on the quantitative analysis of saturation curves of spin label EPR spectra recorded at 77 K, was suggested by Kulikov & Likhtenstein, 1974. This approach has been tested using haemoglobin molecule labeled by SH groups with various nitroxide radicals. Values of the distances between labels and iron ion in haem estimated from the saturation curve parameters (values of ΔT_1^{-1}, ΔT_2^{-1}, and $\Delta T_1^{-1}\bullet\Delta T_2^{-1}$) were compared with distances measured by d_1/d parameter and estimated from the X-ray data of haemoglobin (Kulikov, 1976). Here, T_1 and T_2 are the longitudinal, T_1, and transverse, T_2, relaxation times of the nitroxide electron spin. Results obtained were in reasonable agreement. In the case of rapid spin relaxation of the metal paramagnetic center, this method allows one determine rather long distances up to 2.0 nm. Serious limitation of this method is the following: for structural investigations of haem containing and other proteins, one have to know the exact value of spin relaxation time T_1 of the metal paramagnetic center from independent measurements (Kulikov, 1976).

Theoretical aspects of the method based on spin relaxation of nitroxide radicals in solid matrix (frozen solutions) were considered in the first part of the review by Kulikov & Likhtenstein 1977. The distance between the spins of the radical and the other paramagnetic centre can be determined from the change of the transverse, T_2, and especially from the longitudinal, T_1, relaxation times of the radical due to the dipole-dipole interaction between the radical and the metal ion. The study of structure of several metal-containing proteins: haemoglobin, myosin and nitrogenase, has been carried out by the method of spin labels. The interesting approach, so called method "spin label – paramagnetic probe" has been carefully examined by spin relaxation in solutions of spin labeled proteins in the presence of inert paramagnetics, readily diffused in the solution by measuring T_2 values of the label. The influence of the probes on labels T_2 values depends on the frequency of collisions between a label and a probe, and therefore this approach can be used for quantitative study of the factors, which effect the frequency of collisions: microviscosity, steric hindrances, the presence of electrostatic charges (Kulikov & Likhtenstein 1977).

Later, many scientists began to work on developing various modifications of the relaxation times method for investigation structural peculiarities and conformational dynamics of biological macromolecules, proteins first of all, their aggregates and bio-membranes. One of the most serious and complete reviews in this field to my opinion was written by Eaton & Eaton, 2001a & 2001b. This review contains theoretical and experimental fundamentals of the method in solid and fluid solutions, technical details, a lot of experimental data, their deep analysis, interesting applications.

2.4. Double electron-electron resonance (ELDOR)

A theoretical basis of the method is perfectly described in (Saxena & Freed, 1997). For calculating double quantum two dimensional electron spin resonance spectra in the rigid limit, that correspond to the experimental spectra obtained from a nitroxide biradical, a specific formalism has been developed. The theory includes the dipolar interaction between the nitroxide moieties as well as the fully asymmetric *g* and hyperfine tensors and the angular geometry of the biradical. The effects of arbitrary strong pulses are included by adapting the recently introduced spin-Hamiltonian theory for numerical simulations. Creation of "forbidden" coherence pathways by arbitrary pulses in magnetic resonance, and their role in ELDOR is discussed. The high sensitivity of these ELDOR signals to the strength of the dipolar interaction was demonstrated and rationalized in terms of the orientational selectivity of the "forbidden" pathways. This selectivity also provides constraints on the structural geometry (i.e., the orientations of the nitroxide moieties) of the biradicals. The theory was applied to the double quantum modulation ELDOR experiment on an end-labeled poly-proline peptide biradical. A distance of 1.85 nm between the ends is found for this biradical (Saxena & Freed, 1997).

The results of development of pulse electron-electron double resonance (PELDOR) technique and its applications in structural studies were summarized and described systematically in a review by Tsvetkov et al., 2008. The foundations of the theory of the method were discribed, some experimental features and applications were considered, in particular, determination of the distances between spin labels in the nanometre range for nitroxide biradicals, spin-labeled biological macromolecules, radical-ion pairs, and peptide-membrane complexes. The authors attention was focused on radical systems arising upon self-assembly of nanosized complexes, in particular, from peptides, spatial effects, and radical pairs formation in photolysis and photosynthesis. The position of PELDOR among other structural EPR techniques was analyzed (Tsvetkov et al., 2008).

PELDOR measures via the dipolar electron–electron coupling distances in the nanometre range, currently 1.5–8 nm, with high precision and reliability (Reginsson & Schiemann, 2011). Depending on the quality of the data, the error can be as small as 0.1 nm. Beyond measuring mean distances, PELDOR yields distance distributions, which provide access to conformational distributions and dynamics. The method was also used to count the number of monomers in a complex and allowed determination of the orientations of spin centres with respect to each other. If, in addition to the dipolar through-space coupling, a through-bond exchange coupling mechanism contributes to the overall coupling both mechanisms can be separated and quantified (Reginsson & Schiemann, 2011). This is of principle interest for researchers in many real cases.

Interesting implication of ELDOR to polymer science has been done by Bird et al., 2008.They demonstrated on a series of spin-labeled oligomers quantitative determination the end-to-end lengths and distance distributions. In case of oligomers with well-defined three-dimensional structures (seven different macromolecules, each containing eight monomers) which were labeled with nitroxide radicals, the quantitative information about the shapes and flexibility of the oligomers was obtained, and end-to-end distances were calculared. The shapes of the EPR-derived population distributions allowed authors to compare the flexibility of these spiro-ladder oligomers.

Additional information concerning this very powerful technique one can find in Berliner et al., 2001, Webb, 2006, Reginsson & Schiemann, 2011.

2.5. High frequency/high field EPR spectroscopy

High frequency (high field) EPR spectroscopy opened new approaches for investigating the structure, properties, dynamics, conformational transitions and functioning of many chemical and biological systems. Authors of the recent book considering this method: Möbius & Savitsky, 2009, presented the state-of-the-art capabilities and future perspectives of electron-spin triangulation by high-field/high-frequency dipolar EPR techniques designed for determining the three-dimensional structure of large supra-molecular complexes dissolved in disordered solids. These techniques combine double site-directed spin labeling with orientation-resolving PELDOR spectroscopy. In one of the last publications of this topic, the prospects of angular triangulation, which extends the more familiar distance triangulation was appraise (Savitsky et al., 2011). The three-dimensional structures of two nitroxide biradicals with rather stiff bridging blocks and deuterated nitroxide headgroups have been chosen as a model for spin-labeled proteins. The 95 GHz high-field electron dipolar EPR spectroscopy with the microwave pulse-sequence configurations for PELDOR and relaxation-induced dipolar modulation enhancement (RIDME) has been used. The approach showed good agreement with other structure-determining magnetic-resonance methods, and seems to be one of the most precise orientation-resolving EPR spin triangulation methods for protein structure determination (Savitsky et al., 2011).

To those who want to know about the high field/high frequency approach in detail, I recommend several additional books: Grinberg & Berliner, 2011, Misra, S. K., 2011, Eaton et al., 2010, Hanson & Berliner, 2010.

At the end of this section it is necessary to attract attention to two recent works. The first article considers joint analysis of EPR line shapes and ^{1}H nuclear magnetic relaxation dispersion (NMRD) profiles of DOTA-Gd derivatives by means of the slow motion theory (Kruk et al., 2011). NMRD profiles have been extended to ESR spectral analysis, including in addition g-tensor anisotropy effects. The extended theory has been applied to interpret in a consistent way NMRD profiles and ESR spectra at 95 and 237 GHz for two Gd(III) complexes. The goal was to verify the applicability of the commonly used pseudorotational model of the transient zero field splitting, which was described by a tensor of a constant amplitude, defined in its own principal axes system. The unified interpretation of the EPR and NMRD leads to reasonable agreement with the experimental data. Seems, this approach to the electron spin dynamics can be also effectively used for quantitative description in the case of nitroxide spin probes and labels (Kruk et al., 2011).

One of very important questions in the spin label/probe method is connected with the distance distributions <r> between site-directedly attached spin labels, which obtained by measuring their dipole–dipole interaction in systems under investigation by EPR. As it was shown in (Köhler et al., 2011), the analysis of these distance distributions can be misleading particularly for broad distributions of <r>, because the most probable distance deviates from the distance between the most probable label positions. The authors studied this effect using numerically generated spin label positions, molecular dynamics simulations, and experimental data of a model systems. An approach involving Rice distributions is proposed to overcome this problem (Köhler et al., 2011).

3. The empirical d_1/d parameter

One of the most informative methods for investigating the structure, spatial organization and physical-chemical properties of complex and supramolecular systems on the microscopic, molecular level is EPR spectroscopy in its spin label/probe technique variant (Berliner, 1976; Buchachenko & Wasserman, 1976; Likhtenstein, 1976). Usually, nitroxide radicals of different structure were used for studying of structural peculiarities of spin labeled proteins and the spatial distribution of probes in them. We will discuss shortly the most important results obtained by different authors below.

With the increase of concentration, X-band EPR spectra of nitroxide radicals in frozen solutions reflect not only dipolar broadening of spectral lines and increasing of the M_2 value, but also noticeable changes of the whole spectrum shape which can be characterized with the empirical "shape parameter" d_1/d (Figs. 2, 3). Anisotropic "frozen" EPR spectra of nitroxide radicals at different widths of individual lines have been simulated (Parmon & Kokorin, 1976). They are shown in Fig. 6.

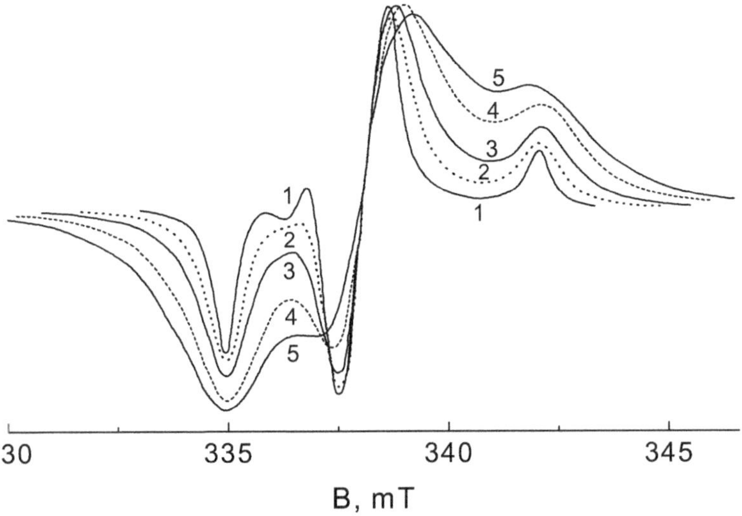

Figure 6. Simulated anisotropic EPR spectra of nitroxide radical at different widths of individual lines: 3 (1), 5 (2), 7 (3), 10 (4), and 13 G (5). The following values were used for simulations: $g_x = 2.0089$, $g_y = 2.0061$, $g_z = 2.0027$; $A_x = 7$, $A_y = 5$, $A_z = 33$ G

It is evidently seen from Figs. 3 and 6 that the line widths and the relative intensities of the center and outer lines in the spectra are changed with the increase of radical concentration. The experimental dependence of d_1/d parameter, characterizing changes of the whole EPR spectrum shape of nitroxides, is not linear (Fig. 7):

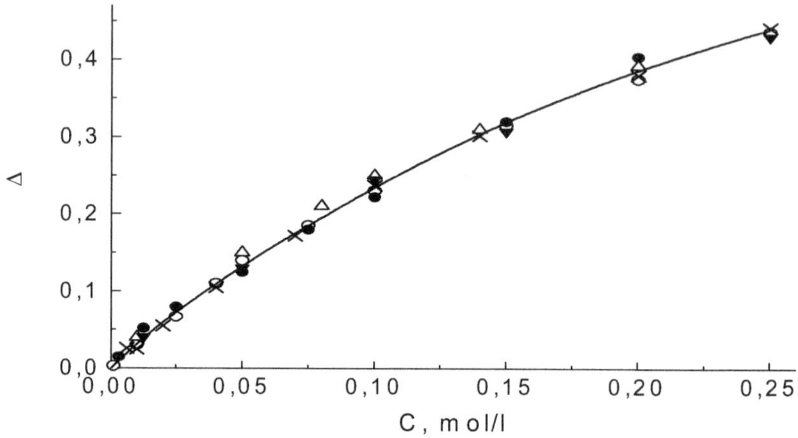

Figure 7. Parameter Δ as a function of nitroxide radical concentration at 77 K: R_6OH in glycerol:H_2O = 1:1 mixture (○) and in ethanol (×), $R_6OCOC_6H_5$ in toluene (•), $R_5{}^{N}CHCOCH_2I$ in ethanol (Δ) and in toluene (▼)

Similar nonlinear dependence have been obtained from simulations of the anisotropic X-band EPR spectra of nitroxides (Parmon & Kokorin, 1976, Kolbanovsky et al., 1992a). For all nitroxide radicals frozen at 77 K in various glazed solvents studied, one can observe, as it was published by Kokorin et al., 1972, 1975, that:

$$d_1/d = (d_1/d)_0 + \Delta \qquad (9)$$

Here Δ is a contribution of the dipole-dipole interaction between radicals, and $(d_1/d)_0$ is a characteristic of the solvent and radical itself (Kokorin, 1974, Kokorin et al., 1975). This equation was verified by precise computer simulations of experimental EPR spectra (Kolbanovsky et al., 1992a). Results obtained for three radicals: R_6OH, $R_6CH_2CH_2Br$ (both in ethanol), and R_6H (in toluene) are given in Table 1. These data confirmed the correctness of usage the d_1/d parameter for quantitative characterization the spatial distribution of spin labels, in the case of their random distribution, in magnetically diluted solid solutions. These results confirmed the possibility of applications of d_1/d parameter for quantitative studies (Kolbanovsky et al., 1992a).

Several experimental correlations between spectral parameters useful for practical applications have been found for nitroxide radicals in solid solutions. They are shown in Figs. 8-10.

C, mol/l	R_6OH		R_6H		$R_6CH_2CH_2Br$	
	Experiment	Theory	Experiment	Theory	Experiment	Theory
0.05	0.12	0.11	0.12	0.12	0.10	0.11
0.1	0.22	0.24	0.23	0.25	0.23	0.23
0.15	0.33	0.37	0.33	0.41	0.31	0.36
0.2	0.38	0.45	0.40	0.49	0.43	0.44
0.25	0.43	0.50	0.43	0.52	0.46	0.48
0.3	0.49	0.54	0.50	0.54	0.51	0.52
0.35	0.55	0.57	0.57	0.56	-	-
0.4	0.58	0.59	0.63	0.57	-	-

Table 1. Experimental and calculated from simulated spectra values of parameter Δ

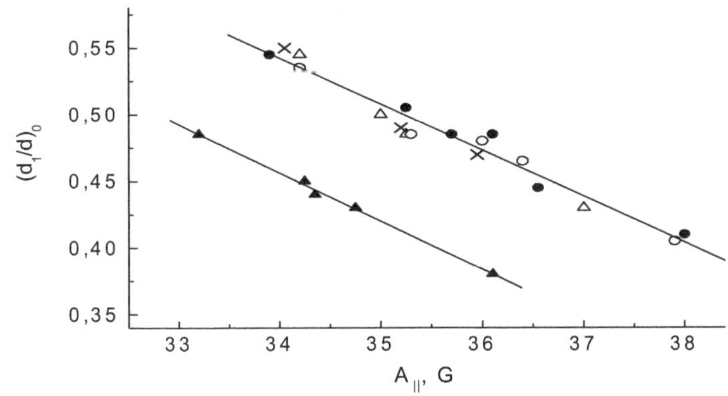

Figure 8. $(d_1/d)_0$ as a function of hyperfine splitting constant $A_{||}$ at 77 K for: R_6=O (▲), R_6H (Δ), R_6OH (○), R_6NH_2 (•), and $R_6OCOC_6H_5$ (×)

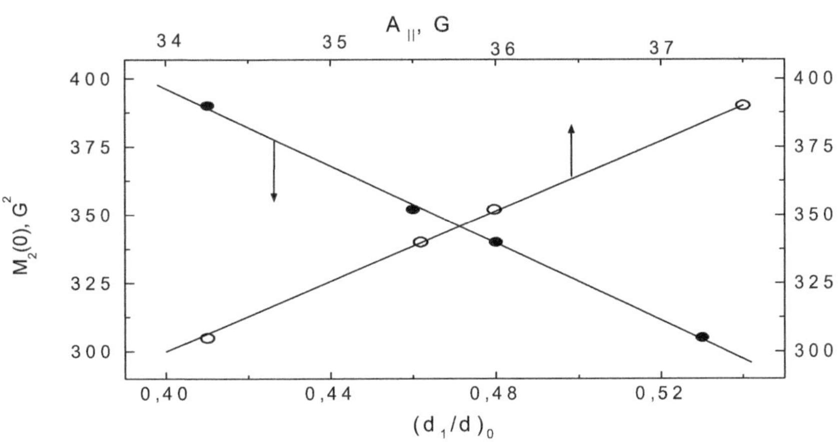

Figure 9. $M_2(0)$ as a function of hyperfine splitting constant $A_{||}$ (○), and of parameter $(d_1/d)_0$ (•) at 77 K for R_6OH in glycerol:H_2O = 1:1 mixture, ethanol, methanol, and $R_6OCOC_6H_5$ in toluene

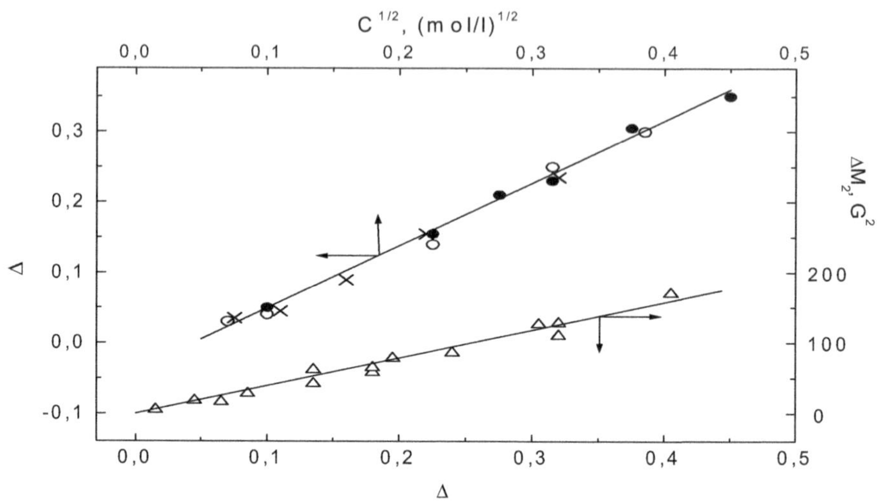

Figure 10. ΔM_2 as a function of parameter Δ (Δ), and parameter Δ as a function of $C^{1/2}$ at 77 K for R_6OH in glycerol:H_2O = 1:1 mixture (×), in ethanol (•), and $R_6OCOC_6H_5$ in toluene (○)

The value of $(d_1/d)_0$ parameter depends on the hyperfine splitting constant $A_{||}$ and on the nitroxide rihg structure (Parmon et al., 1977b). It is seen from Fig. 8 that

$$(d_1/d)_0 = 1.73 - 0.035 \bullet A_{||} \quad (10)$$

These values, 1.73 ± 0.06 and 0.035 ± 0.002, obtained for R_6H, R_6OH, R_6NH_2, and $R_6OCOC_6H_5$ dissolved in various solvents, practically coincide with values of 1.76 and 0.036 correspondingly, published earlier in Parmon et al., 1977b, 1980 and measured for R_6OH radical only. For R_6=O radical, these parameters are equal to 1.70 ± 0.04 and 0.036 ± 0.001, correspondingly.

Fig. 9 presents the value of $M_2(0)$ as a function of $A_{||}$ and of $(d_1/d)_0$ parameter. This can be formalized by corresponding equations:

$$M_2(0) = (25.3 \pm 0.6) \bullet A_{||} - (560 \pm 25) \quad (11)$$

and

$$M_2(0) = (680 \pm 10) - (705 \pm 17) \bullet (d_1/d)_0 \quad (12)$$

Previously (Parmon et al., 1977b, 1980), values of $M_2(0) = 25.5 \bullet A_{||} - 570$ has been reported.

Parameter ΔM_2 also correlates with the values of parameter Δ, as it follows from Fig. 10.

$$\Delta M_2 = (410 \pm 20) \bullet \Delta \quad (13)$$

It was reported (Kokorin, 1986) that parameter Δ, which is not linear on concentration, can be presented as a rather linear plot in other coordinates:

$$\Delta = (0.89 \pm 0.05) \bullet C^{1/2} - (0.025 \pm 0.013) \quad (14)$$

It should be stressed that this linear plot is valid in the concentration range between 0.02 and 0.4 mol/l, i.e., for $0.04 \leq \Delta \leq 0.45$.

Relations presented above were tested with different objects, and can be recommended for application in the case of random or regular distribution of spin probes in chemical and biological systems.

4. Pairwise interaction between two nitroxide spins

Pairwise distribution of the dipolar interacted spins, when a distance between them, r, is noticeably less than a mean distance between pairs, has been investigated in numerous works. This type of spatial distribution contains first of all nitroxide biradicals and double-labeled proteins, peptides and oligomers. In case of short biradicals with r < 1.1 nm, when the dipolar splitting is observed in the EPR spectrum lines, the r_D value, expressed in Eq. (15), can be easily determined from the spectrum simulation (Parmon et al., 1977a, 1980, Kokorin et al., 1984) or just from the experimental value of the dipolar splitting $D_\perp$ by the equation (Kokorin et al., 1972):

$$r_D = 30.3 \bullet D_\perp^{-1/3}, \quad (15)$$

or from the relative intensity α of the half-field ("forbidden", $\Delta M_S = 2$) and normal-field ($\Delta M_S = 1$) EPR transitions, r_S (Lebedev & Muromtsev, 1972, Dubinskii et al., 1974):

$$\alpha = I_2/I_1 = (8/15)(3g\beta/2H_0 r_S^3)^2 \approx 38/r_S^6, \quad (16)$$

where I_2 and I_1 are integrated intensities of EPR spectra of corresponding transitions, $H_0 \approx$ 320 mT is a value of the constant magnetic field in the center of $\Delta M_S = 1$ EPR spectrum, and r_S is measured in Å. In practice, the value of rS one can determine experimentally up to 1.2 nm.

The dipolar interaction impact to second central moment $\Delta M_2 = M_2 - M_2(0)$ of EPR spectrum allows measuring the distance r_M (in Å) in biradicals till ~1.8 nm by the equation (Kokorin et al., 1972, Kulikov et al., 1972):

$$r_M = 23.1 \bullet (\Delta M_2)^{-1/6} \quad (17)$$

Kokorin et al., 1972, reported about 9 biradicals for which distances r were measured independently by methods mentioned above and compared with d_1/d parameter of these biradicals. Later, the number of such "reference" biradicals increased to thirteen, and allowed to plot the experimental dependence of r on Δ presented in Fig. 11 based on the results given in Table 2:

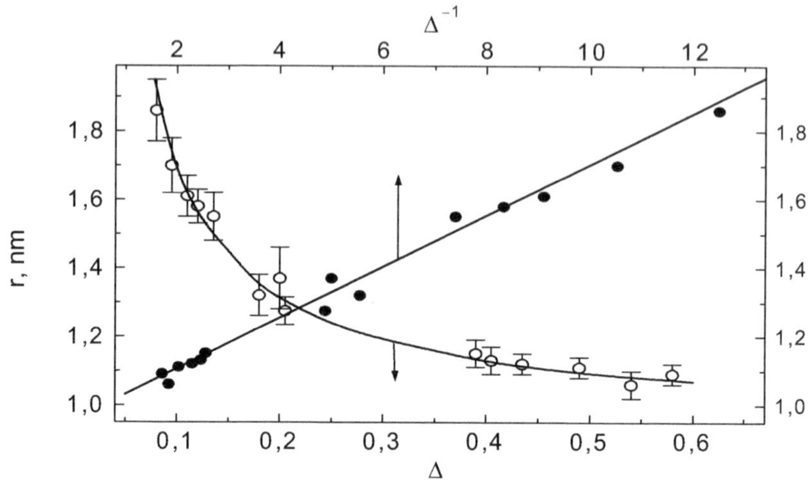

Figure 11. Δ (○) and Δ^{-1} (•) as a function of the distance **r** between unpaired electrons in nitroxide biradicals in frozen solutions at 77 K

Biradical	Solvent	r_M, Å	r_S, Å	r_{calc}, Å
$OC(OR_6)_2$	Toluene	11.2	11.7	11.3*
$S(OR_6)_2$	Toluene	11.1	10.8	11.6
$O_2S(OR_6)_2$	Toluene	10.7	10.3	-
$CH_2CH(O)P(OR_6)_2$	Toluene	10.3	10.2	10.9
$R_6NH(CH_2)_2R_6$	Toluene	10.8	11.5	-
m-$C_6H_4(COOR_6)_2$	Toluene	15.5	-	15.7
o-$C_6H_4(COOR_6)_2$	Toluene	11.5	11.2	11.8
$R_6(CH_2)_4R_6$	Ethanol	12.9	-	13.1*
$(CH_2)_4(HOR_6)_2$	Ethanol	13.6	13.2*	13.8
$(CH_2)_6(COOR_6)_2$	Ethanol	-	18.5*	17.8
m-N_3C_3Cl*$NHCH_2R_6)_2$	Methanol	13.8	-	13.0
m-N_3C_3Cl*$NHR_6)_2$	Methanol	11.5	-	11.3

* measured from the X-ray data as the distance between centres of N–O bonds (Capiomont, 1972); r_M measured from the second central moment ΔM_2 (Kokorin et al., 1972, 1974, 1976, Kulikov et al., 1972); r_S measured from the forbidden transitions $\Delta M_S = 2$ (Dubinskii et al., 1974), r_{calc} calculated from EPR spectra simulation at 77 (K (Kokorin et al., 1984)

Table 2. Distances between N–O groups in nitroxide biradicals measured by different methods

Linear dependence of r on Δ^{-1} obtained in the range $1.0 \le \mathbf{r} \le 1.85$ nm gives a correct, simple and rather precise method for the estimation of **r** values: (Kokorin et al., 1976)

$$\mathbf{r} = (9.6 \pm 0.2) + (0.75 \pm 0.02)/\Delta \qquad (18)$$

Kokorin, 1974, Parmon et al., 1977b, 1980 published similar corresponding parameters equal to 9.3 ± 0.25 and 0.77 ± 0.03. One can see that these values are very close to those in Eq. (18). Therefore, the interval in which d_1/d parameter is recommended for correct distance measurements is $1.2 \leq \mathbf{r} \leq 2.5\text{-}2.7$ nm.

Values of <r> and **r** calculated from the EPR spectra simulation for nitroxide radicals and biradicals in the case of random and pairwise distributions according to recommendations of (Kokorin et al., 1984) were compared with experimental data and shown in Fig. 12. Good correlation between experimental and theoretical data is observed.

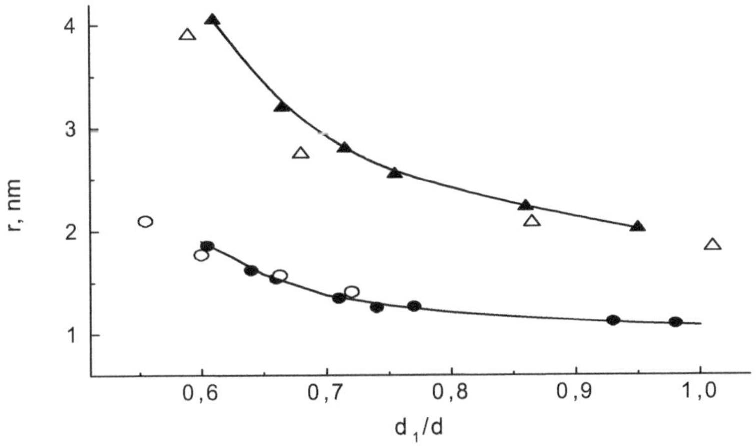

Figure 12. Parameter d_1/d as a function of mean distances <r> for R_6OH radical (Δ, ▲), and distances r for biradicals (○, •) dissolved in frozen at 77 K solutions: experimental (▲, •) and calculated (Δ, ○) from theoretical EPR spectra values

The EPR spectrum shape parameter d_1/d has been used for studying the effect of the solvent on structural organization of nitroxide biradicals.

Results presented in Table 3 show that in glassy solid solutions frozen at 77 K in various solvents the conformational structure of biradicals can be changed by the influence of the solvent in case of non-rigid, rather long flexible molecules such as *m*-$C_6H_4[COO(CH_2)_2R_6]_2$, *o*-$C_6H_4[COO(CH_2)_2R_6]_2$, $S[(CH_2)_2COOR_6]_2$, while for rather short or more rigid molecules ($R_6NH(CH_2)_2R_6$, *m*-$C_6H_4(COOR_6)_2$) the solvent effect is not observed. A long flexible biradical $(CH_2)_4[COO(CH_2)_2R_6]_2$ was not also sensitive to changes in the solvent polarity (Table 3).

It should be stressed that modern EPR techniques allow researchers measuring distances longer 2.5-3.0 nm, which are out of the scale for d_1/d. As an example, it can be illustrated by the work of Bird et al., 2008 in which authors demonstrated the synthesis of a series of spin-labeled curved oligomers and determined their end-to-end lengths and distance distributions using ELDOR technique of EPR spectroscopy. Spin labeled water-soluble, spiro-ladder oligomers with well-defined three-dimensional structures studied with

Biradical	Solvent	$d_1/d \pm 0.01$	r, Å
$R_6NH(CH_2)_2R_6$	Toluene	1.03	11.1 ± 0.2
	Ethanol	0.94	11.0
	H_2O:glycerol = 1:1	0.865	11.0
m-$C_6H_4(COOR_6)_2$	Toluene	0.675	15.5
	1-butanol	0.61	16.0 ± 0.6
	Ethanol	0.60	16.0
	Methanol	0.595	15.7 ± 0.5
o-$C_6H_4[COO(CH_2)_2R_6]_2$	Toluene	0.745	12.7 ± 0.3
	1-butanol	1.01	10.8
	Ethanol	0.92	11.2 ± 0.2
	Methanol	0.82	11.6
m-$C_6H_4[COO(CH_2)_2R_6]_2$	Toluene	0.65	16.1 ± 0.6
	1-butanol	0.55	23.3 ± 2.2
	Ethanol	0.56	19.6 ± 1.4
	Methanol	0.55	19.0 ± 1.2
$(CH_2)_4[COO(CH_2)_2R_6]_2$	Toluene	0.605	20.3 ± 1.6
	Ethanol	0.555	20.3
$S[(CH_2)_2COOR_6]_2$	Toluene	0.66	15.7 ± 0.5
	Ethanol	0.57	18.4 ± 1.0

Table 3. Effect of solvent nature on the distances between unpaired electrons in nitroxide biradicals (Parmon et al., 1980)

ELDOR, provided to obtain quantitative information about the shapes and flexibility of the oligomers. The estimeted end-to-end distance of the oligomers ranges from 23 to 36 Å. The shapes of the EPR-derived population distributions allow the authors to compare the degree of shape persistence and flexibility of spiro-ladder oligomers (Bird et al., 2008).

The last works in this area based on high-frequency pulse EPR technique besides measuring distances allow to determine relative mutual orientation of paramagnetic >N–O groups at distances **r** > 3.0 nm (Savitsky et al., 2011). The 95 GHz high-field electron dipolar EPR spectroscopy with the microwave pulse-sequence configurations for PELDOR has been applied. It was concluded that due to the high detection sensitivity and spectral resolution the combination of site-directed spin labeling with high-field PELDOR stands out as an extremely powerful tool for 3D structure determination of large disordered systems. The authors approach compared with other structure-determining magnetic-resonance methods evidently showed its advantage. Angular constraints were provided in addition to distance constraints obtained for the same sample, and the number of necessary distance constraints was strongly reduced. The reduction of necessary distance constraints became another appealing aspect of orientation-resolving EPR spin triangulation which can be applied for protein structure determination (Savitsky et al., 2011).

Real advantage of d_1/d method in comparison with modern ones is its simplicity, availability and quite good precision within the interval of its correctness (1.2-2.7 nm).

5. Interaction between radicals and paramagnetic metal ions

It was experimentally revealed that d_1/d parameter strongly depends on the longitudinal relaxation time T_1 value, and a new method of measuring distances between spin labels and paramagnetic metal ions in macromolecules suggested (Kokorin & Formazyuk, 1981). The number of paramagnetic metal ions and complexes tested has been enlarged, and possible applications to biological systems discussed (Kokorin, 1986). Some later, this approach has been extended to spin-labeled metal-containing polymers (Kokorin et al., 1989).

Fig. 13 presents typical dependences of d_1/d parameter of R_6OH radical as a function of concentration of R_6OH itself, and of some salts: $Cu(NO_3)_2$, $Ni(en)_2(NO_3)_2$, $CoSO_4$, $MgSO_4$ dissolved in H_2O;glycerol (1:1) mixture, and of $Cr(acac)_3$ in methanol solution. One can see from Fig. 13 that the efficiency of d_1/d increase is different for various metal ions. Indeed, it is known from theory that the dipolar broadening of EPR spectra depends besides concentration of paramagnetic centres on the value of its electron spin and the longitudinal relaxation time T_1 (Abragam, 1961, Molin et al., 1980). Kokorin et al., 1981 suggested to characterize relative efficiency of dipole-dipole interaction between nitroxide radicals and paramagnetic metal ions with parameter α^*:

$$\alpha^* = [4S(S+1)/3]^{-1/2} \bullet \langle \Delta_M/\Delta_R \rangle \tag{19}$$

Here Δ_M and Δ_R are the dipolar impacts to d_1/d parameter measured at the same concentrations in cases of interaction between radicals, Δ_R, or between a radical and a metal ion, Δ_M, $\langle \Delta_M/\Delta_R \rangle$ is the averaging by all concentrations (Fig. 13), and coefficient $\varphi = [4S(S+1)/3]^{-1/2}$ is used for metal ions with the electron spin $S > ½$, and takes into account that a spin probe interacts with several electron spins of the metal, S. In such case the dipolar broadening parameter, A*, will be equal to:

$$A^* = \varphi \bullet \delta H / C_M , \tag{20}$$

analogous to Eq. (1). This correction allows one to determine local concentrations of various paramagnetic metal complexes. A value of $\langle \Delta_M/\Delta_R \rangle$ parameter depends on the T_1 value of the paramagnetic centres under investigation (Kokorin & Formazyuk, 1981). Other approaches for solving this problem as well as a perfect collection of experimentally measured values of T_1 are collected in Eaton & Eaton, 2001a, 2001b.

In case of coupling between a nitroxide radical and a paramagnetic metal complex distributed in the matrix in pairs, interesting results were presented by (Fielding et al., 1986). Low-spin Fe(II1)-tetraphenylporphyrin complexes have been modified with seven nitroxide radicals of different length: $-CONHR_{5,6}$, $-CONHCH_2R_{5,6}$, $-OCH_2CONHR_6$, $-O(CH_2)_2CONHR_6$, $-O(CH_2)_4CONHR_6$. The spin labels were attached by amide or amide and ether linkages to the ortho position of one phenyl ring. The axial ligands were imidazole

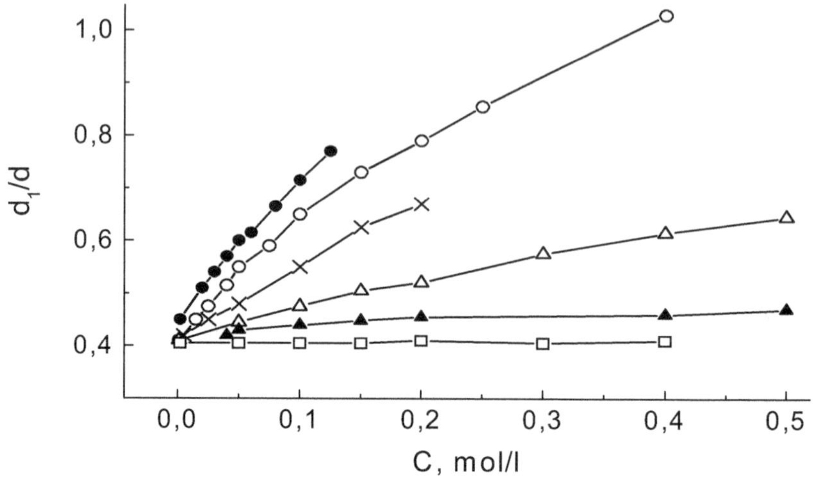

Figure 13. d_1/d as a function of concentration of R_6OH radical (o), $Cu(NO_3)_2$ (×), $Ni(en)_2(NO_3)_2$ (Δ), $CoSO_4$ (▲), $MgSO_4$ (□) in H_2O;glycerol = 1:1, and $Cr(acac)_3$ (•) in methanol at 77 K. en is $NH_2CH_2CH_2NH_2$, and acac is acetylaceronate

or 1-methylimidazole. In frozen solution the complexes with amide linkages adopted two different conformations, and the populations of the conformations were solvent-dependent. Measured value of the exchange integral J was rather high, while the spin-spin interaction in the second conformation was much weaker than in the first conformation. Broadening of the nitroxyl signal in frozen solution was also observed for complexes with longer ether linkages between the phenyl ring and the nitroxyl. Distances between nitroxide N–O groups and Fe(III) ions were estimated by d_1/d parameter and by the Leigh method (Leigh, 1970). The EPR spectra, reported by other authors, of two spin labels coordinated to ferric cytochrome P450 were analyzed with the computer programs developed for the iron porphyrin model systems. The authors showed that electron-electron exchange interaction as well as dipolar interaction must be considered in analyzing the spectra of spin-labeled porphyrin-containing bio-macromolecules (Fielding et al., 1986).

6. Applications to solid solutions and materials

Nitroxide spin probes were successfully used for quantitative investigating the structure and micro-phase organization of frozen two-component solutions (Kokorin & Zamaraev, 1972). R_6OH radical has been chosen as a spin probe to test the homogeneity of heptane-ethanol, carbon tetrachloride-ethanol, and toluene-ethanol mixtures with different ethanol content, frozen at 77 K. EPR spectra of the probe showed non-linear changes of d_1/d parameter, from which local concentrations C_{loc} were calculated. For explanation of the results observed the existence of two different phases in frozen mixed solutions was assumed. By the model suggested in (Khairutdinov & Zamaraev, 1970, Kokorin &

Zamaraev, 1972), radicals were localized not in the whole volume of the sample (V_0, C_0) but only in one mixed phase of total volume V contained both components with local concentration C_{loc}. The second phase was crystallized and did not contain spin probes. Evidently, $C_{loc} = C_0 \bullet V_0/V$ and it was obtained for the coefficient of non-uniformity of probes distribution $\rho = C_{loc}/C_0$:

$$\rho = [(M_{gl}/d_{gl}) + n \bullet (M_{cr}/d_{cr})]^{-1} \bullet C_{Et}^{-1}, \quad (21)$$

where M_{gl}, M_{cr} and d_{gl}, d_{cr} are molecular masses and densities of the glazed (ethanol) and crystallized (heptane, CCL_4) components of the mixture; n is a number of molecules of the crystallized solvent per one molecule of the glazed one (ethanol) in the areas of the spin probe localization. C_{Et} is the concentration of ethanol in the solution. From experimental dependences of ρ on C_{Et}^{-1} for mixtures heptane-ethanol and CCL_4-ethanol a phase of non-polar solvent and the binary mixture of constant composition were observed. The quantitative composition of binary mixtures was determined: 6.5 ± 0.8 ethanol molecules per one heptane molecule, and 2.3 ± 0.3 ethanol molecules per one CCl_4 molecule. Binary toluene-ethanol mixtures were glassy at 77 K at all ratios of components, had complex non-linear dependence of ρ on C_{Et}^{-1} but did not have phases of constant composition (Kokorin & Zamaraev, 1972).

Another interesting and important quantitative application of spin label technique and d_1/d parameter can be illustrated by studies of gold nanoparticles with EPR spectroscopy.

Ionita et al., 2004, investigated the mechanism of a place-exchange reaction of ligand-protected gold nanoparticles using biradical disulfide spin labels which were chemically attached to the surface (Fig. 14). Analysis of reaction mixtures combined GPC and EPR technique allowed authors to determine concentration profile of spin probes and propose a kinetic model for the reaction. Local concentrations of spin labels and mean distances between them were measured using d_1/d parameter. In the model suggested, only one branch of the disulfide ligand was adsorbed on the gold surface during exchange, and the other branch formed mixed disulfide with the outgoing ligand. The two branches of the disulfide ligand therefore did not adsorb in adjacent positions on the surface of gold nanoparticles. This was proven by the powder EPR spectra of frozen exchange reaction mixtures. The data presented also suggested the presence of different binding sites with different reactivity in the exchange reaction. It was assumed that the most-active sites are likely to be nanoparticle surface defects (Ionita et al., 2004).

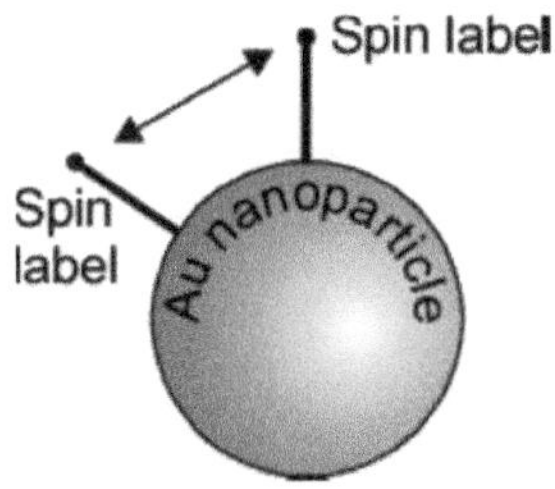

Figure 14. Schematic localization of spin labels on the surface of gold nanoparticles (Ionita et al., 2005)

A series of gold nanoparticles modified with a nitroxide-functionalized ligands was synthesized with a range of spin-label coverage (Ionita et al., 2005). The X-band EPR spectra of frozen solutions of these nanoparticles showed coverage-dependent line-broadening due to dipole-dipole interactions between spin labels, and a noticeable increase of d_1/d parameter. A methodology to analyze such spectra in terms of geometrical features of the nanoparticles (e.g., gold core size and the length of the spin-labeled ligand) was developed. The method was based on the assumption that the spectral line shape was determined by the average distance between nearest-neighboring spin labels adsorbed on the gold particle. Geometrical and statistical analysis related this distance to the line shape parameter d_1/d, which was calibrated using a model system. A calibration curve was suggested as an empirical Eq. (22) (Ionita et al., 2005):

$$d_1/d = a_1\exp[-a_2(r_n - a_3)] + a_4 + a_5/r_n \qquad (22)$$

Here r_n is the average distance between nearest-neighboring nitroxide labels. The values of empirical parameters a_1-a_5 were equal to 0.8050, 3.0150 nm^{-1}, 0.8736 nm, 0.5145, and 0.06824 nm, respectively, as obtained by nonlinear regression. Experimental and calculated values of d_1/d parameter were compared and have been very close to each other. The interspin average distances r_n between nearest-neighboring spins were determined in the range $1.46 \leq r_n \leq 3.3$ nm, and they decreased with increasing coverage of spin labels.

Application of this methodology to the experimental spectra provided information about the conformation of ligands on the gold surface. It was found that, if the spin-labeled ligand was substantially longer than the surrounding protecting layer, it did not adopt a fully stretched conformation but wrapped around the particle immediately above the layer of surrounding ligands. The results obtained also showed that the ligands were not adsorbed cooperatively on the gold surface (Ionita et al., 2005).

The lateral mobility of the thiolate ligands on the surface of gold nanoparticles was also probed by Ionita et al., 2008, using bisnitroxide ligands, which contained a disulfide group in the bridge (to ensure attachment to the gold surface) and a cleavable ester bridge connecting the two spin-labeled branches of the molecule. Upon adsorption of these ligands on the surface of gold particles, the two spin-labeled branches were held next to each other by the ester bridge as evidenced by the spin-spin interactions. Cleavage of the bridge removed the link that kept the branches together. CW and pulsed EPR (ELDOR) experiments showed that the average distance between the adjacent thiolate branches on the gold nanoparticle surface only slightly increased after cleaving the bridge and thermal treatment. This implied that the lateral diffusion of thiolate ligands on the nanoparticle surface was very slow at room temperature and took hours even at elevated temperatures (90°C). The changes in the distance distribution observed at high temperature were likely due to ligands hopping between the nanoparticles rather than diffusing on the particle surface (Ionita et al., 2008).

7. Applications to polymers

Another quantitative application of d_1/d parameter was suggested for determining local concentrations of chain units in macromolecular coils using the spin-label method (Kokorin et al., 1975). Labelling of poly-4-vinyl-pyridine (P4VP) with $R_6CH_2CH_2Br$, R_6OCOCH_2Cl or $R_6NHCOCH_2I$ radicals with the degree of alkylation of pyridine residues from 2 up to 35 mol.% allowed authors to determine such important structural characteristics of the polymer coil as local concentration of pyridine monomers C_N, the effective volume <V> and effective radius R_{eff} of the polymer coil, its local density ρ_{loc}:

$$\langle V \rangle = n/C_{loc};\ R_{eff} = (3n/4\pi C_{loc})^{1/3};\ C_N = [(P - n)/n]\bullet C_{loc};\ \rho_{loc} = P\bullet C_{loc}/n \qquad (23)$$

Here P and n are a degree of polymerization (mean number of monomer units in the chain) and mean number of spin labels in the coil correspondingly. It should be stressed that this approach is correct if: a) distribution of spin labels in the coil is statistically random, b) the mean volume <V> occupied by spin labels is identical to the volume of the coil, and c) spin labeling does not change the conformation of macromolecules.

High values of C_N equal to 0.3 ± 0.1 mol/l, obtained for labeled P4VP molecules in dilute solutions, confirm the theoretical estimations made by Tanford, 1961. Table 4 contains some values obtained for spin labeled P4VP, polyethylenimine (PEI), polyglycidylmethacrylate (PGMA), poly(methacrylic acid) (PMAc), and its sodium salt (PMAcNa). $\langle \hat{h}^2 \rangle = 6\bullet(3\langle V \rangle/4\pi)^{2/3}$ is the mean-square end-to-end distance for a Gaussian chain.

This approach was successfully used for analyzing the conformational state of PGMA macromolecules in diluted and concentrated polymer solutions (Shaulov et al., 1977). Determination of mean distances between monomer units in the chain or their local concentration, C_{loc}, in the effective macromolecular volume <V> as a function of the polymer concentration in solutions of different thermodynamic quality as well as in a solid amorphous powder was carried out. Both labelled and non-labelled polymers were used. It was revealed that within limits of the experimental conditions, the size of the coil considered to be Gaussian, exceeds theta-dimensions, while the coil size in solid polymer is close to θ-dimensions. Models of concentrated PGMA solutions were analyzed and the most probable one was chosen basing on the experiment (Shaulov et al., 1977).

The local density of monomer units of the macromolecule (local density of the host residues, ρ_{host}) in poly(4-vinyl pyridine) solutions in ethanol was determined by the spin label technique and d_1/d parameter (Wasserman et al., 1979). In dilute solutions, ρ_{host} is considerably greater than the mean density of monomer units in the volume of the polymer coil. When the P4VP concentration increases from 0.5 to 65 wt%, the ρ_{host} increase does not exceed 30%. This fact indicates that the differences between polymer coil spatial organization (mutual positions of monomer units close to a labeled unit) in dilute and concentrated solutions are small. The local density of monomer units of neighbouring macromolecule coils (guest macromolecules, ρ_{guest}) is strongly dependent on the polymer concentration in solution. In dilute solutions, $\rho_{host} >> \rho_{guest}$; for polymer concentrations above 2–3 wt%, overlapping and interpenetration of macromolecular coils take place, local density of guest coils, ρ_{guest}; monotonously increases with polymer concentration growing up. The concentration dependence of the local rotational and translational mobility of chain units was also determined for spin-labeled P4VP (Wasserman et al., 1979).

Several works were published in which d_1/d parameter was used for the study of intramolecular dynamics and of local density of P4VP units in diluted and concentrated liquid and frozen solutions with spin label method Wasserman et al., 1980a, 1980b). Temperature dependences of EPR spectra lines of spin labeled P4VP solutions in ethanol allowed estimate dipolar and spin exchange impacts, and to calculate mean local densities ρ_{loc} at 77 and 293 K. Measured parameters: 0.3 ± 0.1 at 77 K and 0.25 ± 0.05 mol/l at 293 K, were several times higher than the average concentrations of the units in solutions. Knowledge of these concentrations allowed to calculate correct diffusion coefficients of spin labels in the P4VP coil. These results are reasonably close to those listed in Table 4.

Polymer	P	C_{loc}, mol/l	ρ, mol/l	<V>, nm³	R_{eff}, nm	$\langle\hat{h}^2\rangle$, nm²
P4VP-25%	430	0.14	0.52	1370	6.8	280
P4VP-35%	270	0.17	0.49	930	6.1	220
P4VP-10%	1330	0.04	0.4	5540	11.0	726
P4VP-15%	1330	0.07	0.49	4570	10.3	636
P4VP-20%	1330	0.11	0.55	4030	9.9	590
P4VP-28%	1330	0.14	0.5	4430	10.2	624
P4VP-40%	1330	0.18	0.45	4930	10.6	670
PEI-5%	120	0.023	0.46	435	4.7	130
PEI-22%	120	0.14	0.62	320	4.25	110
PGMA-10%	690	0.033	0.33	3450	9.4	530
PGMA-16%	690	0.05	0.3	3830	9.7	560
PMAc-23%	1600	0.11	0.48	5580	9.3	520
PMAcNa-23%	1600	0.1	0.43	6130	9.6	550

Table 4. Some parameters characterizing spin-labelled polymers (Kokorin, 1992, Wasserman et al., 1992)

The intramolecular mobility and local density of monomer units in spin labeled styrene co-polymers with maleic anhydride has been studied (Aleksandrova et al, 1986). The estimated value of ρ_{loc} for these co-polymers dissolved in dimethylformamide equal ≈0.03 mol/l was ten-fold less ρ_{loc} values measured for P4VP in ethanol or 50% H_2O:ethanol mixtures (Table 4).

Quantitative measurements of non-crystallized (solvated) water molecules, based on the EPR study of the structure of frozen aqueous solutions of polyvinylpyrrolidone (PVP) and of polyvinylalcohol (PVA) have been reported (Mikhalev et al., 1985). Local concentrations of spin labels and spin probes were determined by d_1/d parameter and C_{loc} values calculated using procedure suggested in Khairutdinov & Zamaraev, 1970, Kokorin & Zamaraev, 1972. In the presence of NaCl salt in frozen solutions of PVP, the formation of strong complexes between PVP links with water has been observed in the areas containing salt, polymer fragments and H_2O molecules.

The application of EPR spin probe and spin label technique for solving two actual problems of polymer physical chemistry was considered in (Wasserman et al., *1996*). The first problem is the determination of conformational state and chain sizes in amorphous solid polymers.

This determination is based on the analysis of the intramolecular dipolar broadening of EPR spectra of spin labelled macromolecules in glassy solvents or in the bulk of unlabeled polymers at 77 K with the use of d_1/d parameter. The second problem is the determination of molecular dynamics and structure of polymer colloid systems: (a) the complex of colloidal silica and synthetic polycation macromolecule, and (b) polymer-surfactant micellar organized systems. Possible approaches were discussed.

Spin labeling was used to investigate the topochemical characteristics of polymer carriers and immobilized metal complexes (Bravaya & Pomogailo, 2000). Functionalized polyethylene (PE) molecules such as PE-grafted-polyallylamine (PE-PAA), PE-grafted-polydiallylamine (PE-PDAA), and PE-grafted-poly-4-vinylpyridine (PE-P4VP), obtained by grafting polymerization of the corresponding monomers, were used as polymer carriers. Metal-containing polymers were synthesized by attaching to polymers either $TiCl_4$ (PE-PDAA-Ti) or $Al(C_2H_5)_2Cl$ (PE-PDAA-Al). 2,2,6,6-tetramethyl-4-(2′-oxy-4′,6′-dichlorotriazine) piperidine-1-oxyl nitroxyl radical (R_1) was used for spin labeling PE-PDAA, while 2,2,5,5,tetramethyl-3-(N-acetoamidiiodine)pyrrollidine-1-oxyl (R_2) was used for spin labeling PE-P4VP, and 2,2,6,6-tetramethyl-4-hydroxy-pyperidine-1-oxyl (R_3) was bound to PE-PDAA-Ti and PE-PDAA-Al respectively. Estimation of the effective distances between the spin labels by d_1/d parameter and the dynamic behavior of nitroxyl radicals in the functionalized polymer matrixes and metal-containing polymers revealed several important features of spin-labeled systems. Metal complex formation of functional polymers made them more accessible for spin labeling and had a considerable effect on the dynamic characteristics of the polymer matrix. Thermodynamic characteristics of the rotational diffusion of the labels were determined (Bravaya & Pomogailo, 2000).

Next serious approach to better understanding structural organization of spin labeled macromolecules in the amorphous solid state and their conformational transitions has been suggested by Khazanovich et al., 1992. The algorithm for EPR spectra computation was developed: it was assumed that molecular weights of labelled linear chains are high enough and their solid solution is diluted. It was shown that the scaling exponent which determines the dependence of mean-square end-to-end distance on molecular weight and stiffness parameter (mean-square length of monomer unit) may be extracted from the spectra. Simulated EPR spectra were compared with experimental ones, measured at 77 K, of diluted solutions of spin-labelled poly(4-vinyl pyridine), P4VP, of different molecular weights in methanol and non-labelled P4VP. The conformational state of the Gaussian coil, parameter of stiffness, and mean square radius of gyration $<R_G^2>^{1/2}$ of spin-labeled P4VP macromolecule in frozen solutions were determined via measuring d_1/d parameter, local density values ρ_{loc} of links, and parameters mentioned above were calculated from it. It was concluded that EPR spectroscopy may become a sensitive tool for studying chain conformation in solid polymers (Khazanovich et al., 1992, Kolbanovsky et al., 1992b).

This approach was successfully used to determine the conformational states of spin labeled P4VP, poly(methacrylic acid), PMAc, and its sodium salt in glassy methanol, ethanol, 1-propanol solutions and in the bulk matrix of unlabeled polymers at 77 K. All macromolecules had near-Gaussian coil conformations. The mean square lengths of the repeating units, the characteristic ratios, and the mean-square end-to-end distances $\langle R^2\rangle^{1/2}$ of the polymers were determined. Typical results are listed in Table 5 with corresponding values of Δ (extracted from Wasserman et al., 1992).

Polymer*	Solvent	Δ	$\langle R^2\rangle^{1/2}$, nm
P4VP-1	Methanol	0.24	22
P4VP-1	1-Propanol	0.29	20
P4VP-2	Methanol	0.14	22
P4VP-4	Ethanol	0.28	36
P4VP-5	Ethanol	0.17	33
P4VP-6	Methanol	0.13	107
PMAc	Methanol	0.25	44
PMAcNa	Methanol	0.23	48

* P4VP-1,2,6 were labeled with $R_6CH_2CH_2Br$, P4VP-4,5 – with $R_5N{=}CHCOCH_2I$, PMAc – with R_6NH_2. P4VP-1,2, P4VP-4,5 and P4VP-6 are of different molecular mass.

Table 5. Parameters Δ and $\langle R^2\rangle^{1/2}$ of spin labeled macromolecules

Analyzing developing of the area of spin probes and labels during a quarter of a century of its application to polymer studies, the author described in details history of the method, investigations of local and segmental mobility in polymers, and paid special attention to the approaches for determining local densities and translational dynamics of monomeric units in a coil (Kovarski, 1996).

It should be noted that EPR data on conformational state and dimensions of the polymer coil can substantially complement the data of other physical methods: neutron scattering, for example.

The spin label method was also used for studying the spatial organization of the labeled linear polyethyleneimine (PEI) macromolecules in glassy 50% water-ethanol solutions in the process of complex formation with transition metal ions (Kokorin et al., 1989). The mean local density of PEI chain units ρ_{loc} was measured by d_1/d parameter, as well as PEI coil volume <V>, the mean coil radius R_{eff}, and the average distance berween spin labels (Table 5). It is known, that if in the sample there are paramagnetic centres of different nature, total dipolar broadening of EPR spectrum lines is the sum of broadenings caused by paramagnetics of each type (Lebedev & Muromtsev, 1972). Analogously, in case of coordination of metal ions (paramagnetic Cu(II), Ni(II), Co(II) and diamagnetic Zn(II) and sodium ions) by the spin-labeled PEI, a procedure for separate determination the impacts of the dipole-dipole interaction between spins of radical labels Δ_{LL} and of radicals with metal ions Δ_{LM} was suggested. These impacts can be expressed as a sum to the experimentally measured value of parameter Δ:

$$\Delta = \Delta_{LL} + \Delta_{LM} \quad (24)$$

The average distances R_{LM} between a label and the nearest paramagnetic complex were determined by the method suggested by Leigh, 1970. The average local concentrations of complexes <C_M> in the PEI coil volume were calculated using the data and coefficients obtained in Kokorin & Formazyuk, 1981. Changes in Δ_{LL} caused by the decrease of the coil volume as a result of polymer-metal complex formation was determined by measuring d_1/d values in the case of PEI interaction with diamagnetic Zn(II) ions in the whole range of [M]/[L] ratios. As an example of this approach Fig. 15 shows the effective local concentration of metal complexes <C_M> in the PEI coil vs. the metal-to-label ratio [M]/[L] for different metal ions at 77 K (Kokorin et al., 1989).

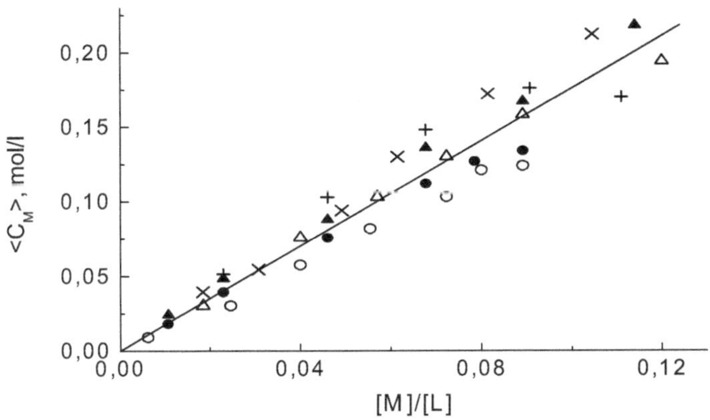

Figure 15. Effective local concentration of metal complexes <C_M> in PEI coil as a function of metal-to-label ratio [M]/[L] at 77 K for: •,o - Cu(II), ▲,Δ - Ni(II) and x,+ - Co(II). Concentration of spin labels [L] = 0.045 (•,▲,x) and 0.012 mol/l (o,Δ,+)

It should be stressed that this procedure is correct only in the assumption that spatial distributions of spin labels and polymer-metal complexes in the coil are the same, random, and there are no areas of their specific localization. A fact that <C_M> plots vs. [M]/[L] ratio are the same for all divalent ions studied allows conclude that the structure of metal-PEI coil for these ions is similar.

Determination of the nanostructure of polymer materials by EPR spectroscopy was considered as one of the few methods that can characterize structural features in the range between 1 and 5 nm in systems that lack long-range order (Jeschke, 2002). Approaches based on various techniques of EPR spectroscopy, such as CW X-band EPR, electron spin echo, ENDOR) provided good structural contrast even in complex materials, because the sites of interest could be selectively labeled or addressed by suitably functionalized spin probes using well established techniques. In the article, experiments on distance measurements on nanoscales in terms of the accessible distance range, precision, and sensitivity were discussed, and recommendations were derived for the proper choice of experiment. Both simple and sophisticated methods for data analysis are described and their limitations are evaluated. The approach of Khazanovich et al., 1992, based on d_1/d and Δ parameters was used for characterization of the chain conformation was described. The conformational organization of the polymer chain and the structure of ionomers based on diblock copolymers were analyzed (Jeschke, 2002).

Bird et al., 2008, demonstrated modern possibilities of ELDOR and computing methods on a series of spin-labeled oligomers to determine their end-to-end lengths, R_{ee}, and distance distributions. Seven different shape-persistent macromolecules from conformationally restricted, asymmetric monomers that are coupled through pairs of amide bonds to create water-soluble, spiro-ladder oligomers with well-defined three-dimensional structures were synthesized and investigated. The ends of these oligomers were labeled with nitroxide radicals. ELDOR experiments were carried out to obtain quantitative information about the shapes and flexibility of the oligomers. The most probable R_{ee} distance of the oligomers ranges from 2.3 to 3.6 nm. The relative distances measured for the oligomers confirm that, by varying the sequence of an oligomer, one can control its shape. The shapes of the EPR-derived population distributions allowed the authors to compare the degree of shape persistence and flexibility of spiro-ladder oligomers to other well-studied nanoscale molecular structures such as *p*-phenylethynylenes (Bird et al., 2008).

Interesting application of spin label method and d_1/d parameter was presented by Kozlov et al., 1981, for investigation the oligomers in solutions where long-chain flexible nitroxide biradicals were used as a model. Measuring distances **r** between N–O groups in oligomers, the dependence of **r** on the number of units in the chain, n, was experimentally obtained for 12 biradicals of different length, and the equation (Flory, 1969):

$$\langle r^2 \rangle = \alpha^2\beta^2 n, \quad (25)$$

where α is a Flory-Fox constant, and β is the characteristic length. It was shown that $\alpha\beta$ = 0.56 nm for hydrocarbon oligomers, $\alpha\beta$ = 0.534 nm for dimethylsiloxane ones, and also β = 0.452 and 0.405 nm for poly(methylene) and poly(dimethylsiloxane) chains correspondingly (Kozlov et al., 1981). Fig. 16 Illustrates Eq. () well.

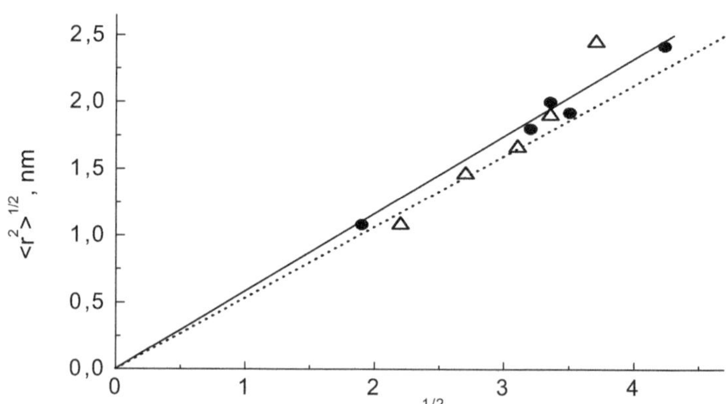

Figure 16. Distance $\langle r^2 \rangle^{1/2}$ as a function of $n^{1/2}$ in toluene solutions at 77 K for $(CH_2)_k(COOR_6)_2$, k = 6-8, 10, 14 (•), and $R_6O\text{–}[Si(CH_3)_2O]_m\text{–}R_6$, m = 2-6 (Δ)

Additional information on applications of spin label technique for investigation structural properties of synthetic polymers is described in detail in monographs by Wasserman & Kovarsky, 1986, Kovarski, 1996, Schlick, 2006.

8. Applications to biological systems

If two nitroxide spin labels are attached to any biological macromolecule, one can measure a distance **r** between their unpaired electrons from the magnitude of dipole-dipole interaction. This approach has been suggested for the first time by Kokorin et al., 1972, Kulikov et al., 1972. Oxy- and met- forms of spin labelled human haemoglobin (Kokorin et al., 1972), egg lysocyme, cachalot myoglobin and myosin from rabbit muscles (Kulikov et al., 1972) were used as probing macromolecules because at that time the X-ray analysis of these proteins was already done and their spatial organization was known. This provided important possibility to compare EPR results with known X-ray structure. The results obtained in these works demonstrated that measuring the second central moment of EPR spectra one can determine distances r with high accuracy in the range of $1.0 < \mathbf{r} < 1.6$ nm, while the experimental EPR spectrum shape parameter d_1/d shifted the upper value of **r** up to 2.5 nm, what is very important for biological systems. Then, the following equation

$$\mathbf{r} = 9.3 + 0.77/\Delta \qquad (26)$$

analogoues to Eq. (18) was suggested (Kokorin, 1974, Parmon et al., 1977b, 1980) for experimental applications. Very often researchers plotted their own graduation curves for using d_1/d parameter. Below, the most interesting results obtained with this parameter in different biological systems are discussed.

8.1. Peptides, proteins, enzymes

The first object to which the approach and d_1/d parameter was applied was D-glyceraldehyde-3-phosphate Dehydrogenase (Elek, et al., 1972), for which authors showed that the distance between spin labels attached to Cys-149 and Cys-153 does not exceed 2.1 nm.

Several works were done on double spin-labelled short proteins – biologically active polypeptides such as Gramicidine S, Bradykinin, etc.

Conformational states of cyclic decapeptide Gramicidine S was studied in (Ivanov et al., 1973). Two ornithine amino acid groups were labeled and at temperatures higher 40°C EPR spectra showed five-component spectra typical for nitroxide biradicals. In frozen solutions the distance **r** equal to 1.25 ± 0.08 and ~1.0 nm was estimated from d_1/d parameter and ΔM_2 value correspondingly, while theoretical calculations of Gramicidine S model estimated the appropriate distance in the range of 1.2 – 1.4 nm.

The most detailed and consecutive study of spatial structure of linear polypeptide Bradykinin was carried out in (Ivanov et al., 1975a, 1975b, Filatova et al., 1977). Attaching by two radicals R_6CH_2COO– or R_5COO– to different amino acid groups as it is shown schematically in Fig. 17:

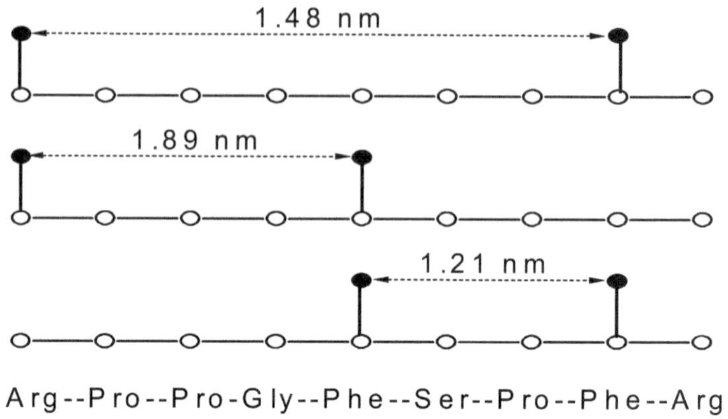

Figure 17. A schematic structure of double spin labelled bradykinin derivatives

The authors could measure a set of distances between various bradykinin analogues. Nine different "biradical" derivatives were synthesized. Some results (parameters d_1/d and **r**) extracted from articles by Ivanov et al., 1975a, 1975b, Filatova et al., 1977, are given in Table 6.

Compound *	d_1/d	r, nm
R_1-Arg^1–Ser^6-R_1	0.64	1.36
R_2-Arg^1–Ser^6-R_2	0.72	1.17
R_1-Arg^1–Tyr^5-R_1	0.57	1.63
R_1-Arg^1–Tyr^8-R_1	0.60	1.48
R_1-Pro^2–Tyr^5-R_1	0.54	1.89
R_1-Pro^2–Tyr^8-R_1	0.58	1.57
BOC-Arg-(R_1)Tyr^5–Tyr^8-R_1	0.65	1.34
BOC-Pro-(R_1)Tyr^5–Tyr^8-R_1	0.56	1.70
BOC-Gly-(R_1)Tyr^5–Tyr^8-R_1	0.65	1.34

* R_1 – R_6CH_2COO–, R_2 – R_5COO–, BOC – *tert*-butyloxycarbonyl

Table 6. Values of d_1/d and interspin distances r in double spin labeled bradykinin derivatives

An important result followed from these data: the bradykinin structure in a solution could not be lenear or chaotically disordered, and the most probable structure was chosen, later confirmed by quantum chemical calculations. It was shown that bradykinin has in solutions a curved, quasi-cyclic structure, which was confirmed by the decay of fluorescence in the case of fluorescent labels.

Study of the interaction between natural and spin labeled steroid hormones and human serum albumin was carried out (Sergeev et al., 1974). The main goal of the work was determination of the relative location of the labeled histidine groups of albumin and a spin labeled steroid. The distance was estimated as **r** > 1.8 nm. Changes of albumin molecule caused by binding steroids had allosteric character and corresponded to the trans-globular effects.

Study of the location of spin-labeled thiol groups relatively the active center of Ca-depending ATP-ase, in which diamagnetic Ca(II) ions were substituted with paramagnetic Mn(II) ions, allowed estimate the distance between the label and manganese ion as 1.1 nm (Maksina et al., 1979.

Later, at the end of 20th and beginning of 21st century, when new, informative, and modern quantitative methods based on double electron-electron resonance (ELDOR) and high frequency EPR spectroscopy were created as well as new methodologies. For example, measurement of the distance between two spin labels in proteins permits distinguish the spatial orientation of elements of defined secondary structure (Hustedt & Beth, 1999). By using site-directed spin labeling, it is possible to determine multiple distance values and thereby build tertiary and quaternary structural models as well as measure the dynamics of structural changes. New analytical methods for determining interspin distances and relative orientations for uniquely oriented spin labels have been developed using global analysis of multifrequency EPR data. New methods have also been developed for determining interprobe distances for randomly oriented spin labels. These methods are being applied to a wide range of structural problems, including peptides, soluble proteins, and membrane proteins, that are not readily characterized by other structural techniques (Hustedt & Beth, 1999). Nevertheless, a simple-measured d_1/d parameter was used rather often during these years.

By using a variety of biochemical and biophysical approaches, a helix packing model for the lactose permease of *Escherichia coli* has been proposed (He et al., 1997). The four residues that are irreplaceable with respect to coupling were paired: Glu269 (helix VIII) with His322 (helix X) and Arg302 (helix XI) with Glu325 (helix X). In addition, the substrate translocation pathway was located at the interface between helices V and VIII, which is in close vicinity to the four essential residues. Based on this structural information and functional studies of mutants in the four irreplaceable residues, a molecular mechanism for energy coupling in the permease has been proposed. It was shown by two methods that Arg302 is also close to Glu269. Glu269-His, Arg302-His, and His322-Phe binds Mn^{2+} with high affinity at pH 7.5, but not at pH 5.5. Site-directed spin-labeling of the double Cys mutant Glu269-Cys / Arg302-Cys exhibited spin-spin interaction with an interspin distance of about 1.4-1.6 nm. The spin-spin interaction was stronger and interspin distance shorter after the permease was reconstituted into proteoliposomes. Taken as a whole, the data were consistent with the idea that Arg302 may interact with either Glu325 or Glu269 during turnover (He et al., 1997).

Hess et al., 2002, have studied the secondary structure, subunit interaction, and molecular orientation of vimentin molecules within intact intermediate filaments and assembly intermediates. Spectroscopy data proved α-helical coiled-coil structures at individual amino acids 316–336 located in rod 2B. Analysis of positions 305, 309, and 312 identify this region as conforming to the helical pattern identified within 316–336 and thus demonstrated that this region is in an α-helical conformation. Varying the position of the spin label, authors could identify both intra- and inter-dimer interactions. With a label attached to the outside of the α-helix, it have been able to measure interactions between positions 348 of separate dimers as they align together in intact filaments, identifying the exact point of overlap. By mixing different spin-labeled proteins, Hess et al. demonstrated that the interaction at position 348 is the result of an anti-parallel arrangement of dimers. This approach provided high resolution structural information (<2 nm resolution), can be used to identify molecular arrangements between subunits in an intact intermediate filament, and should be applicable to other noncrystallizable filamentous systems as well as to the study of protein fibrils (Hess et al., 2002).

Continuing the approach established the utility of site-directed spin labeling and EPR to determine structural relationships among proteins in intact intermediate filaments Hess et al., 2004, have introduced spin labels at 21 residues between amino acids 169 and 193 in rod domain 1 of human vimentin. The EPR spectra provided direct evidence for the coiled coil nature of the vimentin dimer in this region. This result was consistent with predictions but had never been experimentally demonstrated previously. Previously it was identified that residue 348 in the rod domain 2 acted as one point of overlap between adjacent dimers in intact filaments, and a new study was defined residue 191 in the rod domain 1 as a second point of overlap and established that the dimers are arranged in the anti-parallel and staggered orientation at this site. These results are shown in Table 7. By isolating spin-labeled

Position	184	189	190	191	192			
d_1/d [a]	0.33	0.47	0.42	0.45	0.49			
Δ	0.01	0.15	0.1	0.13	0.17			
<r>, Å	> 25	14.5	17.0	15.3	13.9			
Position	281	282	283	284	285	286	287	288
d_1/d [b]	0.43	0.38	0.63	0.44	0.43	0.40	0.45	0.46
Δ	0.11	0.06	0.31	0.12	0.11	0.08	0.13	0.14
<r>, Å	16.3	22.0	11.9	15.7	16.3	18.9	15.3	14.9
Position	289	290	291	292	293	294	295	296
d_1/d [b]	0.39	0.43	0.71	0.38	0.41	0.48	0.44	0.43
Δ	0.07	0.11	0.39	0.06	0.09	0.16	0.12	0.11
<r>, Å	20.2	16.3	11.4	22.0	17.8	14.2	15.7	16.3
Position	297	298	299	300	301	302	304	
d_1/d [b]	0.43	0.51	0.50	0.43	0.48	0.59	0.39	
Δ	0.11	0.19	0.18	0.11	0.16	0.27	0.07	
<r>, Å	16.3	13.4	13.7	16.3	14.2	12.3	20.2	
Position	323	324	325	326	327	328	329	330
d_1/d [c]	0.48	0.33	0.32	0.46	0.34	0.38	0.33	0.49
Δ	0.16	0.01	-	0.14	0.02	0.06	0.01	0.17
<r>, Å	14.2	> 25	-	14.9	> 25	22.0	> 25	13.9
Position	331	332	333	334	335	336		
d_1/d [c]	0.34	0.35	0.46	0.34	0.33	0.33		
Δ	0.02	0.03	0.14	0.02	0.01	0.01		
<r>, Å	> 25	> 25	14.9	> 25	> 25	> 25		

Table 7. Calculation of distances <r> between spin labels at different positions in Vimentin by d_1/d parameter measured from spectra recorded at −100°C in (Hess, [a] 2004, [b] 2006, [c] 2002)

samples at successive stages during the dialysis that lead to filament assembly *in vitro,* authors established a sequence of interactions that occurs during *in vitro* assembly, starting with the α-helix and loose coiled coil dimer formation. Then the formation of tetrameric species centered on residue 191, followed by interactions centered on residue 348 suggestive of octamer or higher order multimer formation. A continuation of this strategy by the authors revealed that both 191–191 and 348–348 interactions were present in low ionic strength Tris buffers when vimentin was maintained at the "protofilament" stage of assembly (Hess et al., 2004).

Mutations in intermediate filament protein genes were responsible for a number of inherited genetic diseases including skin blistering diseases, corneal opacities, and neurological degenerations. It was shown that mutation of the arginine (Arg) residue to be causative in inherited disorders in at least four different intermediate filament (IF) proteins found in skin, cornea, and the central nervous system. Thus this residue is very important to IF assembly and/or function. The impact of mutation at this site in IFs was investigated by spin labeling. Compared with wild type vimentin, the mutant showed normal formation of the coiled coil dimers, with a slight reduction in the stability of the dimer in rod domain 1. Probing the dimer-dimer interactions showed the formation of normal dimer centered on residue 191 but a failure of dimerization at residue 348 in rod domain 2. These data revealed a specific stage of assembly at which a common disease-causing mutation in IF proteins interrupts assembly (Hess et6 al., 2005).

Site-directed spin labeling, EPR and d_1/d parameter were logically used to probe residues 281-304 of human vimentin, a region that has been predicted to be a non-α-helical linker and the beginning of coiled-coil domain 2B (Hess et al., 2006). This region has been hypothesized to be flexible with the polypeptide chains looping away from one another. EPR analysis of spin-labeled mutants indicated that several residues reside in close proximity, suggesting that adjacent linker regions in a dimer run in parallel. Also, the polypeptide backbone was relatively rigid and inflexible in this region. This region did not show the characteristics of a coiled-coil as has been identified elsewhere in the molecule. Within this region, spectra from positions 283 and 291 were unique from all others of the examined. Structural parameters are given in Table 7. These positions displayed a significantly stronger interaction than the contact positions of coiled-coil regions. Analysis of the early stages of assembly by dialysis from 8 M urea and progressive thermal denaturation showed the close apposition and structural rigidity at residues 283 and 291 occurs very early in assembly, well before coiled-coil formation in other parts of the molecule. Spin labels placed further downstream demonstrated EPR spectra suggesting that the first regular heptad of rod domain 2 begins at position 302. In conjunction with previous characterization of region 305-336 by the same authors and the solved structure of rod 2B from 328-405, the full extent of coiled-coil domain in rod 2B became now known, spanning from vimentin positions 302-405 (Hess et al., 2006).

Phosphorylation processes drove the disassembly of the vimentin intermediate filament (IF) cytoskeleton at mitosis. Data of chromatographic analysis have suggested that phosphorylation produced a soluble vimentin tetramer, but little has been determined about the structural changes that were caused by phosphorylation or the structure of the resulting tetramer. Pittenger et al., 2008, have studied site-directed spin labeling and EPR for examining the structural changes resulting from protein kinase A phosphorylation of vimentin IFs in vitro. EPR spectra suggested that the tetrameric species resulting from phosphorylation are the A11 configuration. It was also established that the greatest degree of structural change was connected with the linker 2 and the C-terminal half of the rod domain, despite the fact that most phosphorylation occurs in the N-terminal head domain. The phosphorylation-induced changes notably affected the proposed "trigger sequences" located in the linker 2 region. These data were the first to document specific changes in IF structure resulted from a physiologic regulatory mechanism and provided further evidence that the linker regions play a key role in IF structure and regulation of assembly-disassembly processes (Pittenger et al., 2008).

Four doubly spin-labeled variants of human carbonic anhydrase II and corresponding singly labeled variants were prepared by site-directed spin labeling (Persson et al., 2001). The distances between the spin labels were obtained from CW X-band EPR spectra by analysis of the relative intensity of the half-field transition, Fourier deconvolution of line-shape broadening, d_1/d parameter, and computer simulation of line-shape changes. Distances also were determined by four-pulse double electron-electron resonance. For each variant, at least two methods were applicable and reasonable agreement between methods was obtained. Distances ranged was from 7 to 24 Å. The doubly spin-labeled samples contained some singly labeled protein due to incomplete labeling. The sensitivity of each of the distance determination methods to the non-interacting component was compared (Persson et al., 2001).

The C-terminal end of ubiquitin (Ub) was covalently attached to the amino group of a lysine in a target protein (Steinhoff, 2002). Additional ubiquitin groups were added using Ub-Ub linkages to form a polyubiquitin chain. The accessibility and the molecular dynamics of the target domain for each protein substrate was expected to be distinctive and in this article the author investigated the ubiquitination mediated protein turnover by means of site-directed spin labeling. EPR data were obtained and interpreted in terms of secondary and tertiary structure resolution of proteins and protein complexes. Analysis of the spin labeled side chain mobility, its solvent accessibility, the polarity of the spin label micro-environment and distances between spin labels allowed to model protein domains or protein-protein interaction sites and their conformational changes with a spatial resolution at a reasonable level. The structural changes accompanying protein function or protein-protein interaction were monitored in the millisecond time range (Steinhoff, 2002).

Using modern pulse and multi-frequency techniques combined with site-directed spin labeling and EPR spectroscopy, the protein-protein and protein-oligonucleotide interaction was studied (Steinhoff, 2004). Analysis of the spin label spectra provided information about distances between spin labels and allowed the modeling of protein-protein interaction sites and their conformational changes. Structural changes were detected with millisecond time resolution. Inter- and intra-molecular distances were determined in the range from approximately 0.5 to 8.0 nm by the combination of CW and pulse EPR methods (Steinhoff, 2004).

The elucidation of structure and function of proteins and membrane proteins by EPR spectroscopy has become increasingly important in recent years because of new approaches of spectroscopic methods and in the chemistry of nitroxide spin labels. These new developments have increased the demand for tailor-made amino acids carrying a spin label on the one hand and for reliable methods for their incorporation into proteins on the other. Becker et al., 2005, described methods for site-specific spin labeling of proteins and showed that a combination of recombinant synthesis of proteins with chemically produced peptides (expressed protein ligation) allowed the preparation of site-specifically spin-labeled proteins.

Apolipoprotein A-I (apoA-I) is the major protein constituent of high density lipoprotein (HDL) and plays a central role in phospholipid and cholesterol metabolism. This 243-residue long protein is remarkably flexible and assumes numerous lipiddependent conformations. Using EPR spectroscopy of site-directed spin labels in the N-terminal domain of apoA-I (residues 1–98), Lagerstedt et al., 2007, have mapped a mixture of secondary structural elements, the composition of which was consistent with findings from other methods. Based on side chain mobility, the precise location of secondary elements for amino acids 14–98 was determined for both lipid-free and lipid-bound apoA-I. Based on intermolecular dipolar coupling at positions 26, 44, and 64, and d_1/d measurements, these secondary structural elements were arranged into a tertiary fold to generate a structural model for lipid-free apoA-I in solutions (Lagerstedt et al., 2007).

Site-directed spin labeling and EPR spectroscopy were used for determining the structure of proteins and its conformational changes and dynamics of membrane proteins at physiological conditions. Analysis of these approaches is given in a review written by Czogalla et al., 2007.

β-spectrin is responsible for interactions with ankyrin. Structural studies indicated that this system exhibits a mixed 310/α-helical conformation and is highly amphipathic. The mechanism of its interactions with biological membranes was investigated with a series of singly and doubly spin-labeled erythroid β-spectrin-derived peptides (Czogalla et al., 2008). The spin-label mobility and spin–spin distances were analyzed via EPR spectroscopy, d_1/d parameter, and two different calculation methods. The results indicated that in β-spectrin, the lipid-binding domain, which is part of the 14th segment, has the topology of typical triple-helical spectrin repeat, and it undergoes significant changes when interacting with phospholipids or detergents. A mechanism for these interactions was proposed (Czogalla et al., 2008).

Halorhodopsin from *Natronomonas pharaonis* (pHR) is a light-driven chloride pump that transports a chloride anion across the plasma membrane following light absorption by a retinal chromophore which initiates a photocycle. Analysis of the amino acid sequence of pHR revealed three cysteine (Cys) residues in helices D and E. The Cys residues were labeled with nitroxide radicals and studied using EPR spectroscopy. Labels mobility, accessibility to various reagents, and the distance between the labels have been studied (Mevorat-Kaplan et al., 2006). It was revealed by following the d_1/d parameter that the distance between the spin labels is ca. 13-15 Å. The EPR spectrum suggested that one label had a restricted mobility while the other two were more mobile. Only one label was accessible to hydrophilic paramagnetic broadening reagents leading to the conclusion that this label was exposed to the water phase. All three labels were reduced by ascorbic acid and reoxidized by molecular oxygen. It was found that the protein experiences conformation alterations in the vicinity of the labels during the pigment photocycle. It was suggested that Cys186 is exposed to the bulk medium while Cys184, located close to the retinal ionone ring, exhibits an immobilized EPR signal and is characterized by a hydrophobic environment (Mevorat-Kaplan et al., 2006).

Alliinase, an enzyme found in garlic, catalyzes the synthesis of the well-known chemically and therapeutically active compound allicin (diallyl thiosulfinate). The enzyme is a homodimeric glycoprotein that belongs to the fold-type I family of pyridoxal-50-phosphate-dependent enzymes. There are 10 cysteine residues per alliinase monomer, eight of which form four disulfide bridges and two are free thiols. Cys368 and Cys376 form a SAS-bridge located near the C-terminal and plays an important role in maintaining both the rigidity of the catalytic domain and the substrate-cofactor relative orientation. Weiner et al., 2009, demonstrated that the chemical modification of allinase with the colored ASH reagent yielded chromophore-bearing peptides and showed that the Cys220 and Cys350 thiol groups are accessible in solution. EPR kinetic measurements using disulfide containing a stable nitroxyl biradical showed that the accessibilities of the two ASH groups in Cys220 and Cys350 differ. The enzyme activity and protein structure (measured by circular dichroism) were not affected by the chemical modification of the free thiols. The d_1/d measurements and its calibration curve on distances obtained by authors gave a distance value between Cys220 and Cys350 >2.2 nm; this is in good agreement with known structural data. Modification of the alliinase thiols with biotin and their subsequent binding to immobilized streptavidin enabled the efficient enzymatic production of allicin (Weiner et al., 2009).

New EPR spectroscopy methods allows now measuring distances reasonably larger 3.0 nm which are nor available for d_1/d method. Long-range structural information derived from paramagnetic relaxation enhancement observed in the presence of a paramagnetic nitroxide radical was used for structural characterization of globular, modular and intrinsically disordered proteins, as well as protein–protein and protein-DNA complexes (Gruene et. al., 2011). The authors characterized the conformation of a spin-label attached to the homodimeric protein CylR2 using a combination of X-ray crystallography, EPR and NMR spectroscopy. Close agreement was found between the conformation of the spin label observed in the crystal structure with interspin distances measured by EPR and signal broadening in NMR spectra. It was suggested that the conformation seen in the crystal structure was also preferred in solution. In contrast, conformations of the spin label observed in crystal structures of T4 lysozyme was not in agreement with the paramagnetic relaxation enhancement observed for spin-labeled CylR2 in solution. These data demonstrated that accurate positioning of the paramagnetic center is essential for high-resolution structure determination (Gruene et. al., 2011).

Rabenstein & Shin, 1995, suggested a new elegant, rather precise but a little bit sophisticated EPR "spectroscopic ruler" which was developed using a series of α-helical polypeptides, each modified with two nitroxide spin labels. A synthesized oligopeptide consisted of 21 amino acids with the following a chain: Ac-AAAALAAAALAAAALAAAALA-NH_2, where Ac is acetyl, A is alanine, L - is lysine residue. A series of variants of these peptides in which two alanines were substituted to cysteines with various positions in the chain. In all examples below these numbers are started from the Ac-alanine terminal. The EPR line broadening due to electron-electron dipolar interactions in the frozen state was determined using the Fourier deconvolution method. The dipolar spectra were then used to estimate the distances between the two nitroxides separated by non-labeled amino acids. Results agreed well with a simple α-helical model. The standard deviation from the model system was 0.09 nm in the range of 0.8-2.5 nm. The authors concluded that this technique can be applied to complex systems such as membrane receptors and channels, which are difficult to access with high-resolution NMR or X-ray crystallography, and will be particularly useful for systems for which optical methods are hampered by the presence of light-interfering membranes or chromophores (Rabenstein & Shin, 1995). Indeed, this method was used in several works during last ten years. We have carefully analyzed the data obtained by Rabenstein & Shin, 1995, and compared them with those determined with d_1/d method. Results are shown in Table 8.

Label positions	(6, 12)	(4, 8)	(4, 9)	(4, 11)	(4, 13)	(4, 17)
R, nm	1.46	0.91	1.46	1.00	1.79	2.35
d_1/d	0.49	1.17	0.48	0.69	0.45	0.41
r, nm	1.57	0.99	1.63	1.17	1.89	2.85

Table 8. The interspin distances measured by Rabenstein & Shin, 1995 (R), and from d_1/d parameter (**r**) for different double spin-labeled oligopeptides

The experimental interspin distances R measured by Rabenstein & Shin, 1995, were taken from their Fig. 5 of the article. d_1/d values for all biradicals as well as for a polypeptide with only one spin-labeled cysteine in the 6-th position, $(d_1/d)_0$ value equal to 0.37, we measured from the original EPR spectra shown in Fig. 2 of the article. Interspin distances **r** calculated by Eq. (26) using a parameter Δ, are given in Table 8. We could not use d_1/d parameter for polypeptides (6, 7) and (6, 8) because in their spectra the dipolar splitting of EPR lines are well-defined, and for the (6, 9) system the distance R is too short and a d_1/d value can not be measured. For other six biradical polypeptides (Table 8), one can conclude that R and **r** values are in reasonably good agreement (not worse than ± 0.1 nm) with systematically larger (~0.1 nm) values of **r**, probably because we used EPR spectra printed in the article, and not the original ones. All procedure of estimation **r** values took about one hour, while more precise but more complicated calculations by Rabenstein and Shin method take usually much longer time.

The interspin distances of two or more nitroxide spin labels attached to specific sites in insulins were determined for different conformations with application of EPR by the line broadening due to dipolar interaction (Steinhoff et al., 1997). The procedure was carried out by fitting simulated EPR powder spectra to experimental data, measured at temperatures below 200 K to freeze the protein motion. The experimental spectra were composed of species with different relative nitroxide orientations and interspin distances because of the flexibility of the spin label side chain and the variety of conformational substates of spin labeled insulins in frozen solution. Values for the average distance <r> and for the distance distribution width were determined from the characteristics of the dipolar broadened line shapes and d_1/d parameter. The resulting interspin distances determined for crystallized insulins in the R6 and T6 structure agreed well with structural data obtained by X-ray crystallography and by modeling of the spin-labeled samples. The EPR experiments revealed differences between crystal and frozen solution structures of the B-chain amino termini in the R6 and T6 states of hexameric insulins (Steinhoff et al., 1997). This study of interspin distances between attached spin labels applied to proteins is a nice example how to obtain structural information on proteins under conditions when other methods like two-dimensional NMR spectroscopy or X-ray crystallography are not applicable.

Gramicidin A was studied by CW-EPR and by double-quantum coherence electron paramagnetic resonance (DQC-EPR) in several lipid membranes (Dzikovski et al., 2004). Samples used were macroscopically aligned by isopotential spin-dry ultracentrifugation and vesicles. The nitroxide spin label was attached at the C-terminus yielding the spin-labeled product (GAsl). EPR spectra of aligned membranes containing GAsl showed strong orientation dependence. In DPPC and DSPC membranes at room temperature had the spectral shape consistent with high ordering, which, in conjunction with the observed high polarity of the environment of the nitroxide label, was interpreted in terms of the nitroxide moiety being close to the membrane surface. In contrast, EPR spectra of GAsl in DMPC membranes indicated deeper embedding and tilt of the NO group. The GAsl spectrum in the DPPC membrane at 35°C (the gel to P_β phase transition) exhibited sharp changes, and above this temperature became similar to that of DMPC. The dipolar spectrum from DQC-EPR clearly indicated the presence of pairs in DMPC membranes. This was not the case for DPPC, rapidly frozen from the gel phase but could be a hint of aggregation. The interspin distance in the pairs was determined as 3.09 nm, in good agreement with estimated for the head-to-head GAsl dimer (the channel-forming conformation), which matched the hydrophobic thickness of the DMPC bilayer (Dzikovski et al., 2004). Both DPPC and DSPC, apparently as a result of hydrophobic mismatch between the dimer length and bilayer thickness, did not favor the channel formation in the gel phase. In the P_β and L_α phases of DPPC (above 35°C) the channel dimmer was formed, as evidenced by the DQC-EPR dipolar spectrum after rapid freezing. A comparison with studies of dimer formation by other physical techniques indicated the desirability of using low concentrations of Gramicidin A accessible to the EPR methods for the study (Dzikovski et al., 2004).

Recently, Dzikovski et al., 2011, published experimental results on channel and nonchannel forms of Gramicidin A (GA) studied by EPR in various lipid environments using new mono- and double-spin-labeled compounds. For GA channels, it was demonstrated that pulse dipolar EPR allowed to determine the orientation of the membrane-traversing molecules relative to the membrane normal and to study small effects of lipid environment on the interspin distances in the spin-labeled GA channel. The nonchannel forms of GA were also studied by pulse dipolar EPR for determination of interspin distances corresponding to monomers and double-helical dimers of spin-labeled GA molecules in the organic solvents trifluoroethanol and octanol. The same distances were observed in membranes. Since detection of nonchannel forms in the membrane is complicated by aggregation, the authors suppressed any dipolar spectra from intermolecular interspin distances arising from the aggregates by using double-labeled GA in a mixture with excess unlabeled GA molecules. In hydrophobic mismatching lipids (L-phase of DPPC), GA channels have dissociated into free monomers. The structure of the monomeric form was found similar to a monomeric unit of the channel dimer. The double-helical conformation of gramicidin was also found in some membrane environments. It was revealed that in the gel phase of saturated phosphatidylcholines, the fraction of double-helices increased in the following order: DLPC < DMPC < DSPC < DPPC, and the equilibrium DHD/monomer ratio in DPPC was determined. In membranes, the double-helical form was presented only in aggregates. The effect of N-terminal substitution in the GA molecule upon channel formation was also studied (Dzikovski et al., 2011). This work has demonstrated how pulsed dipolar EPR can be used to study complex equilibria of peptides in membranes.

We would like to attract attention to a recent work by Gordon-Grossman et al., 2009, in which a combined pulse EPR and Monte Carlo simulation study provided the insight on peptide-membrane interactions and the molecular structure of the system. This new approach to obtain details on the distribution and average structure and locations of membrane-associated peptides successfully combined: a) PELDOR to determine intramolecular distances between spin labeled residues in peptides; b) electron spin echo envelope modulation (ESEEM) experiments for measuring water exposure and the direct interaction of spin labeled peptides with deuterium nuclei in the phospholipid molecules, and c) Monte Carlo simulations (MCS) to derive the peptide-membrane populations, energetics, and average conformation of the native peptide and mutants mimicking the spin labeling. The membrane-bound and solution state of the well-known antimicrobial peptide melittin, used as a model system was investigated, and a good agreement between the experimental results and the MCS simulations regarding the distribution of distances between the labeled amino acids, the side chain mobility, and the peptide's orientation was obtained, as well as for the extent of membrane penetration of amino acids in the peptide core. It was shown that the EPR data reported a deeper membrane penetration of the termini compared to the MCS simulations. In case of melittin adsorption on the membrane surface in a monomeric state, it was observed as an amphipatic helix with its hydrophobic residues in the hydrocarbon region of the membrane and its charged and polar residues in the lipid headgroup region (Gordon-Grossman et al., 2009).

8.2. Nucleic acids

EPR spectroscopy has been used broadly for investigating peculiarities of the structural organization and its transformation for DNA, RNA molecules, their models and different protein-nucleic acid complexes. A lot of works were published during last 45 years. Below, we will discuss only several of them relating to the topic of the chapter.

The study of DNA-dye interaction by the spin label method was carried out by Zavriev, et al., 1976. The binding of ethidium bromide and acriflavin dyes with DNA macromolecule modified with spin-labeled analogue of ethylene imine has been studied. These spin labels were shown to bind covalently to DNA, at the same time the number of the dye molecules bound to DNA was decreased without any changes of the binding constant. Analysis of EPR spectra of the samples in the frozen 50% water-glycerol mixture at 77 K for spin-labeled DNA has shown that addition of the dyes increased distances <r> between the labels, that was explained by the increase in DNA length upon formation of the complex with dye molecules. Structural data in the work were obtained from measuring d_1/d values (Zavriev, et al., 1976).

The use of d_1/d measurements for structural characterization of spin labeled DNA and RNA was also described in a review of EPR studies of the structure and dynamic properties of nucleic acids and other biological systems written by Kamzolova & Postnikova, 1981. When spin labels were attached to different sites of a macromolecule, the quantitative information could be obtained about conformational properties of these local regions and, as a result, about the functional behaviour of the systems.

A distance ruler for RNA using EPR and site-directed spin labeling (SDSL) has been suggested by Kim et al., 2004. The site-directed spin-labeled 10-mer RNA duplexes and HIV-1 TAR RNA motifs with various interspin distances were examined. HIV-1 TAR RNA is the binding site of the viral protein Tat, the trans-activator of the HIV-1 LTR. The long terminal repeat, LTR, regulates HIV-1 viral gene expression via its interaction with multiple viral and host factors. It is present at the 5'- end of all HIV-1 spliced and unspliced mRNAs in the nucleus as well as in the cytoplasm. SDSL was applied to RNA structural biology rather rare, despite an importance of knowledge of RNA structure and RNA-protein complex formation. As a model study for measuring distances in RNA molecules using continuous wave (CW) EPR spectroscopy, the spin labels were attached to the 2'-NH_2 positions of appropriately placed uridines in the duplexes, and interspin distances were measured from both molecular dynamics simulations (MDS) and Fourier deconvolution method (FDM) developed by Rabenstein & Shin, 1995. The 10-mer duplexes had interspin distances in the range from 1.0 to 3.0 nm by MDS estimations; however, dipolar line broadening of the CW EPR spectra was observed for the RNAs with interspin distances of 1.0 to 2.1 nm and not for distances over 2.5 nm. Unfortunately, the authors did not use the d_1/d method for the distance measurement, probably because this approach was described only in Russian language literature. Its application could add the necessary information to the subject, the more so the appropriate EPR spectra at low temperature were recorded. The conformational changes in TAR (transactivating responsive region) RNA in the presence and in the absence of different divalent metal ions were monitored by measuring distances between two nucleotides in the bulge region. The predicted interspin distances obtained from the FDM method and those from MDS calculations match well for both the model RNA duplexes and the structural changes predicted for TAR RNA. These results demonstrate that distance measurement using EPR spectroscopy is a potentially powerful method to help predict the structures of RNA molecules (Kim et al., 2004).

Site-directed spin labeling measurements of nanometer distances in nucleic acids using a sequence-independent nitroxide probe has been carried out also in (Cai et al., 2006). Analysis of electron spin dipolar interactions between pairs of nitroxides yields the inter-nitroxide distance, which provides quantitative structural information. The PELDOR and NMR methods had enabled such distance measurements up to 7.0 nm in bio-molecules, thus opening up the possibility of SDSL global structural mapping. The study evaluated SDSL distance measurement using a nitroxide spin label that was attached, in an efficient manner, to a phosphorothioate backbone position at arbitrary DNA or RNA sequences. Radical pairs were attached to selected positions of a dodecamer DNA duplex with a known NMR structure, and eight distances, ranging from 2.0 to 4.0 nm, were measured using PELDOR technique. The measured distances correlated strongly ($R^2 = 0.98$) with the predicted values calculated based on a search of sterically allowable radical conformations in the NMR structure, and the accurate distance measurements was demonstrated. The method was proposed for global structural mapping of DNA and DNA–protein complexes (Cai et al., 2006).

The molecular chaperone DnaK recognizes and binds substrate proteins via a stretch of seven amino acid residues that is usually only exposed in unfolded proteins. The binding kinetics is regulated by the nucleotide state of DnaK, which alternates between DnaK and ATP (fast exchange) and DnaK and ADP (slow exchange). These two forms cycle with a rate mainly determined by the ATPase activity of DnaK and nucleotide exchange. The different substrate binding properties of DnaK were mainly attributed to changes of the position and mobility of a helical region in the C-terminal peptide-binding domain, the so-called LID (Popp et al., 2005). Authors investigated the nucleotide-dependent structural changes in the peptide-binding region and the question: could they induce structural changes in peptide stretches using the energy available from ATP hydrolysis. Model peptides contained two cysteine residues at varying positions and were derived from the structurally well-studied peptide NRLLLTG and labelled with spin probes. Measurements of distances between spin labels were carried out by EPR for free peptides or peptides bound to the ATP and ADP-state of DnaK, respectively. No significant change of distances between labels was observed, hence, no structural changes that could be sensed by the probes at the position of central leucine residues located in the center of the binding region occur due to different nucleotide states. It was concluded that the ATPase activity of DnaK is not connected to structural changes of the peptide-binding pocket but has an effect on the LID domain or other further remote residues (Popp et al., 2005).

A rigid, spin-labeled nucleoside was prepared using a convergent synthetic strategy that could also be applied for the synthesis of the corresponding ribonucleoside (Barhate et al., 2007). EPR spectroscopic analysis of a DNA that contained the rigid spin label verified its limited mobility within a DNA duplex. The rigid spin label had several advantages over previously reported spin labels for nucleic acids: a) distance measurements between two rigid spin labels can be done more accurate than between flexible nitroxides; b) possibility of determination of relative orientations of the two rigid labels, that thereby provided more detailed structural information; c) the nucleoside became fluorescent upon reduction of the nitroxide with a mild reducing agent, that was the first example of a spectroscopic probe that could be used for structural studies by both EPR and fluorescence spectroscopy. The dual spectroscopic activity of the spin label enabled the preparation of nucleic acids that contain a redox-active sensor in their structure. More detailed characterization of the new bifunctional spectroscopic probe and its application for the studies of the structure and dynamics of nucleic acids will be reported in due course (Barhate et al., 2007).

Schiemann et al., 2007, described the facile synthesis of the nitroxide spin-label 2,2,5,5-tetramethyl-pyrrolin-1-oxyl-3-acetylene, TPA, and its binding to DNA/RNA through Sonogashira cross-coupling during automated solid-phase synthesis. They also have measured distance between two such spin-labels on RNA/DNA using PELDOR, and suggested to use this approach for studying global structure elements of oligonucleotides in frozen solutions at RNA/DNA amounts of ~10 nmol. The procedure suggested by authors should be applicable to RNA/DNA strands of up to ~80 bases in length and PELDOR yields reliably spin–spin distances up to ~6.5 nm (Schiemann et al., 2007).

Indeed, over the last 10 years PELDOR has emerged as a powerful new biophysical method without size restriction to the biomolecule under studying, and has been applied to a large variety of nucleic acids as well as proteins and protein complexes in solution or within membranes. Small nitroxide spin labels, paramagnetic metal ions, amino acid radicals or intrinsic clusters and cofactor radicals have been already used as spin centres (Reginsson & Schiemann, 2011).

8.3. Biomembranes and lipid-protein complexes

Native biological membranes and their chemical models such as vesicles and lipid double-layered membranes (emulsions) as well as various lipid-protein complexes are a huge class of objects suitable for investigation by spin probe/label technique. Structural results obtained by EPR for the third group (lipid-protein complexes) were discussed in Section 8.1 of this review. Numerous articles were published during last 50 years concerning biomembranes but only very few were related to quantitative measurements of the local concentration of spin probes and interspin distances. The main problem in such studies of the structural organization of biological membranes is that when spin probes (nitroxide radicals) are penetrated into outer or inner layer of the double-layered membrane, they promptly become distributed in both layers of the membrane because of flip-flop transitions

(Berliner, 1976, Kuznetsov, 1976, Likhtenshtein et al., 2008, Hemminga & Berliner, 2007, Webb, 2006). Therefore, it is usually quite difficult to distinguish between dipolar coupling of spins inside one lipid layer or between spins localized in two lipid monolayers.

9. Conclusion

Measurement of distances or local concentrations in physical and macromolecular chemistry, solid state chemistry, molecular biology and biophysics is an important quantitative tool for investigating structure, spatial organization and conformational transitions in solids, solid solutions, polymers, biological macromolecules, and complex supramolecular systems using site-directed spin labeling and various techniques of EPR spectroscopy. Modern approaches of EPR such as high frequency/high field EPR, pulce technique and double resonances, dipolar EPR spectroscopy allow researchers determine not only interspin distances but also their relative 3-D orientation and the behaviour of these complex systems under their functioning using stable nitroxide radicals. Analysis of the results obtained shows that became a method making possible controlling quantitatively spatial structure and properties of chemical and biological systems in conditions the most close to natural. And this is very important for correct understanding of the mechanisms of these processes.

Scientific progress is irreversible. During the last forty years three generations of researchers were changed, and naturally the development of new modern methods of quantitative investigation is continuously in progress. These methods are usually correct, informative but very technically complex and need a lot of theoretical and computer calculations, i.e. take a lot of time. Therefore, such simple method as d_1/d parameter for estimation distances **r** in the case of pairwise distribution of nitroxide radicals and local concentrations C_{loc} or mean local distances <r> at their random (chaotic) distribution of paramagnetic centres can provide valuable structural information at the beginning serious complex and long-time investigation. Measurements of d_1/d values are simple and do not need much time especially considering important information provided by it. Evidently, d_1/d parameter should be used in the distance or concentration intervals and experimental conditions in which it determination is correct.

Author details

Alexander I. Kokorin
N.Semenov Institute of Chemical Physics RAS, Moscow, Russian Federation

Acknowledgement

The author thanks the Russian Foundation for Basic Research (grant 12-03-00623_a) for financial support of the work. I thank also Mrs. A. A. Goncharova, Mr. O. I. Gromov and Dr. I. A. Kokorin for their technical assistance in Chapter preparation.

10. References

Abragam, A. (1961). *The principles of nuclear magnetism*, Clarendon Press, Oxford

Aleksandrova, T. A.; Wasserman, A. M.; Medvedeva, T. V.; Shapiro, A. M. & Korshak, Yu. V. (1986). The intramolecular mobility and local density of links in nitroxide polyradicals based on styrene co-polymers with maleic anhydride. *Vysokomolec. Soed. B*, Vol. 28, No. 11, 832-835, ISSN: 0507-5475

Altshuler, S. A. & Kozirev, B. M. (1964). *Electron Paramagnetic Resonance*, Academic Press, New York

Barhate, N.; Cekan, P.; Massey, A. P. & Sigurdsson, S. Th. (2007). A nucleoside that contains a rigid nitroxide spin label: A fluorophore in disguise, *Angew. Chem. Int. Ed.*, Vol. 46, No. 15, 2655 –2658, ISSN 1433-7851

Bender, C. J. & Berliner, L. J., Eds. (2006). *Computational and Instrumental Methods in EPR*, in *Biological Magnetic Resonance*, Vol.25, Springer Verlag, ISBN: 0-387-33145-X

Becker, C. F. W.; Lausecker, K.; Balog, M.; Kalai, T.; Hideg, K.; Steinhoff, H.-J. & Engelhard, M. (2005). Incorporation of spin-labelled amino acids into proteins. *Magn. Reson. Chem.*, Vol. 43, S34–S39, ISSN 0749-1581

Berliner, L. J. (1976). *Spin Labeling: Theory and Applications*. Academic Press, ISBN 0-120-92350-5, New York

Berliner, L. J. (1979). *Spin Labeling II: Theory and Applications*. Academic Press, ISBN 0120923505, New York

Berliner, L. J.; Eaton, S. S. & Eaton, G. R., Eds (2001). *Distance Measurements in Biological Systems by EPR*. In *Biological Magnetic Resonance*, Vol. 19. Springer Verlag, ISBN: 0-306-46533-7, New York, London

Bird, G. H.; Pornsuwan, S.; Saxena, S. & Schafmeister, C. E. (2008). Distance distributions of end-labeled curved bispeptide oligomers by E. S. R. *ACS Nano*, Vol. 2, No. 9, 1857–1864, ISSN 1936-0851

Blumenfeld, L. A.; Voevodskii, V. V. & Semenov, A. G. (1962). *Application of Electron Paramagnetic Resonance in Chemistry*, Nauka, Novosibirsk

Bravaya, N. M. & Pomogailo, A. D. (2000). Spin labels as the instrument for analysis of topochemistry of polymer-immobilized Ziegler catalytic systems. *J. Inorg. Organomet. Polymers*, Vol. 10, No. 1, 1-22, ISSN 1053-0495

Brustolon, M. R. & Giamello E., Eds. (2009). *Electron Paramagnetic Resonance Spectroscopy: A Practitioner's Toolkit*, Wiley, ISBN-10: 0-4702-5882-9

Buchachenko, A. L. & Wasserman, A. M. (1973). *Stable Radicals*, Khimiya, Moscow. (1976). *Stable Radicals*. Wiley, ISBN 9067642592, *London*

Cai, Q.; Kusnetzow, A. K.; Hubbell, W. L.; Haworth, I. S.; Gacho, G. P. C.; Van Eps, N.; Hideg, K.; Chambers, E. J. & Qin, P. Z. (2006). Site-directed spin labeling measurements of nanometer distances in nucleic acids using a sequence-independent nitroxide probe. *Nucleic Acids Research*, Vol. 34, No. 17, 4722-4730, ISSN 0305-1048

Capiomont, A. (1972). Structure cristalline du radical nitroxyde: subérate de di(tétraméthyl-2,2,6,6 pipéridinyl-4 oxyle-1). *Acta Cryst. B*, Vol. B28, No. 7, 2298-2301, ISSN 0567-7408

Czogalla1, A.; Pieciul, A.; Jezierski, A. & Sikorski, A. F. (2007a). Attaching a spin to a protein – site-directed spin labeling in structural biology, *Acta Bioch. Polonica*, Vol. 54, No. 2, 235–244, ISSN 0001-527X

Czogalla a, A.; Grzymajło a, K.; Jezierski, A. & Sikorski, A. F. (2008). Phospholipid-induced structural changes to an erythroid β spectrin ankyrin-dependent lipid-binding site Biochimica et Biophysica Acta, Vol. 1778, No. 11, 2612–2620, ISSN 0005-2736

Dubinskii, A. A.; Grinberg, O. Ya.; Tabachnik, A. A.; Shapiro, A. B.; Ivanov, V. P.; Rozantsev, E. G. & Lebedev, Ya. S. (1974). The measurement of distances between paramagnetic fragments in biradicals by the forbidden transition ΔM_S=2. *Biofizika*, Vol. 19., No. 5, 840-842, ISSN 0006-3029

Dzikovski, B. G.; Borbat, P. P. & Freed, J. H. (2004). Spin-labeled Gramicidin A: Channel formation and dissociation. *Biophys. J.*, Vol. 87, No. 5, 3504-3517, ISSN 0006-3495

Dzikovski, B. G.; Borbat, P. P. & Freed, J. H. (2011). Channel and Nonchannel Forms of Spin-Labeled Gramicidin in Membranes and Their Equilibria. *J. Phys. Chem. B*, Vol. 115, No. 1, 176-185, ISSN 1520-6106

Eaton, S. S. & Eaton, G. R. (2001a). Relaxation times of organic radicals and transition metal ions. In: *Distance Measurements in Biological Systems by EPR*. In *Biological Magnetic Resonance*, (2001), Berliner, L. J.; Eaton, S. S. & Eaton, G. R., Eds. Vol. 19, 29-154, Springer Verlag, ISBN: 0-306-46533-7, New York, London

Eaton, S. S. & Eaton, G. R. (2001b). Determination of distances based on T_1 and T_m effects. In: *Distance Measurements in Biological Systems by EPR*. In *Biological Magnetic Resonance*, (2001), Berliner, L. J.; Eaton, S. S. & Eaton, G. R., Eds. Vol. 19, 348-382, Springer Verlag, ISBN: 0-306-46533-7, New York, London

Eaton, S. S. & Eaton, G. R. (2004). *Measurements of Interspin Distances by EPR*. In: *Electron Paramagnetic Resonance*, Vol. 19, 318-337, ISBN: 0-306-46533-7

Eaton, G. R.; Eaton, S. S.; Barr, D. P. & Weber, R. T. (2010). *Quantitative EPR*. Springer, ISBN 978-3211929476,

Elek, G.; Sajgo, M.; Grigorian, G. L.; Chibrikin, V. M. &Keleti, T. (1972). Spin labelling of D-glyceraldehyde-3-phosphate Dehydrogenase with specific reagents. *Acta Biochim. et Biophys. Acad. Sci. Hung.*, Vol. 7, No. 2, 119-131, ISSN 0237-6261

Fielding, L.; More, K. M.; Eaton, G. R. & Eaton, S. S. (1986). Metal-nitroxyl interactions. 46. Spectra of low-spin Iron(III) complexes of spin-labeled tetraphenylporphyrines and their implications for the interpretation of EPR spectra of spin-labeled Cytochrome P450. *J. Am. Chem. Soc.*, Vol. 108, No. 3, 618-625, *ISSN* 0002-7863

Filatova, M. P.; Reissmann, Z.; Reutova, T. O.; Ivanov, V. T.; Grigoryan, G. L.; Shapiro, A. M. & Rozantsev, E. G. (1977). Conformational states of bradykinin and its analogs in solution. III. ESR spectra of spin-labeled analogs. *Bioorg. Khimiya*, Vol. 3, No. 9, 1181-1189, ISSN 0132-3423

Flory, P. J. (1969). *Statistical Mechanics of Chain Molecules*. Wiley, ISBN 0-470-26495-0; reissued 1989, ISBN 1-56990-019-1, New York

Gordon-Grossman, M.; Gofman, Y.; Zimmermann, H.; Frydman, V.; Shai, Y.; Ben-Tal, N. & Goldfarb, D. (2009). A combined pulse EPR and Monte Carlo simulation study provides molecular insight on peptide-membrane interactions. *J. Phys. Chem. B*, Vol. 113, No. 38, 12687–12695, ISSN 1089-5647

Grinberg, O. & Berliner, L. J., Eds. (2011). *Very High Frequency (VHF) ESR/EPR*, Springer, ISBN 978-1441934420,

Grinberg, O. Ya.; Nikitaev, A. T.; Zamaraev, K. I. & Lebedev, Ya. S. (1969). Influence of the concentration on the EPR line width in solid solutions of VO^{2+} and MoO^{3+}. *Zh. Strukt. Khimii*, Vol. 10, No. 2, 230-233, ISSN 0136-7463

Gruene, T.; Cho, M.-K.; Karyagina, I.; Kim, H.-Y.; Grosse, C.; Giller, K.; Zweckstetter, M. & Becker, S. (2011). Integrated analysis of the conformation of a protein-linked spin label by crystallography, EPR and NMR spectroscopy. *Biomol. NMR*, Vol. 49, No. 1, 111–119, ISSN 0925-2738

Hanson, G. & Berliner, L. J., Eds., (2010). *Metals in Biology: Applications of High-Resolution EPR to Metalloenzymes*, in: *Biological Magnetic Resonance*, Vol. 29, Springer, ISBN 978-1441911384,

He, M. M.; Voss, J.; Hubbell, W. L. & Kaback, H. R. (1997). Arginine 302 (Helix IX) in the lactose permease of *Escherichia coli* is in close proximity to Glutamate 269 (Helix VIII) as well as Glutamate 325 (Helix X). *Biochemistry*, Vol. 36, No. 44, 13682-13687, ISSN 0001527X

Hemminga, M. A. & Berliner, L. J., Eds. (2007). *ESR Spectroscopy in Membrane Biophysics*, in *Biological Magnetic Resonance*. Vol. 27, Springer Verlag 2007. ISBN: 0-387-25066-2.

Hess, J. F.; Voss, J. C. & FitzGerald, P. G. (2002). Real-time observation of coiled-coil domains and subunit assembly in intermediate filaments. *J. Biol. CChem.*, Vol. 277, No. 38, 35516–35522, ISSN 0021-. 9258

Hess, J. F.; Budamagunta, M. S.; Voss, J. C. & FitzGerald, P. G. (2004). Structural characterization of Human Vimentin Rod 1 and the sequencing of assembly steps in intermediate filament formation *in vitro* using site-directed spin labeling and EPR. *J. Biol. CChem.*, Vol. 279, No. 43, 44841–44846, ISSN 0021-. 9258

Hess, J., Budamagunta, M., FitzGerald, P., Voss, J. (2005). Characterization of structural changes in vimentin bearing an EBS-like mutation using site directed spin labeling and electron paramagnetic resonance, *J. Biol. Chem.*, Vol. 280, No. 3, 2141-2146, ISSN 0021-. 9258

Hess, J. F.; Budamagunta, M. S.; Shipman, R. L.; FitzGerald, P. G. & Voss, J. C. (2006). Characterization of the linker 2 region in human Vimentin using site-directed spin labeling and EPR. *Biochemistry*, Vol. 45, No. 39, 11737-11743, ISSN 0001527X

Hustedt, E. J. & Beth, A. H. (1999). Nitroxide spin-spin interactions: Applications to protein structure and dynamics. *Ann. Rev. Biophys. Biomol. Struct.*, Vol. 28, No. 1, 129-153, ISSN 1056-8700

Ionita, P.; Caragheorgheopol, A.; Gilbert, B. C. & Chechik, V. (2004). Mechanistic study of a place-exchange reaction of Au nanoparticles with spin-labeled disulfides. *Langmuir*, Vol. 20, 11536-11544. *ISSN* 0743-7463

Ionita, P.; Caragheorgheopol, A.; Gilbert, B. C. & Chechik, V. (2005). Dipole-dipole interactions in spin-labeled Au nanoparticles as a measure of interspin distances. *J. Phys. Chem. B*, Vol. 109, 3734-3742 *ISSN* 1089-5647

Ionita, P.; Volkov, A.; Jeschke, G. & Chechik, V. (2008). Lateral diffusion of thiol ligands on the surface of Au nanoparticles: An EPR study. *Anal. Chem.*, Vol. 80, No. 1, 95-106, *ISSN* 0003-2700

Ivanov, V. T.; Filatova, M. P.; Reissmann, Z.; Reutova, T. O.; Kogan, U. A.; Efremov, E. S.; Ivanov, V. S.; Galaktionov, S. G.; Grigoryan, G. L. & Bystrov, V. F. (1975a). The conformational states of bradykinin in solutions. *Bioorg. Khimiya*, Vol. 1, No. 8, 1241-1244, ISSN 0132-3423

Ivanov, V. T.; Filatova, M. P.; Reissmann, S.; Reutova, T. O.; Efremov, E. S.; Pashkov, V. S.; Galaktionov, S. G.; Grigoryan, G. L. & Ovchinnikov, Yu. A. (1975b). In: *Peptides: chemistry, structure and biology*, Walter, R. & Meienhofer, J., Eds. Ann Arbor Science, New York, 151-157.

Ivanov, V. T.; Miroshnikov, A. I.; Snezhkova, L. G.; Ovchinnikov, Yu. A.; Kulikov, A. V. & Likhtenstein, G. I. (1973). The application of EPR spectroscopy for studying conformational states of peptides. Gramicidine S. *Khimiya Prirod. Soed.* (Chemistry of Natural Compounds), Vol. 9, No. 1, 91-98, ISSN 0023-1150

Jeschke, G. (2002). Determination of the nanostructure of polymer materials by EPR spectroscopy. *Macromol Rapid Commun.*, Vol. 23, No. 2, 227-246, ISSN: 1521-3927

Kamzolova, S. G. & Postnikova, G. B. (1981). Spin-labelled nucleic acids. *Quart. Rev. Bioph.*, Vol. 14, No. 2, 223-288, ISSN 0033-5835

Khairutdinov, R. F. & Zamaraev, K. I. (1970). Study of the structure of frozen solutions with paramagnetic probes. Bull. Acad. Sci. USSR, Div. chemistry, No. 7, 1524-1528, ISSN 0002 3353

Khazanovich, T. N.; Kolbanovsky, A. D.; Kokorin, A. I.; Medvedeva, T. V. & Wasserman, A. M. (1992). EPR spectroscopy of spin-labeled macromolecules as a tool for determining chain conformations in amorphous solid polymers. *Polymer*, Vol. 33, No. 24, 5208-5214, ISSN: 0032-3861

Kim, N.-K.; Murall, A. & DeRose, V. J. (2004). A distance ruler for RNA using EPR and site-directed spin labeling. Chemistry & Biology, Vol. 11, No. 7, 939-948, *ISSN*: 1074-5521

Kokorin, A. I. (1974). *The measurement of distances between spin labels as a method of studying the structure of macromolecules and solid solutions*. Ph.D. Thesis, ICP AN SSSR, Moscow

Kokorin, A. I. (1986). Application of nitroxyle biradicals in medical-biological studies. In: *Method of Spin Labels and Probes. Problems and Perspectives*, N. M. Emanuel & R. I. Zhdanov (Eds.), 61-79, Nauka, Moscow

Kokorin, A. I. (1992). *The structure of coordination compounds with macromolecular ligands*. Dr.Sci. Thesis, ICP RAS, Moscow

Kokorin, A. I. & Formazyuk, V. E. (1981). New method of measuring distances between spin-label and paramagnetic ions in macromolecules. *Russ. Molek. Biol.*, Vol. 15, No. 4, 930-938 (p. 722-728 in transl.) ISSN 0026-8984

Kokorin, A. I. & Zamaraev, K. I. (1972). Investigation of the structure of frozen two-component solutions with the aid of iminoxy-radicals. *Russian J. Phys. Chem.*, Vol. 46, No. 11, 1658-1659, ISSN 0044-4537

Kokorin, A. I.; Kirsh, Yu. E. & Zamaraev, K. I. Determining local concentrations of units in macromolecular coils using the spin-labelling method. (1975). *Vysokomol. Soed.*, Vol. a17, No. 7, 1618-1621, (p. 1864-1868 in transl.) ISSN: 0507-5475

Kokorin, A. I.; Parmon, V. N. & Shubin, A. A. (1984). *Atlas of the anisotropic EPR spectra of nitroxide biradicals*. Nauka, Moscow

Kokorin, A. I.; Bogach, L. S.; Shapiro, A. B. & Rozantsev, E. G. (1976). Study of conformational peculiarities of triazine nitroxide biradicals by EPR technique. *Bull. Acad. Sci. USSR, Chemistry*, No. 9, 1994-1999, ISSN 0002 3353

Kokorin, A. I.; Molochnikov, L. S.; Yakovleva, I. V.; Shapiro, A. B. & Gembitskii, P. A. (1989). Study of the interaction of transition metal ions with polyethyleneimine using the spil-label method. *Vysokomol. Soed. A*, Vol. 31, No. 3, 546-551, ISSN: 0507-5475 (Russ. Ed.); *Polymer Sci. USSR* (1990), Vol. 31, No. 3, 597-603, ISSN: 0507-5475 (Engl. Ed.)

Kokorin, A. I.; Parmon, V. N.; Suskina, V. I.; Ivanov, Yu. A.; Rozantsev, E. G. & Zamaraev, K. I. (1974). Intramolecular exchange and dipole-dipole interactions in solutions of some iminoxyl biradicals. *Russ. J. Phys. Chem.*, Vol. 48, No. 4, 548-551, ISSN 0036-0244

Kokorin, A. I.; Zamaraev, K. I.; Grigoryan, G. L.; Ivanov, V. P. & Rozantsev, E. G. (1972). Measurement of the distances between the paramagnetic centres in solid solutions of nitroxide radicals, biradicals and spin-labeled proteins. *Biofizika*, Vol. 17, No. 1, 34-41 (p. 31-39 in transl.). ISSN 0006-3029

Kolbanovsky, A. D.; Wasserman A. M.; Kokorin, A. I. & Khazanovich, T. N. (1992a). Experimental verification of the theory of the dipole-dipole broadening of the EPR spectra of solid solutions of radicals. *Russ. J. Chem. Phys.*, Vol. 11, No. 1, 94-98, ISSN 0207-401X

Kolbanovsky, A. D.; Wasserman A. M.; Medvedeva, T. V. & Khazanovich, T. N. (1992b). EPR as a method for the determination of the spin-labeled macromolecule conformation in a solid state. *Russ. J. Chem. Phys.*, Vol. 11, No. 8, 1129-1135, ISSN 0207-401X

Kovarski, A. L. (1996). Spin probes and labels. A quarter of a century of application to polymer studies, in: *Polymer Yearbook*, R. A. Pethrick, Ed. Vol. 13, 113-139, ISBN 3-7186-5712-0

Kozlov, S. V.; Kokorin, A. I.; Shapiro, A. B. & Rozantsev, E. G. (1981). Chain nitroxide biradicals – a model for investigation the oligomers in solutions. *Vysokomolec. Soed. B*, Vol. 23, No. 5, 323-327, ISSN 0507-5475

Köhler, S. D.; Spitzbarth, M.; Diederichs, K.; Exner, T. E. & Drescher, M. (2011). A short note on the analysis of distance measurements by electron paramagnetic resonance. *J. Magn. Reson.*, Vol. 208, No. 1, 167–170, ISSN 1090-7807

Kruk, D.; Kowalewski, J.; Tipikin, D. S.; Freed, J. H.; Mo´scicki, M.; Mielczarek, A. & Port, M. (2011). Joint analysis of ESR lineshapes and ^{1}H NMRD profiles of DOTA-Gd derivatives by means of the slow motion theory. *J. Chem. Phys.*, Vol. 134(2), 024508, ISSN 0021-9606

Kulikov, A. V. (1976). Evaluation of the distance between spins of the spin label and paramagnetic centre in spin-labeled proteins from the parameters of saturation curve of EPR spectra of labels at 77 K. *Russ. Molek. Biol.*, Vol. 10, No. 1, 132-141, ISSN: 0026-8984

Kulikov, A. V. & Likhtenstein, G. I. (1974). The use of the saturation curves for estimation of distances in biological systems by the method of double spin labels. *Biofizika*, Vol. 19, No. , 420-423, ISSN 0006-3029

Kulikov, A. V. & Likhtenstein, G. I. (1977). The use of spin relaxation phenomena in the investigation of the structure of model and biological systems by the method of spin labels. *Adv. in Molec. Relax. and Interaction Processes*. Vol. 10, No. 1, 47-79, ISSN 0378-4487

Kulikov, A. V.; Likhtenstein, G. I.; Rozantsev, E. G.; Suskina, V. I. & Shapiro, A. B. (1972). On possible determination of distances between functional groups of protein by the method of spin labels. *Biofizika*, Vol. 17, No. 1, 42-48, ISSN 0006-3029

Kuznetsov, A. N. (1976). *The Method of Spin Probes*, Nauka, Moscow

Lagerstedt, J. O.; Budamagunta, M. S.; Oda, M. N. & Voss, J. C. (2007). EPR spectroscopy of site-directed spin labels reveals the structural heterogeneity in the N-terminal domain of ApoA-I in solution. *J. Biol. Chem.*, Vol. 282, No. 12, 9143–9149, ISSN 0021-. 9258

Lebedev, Ya. S. (1969). *Free radicals in the Solid State*, Dr. Sci. Thesis, ICP AN USSR, Moscow

Lebedev, Ya. S. & Muromtsev, V. I. (1971). *EPR and relaxation of the stabilized radicals*, Khimiya, Moscow

Leigh, J. S. (1970). ESR rigid-lattice line shape in a system of two interacting spins. *J. Chem. Phys.*, Vol. 52, No. 5, 2608-2612, ISSN 0021-9606

Likhtenshtein, G. I. (1974). *The Method of Spin Labels in Molecular Biology*, Nauka, Moscow; (1976). *Spin Labeling Methods in Moleculat Biology*. Wiley, *ISBN*-10: 0-4702-5882-9 New York

Likhtenshtein, G.; Yamauchi, J.; Nakatsuji, S.; Smirnov, A. I. & Tamura, R. (2008). *Nitroxides*: Applications in Chemistry, Biomedicine, a*nd Materials Science*, Wiley-VCH, ISBN: 978-3-527-31889-6, New York

Maksina, A. G.; Azizova, O. A.; Artemova, L. G.; Vladimirov, Yu. A. & Kokorin, A. I. (1979). Study of the location of spin-labeled thiol groups relatively the active center of Ca-depending ATP-ase. *Proc. Acad. Sci. USSR*, Vol. 247, No. 4, 982-985, ISSN 0891-5571

Mevorat-Kaplan, K.; Weiner, L. & Sheves, M. (2006). Spin labeling of *Natronomonas pharaonis* halorhodopsin: Probing the cysteine residues environment. *J. Phys. Chem. B*, Vol. 110, No. 17, 8825-8831, ISSN 1089-5647

Mikhalev, O. I.; Yakovleva, I. V.; Trofimov, V. J. & Shapiro, A. B. (1985). The ESR study of the structure of frozen aqueous solutions of Polyvinylpyrrolidone and Polyvinylalcohol. *Cryo-Letters*, No. 6, 245-256, ISSN 0143-2044

Misra, S. K., Ed. (2011). *Multifrequency Electron Paramagnetic Resonance: Theory and Applications*, Wiley-VCH, ISBN 978-3527407798,

Molin, Yu. N.; Salikov, K. M. & Zamaraev, K. I. (1980). *Spin Exchange*, Springer-Verlag, ISBN 3-540-10095-4, Berlin, New York

Möbius, K. & Savitsky, A. (2009). *High-field EPR Spectroscopy on Proteins and their Model Systems*, RSC Publishing, ISBN: 0-8540-4368-3

Neiman, M. B.; Rozantsev, E. G. & Mamedova, Yu. G. (1962). Free radical reactions involving no unpaired electrons. *Nature*, Vol. 196, 472-474, ISSN 0028-0836

Parmon, V. N. & Kokorin, A. I. (1976), unpublished results.

Parmon, V. N.; Kokorin, A. I. & Zhidomirov, G. M. (1977a). The interpretation of the polycrystalline ESR spectra of nitroxide biradicals. *J. Magn. Res.*, Vol. 28, No. 2, 339-349, ISSN 1090-7807

Parmon, V. N.; Kokorin, A. I. & Zhidomirov, G. M. (1977b). Conformational structure of nitroxide biradicals. Use of biradicals as spin probes. *Russ. J. Struct. Chem.*, Vol. 18, No. 1, 104-147, ISSN 0022-4766

Parmon, V. N.; Kokorin, A. I. & Zhidomirov, G. M. (1980). *Stable Biradicals.* Nauka, Moscow

Persson, M.; Harbridge, J. R.; Hammarström, P.; Mitri, R.; Mårtensson, L.-G.; Carlsson, U.; Eaton, G. R. & Eaton, S. S. (2001). Comparison of EPR methods to determine distances between spin labels on Human Carbonic Anhydrase II. *Biophys. J.*, Vol. 80, No. 6, 2886–2897, ISSN 0006-3495

Pittenger, J. T.; Hess, J. F.; Budamagunta, M. S.; Voss, J. C. & FitzGerald, P. G. (2008). Identification of phosphorylation-induced changes in Vimentin intermediate filaments by site-directed spin labeling and EPR. *Biochemistry*, Vol. 47, No. 41, 10863–10870, ISSN 0001527X

Popp, S.; Packschies, L.; Radzwill, N.; Vogel, K. P.; Steinhoff, H.-J. & Reinstein, J. (2005). Structural dynamics of the DnaK–peptide complex. *J. Mol. Biol.*, Vol. 347, No. 4, 1039–1052, ISSN 0022-2836

Pryce, M.H.L. & Stevens, K.W.H. (*1950*). The Theory of magnetic resonance-line widths in crystals. Proc. Phys. Soc. (London), Vol. A63, No. 1, 36-, ISSN 0370-1301

Rabenstein, M. D. & Shin, Y.-K. (1995). Determination of the distance between two spin labels attached to a macromolecule. *Proc. Nat. Acad. Sci. USA*, Vol. 92, No. , 8239-8243, ISSN 0027-8424

Reginsson, G. W. & Schiemann, O. (2011). Pulsed electron–electron double resonance: beyond nanometre distance measurements on biomacromolecules. *Biochem. J.*, Vol. 434, No. 2, 353–363, ISSN 0264-6021

Rozantsev, E. G. (1964). On free organic radicals with a hydroxy group. Izv. AN SSSR, Ser. Khim., No. 12, 2187-2191, *ISSN* 0002 3353

Rozantsev, E. G. (1970). *Free Iminoxyl Radicals*, Khimiya, Moscow. *Free Nitroxyl Radicals*, Plenum Press, *ISBN* 0-608-05755-X, New York

Rozantzev, E. G. & Neiman, M. B. (1964). Organic radical reactions involving no free valence. *Tetrahedron*, Vol. 20, No. 1, 131-137, ISSN: 00404020

Salikhov, K. M. (2010). Contributions of exchange and dipole–dipole interactions to the shape of EPR spectra of free radicals in diluted solutions. *Appl. Magn. Reson.*, Vol. 38, No. 2, 237-256, ISSN 0937-9347

Savitsky, A.; Dubinskii, A. A.; Zimmermann, H.; Lubitz, W. & M€obius, K. (2011). High-field dipolar EPR spectroscopy of nitroxide biradicals for determining three-dimensional structures of biomacromolecules in disordered solids. *J. Phys. Chem. B*, Vol. 115, No. 41, 11950-11963, ISSN 1089-5647

Saxena, S. & Freed, J. H. (1997). Theory of Double Quantum Two-Dimensional ESR with Application to Distance Measurements. *J. Chem. Phys.*, Vol. 107, No. 5, 1317–1340, *ISSN* 0021-9606

Schiemann, O.; Piton, N.; Plackmeyer, J.; Bode, B. E.; Prisner, T. F. & Engels, J. W. (2007). Spin labeling of oligonucleotides with the nitroxide TPA and use of PELDOR, a pulse EPR method, to measure intramolecular distances. *Nature Protocols, No.* 2, 904 – 923, ISSN 1754-2189

Schlick, S., Ed. (2006). *Advanced ESR Methods in Polymer Research*. Wiley, ISBN: 0-471-73189-7, Hoboken

Schweiger, A. & Jeschke, G. (2001). *Principles of Pulse Electron Paramagnetic Resonance*, Oxford University Press, *ISBN* 13: 978-0-19-850634-8, Oxford

Sergeev, P. V.; Ul'yankina, T. I.; Seifulla, R. D.; Grebenshchikov Yu. B. & Likhtenstein, G. I. (1974). Study of the interaction of steroids with human serum albumin by the spin label method. *Molec. Biol.*, Vol. 8, No. 2, 206-217, ISSN 0026-8984

Shaulov, A. Yu.; Kharitonov, A. S. & Kokorin, A. I. (1977). Analysis of the conformational state of the macromolecules in dilute and concentrated polymer solutions using spin labelling. *Polymer Sci. USSR*, Vol. 19, No. 8, 2075-2085, ISSN: 0507-5475

Steinhoff, H.-J. (2002). Methods for study of protein dynamics and protein-protein interaction in protein-ubiquitination by EPR spectroscopy. Frontiers in Bioscience, Vol. 7, c97-110, *ISSN* 1093-9946

Steinhoff, H.-J. (2004), Inter- and intra-molecular distances determined by EPR spectroscopy and site-directed spin labeling reveal protein-protein and protein-oligonucleotide interaction. *J. Biol. Chem.*, Vol. 385, No. 10, 913-920, ISSN 1431-6730

Steinhoff, H. J.; Radzwill, N.; Thevis, W.; Lenz, V.; Brandenburg, D.; Antson, A.; Dodson, G. & Wollmer, A. (1997). Determination of interspin distances between spin labels attached to insulin: comparison of EPR data with the X-ray structure. *Biophys. J.*, Vol. 73, No. 6, 3287-3298, ISSN 0006-3495

Tanford, C. (1961). *Physical Chemistry of Macromolecules*. Wiley, ISBN 0-19-850466-7, New York

Tsvetkov, Yu. D.; Milov, A. D. & Mar'yasov, A. G. (2008), Pulse electron-electron double resonance (PELDOR) as nanometre range EPR spectroscopy. *Usp. Khimii*, Vol. 77, No. 6 515-550 ISSN 0042-1308

Van Vleck, J. H. (1948). The dipolar broadening of magnetic resonance lines in crystals. *Phys. Rev.*, Vol. 74, No. ,1168-1183

Wasserman, A. M. & Kovarsky, A. L. (1986). *Spin Labels and Probes in Physical Chemistry of Polymers*, Nauka, Moscow

Wasserman, A. M.; Aleksandrova, T. A. & Kirsh, Yu. E. (1980a). The study of intramolecular mobility and of local density of poly-4-vinylpyridine units in solution with spin label method. *Vysokomolec. Soed. A*, Vol. 22, No. 2, 275-281, ISSN: 0507-5475

Wasserman, A. M.; Aleksandrova, T. A. & Kirsh, Yu. E. (1980b). The study of of local units density and of molecular dynamics in concentrated poly-4-vinylpyridine solutions with spin label method. *Vysokomolec. Soed. A*, Vol. 22, No. 2, 282-291, ISSN: 0507-5475

Wasserman, A. M.; Khazanovich, T. N. & Kasaikin, V. A. (1996). Some EPR spin probe and spin label studies of polymer systems. *Appl.* Magn. *Reson.*, Vol. 10, No. 1-3, 413-429, ISSN 0937-9347

Wasserman, A. M.; Aleksandrova, T. A.; Kirsch, Yu. E. & Buchachenko, A. L. (1979). Investigation of local density of monomer units and molecular dynamics in solutions of poly-4-vinyl-pyridine by the spin label technique. *Eur. Polymer J.*, Vol. 15, No. 11, 1051–1057, ISSN: 0014-3057

Wasserman, A. M.; Kolbanovsky, A. D.; Kokorin, A. I.; Medvedeva, T. V. & Khazanovich, T. N. (1992). Assessing the conformations of spin-labeled macromolecules in solid amorphous polymers using ESR. *Polymer SScience*, Vol. 34, No. 10, 858-862, ISSN: 0507-5475

Webb, G. A., Ed. (2006). *Modern Magnetic Resonance: I. Applications in Chemistry; II. Applications in Biological, Medical and Pharmaceutical Sciences Volume; III. Applications in Materials, Food and Marine Sciences*, Springer Verlag, ISBN 1-402-03894-1

Weil, J. A. & Bolton, J. R. (2007). *Electron Paramagnetic Resonance: Elementary Theory and Practical Applications*. Wiley & Sons, Inc., ISBN 978-0471754961, Hoboken

Weiner, L.; Shin, I.; Shimon, L. J. W. ; Miron, T.; Wilchek, M.; Mirelman, D.; Frolow, F. & Rabinkov, A. (2009). Thiol-disulfide organization in alliin lyase (alliinase) from garlic (*Allium sativum*). *Protein Sci.*, Vol. 18, No. 1, 196-205, ISSN 0961-8368

Zavriev, S. K.; Grigoryan, G. L. & Minchenkova, L. E. (1976). The study of DNA-Dye interaction by the method of spin labels. *Russ. Molek. Biol.*, Vol. 10, No. 5, 1387-1393, ISSN: 0026-8984

Simulation of Rigid-Limit and Slow-Motion EPR Spectra for Extraction of Quantitative Dynamic and Orientational Information

Andrey Kh. Vorobiev and Natalia A. Chumakova

Additional information is available at the end of the chapter

1. Introduction

The present chapter is devoted to methods of extraction of quantitative data from slow-motion EPR spectra of nitroxides in viscous and rigid media. Slow-motion spectra correspond to the range of rotation relaxation times in which the EPR spectra cannot be reduced to superposition of the Lorentzian lines, which is typical of the range of fast rotation. In the case of the X-band EPR spectra of nitroxides, this range lies approximately from 10^{-6} to 10^{-9} s. In the limit of slow motions the movements of the probe do not influence the EPR spectrum shape (rigid-limit conditions). The aim of the present chapter is to describe the methods of yielding quantitative data on molecular mobility and orientation alignment. Such data can be obtained most reliably by means of numerical simulation of the spectra. In the appendix of this chapter, we describe the homemade software used for spectra simulation.

The methods of extraction of quantitative data on molecular mobility from EPR spectra have been developed rather intensively during the last decades. In the case of fast rotation (characteristic time shorter than approximately 10^{-9} s), the simple measurements of line widths and intensities of spectral components are enough to estimate the isotropic rotation correlation time. The corresponding procedures and formulas can be found elsewhere [1-3]. When more exact data are desirable and in the case of anisotropic rotational mobility, the well-known method developed by Freed and collaborators for numerical simulation of the spectra can be used [4, 5]. Examples of applications of this technique to different systems can be found in works [6-11].

In the present chapter, we focus the reader's attention on obtaining quantitative data when the conventional methods produce incorrect or unreliable results. Section II describes the method of determining the translational diffusion coefficient. The technique is simple and does not require spectra simulation. It was suggested in [12] about 20 years ago but has not found wide application. We illustrate this technique on samples of ordinary solvents and ionic liquids. The methods of spectra simulation are considered in section III. First, in the rigid-limit spectra of the isotropic samples, we showed the possibility to achieve agreement between experimental and calculated spectra within errors of spectrum recording. The magnetic and line width characteristics of the nitroxide probe are determined in the course of this simulation. The obtained values are necessary for the analysis of spectra recorded in more complicated conditions. The spectra simulation and determination of the rotational characteristics of the probes in polymer matrices are presented as examples. The quantitative description of EPR spectra in a wide temperature range up to the glass transition point was found to require consideration of quasi-libration movements. Section IV deals with the study of orientation alignment of paramagnetic probes. The section describes the orientation distribution functions that can be determined by simulation of spectrum angular dependencies. Liquid crystalline, polymeric, and low-molecular glassy systems are considered. The model-free method of characterizing orientation distribution function is compared with the mean-force approach realized in the known software [4, 5]. The advantages and drawbacks of both approaches are discussed.

2. Determination of the translation diffusion coefficients of the paramagnetic molecules from the analysis of temperature dependence of EPR spectra concentration broadening

It is well known that the spin probe technique can be used to measure molecular rotational mobility, as the rotational movements are reflected in the width and shape of EPR spectrum lines [1-5]. On the contrary, translational molecular mobility of paramagnetic molecules is studied by EPR technique very rarely. Meanwhile, it is known that translation of radicals influences EPR spectrum. There are two underlying mechanisms of this influence: dipole-dipole interaction and spin (Heisenberg) exchange. Both of these mechanisms lead to broadening of the spectral lines. At low temperatures, when translational mobility is substantially hindered, the main cause of spectral line broadening is the dipole-dipole interaction of paramagnetic molecules. At high temperatures, the intensive translational movements average the dipole-dipole interaction of radicals while increasing spin exchange. It is obvious that there exists a temperature region where the contributions of dipole-dipole and exchange interactions to line broadening are comparable. A theoretic research [13] showed earnestly the difficulty of a direct analysis of the EPR spectra with the purpose of determining the translational diffusion coefficients. Nevertheless, the translational diffusion coefficients may be estimated by an analysis of the broadening of EPR spectra. For this purpose, the method described in [12, 14-16] can be used. This method allows estimating contributions of dipole-dipole and exchange interactions to the line width by means of analyzing the temperature dependence of the concentration broadening. The concentration broadening can be represented as follows:

$$\delta H = [A\exp(-E_{tr}^{a} / kT) + B\exp(E_{tr}^{a} / kT)]\delta c\ , \qquad (1)$$

where δH is the line broadening, E^{a}_{tr} is the effective activation energy of translational movement, and δc is the difference in concentrations of two radical solutions. The first summand in Eq. (1) describes the effect of the exchange broadening, and the second term describes the effect of the dipole-dipole interaction.

The widths of the spectral lines are determined as the distances between the points of the maximal slope of the absorption lines (peak-to-peak distances of the first derivatives of the absorption lines). In the paper [12], the application of this method in determining the translational diffusion coefficients of the various spin probes in liquid crystal matrices was demonstrated. In the joint work with our colleagues from the University of Graz (Austria), we showed the possibility of using the method to investigate the translation of radicals in standard low-molecular-weight solvents and room-temperature ionic liquids [17].

In Figure 1, one can see the EPR spectra recorded for two different concentrations of radical TEMPOL d17 in glycerol at various temperatures. It is obvious that the spectra of the solution with a larger concentration are broadened noticeably in comparison with the spectra of the less concentrated solution. Because the rotational mobility of the radicals does not depend on concentration, the difference observed was therefore caused by the dipole-

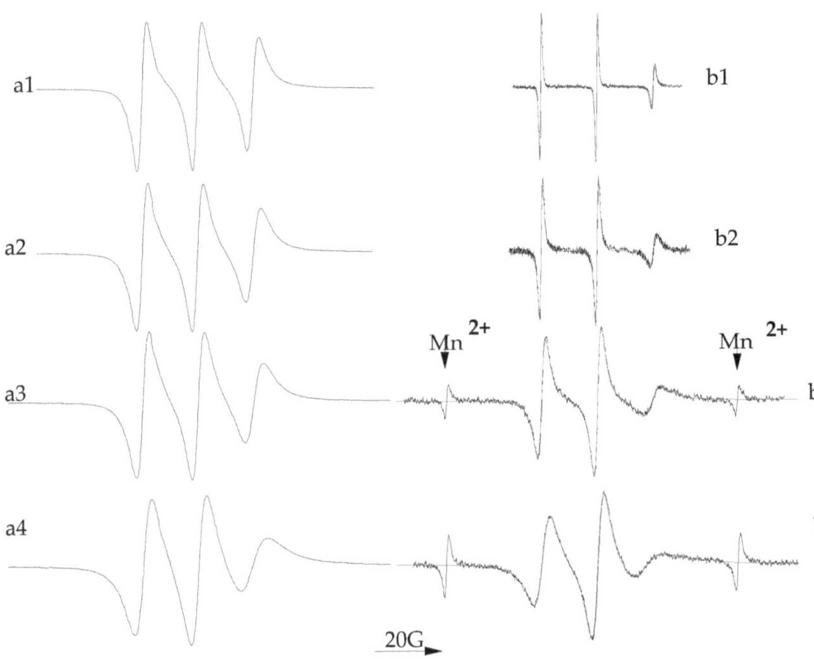

Figure 1. The EPR spectra of TEMPOL-d17 in glycerol: (a) 7.1·× 10^{-2} mol/L and (b) 3.0·× 10^{-3} mol/L, recorded at temperatures 333K (a1, b1), 318K (a2, b2), 303K (a3, b3), and 295K (a4, b4). The red lines are the results of the simulation of the spectra according to the method described in [4,5].

dipole and exchange interactions of the paramagnetic molecules. The result of the computer simulation of the spectra for the solution with low concentration of the radicals is also presented in Figure 1. The simulation was performed according to [5] and was used to determine the rotation diffusion coefficients of TEMPOL-d17 in glycerol at different temperatures and estimation of the effective activation energy of the rotational movements.

Figure 2(a) presents the temperature dependence of the line broadening δH normalized on the difference of concentrations δc for radical TEMPOL-d17 in glycerol and the result of fitting the experimental data according to Eq. (1). The modeling was performed by the least squares method, with simultaneous variation of three parameters: A, B, and E^{a}_{tr}. In this figure, one can also see the calculated contributions of the exchange and dipole-dipole broadening. Obviously, at temperatures 320–330K, the line broadening is caused mainly by the dipole-dipole interactions of the radicals, whereas at 360–370K, the lines are broadened basically by spin exchange. At the intermediate temperature range 340–360K, the contributions of dipole-dipole and exchange broadening are comparable. The effective activation energies of the rotational and translational movements of the radicals TEMPOL-d17 in glycerol are presented in Table 1.

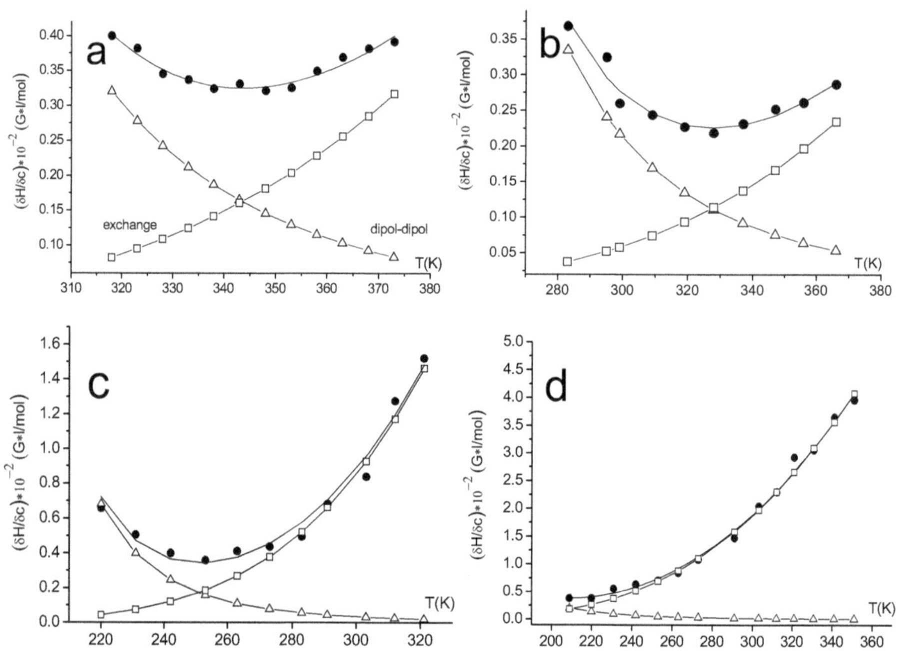

Figure 2. $\delta H/\delta c$ as a function of temperature for TEMPOL-d17 in glycerol (a), TEMPOL in omimBF4 (b), 1-propanol (c), and cumene (d); experimental broadening (solid circles); contribution of spin exchange (open squares); and contribution of dipole-dipole interaction (open triangles).

Matrix	E^a_{rot}/kJ·mol^{-1}	E^a_{tr}/kJ·mol^{-1}
emimBF$_4$	26.8 ± 0.6 (280–360K)	8.4 ± 1.1 (295–380K)
bmimBF$_4$	37.5 ± 0.8 (260–350K)	15.8 ± 1.1 (295–380K)
omimBF$_4$	36.1 ± 0.4 (280–400K)	19.0 ± 1.2 (295–380K)
omimPF$_6$	47.6 ± 1.3 (280–370K)	15.3 ± 1.0 (295–380K)
omimCl	38.7 ± 1.4 (290–360K)	20.4 ± 0.5 (295–380K)
glycerol*	49.4 ± 0.8 (295–345K)	24.1 ± 1.0 (320–380K)
cumene	18.4± 0.9 (200–290K)	13.4 ± 0.6 (210–360K)
n-propanol	18.0 ± 0.6 (200–290K)	20.5 ± 1.4 (220–320K)

* Radical TEMPOL-d17

Table 1. The effective activation energies of the rotational and translational moves of the radicals TEMPOL in different matrices

Consideration of spin-exchange contribution to spectral broadening makes it possible to calculate the spin-exchange constant as follows [18]:

$$k_{exch} = [1.52 \cdot 10^7 (g/2)(2/3)(\delta H)_{exch}] / \delta c \, , \tag{2}$$

where g is the average value of the g-factor of the radical, δH_{exch} is the contribution of spin exchange to the broadening of the spectra.

On the basis of the spin-exchange constant, it is easy to calculate the translation diffusion coefficient [18]:

$$D_{tr} = k_{exch} / 16 f \pi r \, , \tag{3}$$

where f is the steric factor reflecting different probability of spin exchange upon collisions with different mutual orientations of the radicals (for TEMPOL, f = 0.8 [19]) and r is the radius of the paramagnetic molecule.

The translational diffusion coefficients of the TEMPOL-d17 in glycerol, calculated at some temperatures, are presented in Table 2.

In Figure 2(b–d), the examples presented illustrate the application of the method [12] in determining the translational diffusion coefficients of the undeuterated probe TEMPOL in the ionic liquid omimBF$_4$ and standard molecular solvents *n*-propanol and cumene. The high temperature spectra of the diluted solutions of undeuterated TEMPOL demonstrate the additional hyperfine structure on protons. In these cases, we measured the width of the envelope of the spectral line. It can be seen that such method does not lead to distortion of the temperature dependence of the spectral line broadening. Indeed, intensive spin exchange at high temperatures and large concentration leads to substantial spectral line broadening. As a result, subtraction of a small value measured with some error from a big value measured exactly does not lead to a significant error in the result. The values of the effective activation energies for translation and rotation of the radical TEMPOL in different matrices are presented in Table 1.

In Figure 2(c, d), one can see the temperature dependence of the EPR spectra broadening for TEMPOL in the low-viscous solvents *n*-propanol and cumene. It is seen that in these cases, the temperature range in which the dipole-dipole broadening can be observed is confined significantly. At lower temperatures, the rotational mobility of the probe is so small that measurement of the spectral line widths becomes impossible. In the case of *n*-propanol, the parabolic dependence (1) has feebly marked the left dipole-dipole branch (viscosity at room temperature, η_{295} = 1.8 sP). In the case of cumene (η_{295} = 1.0 sP), we can observe practically only the exchange branch of the concentration broadening. However, even in this case, it is possible to distinguish the significantly different contributions of dipole-dipole interaction and spin exchange to broadening of the spectral lines by means of the method [12].

	220K	240K	260K	280K	295K	300K	320K	340K	360K	380K
emimBF_4	–	–	–	–	1.9	2.0	2.5	3.0	3.6	4.2
bmimBF_4	–	–	–	–	1.1 ± 0.4 0.8 ± 0.3* 0.9 ± 0.3**	1.3	1.9	2.7	3.7	4.8
omimBF_4	–	–	–	–	0.6	0.7	1.0	1.6	2.4	3.3
omimPF_6	–	–	–	–	0.9	1.0	1.5	2.1	2.8	3.7
omimCl	–	–	–	–	0.3	0.4	0.6	0.9	1.4	1.9
glycerol***	–	–	–	–	–	–	1.0	1.7	2.7	4.1
cumene	3.0	5.5	9.2	14.3	19.2	21.0	29.4	39.6	51.5	–
n-propanol	0.5	1.2	2.7	5.4	8.4	9.6	16.1	–	–	–

*The data were obtained by means of cyclic voltammetry
**The data were obtained by means of chronoammetry
***Radical TEMPOL-d17

Table 2. The translation diffusion coefficients ($D_{tr}\cdot10^7$) of the radical TEMPOL at various temperatures

Recently it was supposed [20-22] that paramagnetic molecules in the solvent cage repeatedly collide and exchange their spin states. In such case the spin exchange does not reflect adequately the rate of translational diffusion of molecules in the medium. To check the correctness of the obtained data we compared the value of translation diffusion coefficient of TEMPOL in ionic liquid bmimBF4 with the values measured for the same system by two independent electrochemical methods - cyclic voltammetry and chrono-ammetry [23]. The results of all types of measurements are in agreement within experimental errors (Table 2).

From Table 2, it can be seen that in cases of viscous solvents, the effective values of activation energy for rotational mobility exceed noticeably the effective values of activation energy for translational movements. The reason of this phenomenon is not clear at the present time. Perhaps it is a result of microstructure of viscous solvents, such as glycerol, and all ionic liquids.

We hope that the method of determining the translation diffusion coefficients of paramagnetic molecules [12], which possibly does not possess high accuracy but is very simple to use and does not demand computer simulation of the spectra, will be widely applicable.

3. Simulation of EPR spectra in glassy polymer matrices: Rigid-limit and quasi-libration model

The example presented earlier demonstrates that simple semi-empirical treatment of EPR spectra can provide useful information concerning the dynamic properties of condensed media. Numerous examples prove, however, that numerical modeling of experimental EPR spectra leads to more unambiguous and reliable results. Therefore, methods of simulation of EPR spectra are the main subject of the present chapter. In earlier works, simulation of spectra was based on the trial-and-error method. The researcher chose the parameters of the EPR spectrum, taking into account theoretical consideration or analogy with known results, and calculated the spectrum. The quality of the calculated spectrum was determined by visual comparison with the experimental one. The parameter values were then adjusted to improve the agreement with the experiment. This approach is often used up to the present. However, most investigators currently use the numerical fitting of experimental spectra with variation of parameters. In this technique, the desirable values are determined in the course minimizing the discrepancy between the calculated and experimental spectra by means of nonlinear least-squares fitting procedures. The higher objectivity of this technique and the possibility of determining errors make this approach preferential. The different fitting algorithms are compared in [24]. The minimizing program NL2SOL [25] was used in the present work.

The shape of the EPR spectrum can be calculated using a different software, for example, EasySpin (http://www.easyspin.org/) and SimFonia (http://www.bruker-biospin.com/xsophe.html). We used the homemade software described in the appendix of the present chapter. Unweighted differences between the experimental and simulated spectra, which are calculated in each point of the spectrum, are used as minimized residuals r_i. The resulting discrepancy is calculated as follows:

$$D = \frac{1}{2}\sum_i \frac{r_i^2}{n}, \qquad (4)$$

where n is the number of calculated points in the spectrum.

Since the simulation of the ESR spectra via the determination of required parameters is an inverse problem, some restrictions should be imposed on the simulation results to avoid ambiguity. We consider the description of the ESR spectra as satisfactory when the following requirements are met:

1. The discrepancies between the simulated and experimental spectra should be within the experimental errors.
2. The resulting optimal set of parameters should be stable. The simulation procedure should converge to the same optimal set of parameters, independent of the choice of initial parameter values within physically reasonable limits.
3. The resulting values of the magnetic and dynamic parameters and the values of the parameters of individual line width should be physically meaningful.

Numerical simulation of the EPR spectra is used mainly with two aims: to determine the structural, dynamic, or chemical properties of the paramagnetic particle studied and to estimate the characteristics of the medium under consideration (spin probe technique). The determination of magnetic parameters of paramagnetic species is a necessary step in the study in both cases. The determination of magnetic parameters is more reliable when complicated factors, such as molecular mobility, intramolecular transitions, dipole-dipole broadening, and orientation alignment, are absent. Hence, the measurements of magnetic parameters are often performed using diluted glassy solutions at low temperatures, when molecular mobility is frozen (rigid-limit conditions). The simulation of rigid-limit spectra is considered in the next section in more detail. This problem will be used to illustrate the strategy of fitting procedure and to discuss the possible troubles and errors. The structures of the nitroxide probes used are shown in Figure 3.

3.1. Quality of the experimental spectra

The requirement of coincidence of experimental and calculated spectra within experimental errors impose the following additional conditions on the procedure of spectra recording:

1. Modulation amplitudes should be less than the characteristic features of the spectrum shape. The most reliable way to check the fulfillment of this condition is recording the spectrum with different modulation amplitudes and comparing the spectra obtained.
2. The microwave power should not induce saturation. To check this condition, the recording spectrum at a different microwave power is necessary. The signal amplitude then is plotted versus the square root of power. The appropriate power is chosen within the linear part of this dependency. This check is particularly important when low-temperature spectra disposed to saturation are recorded.
3. The field range for the recorded spectra should contain sufficiently long left and right "tails," where the signal is negligibly small. The baseline should be carefully subtracted from the spectrum using the tail fragments.
4. The recorded spectrum should be checked for the absence of the "fast passage" effect [26], which is seen more often in solids at low temperatures. This effect leads to superposition of the integral signal of dispersion and the first derivative EPR lines. To check the absence of this effect, the calculation of the first moment should be used. If the

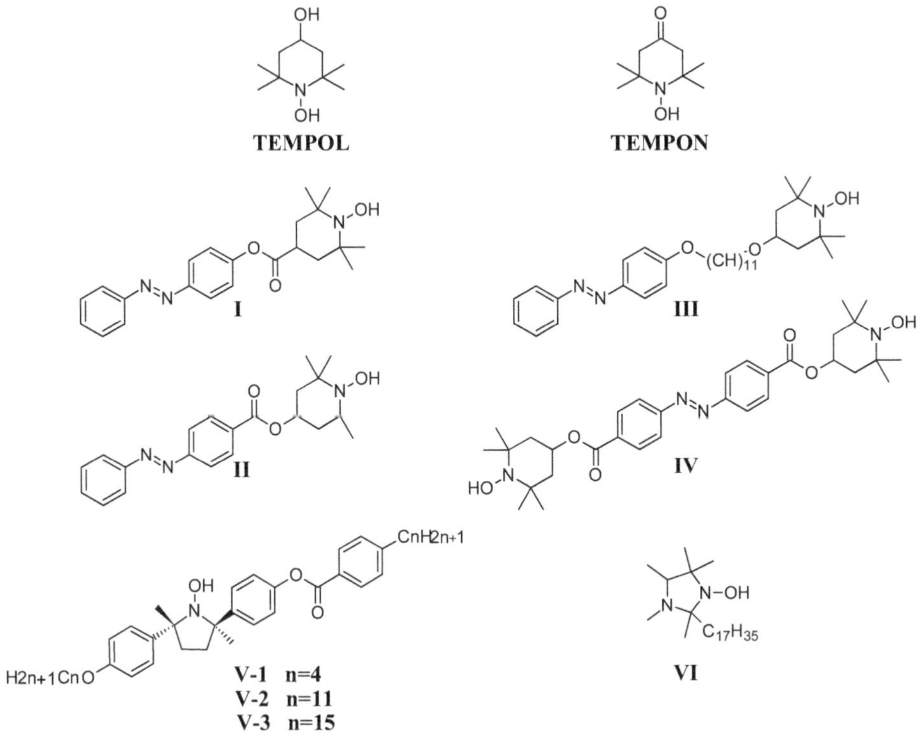

Figure 3. The structures of the nitroxide probes used.

center of the spectrum is taken as zero point, the value of the first moment calculated for the left part of spectrum should be equal to the value of the first moment calculated for the right part. This calculation is useful for controlling baseline subtraction as well. Our practice showed that the spectra with differences of less than 10% between the values of the left and right first moments are acceptable for simulation.

Before the simulation, the experimental spectra are normalized so that the spectrum area is equal to unity. In this case, the final discrepancies between the experimental and calculated spectra for the different experiments can be compared.

For these checks, subtractions, and other manipulations with the experimental spectra, the homemade program esrD is used in our laboratory. The short description of the program is presented in the appendix of this chapter.

3.2. Simulation of rigid-limit EPR spectra

The shape of the EPR spectrum of the disordered sample in rigid-limit conditions is described by the following expression:

$$F(H) = \frac{1}{4\pi}\int_0^{2\pi} d\varphi \int_0^{\pi} f(H,g,A,\theta,\varphi)\sin\theta d\theta\,, \tag{5}$$

where angles θ, φ define the magnetic field direction in the frame of paramagnetic species, $f(H,g,A,\theta,\varphi)$ is the shape of the individual resonance line, the position of the resonance line is defined by the g value $g(\theta,\varphi)$ and by the hyperfine interaction constant $A(\theta,\varphi)$, and the shape of the resonance line is described by the width and type of function (Gaussian, Lorentzian, or mixed function).

Sufficient description of the magnetic properties of the paramagnetic probe consists in determining three g-tensor components, three components of hfi tensor for each magnetic nucleus, three Euler angles connecting each hfi frame with g-tensor frame, and the characteristics of an individual resonance line. Determination of the magnetic parameters of a paramagnetic particle in case of several nuclei with noticeable hyperfine interaction is a rather complicated problem that remains unsolved in full measure up to the present. Fortunately, only the hfi on a nitrogen nucleus is usually apparent in the EPR spectra of nitroxides. The hyperfine interaction with the protons in nitroxides is weaker by two orders of magnitude and thus can be neglected. For further simplification of the EPR spectrum, the isotope-substituted compounds are used. In this case, the hydrogen atoms in the probe molecule are replaced by deuterium and/or ^{14}N nucleus (spin 1) is replaced by ^{15}N (spin 1/2).

3.2.1. *Shape and width of the individual line*

The condition of coincidence of the calculated and experimental spectra within the errors of the experiment requires application of the most comprehensive function of line shape. It is the convolution of Gaussian and Lorentzian functions (Voigt profile). Such convolution is calculated mostly using fast Fourier transformation. We have found, however, that this procedure is insufficiently stable in the course of line width variation. Using equation (6)[27],

$$\int_{-\infty}^{+\infty} \frac{t\cdot \exp(-y^2)dy}{(x-y)^2+t^2} = \pi\cdot \mathrm{Re}\,\omega(x+it)\,, \tag{6}$$

where $\omega(z)=\exp(-z^2)\cdot erfc(-iz)$ and $erfc(x)=\frac{2}{\sqrt{\pi}}\int_x^{\infty} e^{-t^2}dt$ are the additional probability integral, the following explicit expression for the first derivative of the absorption line can be obtained:

$$F'(H) = \frac{4}{h_G^2\sqrt{\pi}}\mathrm{Re}[z\cdot \exp(-z^2)\cdot erfc(-iz)] \tag{7}$$

$$z = \frac{(H-H_0)\sqrt{2}}{h_G} + i\sqrt{\frac{3}{2}}\frac{h_L}{h_G}$$

where $H\text{-}H_0$ is distance from the center of the line, h_G is the Gaussian line width, and h_L is the Lorentzian line width.

Expression (7) should be used in the course of fitting despite the fact that numerical procedure becomes sufficiently slower.

It is known also that the line width of the EPR spectrum depends on the orientation of the paramagnetic particle relative to the magnetic field of the spectrometer. It means that both line width values, $h_G = h_G(\theta, \varphi)$ and $h_L = h_L(\theta, \varphi)$, are orientation dependent. To take into account this dependence, we describe the Gaussian and Lorentzian line widths as second-rank tensors, which can be tilted relative to g-tensor axes.

3.2.2. *Number of fitting parameters and the uniqueness of their determination*

The fitting parameters that are used in the course of the simulation of rigid-limit EPR spectrum are listed in Table 3. It is seen from Table 3 that the number of adjustable parameters for the rigid-limit simulation in the general case can reach 23. We do not know the examples of determination of all these values from the simulation of the EPR spectra. As a rule, some of the indicated parameters do not influence significantly the agreement between the calculated and experimental spectra. The attempt to determine such values by the fitting procedure leads to false or singular convergence in the course of minimization. In that case, spectrum fitting should be repeated after elimination of the indefinable values from the set of varied parameters. Other varied values can come to zero in the course of minimization. It means that this parameter is well defined by the EPR spectrum, but in the course of minimization, this parameter also can be removed from the set of variables.

The ambiguity of the values determined and several minima with the comparable description of the experimental spectrum can be observed when two or more parameters equally influence the spectrum. Such improper parameters can be revealed by an analysis of covariance matrix calculated in the final point of minimization. The values that demonstrate the covariance coefficients more than 0.7–0.8 are not sufficiently independent and possibly cannot be determined simultaneously.

In the course of simulation of the EPR spectra of nitroxides, such troubles appear first at the determination of the Euler angles that describe the relative orientation of different molecular frames related with the tensors shown in Table 3. In particular, the tilt of the axes of the Gaussian and Lorentzian line width tensors are determined from the rigid-limit EPR spectra exceptionally rarely. Commonly, it is enough to assume that the principal frames of these tensors coincide with the g-tensor frame. The determination of the orientation of the hfi-tensor frame relative to the g-tensor frame is also a rare case, as the structure of nitroxides usually dictates almost complete coincidence of these frames.

Sometimes the interdependence is observed between the values of the hfi components and the corresponding components of the Gaussian and Lorentzian line width, namely, between values Ax, h_{Lx}, and h_{Gx} and between values Ay, h_{Ly} and h_{Gy}. This interdependence is a result of insufficient resolution of the spectra when the value of the line width is comparable

Fitting parameters	Description	Number of parameters and their determinability in the course of simulation of the nitroxide EPR spectrum
I	Spectrum intensity	1*
Field shift	Microwave frequency or $\langle g \rangle$ value or field shift is varied to adjust the field position of the calculated and experimental spectra	1*
gx, gy, gz	Principal values of the electronic g-tensor Only two values are varied simultaneously; the third one is defined by isotropic value, if it is known: $\langle g \rangle = (gx + gy + gz) / 3$	2, 3*
Ax, Ay, Az	Principal values of the nuclear hfi tensor Only two value are varied simultaneously; the third one is defined by isotropic value, if it is known: $\langle A \rangle = (Ax + Ay + Az) / 3$	2, 3**
$\Omega(A \rightarrow g)$	Euler angles connecting the own frame of the hfi tensor with the own frame of the g-tensor	3**
h_{Lx}, h_{Ly}, h_{Lz}	Principal values of the orientation-dependent Lorentzian tensor of line width	3*
$\Omega(h_L \rightarrow g)$	Euler angles connecting the own frame of the h_L tensor with the own frame of the g-tensor	3**
h_{Gx}, h_{Gy}, h_{Gz}	Principal values of the orientation-dependent Gaussian tensor of line width	3*
$\Omega(h_G \rightarrow g)$	Euler angles connecting the own frame of the h_G tensor with the own frame of the g-tensor	3**

*Parameters are easily determined

**Some troubles are observed in the course of parameter determination

Table 3. Fitting parameters

with the values of hyperfine splitting. This interdependence is vanished when the deuterated probes are used. To obtain more resolved spectra and more precise determination of Ax and Ay values, the spectra recording at higher temperatures is often used. For example, the magnetic parameters for a number of nitroxides presented in [28] were determined from the spectra recorded at a temperature below glass transition point but higher than 77K. It should be taken into account that narrowing of the spectrum at higher temperatures indicates appearance of some intramolecular or intermolecular mobility. Thus, the obtained values can be slightly averaged by molecular movements. Partial averaging of magnetic parameters by molecular mobility in low-temperature glassy matrices will be considered in detail in section III.3.

The number of varied parameters diminishes when tensors taken into account have uniaxial or isotropic symmetry. The number of parameters that can be varied simultaneously commonly does not exceed 10–15.

3.2.3. *Examples of rigid-limit EPR spectra simulations*

The examples of X-band EPR spectra described by the fitting procedure are presented in Figure 4.

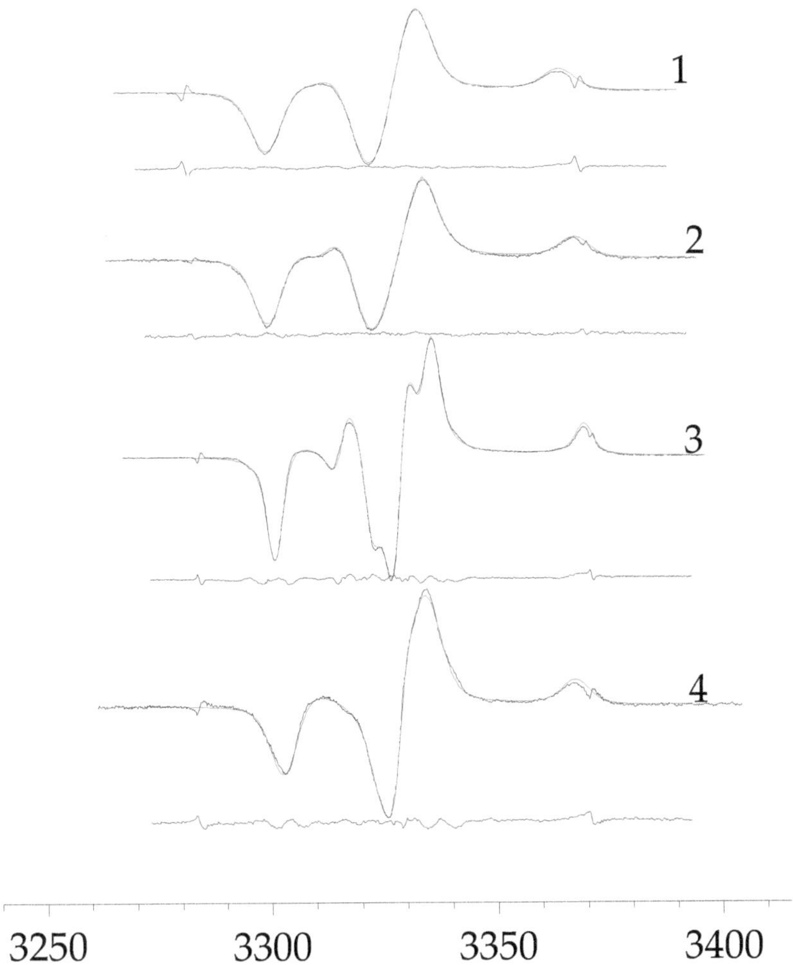

Figure 4. Figure 4. Experimental (black lines) and calculated EPR spectra (red lines) in rigid limit for the systems indicated in Table 4. Below every spectrum, the difference between the experimental and calculated spectra is plotted.

Figure 4 shows the coincidence of the calculated and experimental spectra within the errors of recording. The parameter values determined by means of fitting are collected in Table 4. One can see from that table that the Lorentzian line width at 77K is ordinarily smaller than the Gaussian line width, but the former is not negligible. When the temperature is higher, the Gaussian line width diminishes but the Lorentzian line width rises. It is seen also that the values of Ax and Ay are comparable with the values of the corresponding Gaussian line width; and as a result, they determined with larger errors. In the case of deuterated TEMPOL (row 3 in Table 4), the accuracy of Ax and Ay determination is noticeably higher.

	Systems	g	A	Gaussian linewidth	Lorentzian Linewidth	Discrepancy and error levels
1	TEMPON in AF2400, 77K	2.00885±0.00009 2.00637±0.00008 2.00227	8.13±0.17 3.01±0.40 32.03±0.05	7.38±0.30 7.82±0.16 5.88±0.15	1.66±0.26 0.00±0.27 1.59±0.11	D = 1.4 10^{-9} Dn = 1.3 10^{-10} Dr = 5.0 10^{-10}
2	Probe I in polystyrene, 77K	2.00960±0.00006 2.00638±0.00005 2.002018	7.12±0.11 5.77±0.12 33.61±0.04	5.99±0.16 7.54±0.14 5.01±0.11	1.25±0.16 0.00±0.13 2.01±0.09	D = 9.2 10^{-10} Dn = 4.3 10^{-10} Dr = 1.5 10^{-9}
3	Deuterated TEMPOL in m-tetrahydrofuran, 77K	2.00985±0.00008 2.00612±0.00006 2.002227	6.53±0.04 6.03±0.03 34.15	2.47±0.08 2.97±0.06 2.09±0.07	0.90±0.05 0.30±0.07 1.98±0.04	D = 1.9 10^{-9} Dn = 7.7 10^{-11} Dr = 1.0 10^{-9}
4	Probe V (n=11) in pentyl-cyanobiphenyl (5CB) at 111K	2.00871±0.00006 2.00614±0.00006 2.002144	2.85±0.18 3.70±0.11 32.27±0.06	4.99±0.13 1.86±0.33 4.68±0.17	0.00±0.29 2.57±0.18 1.88±0.11	D = 3.6 10^{-9} Dn = 6.3 10^{-10} Dr = 1.5 10^{-9}

Table 4. The magnetic parameters and line width characteristics obtained by simulation of some EPR spectra in rigid limit

3.2.4. Quality of simulation and errors of the values defined

The correct measure of the acceptable deviations of the calculated spectrum from the experimental one is the errors of the spectrum recording. In general, there are several sources of experimental errors related to recording of EPR spectra: noise of spectrometer, nonlinear baseline, presence of paramagnetic impurities, etc. Commonly only the noise level of the spectrometer is estimated in experiments and used in the analysis of EPR spectra. This value is the obtained variance of a linear fit to the two baseline segments at either end of the spectrum [5]. Such a value will be designated below as Dn. The more reliable way of estimating error level is reproducing the experiment and calculating the standard deviation between the two spectra according to formula (4). Such value will be denoted as Dr. The value of the experimental errors estimated by reproducing experiment Dr is often up to 10 times larger than the noise level Dn estimated using the outside fragments of the spectrum (last column on Table 4). Unfortunately, some errors of spectrum recording (presence of paramagnetic impurities, nonlinear field sweep, etc.) are hardly estimated quantitatively. Obviously, ignoring some error sources is the reason why discrepancy in the final point of minimization is somewhat larger than the estimated level of the recording errors.

Comparison of the discrepancy with the error level is a sensitive and reliable characteristic of validity of parameters that vary in the course of the fitting. It is clear that using of an additional fitting parameter is reasonable when the achieved decrease in discrepancy is more than the level of the experimental errors. If the achieved improvement of fitting is less than the error level, the additional parameter should be considered as redundant.

Standard deviation and confidence range for parameter value can be estimated on the basis of the covariance matrix at the minimum point as follows [25]:

$$\sigma_i = \sqrt{c_{ii}}; \quad \delta x_i = \pm t_{n-p}^{\alpha/2}\sqrt{c_{ii}}, \tag{8}$$

where t is the Student coefficient, α is the confidence probability, c_{ii} is the covariance coefficient for the varied parameter x_i.

Such estimation of errors means that the difference between the experimental and calculated spectra, which is caused by imperfection of calculation model, is assumed to be equivalent to noises and other errors of the spectra recording. For unweighted residual minimization, the t distribution statistics is used to estimate the confidence bounds for each parameter [29]. Values of the standard deviations of the determined values are used in what follows as errors of defined values. It should be noted that repeated performance of the experiment and spectrum simulation shows that the errors calculated as described earlier are somewhat underestimated. The calculations of the errors using χ^2 statistics produce the similar underestimated values. The more realistic error of the defined value seems to be the confidence interval for a 99% confidence level (Student's t-distribution coefficient is equal to 2.57).

3.3. EPR spectra in case of librational molecular movements

Rotational mobility is usually considered as one of the simple models: Brownian rotation diffusion, free rotation, or rotational jumps [4,30]. It is known, however, that these models are not sufficient for satisfactory description of EPR spectra in some media. The largest differences between predicted and experimental spectra are observed when the paramagnetic probe is in media with inhomogeneous microstructure, for example, in the liquid crystalline media, on the surface of adsorbents or in polymer matrices. The last case is better studied and is considered in the following section in detail.

3.3.1. Inadequacy of the simple models of motion

The temperature dependence of the EPR spectrum for spin probe 2,2,6,6-tetramethyl-4-oxopiperidine-1-oxyl (TEMPON) in polystyrene (PS) matrix is presented in Figure 5 as an example.

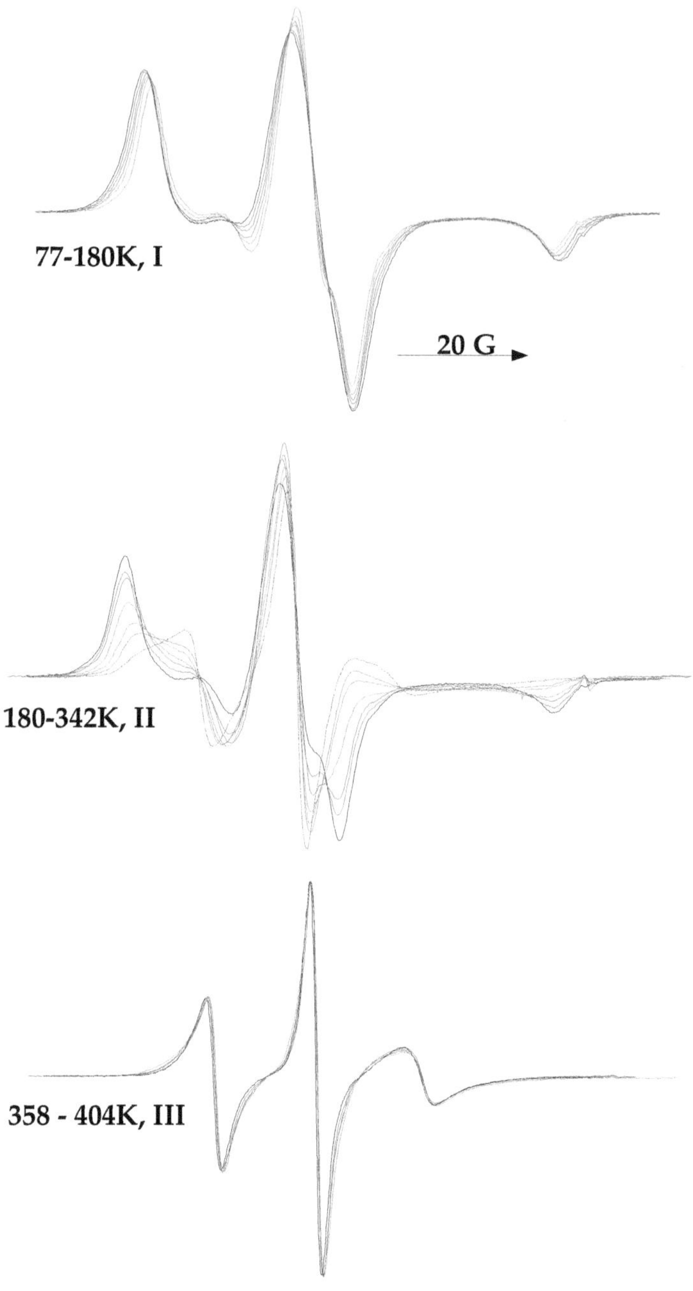

Figure 5. ESR spectra of TEMPON in polystyrene [31].

One can see that when temperature rises, a broad asymmetric spectrum transforms into a narrowed spectrum characterized by motion-averaged magnetic parameters. Increase in rotational mobility results in decrease in separation of the outermost spectral components and diminishing line width. For clarity, all the spectra in Figure 5 are normalized to make the amplitude of the central peak of the spectrum equal to unity. It is seen that the whole temperature dependence of the EPR spectrum in polymer can be divided into three ranges with different specific spectral changes. In the low temperature range (range I, <180 K), the outer extrema of the spectrum smoothly move toward the center as temperature increases. The second range (range II, 180–350K) is characterized by qualitative changes in the shape of the EPR spectrum. A new phenomenon specific for rigid glassy polymers near glass transition temperature is seen in temperature range III (350–405K). In this range, spectral lines narrow, whereas the ratios of amplitudes of different components vary insignificantly. Such spectral behavior is in contradiction with the results of the Redfield relaxation theory [32] and semi-empirical formulas used for analysis of the EPR spectra of paramagnetic probes in liquids [1-3, 33, 34].

The best results of spectra simulation for TEMPON in polystyrene using the model of Brownian diffusion are presented in Figure 6. The deviations far exceeding the experimental errors are seen in this figure for almost the whole temperature range presented. Low-temperature spectra are reproduced rather well, but the rotation correlation times obtained are of the order of 10^{-7} s. This value is in the sharp disagreement with the other measurements, as it will be shown below. The spectra for the middle of the presented temperature range show the qualitative difference with the experimental ones. The above-described specific feature of high temperature spectra, that is, the constancy of the components ratio, is not obtained in the course of fitting.

There are two causes for this disagreement. The first one is the oversimplified model of rotational movements. Today, it is clear that the particle in the polymer medium should be characterized by a wide spectrum of rotations with different frequencies. The most comprehensive model of molecular mobility of the EPR probe in condensed media is the model of slowly relaxing local structure (SRLS) [7, 35], which assumes the fast motions of nitroxide in a matrix cage and the simultaneous slow cage reorientation. If the cage motion is slow enough to be disregarded, SRLS is reduced to the microscopic-order macroscopic-disorder model, which considers nitroxide motion in the potential produced by the cage [36, 37]. In the rigid matrices at low temperatures, the cage barrier exceeds significantly the thermal energy $E_{barrier} \gg kT$. Then, the values of the angular displacements of nitroxides are restricted, and the motion is turned into librations near equilibrium position.

The second feature that produces troubles in the course of spectra simulation is the distribution of probe molecules according to their mobility. The local surrounding of different probe molecules differs in free volume, molecular alignment, and other conditions. One of the most known examples are the polymers containing crystalline and amorphous areas. As a result, different molecules demonstrate different rotational mobility.

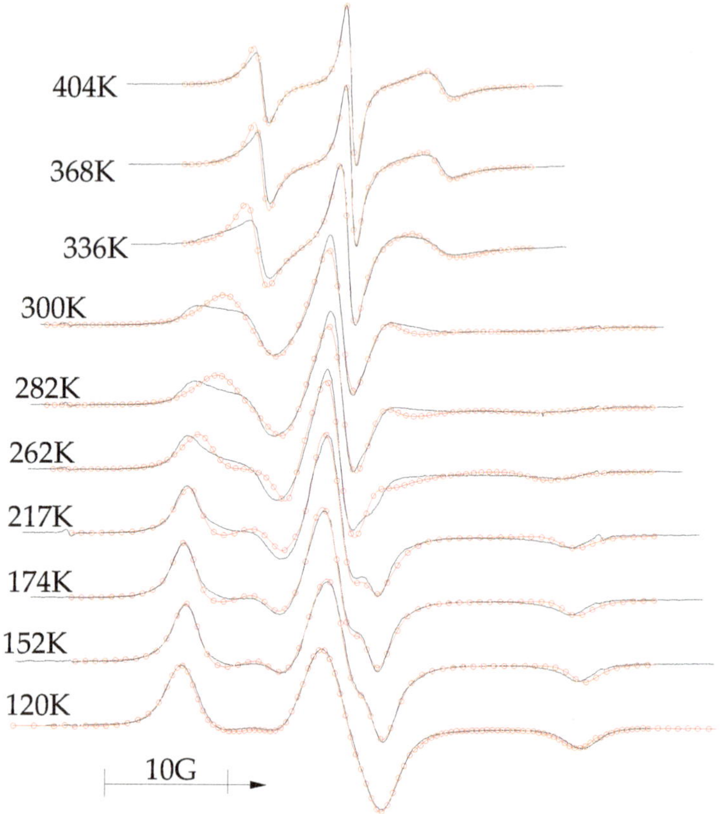

Figure 6. The best fitting of the spectra for TEMPON in polystyrene within the framework of the anisotropic diffusion model.

3.3.2. Concept of molecular quasi-librations [38]

The term "libration" is commonly used for harmonic angular oscillations of molecules in crystals with frequencies of 10^{11}–10^{12} s^{-1} and amplitudes of approximately 2–3°. A similar type of motion in glasses has been evidenced by high-frequency EPR [39], magnetization transfer [40], and spin-echo experiments [41-44]. Librations are averaging motions in time scale of EPR. The sensitivity of the EPR method to angular displacements with different frequencies is qualitatively illustrated in Figure 7.

The figure presents the time required for the rotational displacement on the angle specified on the abscissa. This dependence characterizes the molecular mobility in the medium. The range of frequencies and corresponding amplitudes that influence the X-band EPR spectrum are shown in Figure 7 as a sensitivity window. The window is shifted to higher frequencies when high-frequency EPR spectroscopy is used.

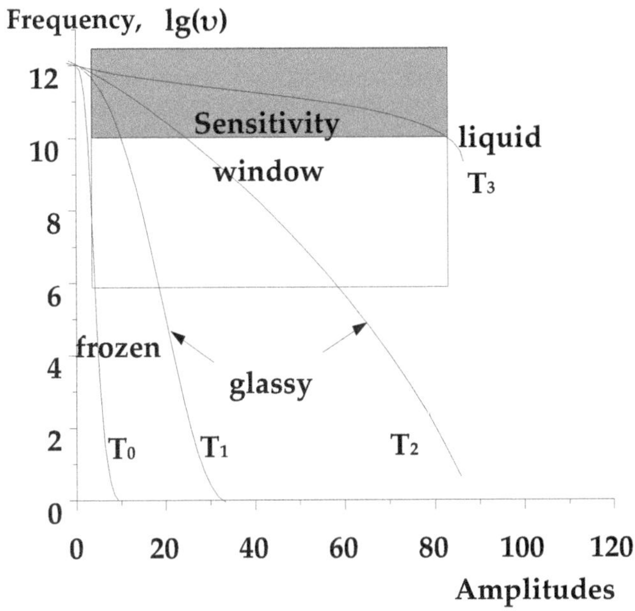

Figure 7. Relaxation curves typical for different states of matter ($T_0 < T_1 < T_2 < T_3$).

In Figure 7, the curve marked T_0 corresponds to a solid sample at a low temperature. The molecular motions possible in such conditions do not fall into the sensitivity window and, thus, do not change the EPR spectrum (rigid-limit conditions). The curve T_3 corresponds to the liquid sample with low viscosity. Any angular displacement of the probe in such medium requires the time less than 10^{-10} s. It is known that such rotations fully average the magnetic parameters of the nitroxide probe. It means that EPR spectrum recorded in these conditions consists of components corresponding to averaged g value ($\overline{g}$) and averaged hyperfine constant $\overline{A}$. Rotational mobility shows itself in such spectra in width and amplitudes of spectral components and can be estimated using the Redfield relaxation theory [32, 33]. The area of averaging movements is marked in Figure 7 by the gray color.

There are many systems in which small angular displacements proceed with averaging frequencies in EPR time scale, but greater displacements that require larger time do not average magnetic parameters (curves T_1 and T_2 in Figure 7). These conditions are ordinary, for example, in cases of glassy polymers or paramagnetic labels attached to large biomolecules. In these cases, the magnetic parameters of probe or label are averaged only partially. The set of movements with frequencies that are averaging in the EPR time scale are not limited to harmonic vibrations near the equilibrium position but comprises combined movements of the local probe surrounding. It means that these movements can be stochastic and of higher amplitude than ordinary librations in crystals. To draw a distinction between ordinary librations in crystals and stochastic librations of spin probe in glasses, the latter should be referred to as "quasi-librations."

Partial averaging of the probe magnetic parameters by the movement with high-frequency but limited amplitude was considered earlier in [45, 46]. On the basis of these works, it can be shown [47] that the following averaging formulas are valid in the case of quasi-librations around three g-tensor axes simultaneously:

$$\begin{aligned} < A_x > &= A_x + 0.5(A_z - A_x)(1 - P_y) + 0.5(A_y - A_x)(1 - P_z) \\ < A_y > &= A_y + 0.5(A_z - A_y)(1 - P_x) + 0.5(A_x - A_y)(1 - P_z) \\ < A_z > &= A_z + 0.5(A_x - A_z)(1 - P_y) + 0.5(A_y - A_z)(1 - P_x) \end{aligned} \qquad (9)$$

where $P_x = (\sin L_x \cdot \cos L_x) / L_x$, $P_y = (\sin L_y \cdot \cos L_y) / L_y$, $P_z = (\sin L_z \cdot \cos L_z) / L_z$;

$<A_x>$, $<A_y>$, $<A_z>$ and A_x, A_y, A_z are averaged and intrinsic hyperfine constants, respectively;

L_x, L_y, and L_z are half-amplitudes of the motion around the X, Y, and Z axes, respectively.

It should be noted that equation (9) uses the assumption of independency of quasi-librations around different axes. As a result, the equation is applicable for any quasi-libration amplitudes in the case of movements around a single axis. In the case of simultaneous quasi-librations around three axes, this assumption is valid at small amplitudes only (less than approximately 45°).

Some authors use mean-squared sine of the displacement angle averaged over all the paramagnetic particles, $<\sin^2\alpha_x>$, $<\sin^2\alpha_y>$, and $<\sin^2\alpha_z>$, for characterization of quasi-libration motion [43, 44, 48-50]. The relations between these characteristics and the quasi-libration amplitudes mentioned earlier are given by

$$\begin{aligned} < \sin^2 a_x > &= 0.5(1 - P_x) = 0.5\left[1 - (\sin L_x \cdot \cos L_x) / L_x\right] \\ < \sin^2 a_y > &= 0.5(1 - P_y) = 0.5\left[1 - (\sin L_y \cdot \cos L_y) / L_y\right] \\ < \sin^2 a_z > &= 0.5(1 - P_z) = 0.5\left[1 - (\sin L_z \cdot \cos L_z) / L_z\right] \end{aligned} \qquad (10)$$

By using the averaged mean-squared sine of displacements, equations (9) transforms into the following:

$$\begin{aligned} < A_x > &= A_x + (A_z - A_x) < \sin^2 a_y > + (A_y - A_x) < \sin^2 a_z > \\ < A_y > &= A_y + (A_z - A_y) < \sin^2 a_x > + (A_x - A_y) < \sin^2 a_z > \\ < A_z > &= A_z + (A_x - A_z) < \sin^2 a_y > + (A_y - A_z) < \sin_2 a_x > \end{aligned} \qquad (11)$$

The average expressions for the case of arbitrary direction of quasi-libration axis in g-tensor frame are presented in [51, 52]:

$$
\begin{aligned}
< A_x > &= k_1 A_x + 2k_2 c_1^2 A_x + k_3\left(c_1^2 A_x + c_2^2 A_y + c_3^2 A_z\right)c_1^2 + k_4\left(c_2^2 A_z + c_3^2 A_y\right) \\
< A_y > &= k_1 A_y + 2k_2 c_2^2 A_y + k_3\left(c_1^2 A_x + c_2^2 A_y + c_3^2 A_z\right)c_2^2 + k_4\left(c_3^2 A_x + c_1^2 A_z\right) \\
< A_z > &= k_1 A_z + 2k_2 c_3^2 A_z + k_3\left(c_1^2 A_x + c_2^2 A_y + c_3^2 A_z\right)c_3^2 + k_4\left(c_2^2 A_x + c_1^2 A_y\right),
\end{aligned} \quad (12)
$$

where c_1, c_2, and c_3 are the direction cosines of the quasi-libration axis, $k_1 = 0.5(1 + \sin 2L / 2L)$, $k_2 = \sin L / L - k_1$, $k_3 = 1 - 2\sin L / L + k_1$, $k_4 = 0.5(1 - \sin 2L / 2L)$, and L is the half amplitudes of motion around the libration axis.

The expressions for the averaged g-tensor components are analogous to that presented for hfi-tensor.

The quasi-libration concept essentially divides molecular rotational movements into two kinds: high and low frequencies. Similar differentiation lies on the basis of the known SRLS model [7, 35]. Thus, the quasi-libration concept is the simplified version of the SRLS model. Both models describe the frequency and amplitude distribution of rotational movements of one paramagnetic center and assume that all probe molecules in the sample demonstrate the identical molecular mobility. On the other hand, it is known that real systems often show clear microscopic inhomogeneity that induces the difference in mobility of different particles as a result of variation of local structure [34, 39, 53-56]. Thus, generally, both distributions: distribution characterizing the different movements of one particle and distribution of particles, should be taken into account for adequate description of EPR spectra.

3.3.3. Low temperature range of temperature dependence

Figure 5 shows that at low temperatures, the EPR spectra of nitroxides (temperature range I) slowly change as the temperature increases and retain shape typical for rigid limit. This set of spectra can be successfully described using different models of rotation mobility, particularly within the model of Brownian diffusion, jump rotation, quasi-librations, and the rigid-limit model with some changed magnetic parameters. Thus, the changes in spectrum in this temperature range are model insensitive. The choice of an adequate model in this situation should be based on additional data obtained using other experimental techniques. The simulation of spectra in range I within the models of Brownian rotation diffusion and jump rotation leads to the conclusion that the rotation correlation time of admixture molecules in glassy polymers noticeably below the glass transition point lie in the range 10^{-6}–10^{-7} s. Such rapid rotation in solid media at a low temperature seems to be quite unrealistic. Experimental measurements of rotational relaxation times by means of light induced alignment of nitroxides [48, 49] give for glassy polystyrene characteristic values approximately 10^{1}–10^{3} s for room temperature. It means that rotation correlation time estimated from the analysis of the EPR spectrum within the Brownian rotation model exceeds the real value by 10 orders of magnitude. This contradiction leads us to conclude that only the model of quasi-librations with limited amplitudes adequately describes the real molecular rotations in range I.

The fitting of experimental spectra in the framework of the quasi-libration model included the variation of quasi-libration amplitudes and width of individual resonance line. The magnetic parameters of the probes were obtained by simulation of rigid-limit spectra and were not further changed. The results of the fitting of the EPR spectra in the temperature range I are presented in Figure 8(a).

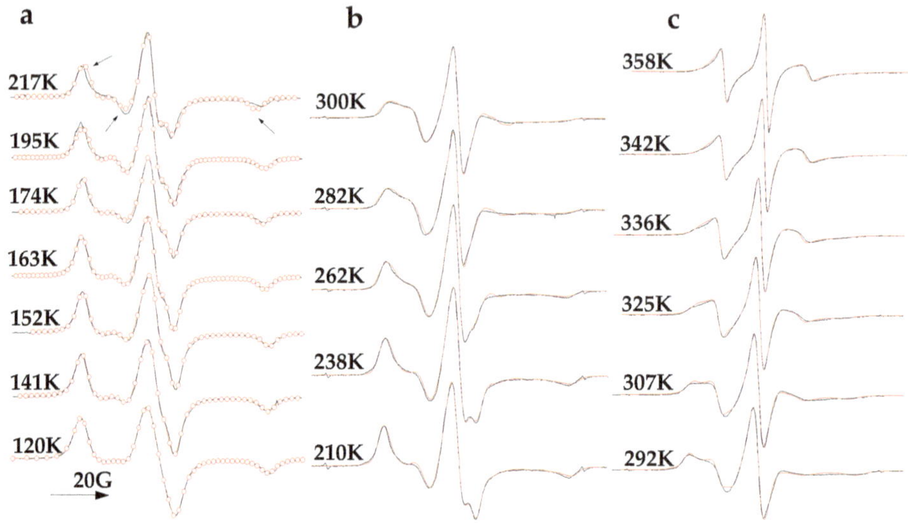

Figure 8. Experimental (black lines) and calculated EPR spectra (red lines) for TEMPON in polystyrene: (a) quasi-libration model, (b) Gaussian distribution of quasi-libration amplitudes, and (c) quasi-librations and lognormal distribution of rotation correlation times.

One can see from Figure 8a that satisfactory agreement between the experimental and calculated spectra holds up to a temperature of approximately 200K. One can see that the spectrum at 217K cannot be qualitatively reproduced taking into account the simple quasi-librations. At higher temperatures, the deviations of the calculated spectra from the experimental ones become larger still.

In the course of fitting, it was found that the amplitudes of quasi-librations around different molecular axes are determined with different accuracy. In particular, the amplitudes of movements around the *z*-axis are defined with uncertainty as approximately 10°. In this connection, for description of the presented experimental spectra, it was possible to assume the uniaxial symmetry of quasi-librations, namely, amplitudes for the *y* and *z* axes being equal: Ly = Lz = Lyz.

3.3.4. Distribution of quasi-libration amplitudes [47]

For quantitative description of the spectra recorded at higher temperatures, the distribution of the probe molecules according to their libration amplitudes should be taken into account.

We tried to describe the experimental spectra using three different distribution functions: rectangular, bimodal, and Gaussian distributions. The fitting procedure was found to be rather sensitive to the distribution shape. We have not achieved positive results when the rectangular distribution was used. The bimodal distribution was found to be useful only in the case of porous Teflon AF2400. Obviously, it is a result of the specific structure of this polymer, which is characterized by bimodal distribution of microstructure. An annihilation positron study shows that free volume distribution in AF2400 has a bimodal shape [57]. In the cases of conventional polymers, using the Gaussian distribution of quasi-libration amplitudes permits to expand the temperature range of quantitative simulation of the spectra shape. Figure 8(b) demonstrates as an example the coincidence of experimental and calculated spectra of TEMPON in polystyrene up to room temperature.

The temperature evolution of the distribution is presented in Figure 9. One can see a graduate shift of the distribution to higher quasi-libration amplitudes and narrowing of the distribution when the temperature increases. The temperature dependencies of averaged quasi-libration amplitudes for TEMPON in different polymers are presented in Figure 10. The difference in these dependencies reflects different properties of polymer matrices. The clear dependence of quasi-libration amplitudes on the probe molecular size is detected as well.

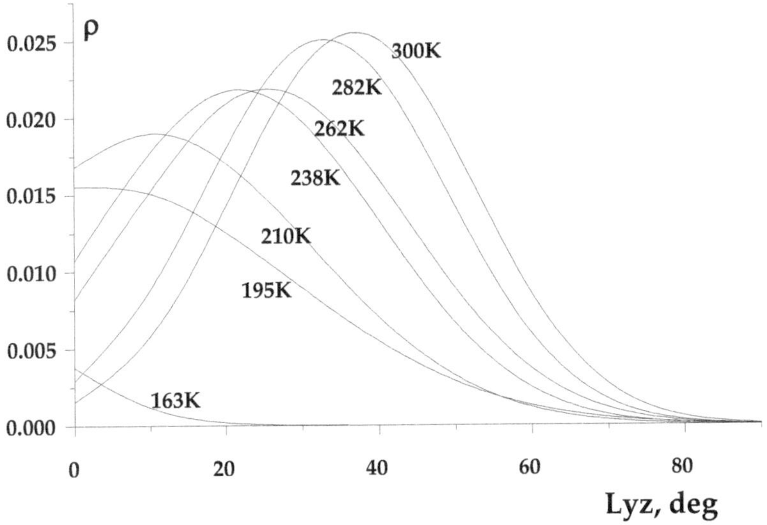

Figure 9. Shape of the Gauss distribution for TEMPON in PS.

It is seen from Figures 9 and 10 that when the averaged quasi-libration amplitude reaches 40–50° the high-amplitude tail of the distribution spread to 90°. The quasi-librations with this amplitude cannot be distinguished from rotational movements. It is not surprising, therefore, that EPR spectra at higher temperatures cannot be simulated within the framework of the quasi-libration model only.

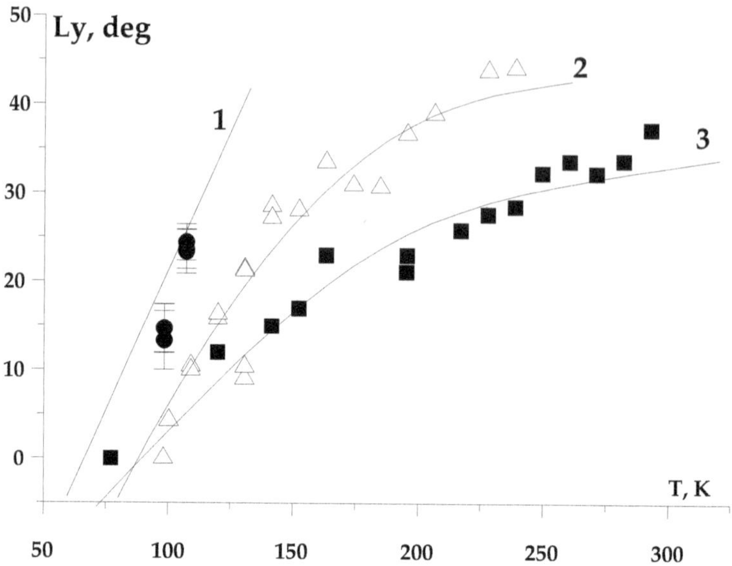

Figure 10. Quasi-libration amplitudes of TEMPON in AF-2400 (1), PVTMS (2), and PS (3).

3.3.5. Quasi-libration and rotational diffusion

At a temperature higher than 300K, the EPR spectra of TEMPON in polystyrene cannot be qualitatively described using the Brownian rotation diffusion (Figure 6) or quasi-librations only. Both these movements should be taken into account simultaneously. This model assumes that the probe takes part in two types of motion: fast quasi-librations restricted by the matrix cage and slow Brownian diffusion caused by the cage rearrangement. The difference in the cage properties can be taken into account by means of log-normal distribution of correlation times for cage reorientation:

$$\rho(R)=\begin{cases} 0, & R<0 \\ \dfrac{1}{\sigma\sqrt{2\pi}}\exp\left(-\dfrac{\left(\ln R-\ln R_0\right)^2}{2\sigma^2}\right), & R>0' \end{cases} \tag{13}$$

where R is the rotational diffusion coefficient and $\ln R_0$ and σ are the center and dispersion of the distribution, respectively.

Application of this combined model gives satisfactory agreement between calculated and experimental spectra in the high temperature range (Figure 6(c)). It should be pointed out that taking into account quasi-librations is a necessary requirement for satisfactory description of these spectra. The average correlation time for cage reorientation obtained as results of spectra simulations was found to be approximately 10^{-7} s at 360K. Characteristic widths of the distributions lie in the range of 0.6–1.0 decades. The obtained data conform to the literature data on the width of distribution of correlation times in polymers [56, 58-60]. The distribution narrows when temperature rises. Our calculations show that the distribution with a width of 0.4 decades or smaller does not influence the simulated ESR spectra.

The presented consideration shows that by using the quasi-libration model and taking into account the molecular distributions, it is possible to describe within experimental accuracy the EPR spectra of paramagnetic probe in polymer matrix in the whole temperature range near and below the glass transition point. The examples of some other polymers are presented in [38, 47].

3.3.6. Correlation between molecular mobility and reaction rate [61]

Having the opportunity to obtain detailed characteristics of rotational mobility of admixture molecules in polymer matrix, it is interesting to examine the influence of rotation mobility on the molecular reactivity. For this purpose, we used four bifunctional probe molecules carrying a paramagnetic nitroxide fragment and a photochromic azobenzene moiety (Figure 3). The azobenzene fragment of these molecules is known to undergo photochemical *trans*-to-*cis* isomerization [62]. The kinetic curves of photo-isomerization of probe II in polystyrene matrix are presented in Figure 11 as an example.

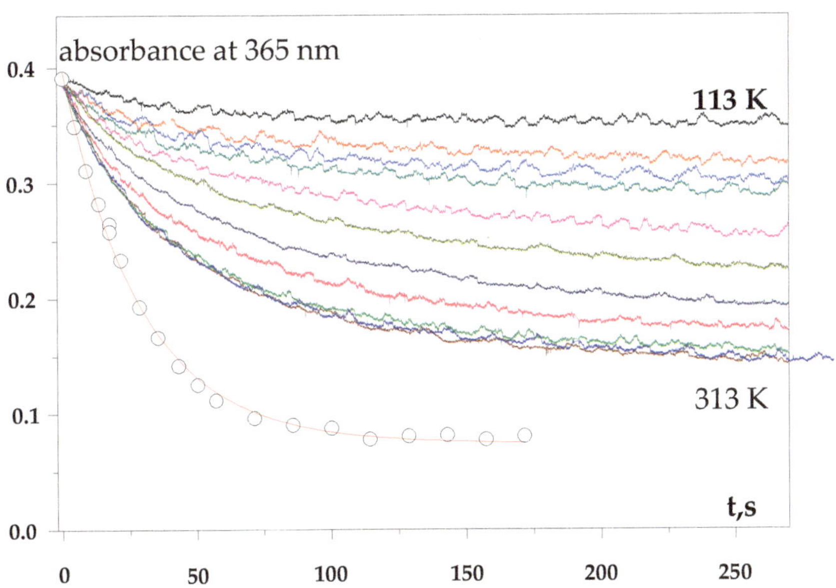

Figure 11. Kinetic curves of *trans-cis* photo-isomerization of probe II in glassy polystyrene and in solution (points).

As one can see, the influence of the polymer media consists in diminishing photoreaction rate (reduction of reaction quantum yield) and lowering the reaction yield at prolonged irradiation. Both these effects are the results of restriction imposed by the matrix on molecular rearrangement. The less extent of the photoreaction indicates that some fraction of initial trans-isomer is incapable of the reaction as a result of the rigid matrix surrounding. This conclusion was proven by the following experiment (Figure 12). The sample was irradiated at a low temperature until photoreaction has almost stopped. Then the sample was annealed at room temperature in the dark and cooled to the initial low temperature. Such annealed sample demonstrated recovered high rate of photoreaction at irradiation.

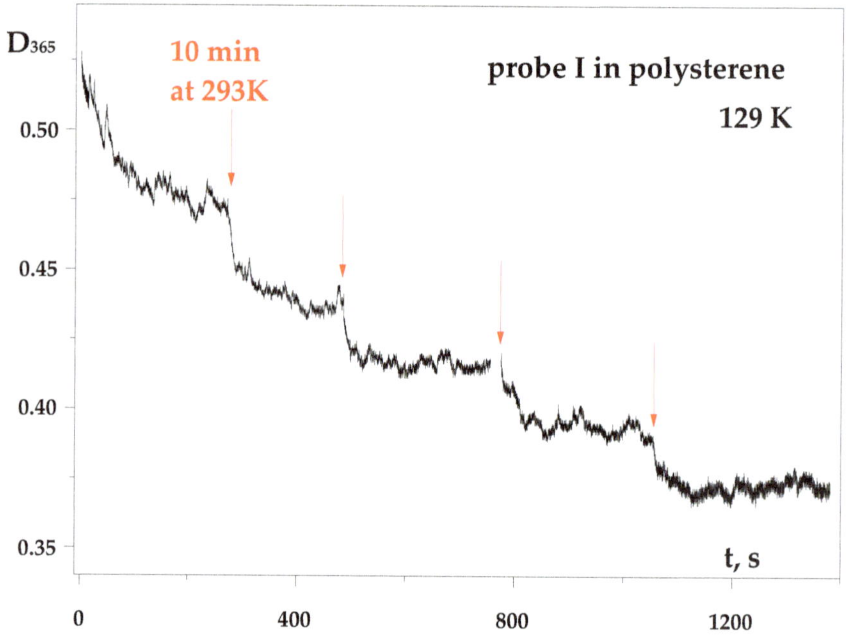

Figure 12. The kinetics of photoreaction of probe I in polymer media at 129K, which was interrupted in pointed moments, and the sample was annealed 10 minutes at 293K without light irradiation.

Determination of the quasi-libration amplitudes was performed by numerical simulation of the EPR spectra as described earlier. The obtained amplitudes are collected in Figure 13. In accordance with the molecular geometry, the largest amplitudes are detected for quasi-libration around the longest molecular axis (x-axis of the g-tensor).

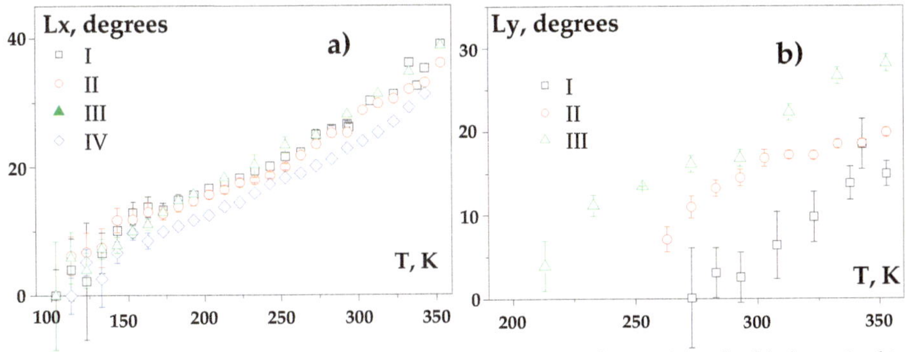

Figure 13. Averaged values of the quasi-libration amplitudes around axes x (a) and y (b), determined in the course of the numerical simulation of ESR spectra, probes I–IV.

The obtained values of quasi-libration amplitudes can be confronted with the quantum yields of photo-isomerization. The very good correlation of these values is illustrated by Figure 14.

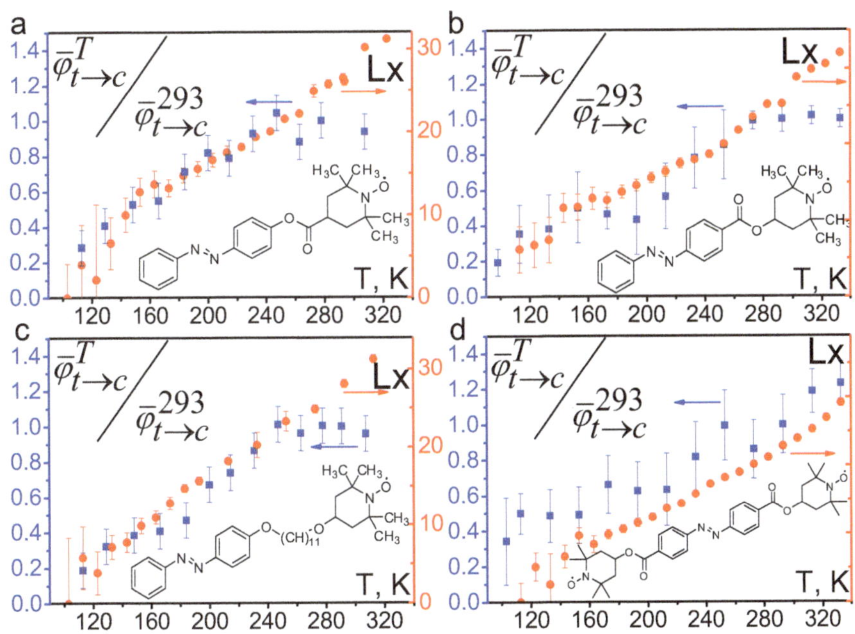

Figure 14. Quantum yields of *trans-cis* photo-isomerization and amplitudes of quasi-librations versus temperature.

Even more interesting is the comparison of quasi-libration distributions with the extent of reaction. The Gaussian distribution functions of quasi-libration amplitudes are presented in

Figure 15. Hatched areas denote the fraction of molecules, which are able to undergo photochemical isomerization. The fraction of active molecules in this figure was determined from the results of photochemical experiments. It is seen that the threshold of reactivity for different temperatures is observed at the same quasi-libration amplitude. This value amounts to approximately 10° and can be rationalized by consideration of displacements required for the elementary act of reaction.

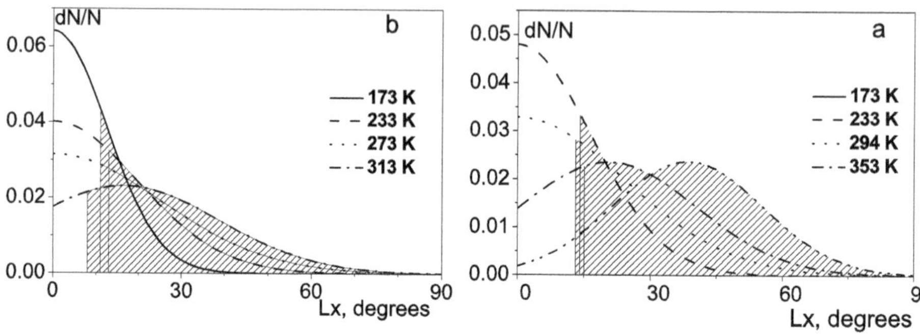

Figure 15. Temperature dependence of the distribution density of the quasi-libration amplitudes for probes I (a) and IV (b), determined as a result of the numerical simulation of the ESR spectra; hatching indicates photo-chemically active molecules.

Thus, the model of quasi-librations and characteristics of molecular displacements in polymer matrix determined by simulation of the EPR spectra are confirmed by kinetic chemical experiments.

4. Orientational alignment of paramagnetic molecules in a sample

Properties of many materials, such as stretched polymers, liquid crystals, and LB-films, are determined by the orientational order of molecules. The most precise characteristic of the molecular order is the orientation distribution function $\rho(\alpha,\beta,\gamma)$, which shows the number (or fraction) of molecules oriented in the angular interval $\alpha + d\alpha$, $\beta + d\beta$, $\gamma + d\gamma$, where α,β,γ are Euler angles connecting the molecular reference frame with the sample frame. Information about the characteristics of orientation distribution function is contained in the EPR spectra of nitroxide probe introduced into the ordered medium. The anisotropy of the medium manifests itself in the dependence of spectrum shape on the orientation of the sample respective to the direction of the magnetic field of the spectrometer. Most often, this information was obtained using any assumption about the shape of orientation distribution function. By means of comparison of the calculated and experimental spectra, researchers found the parameters of a priori defined function [4-11, 63-69].

In the present chapter, we describe the model-free method of determining of orientation distribution function [70-76]. The method is based on the expansion of the determined function in a series of orthonormal functions with variable coefficients. The expansion coefficients are determined through minimization of discrepancies between simulated and experimental spectra.

4.1. Orientation distribution function

In the general case, the orientation distribution function is represented as a series of Wigner functions:

$$\rho(\alpha,\beta,\gamma) = \sum_{j,m',m} \frac{2j+1}{8\pi^2} \left\langle D^{j\;*}_{m'm} \right\rangle D^{j}_{m'm}(\alpha,\beta,\gamma) \tag{14}$$

However, depending on the symmetry of the system under consideration, the representation of the function can be simplified. When uniaxial molecules form a uniaxial sample, the orientation of each molecule in the sample is uniquely determined by the angle between the anisotropy axis of this molecule and the symmetry axis of the sample. In this case, the orientation function is the function of only one angle $\rho = \rho(\beta)$ and can be represented as a series of Legendre polynomials. When uniaxial paramagnetic particles are arbitrarily distributed in the sample, the orientation of each particle is determined by two angles that characterize the orientation of the anisotropy axis of the particle in the reference frame associated with the sample. In this case, orientation function is a function of two angles $\rho = \rho(\beta,\gamma)$. In a uniaxial sample, the distribution function of particles characterized by three different principal values of g-tensor and/or hfi tensor is also a function of two angles. In this case, the angles β,γ characterize the orientation of the symmetry axis of the sample in the reference frame associated with a paramagnetic particles. The orientation distribution function in these two cases can be presented as a series of spherical harmonics:

$$\rho(\beta,\gamma) = \frac{1}{2\pi} \sum_{j=0}^{\infty} \left(\frac{1}{2} a_{j0} P_j(\cos\beta) + \sum_{m=1}^{j} P_{jm}(\cos\beta) \left[a_{jm} \cos m\gamma + b_{jm} \sin m\gamma \right] \right) \tag{15}$$

where P_j are Legendre polynomials and P_{jm} are associated Legendre functions.

In practice, axially symmetrical samples are studied most often, but magnetic parameters of spin probes possess ordinary orthorhombic symmetry. Hence, the orientation distribution function usually can be represented as series (15). Coefficient a_{00} reflects the full number of radicals; for a normalized orientation function, it is equal to unity.

To characterize the orientation alignment of the uniaxial system, the order parameters are used:

$$\begin{aligned} A^j_m \equiv S_{jm} &= \frac{\left\langle D^{j\;*}_{0m} \right\rangle + \left\langle D^{j\;*}_{0-m} \right\rangle}{2} = \frac{a_{jm}}{2j+1} \sqrt{\frac{(j+m)!}{(j-m)!}} \\ B^j_m &= \frac{\left\langle D^{j\;*}_{0m} \right\rangle - \left\langle D^{j\;*}_{0-m} \right\rangle}{2i} = \frac{b_{jm}}{2j+1} \sqrt{\frac{(j+m)!}{(j-m)!}} \end{aligned} \tag{16}$$

Specification of all the order parameters or all the coefficients in expansions (14) and (15) gives a complete specification of the orientation distribution function. To date, there is no technique for complete experimental determination of an orientation distribution function for soft matter. Only the second moments of orientation function (order parameters of rank two) are determined usually, as they can be obtained using one-photon optical methods. Orientation characteristics of rank four and higher are determined extremely rarely. The EPR spectroscopy of the spin probes and labels provides, in principle, the possibility of complete determination of an orientation distribution function.

For computer realization of the suggested method, the homemade program ODF3 was worked out. This program allows to calculate the series of EPR spectra recorded at various orientations of the sample in the magnetic field of the spectrometer and to determine the characteristics of orientation distribution function in the course of the minimization procedure. A brief description of the program ODF3 is presented in the appendix.

It is necessary to note that orientation function determined by the analysis of EPR spectra is always symmetric with respect to the center of coordinates. Indeed, at turning the magnetic field direction by 180°, the EPR spectrum does not change, and so some unequal orientations of the paramagnetic particle are indistinguishable by EPR. This limitation equally concerns all methods, which use EPR spectroscopy for the study of orientation order. Unfortunately, researchers often do not take this circumstance into account at interpretation of their results. This limitation imposes constraints on the expansion coefficients. When orientation distribution function is represented in the form of Eq. (15), all the coefficients a_{jm}, b_{jm} with odd j are equal to zero.

4.2. Determination of orthorhombic orientation distribution function

The application of the suggested method is demonstrated below on the examples of spin probes in liquid crystals, polymer matrices, and supercooled glasses [73-77].

In Figure 16(a), one can see some EPR spectra of the standard spin probe TEMPOL (Figure 3) in nematic liquid crystal 5CB (4-*n*-amyl-4′-cyanobiphenyl) aligned by the magnetic field of the EPR spectrometer [73,74]. The spectra were recorded at T = 77K (in liquid nitrogen) at various angles between the symmetry axis of the sample and the magnetic field of the EPR spectrometer. The angular dependence of the spectrum shape proves that the paramagnetic molecules are partially ordered by the liquid crystal. In Figure 16(b), the result of the joint simulation of these spectra is presented. Here, as well as in other examples, the magnetic parameters of the radical were determined previously by means of simulation of the spectrum of isotropic sample. Hence, in the course of simulation of the angular dependence, only expansion coefficients of the orientation function were varied.

Because in EPR spectroscopy there is no principal prohibition for determination of the high–rank-order parameters of the orientation function, we simulated the series of EPR spectra several times with consecutive increases in the expansion order. When adding expansion members of the next order did not lead to improvement of the description of the

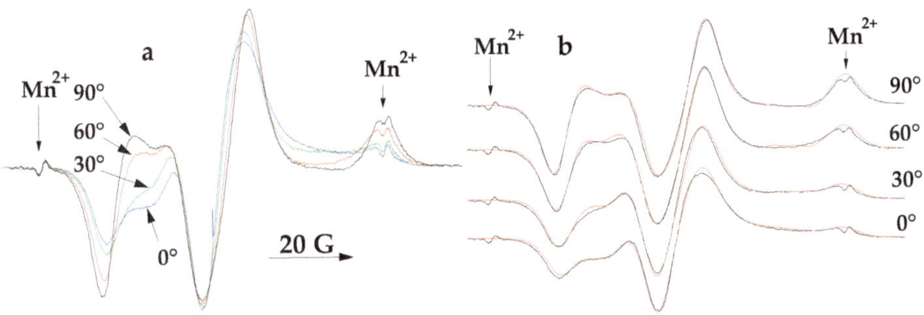

Figure 16. EPR spectra of TEMPOL in aligned liquid crystal 5CB recorded at different angles between the magnetic field vector and the sample director.

experimental spectra we supposed that the next coefficients are close to zero within the errors of determination. The result of simulation of the spectra is shown in Figure 16(b). The values of the coefficients are presented in Table 5. It was found that the expansion coefficients of the second and fourth orders are determined reliably.

The orientation distribution function of TEMPOL in 5CB is presented in Figure 17(a). The function describes the orientation distribution of the sample director in the frame of paramagnetic molecule. The principal axes of the g-tensor are used as coordinate axes. The orientation of these axes in the nitroxide radical is shown in Figure 17(b). From Figure 17(a), one can see that the TEMPOL magnetic axis Z is oriented predominantly perpendicular to the sample symmetry axis. Possibly, it is a result of the interaction of the electron pairs of nitrogen and oxygen atoms with the π-system of the benzene ring of the liquid crystal molecules. The X and Y axes of the TEMPOL molecules are predominantly directed at 50º and 40º to the sample anisotropy axis, respectively. The corresponding orientation of the molecules of spin probe and liquid crystal is shown in Figure 17(b).

It is necessary to emphasize that in the presented case, the orientation distribution function possesses orthorhombic symmetry. The principal axes of the g-tensor and hfi-tensor practically coincide in the case of TEMPOL and other nitroxide radicals. It means that nitroxide probes have orthorhombic symmetry. Since the effective values of the magnetic parameters are defined by squares of the directional cosines of the magnetic field vector in the g-frame and hfi-frame correspondently (17), eight different molecular orientations are undistinguishable by EPR.

$$g_{eff}^2 = g_{xx}^2 \cos^2(HX_g) + g_{yy}^2 \cos^2(HY_g) + g_{zz}^2 \cos^2(HZ_g) \tag{17}$$

$$A_{eff}^2 = A_{xx}^2 \cos^2(HX_A) + A_{yy}^2 \cos^2(HY_A) + A_{zz}^2 \cos^2(HZ_A) \tag{18}$$

System	a_{20} a_{40} a_{60}	a_{21} b_{21}	a_{22} b_{22}	a_{42} a_{44}	a_{62} a_{64} a_{66}
TEMPOL in 5B	−0.872 ± 0.007 0.247 ± 0.015 –	– –	0.082 ± 0.004 –	−0.0004 ± 0.0012 −0.0047 ± 0.0003	– – –
V(n=11) in 5CB	1.91 0.11 −0.70	– –	−0.133 –	−0.065 ~0	−0.024 ~0 ~0
V(n=11) in 5CB/pores	1.35 0.21 –	– –	−0.18 –	−0.010 −0.0006	– – –
V(n=4) in stretched PE	0.82 – –	– –	−0.029 –	– –	– – –
V(n=11) in stretched PE	1.43 0.37 –	– –	−0.106 –	−0.018 −0.0016	– – –
V(n=15) in stretched PE	2.42 1.06 0.144	– –	−0.103 –	−0.056 0.0011	−0.0016 ~0 ~0
VI in stretched PE	0.49 0.22 –	– –	−0.056 –	−0.010 ~0	– – –
HO_2 in glassy H_2O_2	0.25 – –	0.073 0.181	0.025 0.038	– –	– – –
Cl_2^- in glassy LiCl (5mol/l)	−0.25 – –	– –	– –	– –	– – –

Typical errors of values determined are presented in the first row of the table.

Table 5. The expansion coefficients of series (15) for the orientation distribution functions of some radicals in different matrices

The function possessing orthorhombic symmetry consists of eight equal "petals," which have maxima at angles (βmax,γmax), (βmax,−γmax), (βmax,π + γmax), (βmax,π − γmax), (π − βmax,γmax), (π − βmax,−γmax), (π − βmax,π + γmax), (π − βmax,π − γmax). For the function presented in Figure 17(a), βmax is approximately 90°, as the petals corresponding to the angles (βmax,γmax) and (π − βmax,γmax), (βmax,−γmax) and (π − βmax,−γmax), etc. overlap in pairs.

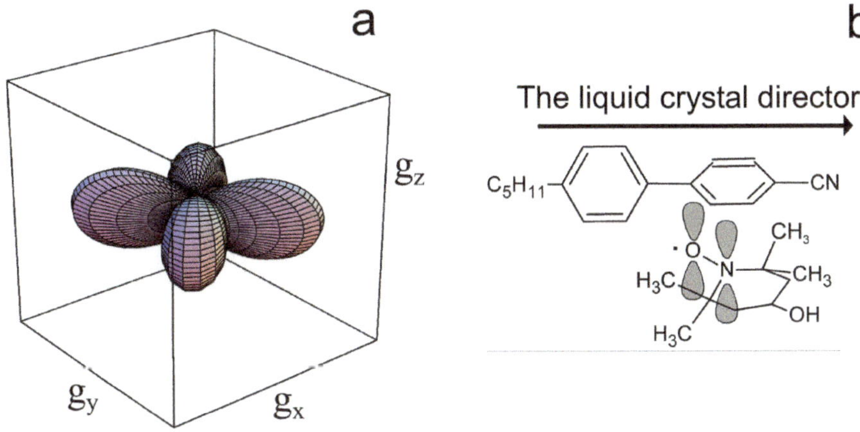

Figure 17. The orientation distribution function of TEMPOL in 5CB aligned by the magnetic field (a) and the reciprocal orientation of the molecules of the spin probe and liquid crystal (b).

In Figure 18, one can see the angular dependence of the EPR spectrum for the nitroxide radical *V*(n = 11) in supercooled 5CB aligned by magnetic field and the result of the computer simulation of spectra [76]. It is seen that angular dependence (the difference between the spectra recorded at different angles) in the case of the spin probe *V*(n = 11) is much larger than in the case of TEMPOL. Obviously, molecules of probe V, which have rigid central fragments, are built into a liquid crystal structure better than the molecules of TEMPOL. It was found out that orientation function in this case is determined reliably up to the sixth order of expansion. The values of the coefficients are presented in Table 5. The function is shown in Figure 19. It can be seen that the central rigid fragments of the paramagnetic molecules preferably order along the liquid crystal molecules. The magnetic axis *X* is situated perpendicular to the sample director.

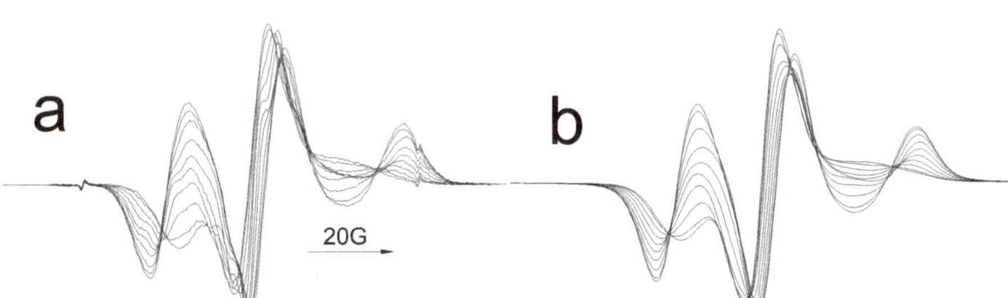

Figure 18. The angular dependence of the EPR spectrum of the nitroxide radical *V*(n = 11) in 5CB aligned by magnetic field (a) and the result of its computer simulation (b).

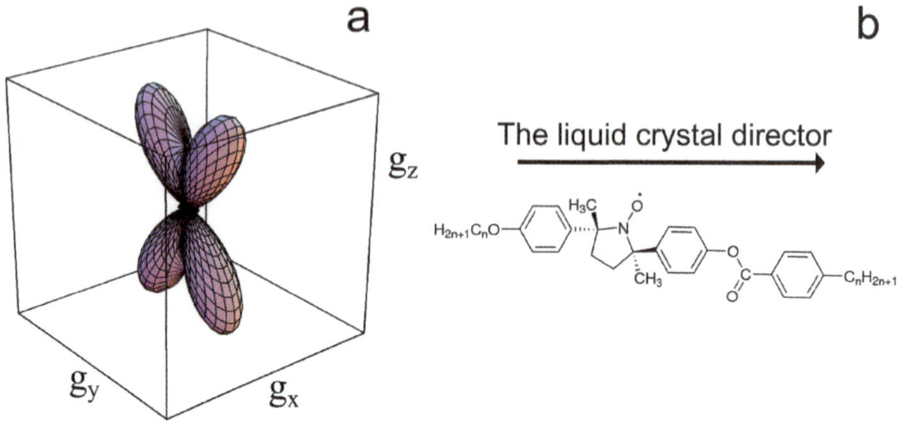

Figure 19. The orientation distribution function of radical *V*(n = 11) in 5CB aligned by the magnetic field (a) and orientation of the molecules of spin probe and liquid crystal (b).

In the examples presented earlier, the uniaxial samples were studied. However, the method described can be used for investigating more complicated systems. In Figure 20, one can see the scheme of the following experiment. The sample of liquid crystal 5CB with embedded the spin probe TEMPOL was aligned by the magnetic field of the EPR spectrometer along the direction D_1 at T = 295K and then was quickly cooled to T = 77K as it was performed in the previous examples. After that, the sample was heated in the spectrometer resonator to T = 220K in such position that the axis D_1 was approximately perpendicular to the magnetic field direction. In the course of this annealing, the new axis D_2 directed along the new magnetic field vector arose. After realignment, the sample was quickly cooled in the magnetic field to T = 77K, and the angular dependence of EPR spectra was recorded. The orientation distribution for this sample was presented as a sum of two different uniaxial functions with different directors, D_1 and D_2. The varied parameters in this case were the expansion coefficients of these functions in series (15) and the fractions of particles oriented along the D_1 and D_2 axes. In this assumption, it was possible to simulate the angular dependence of the EPR spectrum within the experimental errors. The orientation distributions of the probe molecules respective to D_1 and D_2 axes are shown in Figure 20 (functions F1 and F2, correspondingly). One can see that the function F1 is practically equal to the function presented in Figure 17(a). Hence, a part of the liquid crystal is not realigned by magnetic field at T = 220K. Another part of the sample possesses mobility at T = 220K, which is sufficient to change the orientation according to the new director D_2; thus, this part of the probe molecules are turned at an angle of 90°. One can see that the mobility of the liquid crystal particles at T = 220K is sufficient for realization of realignment. The ratio of the number of radicals oriented axially along D_1 to those oriented axially along D_2 is 2.5:1. As a whole, the result of this experiment provide evidence of considerable distribution of molecular mobility in supercooled liquid crystal.

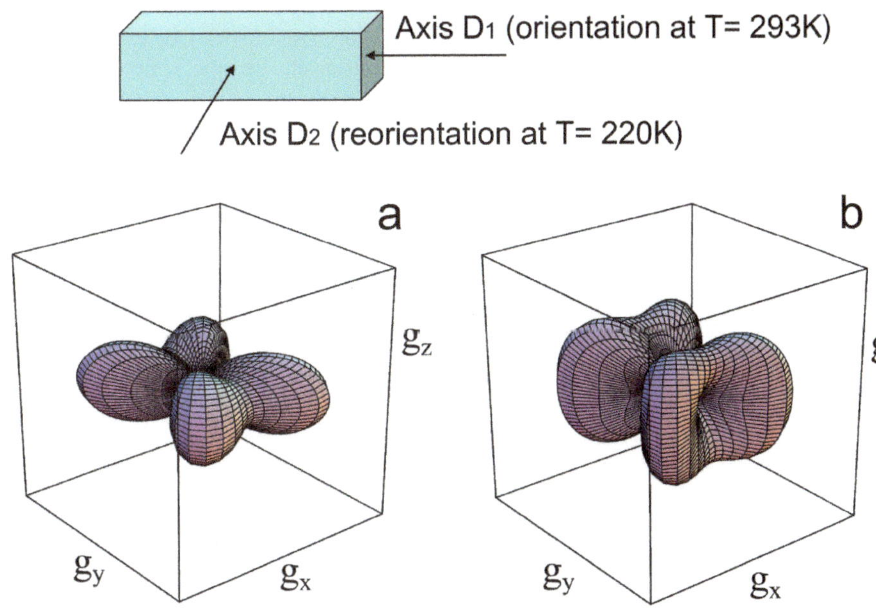

Figure 20. The scheme of the experiment of the liquid crystal realignment (see text). The distribution functions for axis D_1 (a) and axis D_2 (b) in the molecular reference frame.

4.3. Non-orthorhombic spin probe

In all the previous examples, nitroxides were used as spin probes. For these radicals, the directions of the main axes of the g- and hfi-tensors coincide, and the orientation distribution possesses orthorhombic symmetry. In series (15) in this case, only coefficients a_{jm} with even *j и m* are nonzero. This feature is a consequence of the existence of eight molecular orientations undistinguishable by EPR. To characterize molecular order more definitely, the paramagnetic probe with lower symmetry is necessary. In the following text, we present an example of orientation distribution of HO_2 radicals with distinct axes of g- and hfi-tensors [77]. The HO_2·radicals were generated in the matrix of glassy hydrogen peroxide at 77K by the light irradiation with wavelength $\lambda = 254$ nm. The long irradiation by the collimated beam of the nonpolarized light leads to orientational alignment of the radicals as a result of the photo-orientation process. In the course of photo orientation, the radicals are aligned in such a way that the vectors of their optical dipole transition moments are directed along the symmetry axis of the sample (direction of light beam). By means of simulation of the spectrum of the isotropic sample, it was established that the Euler angles connecting the frames of the g- and hfi-tensors come to $\zeta = -70°$, $\xi = 47°$, and $\varsigma = 35°$. Hence, in this case, the magnetic properties of paramagnetic particles cannot be described by tensor rank two, and each direction in the radical coordinate frame is individual and can be found from the analysis of the EPR spectra. Series (15) in this case includes nonzero coefficients a_{jm} и b_{jm} with both even and odd *m*. The orientation distribution function of the HO_2·radicals is presented in Figure 21; the expansion coefficients are shown in Table 5. One can see that in this case, one predominant orientation of the radicals respective to the sample symmetry axis is observed. Obviously, this orientation is dictated by the optical dipole transition moment of HO_2. Hence, determination of the orientation distribution function of the paramagnetic molecules allows establishing experimentally the direction of optical dipole transition moment in the molecular coordinate frame.

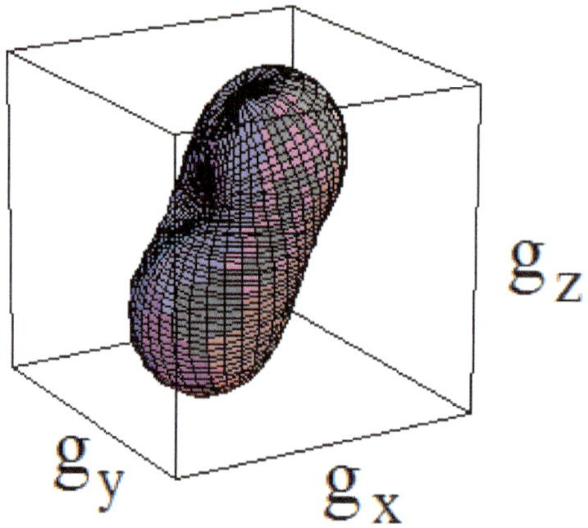

Figure 21. The orientation distribution function of the HO_2·radicals in supercooled hydrogen peroxide.

4.4. Molecular orientation frame

In the example presented previously, the data concerning the properties of the paramagnetic molecules were obtained. However, most often, orientation distribution of the spin probe is studied to obtain information about the structure and dynamics of the matrix. In this case, the choice of the spin probe is of fundamental importance. It was shown earlier that radical V reflect the alignment of liquid crystals more effectively than the standard spin probes of the piperidine series. As another illustration of the sensitivity of the method to the structure of the probe, the orientation distributions for different radicals in the thin film of polyethylene stretched by five times are shown in Figure 22. It is seen that radicals that have rigid central fragment *V*(n = 4), *V*(n = 11), and *V*(n = 15) are ordered in this matrix more effectively than the imidazolidine derivative VI (Figure 3). At the same time, ordering of radical V depends on the length of the saturated substituents, which evidently align in the polymer matrix along the macromolecules of polyethylene.

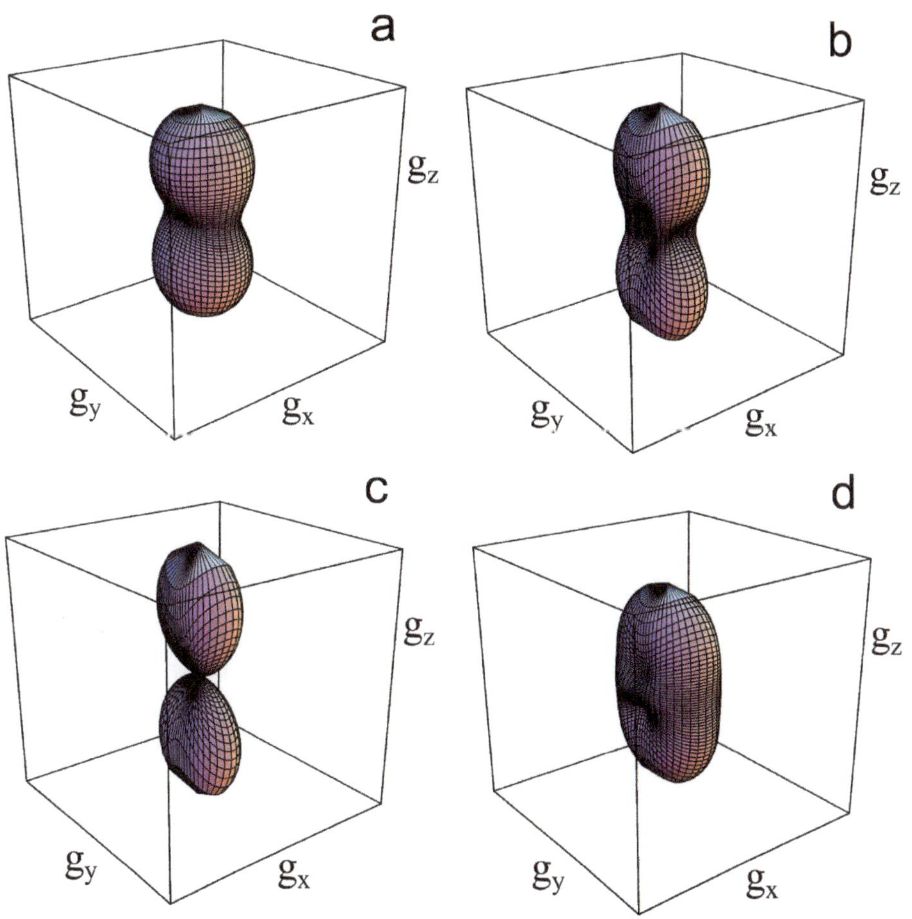

Figure 22. The orientation distribution function of the radicals *V*(n = 4) (a), *V*(n = 11) (b), *V*(n = 15) (c), and VI (d) in the polyethylene film stretched five times.

It is obvious that orientation of a paramagnetic molecule in the ordered matrix is defined not by magnetic axes but by other molecular properties (geometric shape, interaction with the medium molecules, etc.). In every case, there is a molecular axis that is ordered to the maximum extent. This axis will be referred to as main molecular orientation axis Z_t. The second most ordered axis, which is orthogonal to the first one, defines completely the molecular orientation frame (X_t, Y_t, Z_t). A practically important problem is to establish the direction of the orientation axes of the paramagnetic molecules relatively to known axes, for example, principal g-tensor axes. In addition, it is necessary to determine the order parameters (orientation factors) for the orientation axes because such parameters more adequately reflect orientation alignment of media.

As it was noted previously, uniaxial samples are studied commonly in practice. At the same time, most spin probes have three different principal values of g-tensor. The formation of the uniaxial sample by orthorhombic particles can be done in two ways, which are illustrated in Figure 23. First, such situation can be realized if all three orientation axes are ordered axially in the sample (they form the cones around the symmetry axis of the sample; Figure 23a). Let us call this type of the uniaxial sample A1. The second possibility is shown in Figure 23b. In this case, one orientation axis (denote it as Z_t) is ordered axially in the sample (forms the cones around the symmetry axis) and two other orientation axes are directed statistically in the plane (X_tY_t). In this type of axial sample (A2), we will name the case of hidden molecular axiality.

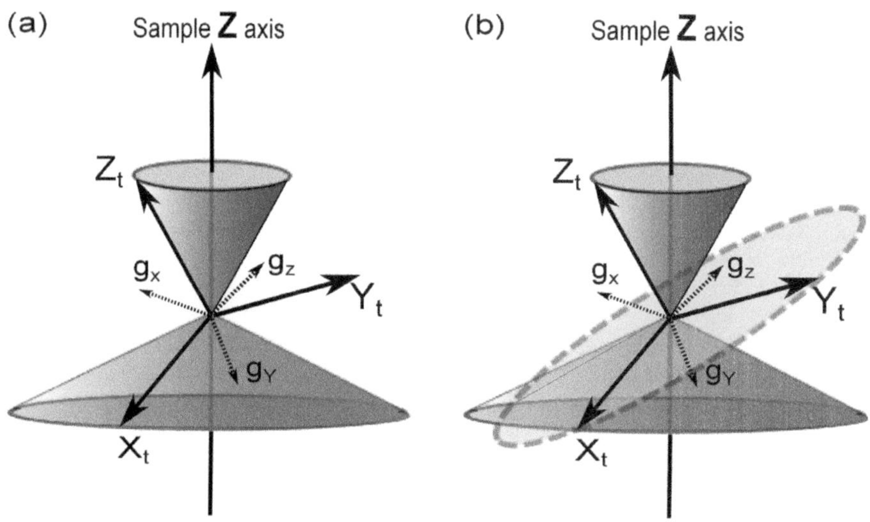

Figure 23. Mutual disposition of the magnetic and orientation frames of the spin probe and the sample director: A1 (a) and A2 (b) (see text).

It can be shown that the expansion coefficients of the distribution function presented in the molecular orientation frame and the coefficients for the same function described in the g-tensor frame in the case of the A1 sample are connected by the following expression:

$$\left\langle D^j_{m'm} \right\rangle_{gF} = \sum_k \left\langle D^j_{m'k} \right\rangle_t D^j_{mk}(\varphi,\theta,\psi), \tag{19}$$

where angles φ, θ, and ψ are Euler angles connecting the molecular orientation frame with the g-tensor frame.

Orientation distribution function of the sample symmetry axis in the molecular orientation frame of the spin probe has a maximum along the axis Z_t. In the case of A2, this function possesses uniaxial symmetry and can be described by series (15), in which only members with zero second indices are nonzero:

$$\rho(\beta_t)=\frac{1}{4\pi}\sum_{j=0}^{\infty}\left(a_{j0}\right)_t P_j\left(\cos\beta_t\right) \tag{20}$$

where β_t is the angle between the symmetry axis of the sample and the axis Z_t and $(a_{j0})_t$ are the expansion coefficients in series (15) for the distribution function presented in the molecular orientation frame (X_t, Y_t, Z_t).

Let the angles (θ, φ) determine the direction of Z_t in the g-tensor frame. Using Eq. (19), it can be shown that the expansion coefficients of the function in the g-frame $(a_{j0})_{gF}$ are connected with the coefficients $(a_{j0})_t$ and angles (θ, φ) by the following expression:

$$\left\langle D_{0m}^{j\ *}\right\rangle_{gF}=\left(A_0^j\right)_t D_{m0}^j(\varphi,\theta,0)$$

$$\begin{aligned} a_{jm}&=\left(a_{j0}\right)_t\frac{(-1)^m(j-m)!}{(j+m)!}\cos m\varphi P_{jm}(\cos\theta)\\ b_{jm}&=-\left(a_{j0}\right)_t\frac{(-1)^m(j-m)!}{(j+m)!}\sin m\varphi P_{jm}(\cos\theta)\end{aligned} \tag{21}$$

Using Eq. (21) in the course of the computer simulation of the spectrum angular dependence, one can vary coefficients $(a_{j0})_t$ and angles (θ, φ) and simultaneously find the order parameters and direction of the main molecular orientation axis in the g-tensor frame of the radical. It is necessary to note that in the case of orthorhombic probe, one of the eight undistinguishable pairs (θ, φ) is determined in the course of such calculations.

In the case A1, the orientation distribution function described in the molecular orientation frame does not possess uniaxial symmetry and can be described by angles (φ, θ, ψ) and coefficients $(a_{jm})_t$, including the members with nonzero m in accordance with Eq. (19).

The possibilities of simulation of the spectrum angular dependence according to assumptions A1 and A2 are contained in the program ODF3.

In Figure 24, one can see the orientation functions of the radical *V*(n = 11) in supercooled 5CB aligned by magnetic field of the EPR spectrometer. The function in Figure 24(a) was obtained in orthorhombic assumption; the function in Figure 24(b) corresponds to hidden axiality **A2**. It is seen that Eq. (21) permit to separate one petal from orthorhombic function. The discrepancy of the calculated spectra from the experimental ones for these two functions is the same. This calculation leads to the determination of the main molecular orientation axis for the probe V. This axis was found to be directed with the angles θ = 30°, φ = 0° to the g-tensor axes. It was verified that the other seven pairs of angles mentioned earlier lead to the same calculated spectra. Hence, the choice between these possible directions can be done only by using additional data obtained by other experimental techniques.

The analogous functions are presented in Figure 25(a, b), but in this case, the liquid crystal was aligned in uniaxially ordered cylindrical pores of porous polyethylene [78]. Figures

24(b) and 25(b) show that the obtained functions are very close. The difference between the functions consists in the extent of anisotropy. It is seen that liquid crystal 5CB is aligned by magnetic field more effectively than by the pores of polyethylene. The significant visual difference between the orthorhombic functions 24(a) and 25(a) is explained by the fact that individual petals that are well divided in function 24(a) become wider and flow together in function 25(a).

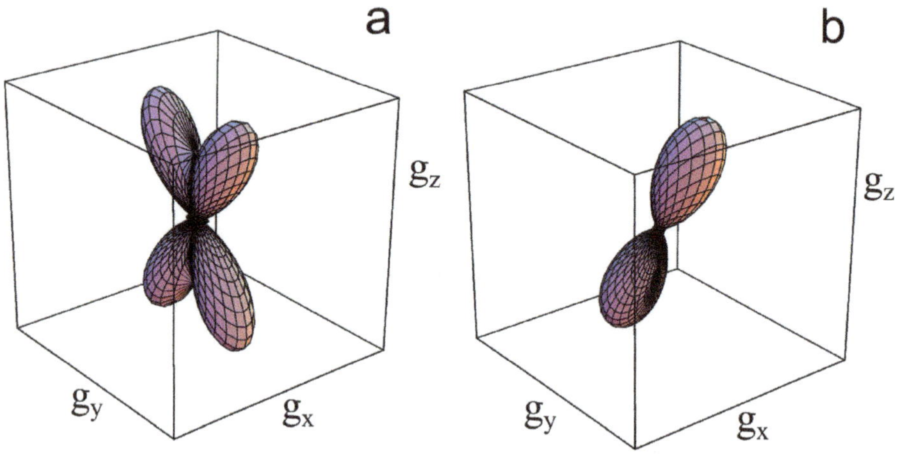

Figure 24. The orientation distribution functions of radical V(n = 11) in 5CB aligned by the magnetic field: the orthorhombic function (a) and the function calculated with the assumption A2 (b).

Almost all the cases presented previously showed that the simulations within the uniaxial models A1 and A2 lead to the very close values of discrepancy between the calculated and experimental spectra. It means that the model of hidden axiality is often confirmed by experiment. The exception was found in the case of liquid crystal aligned in porous polyethylene. In that case, the simulation of the angular dependence for V(n = 11) in the assumption A1 leads to diminishing of discrepancy by 14% in comparison with the model A2. The found distortion of uniaxial symmetry of the distribution is shown in Figure 25(c, d). This example demonstrates that the biaxiality of nematic media can be studied by spin probe technique.

4.5. Order parameters

On the basis of orientation distribution function, one can calculate the order parameters for any molecular axis of the radical in the matrix under consideration. For this aim, orientation distribution functions are transformed to the chosen molecular frame using Eq. (19) or (21). The order parameters for new axes are calculated according to Eq. (16).

The second-rank order parameters for the g-tensor axes can be calculated as follows:

$$A_0^2\left(g_x\right) = -\frac{a_{20}}{10} + \frac{6a_{22}}{5} \tag{22}$$

$$A_0^2\left(g_y\right) = -\frac{a_{20}}{10} - \frac{6a_{22}}{5} \tag{23}$$

$$A_0^2\left(g_z\right) = \frac{a_{20}}{5}, \tag{24}$$

where a_{20} and a_{22} are obtained in orthorhombic approximation in the g-tensor frame.

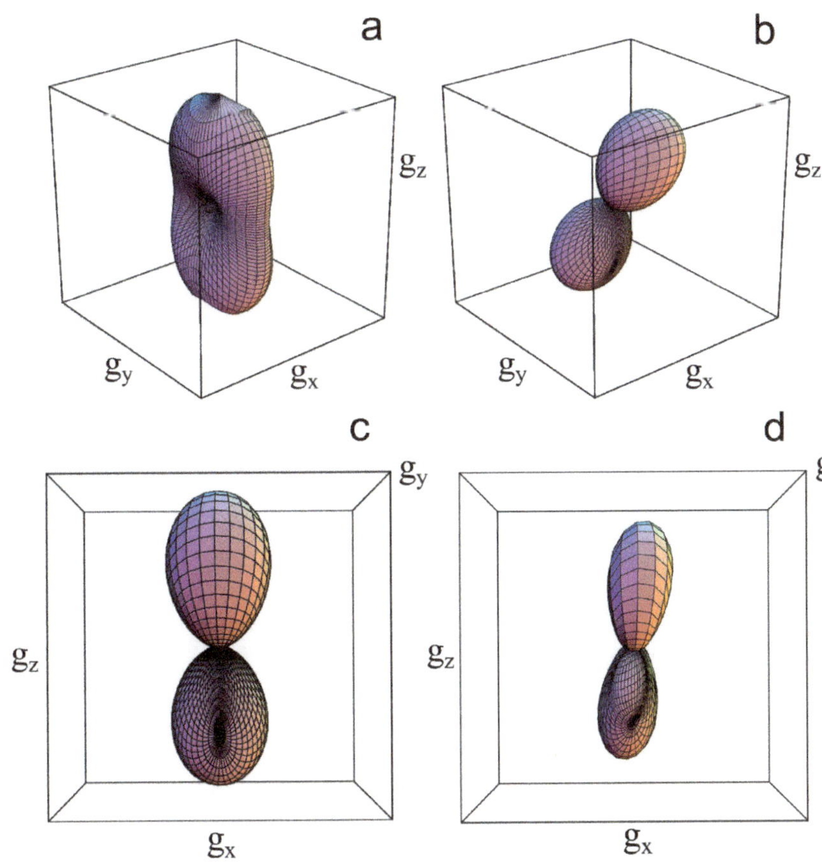

Figure 25. The orientation distribution functions of radical V(n = 11) in 5CB aligned in the pores of the porous polyethylene: orthorhombic function (a), the function calculated with the assumption A2 (b,c), and function calculated with the assumption A1 (d).

The order parameters for the magnetic axes of the radical *V*(n = 11) in 5CB aligned by magnetic field (Figure 24(a)) arrive at the values $f(g_x) = -0.35$, $f(g_y) = -0.031$, and $f(g_z) = 0.38$. These values show that the axis *X* is ordered predominantly perpendicular to the symmetry axis of the sample. The order parameters for axes *Y* and *Z* are determined not only by their extent of ordering but also by the angles between these axes and the director of the sample. Rather small values of the order parameters for the magnetic *Y* and *Z* axes reflect the tilt of these axes relative to sample director. Thus, the values of the order parameters for magnetic axes are useful in defining the orientation of the probe relative to the medium but do not characterize clearly the medium alignment.

The more adequate characteristics of the medium order are the order parameters of the probe orientation axes. The values calculated using Eqs. (21) and (16) for some studied liquid crystalline systems are collected in Table 6. One can see from the table that the second-rank order parameters achieve the values 0.6–0.7, which are in agreement with the values for nematic mesophase obtained by other methods. The important advantage of the presented method is the possibility to estimate order parameters of higher rank. Some of these values are presented in Table 6 as well.

Systems	$(A_{20})_t$	$(A_{40})_t$	$(A_{60})_t$
V(n=11) in 5CB	0.626	0.347	0.148
V(n=15) in stretched PE	0.673	0.357	0.088
V(n=11) in 5CB/pores	0.522	0.089	–
I-trans in 5CB/pores	0.376	0.059	–
I-cis in 5CB/pores	0.206	0.013	–
Errors	~3%	~12%	~15%

Table 6. Order parameters obtained within the assumption of A2 axiality

At present, the most widespread technique for the study of orientational alignment is optical spectroscopy, particularly the measuring of the linear dichroism in UV-vis range. It is known that such measurements give only order parameters of rank 2 [79]. To verify the discussed EPR technique, it is important to compare the values obtained by EPR and the optical measurements for the same samples. Such comparison is presented in Table 7 using four systems.

The orientation distribution of anion-radicals Cl_2^- in low temperature glass formed by 5M aqueous solution LiCl at 77K is studied in [80]. The orientational alignment in this system was induced by irradiation of the sample with parallel beam of the linearly polarized light using photo-orientation phenomenon. In this case, the anion-radicals are predominantly oriented perpendicular to the electric vector of light wave. Taking into account that optical transition moment directed along the axis of anion-radical, the order parameter of rank two can be measured as linear dichroism:

$$A_{0\,opt}^{2} = d = \frac{D_{\parallel} - D_{\perp}}{D_{\parallel} + 2D_{\perp}}, \tag{25}$$

where $D_{\parallel}$ and $D_{\perp}$ are the values of optical absorbance at mutually parallel and perpendicular polarizations of the irradiating and probing beams. The dichroism of the sample presented in the table was averaged over the range of wavelengths from 349 to 390 nm.

	Anion-radical Cl_2^- in glassy aqueous solution of LiCl	*Trans*-I in 5CB embedded in pores of polyethylene	*Cis*-I in 5CB embedded in pores of polyethylene	*V*(n = 4) in liquid crystalline comblike polymer
$A_0^2{}_{EPR}$	−0.10 ± 0.01	0.38 ± 0.01	0.21 ± 0.01	0.55 ± 0.03*
$A_0^2{}_{opt}$	−0.12 ± 0.01	0.35 ± 0.04	0.23 ± 0.03	0.36 ± 0.01**

*Orientation factor of the orientation axis of radical *V*(n = 4)
**Orientation factor of the optical transition dipole moment of merocyanine dye (Figure 26)

Table 7. The order parameters determined by means of optical spectroscopy and EPR

The dichroism of the samples containing the ordered radical I in *cis* and *trans* forms was measured similarly using the recording of UV-vis spectra in polarized light. The radicals were ordered in 5CB aligned in pores of polyethylene.

To obtain the order parameters for ordered comblike liquid crystalline polymer (Figure 26) by means of EPR spectroscopy, the spin probe *V*(n = 4) was used. For measurements of linear dichroism, merocyanine dye Ash253a was embedded in the sample.

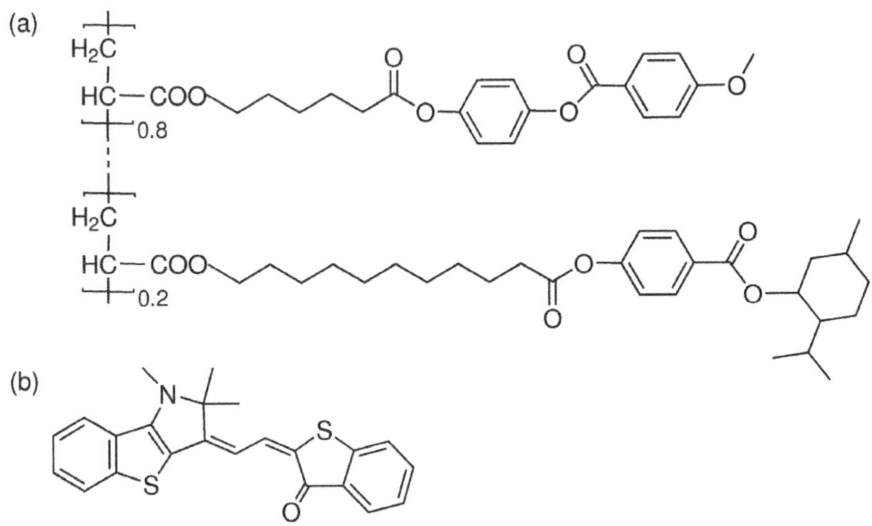

Figure 26. The structures of comblike polymer containing racemic menthyl moieties and nematogenic phenylbenzoate moieties (a) and merocyanine dye Ash253a (b).

One can see from Table 7 that the EPR technique produces the values of the order parameters, which are in agreement with the optical measurements. The noticeable difference is observed only in the case of comblike polymer when the dichroism of specially introduced dye was measured. Possibly, the optical measurement in this case gives the understated value as a result of the tilt of the transition dipole moment with respect to the director of liquid crystal matrix. Thus, the proposed method is efficient for determining order parameter.

4.6. Determination of orientation distribution function of rotating radicals

The serious disadvantage of the method described previously is the impossibility of using the EPR spectra if probe molecules rotate in the time scale of EPR. Many systems of great interest can exist only in temperature ranges in which the intensive molecular rotational mobility takes place. Orientational alignment of the nematic phases presented previously was studied by means of quick cooling of the samples to $T = 77$K and subsequent recording of the spectra of fixed particles. We supposed that the structure of the material does not change considerably at such cooling. However, such an approach cannot always be used. For example, the temperature dependence of the liquid crystal structure or alignment cannot be studied using this approach.

At present, the most widespread method for simulation of EPR spectra of rotating spin probes with determination of orientation distribution function is the method described in [4, 5]. This method is based on the assumption that each paramagnetic molecule is situated in the field of ordering potential, which is induced by the aligned matrix $U(\alpha, \beta, \gamma)$. This approach will be referred henceforth as ordering potential (OP) method to distinguish from the method described earlier, the orientation function (OF) method. The orientation distribution of the molecules in the potential is determined by the Boltzmann equilibrium:

$$\rho(\beta,\gamma) = \frac{e^{-U(\beta,\gamma)/k_b T}}{\int e^{-U(\beta,\gamma)/k_b T} \sin\beta d\beta d\gamma} \tag{26}$$

The ordering potential is represented as a series of spherical functions:

$$\frac{U(\beta,\gamma)}{k_b T} = -\sum_{j,m} c_{jm} D^j_{0m}(\beta,\gamma) \tag{27}$$

In practice, the expression (24) is presented as follows:

$$\frac{U(\beta,\gamma)}{k_b T} = -\sum_{j} c_{j0} D^j_{00}(\beta,\gamma) - \sum_{j,m} c_{jm}\left[D^j_{0m}(\beta,\gamma) + D^j_{0-m}(\beta,\gamma)\right], \tag{28}$$

where $j, m = 2, 4$.

The program realization of the OP method is based on the stochastic Liouville equation; hence, it can be used for analysis of the orientation ordering of the rotating molecules. The available software for the OP method utilizes the assumption that the axis of fastest rotation of the paramagnetic molecule is directed along the orienting potential. A lot of useful and important information concerning different oriented media was obtained using this method [6-11].

For comparison of the results obtained by the OP and OF methods, we determined the orientation distribution functions of some spin probes in the liquid crystal 5CB, aligned in pores of the porous polyethylene, at the temperature of liquid nitrogen, $T = 77K$ (OF), and at the temperature of existence of the nematic phase, $T = 298K$ (OP). It was found that 5CB ordered in the pores is not affected by the orienting action of the magnetic field of the EPR spectrometer. Therefore, it is possible to record the angular dependence of the EPR spectrum of such sample at $T = 298K$.

In Figure 27(b), one can see the orientation distribution function determined by the OP method at $T = 298K$. Visually, this function is quite identical to the function obtained for the same system by means of the OF method at $T = 77K$ (Figure 27(a)). The order parameter of the axis of the fastest rotation of the paramagnetic molecule in this case comes to $(A_0^2)_{rot} = 0.48$ and agrees within the experimental errors, with the value of the order parameter obtained by means of the OF method, $(A_0^2)_t = 0.50$. The direction of the rotation axis in the magnetic frame is described by the angles $\theta_{rot} = 39°$ and $\varphi_{rot} = 0°$; it is close to the direction of the orientation axis $\theta_t = 30°$, $\varphi_t = 0°$. Evidently, the molecular orientation axis in this case coincides with the axis of the fastest rotation, and the methods OP and OF give the same results. This example shows as well that the structure of the matrix really does not noticeably change at rapid cooling of the sample by immersing it into liquid nitrogen.

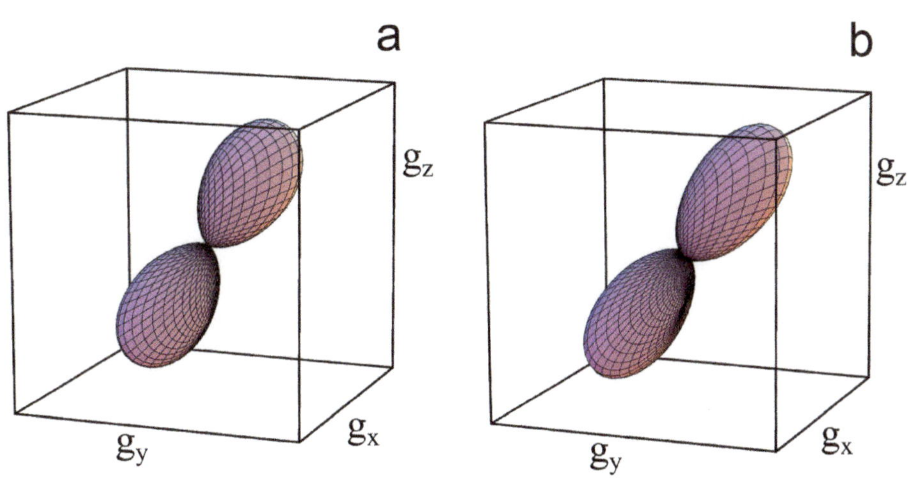

Figure 27. The orientation distribution function of radical V(n=11) in 5CB defined from the room temperature spectra by the OP method.

The similar comparison of the methods OP and OF was performed by determination of the orientation functions of the nitroxide radicals containing azobenzene fragment (Figure 28) in 5CB aligned in the pores of polyethylene. Conversion of the azobenzene moiety from *trans* to *cis* form was realized by irradiation of the samples with light [62].

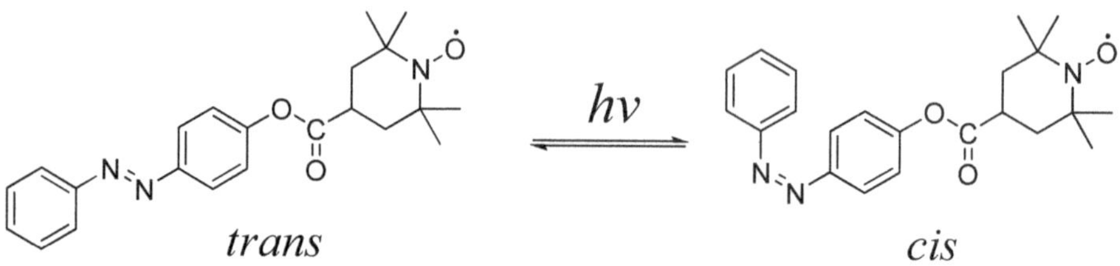

Figure 28. Photochemical conversion of the azobenzen fragment of radical I.

In Figures 29 and 30, one can see the orientation distribution functions of radical I in *trans* and *cis* configurations. The functions were determined at temperatures $T = 77K$ (OF method; Figures 29(a) and 30(a)) and $T = 298K$ (OP method; Figures 29(b) and 30(b)). The obtained characteristics are collected in Table 8.

System	$(a_{20})_t$, $(a_{40})_t$	θ_t, φ_t, degree	$(A_0^2)_t$	c_{20}, c_{22}	c_{40}, c_{40}, c_{40}	θ_{rot}, φ_{rott}, degree	$(A_0^2)_{rot}$
V(n=11) in 5CB/pores	2.61±0.02 0.80±0.08	33.6±0.2, 90.3±1.4	0.52	2.22±0.15 0.32±0.05	–	40.7±0.4 109.3±1.0	0.48
I-trans in 5CB/pores	1.88±0.02 0.53±0.03	86.8±2.2 11.9±0.7	0.38	1.24±0.01 -0.080±0.002	–	106.4±0.2 10.2±0.3	0.28
I-cis in 5CB/pores	1.03±0.02 0.17±0.02	86.1±2.8 -42.4±0.3	0.21	0.23±0.01 -0.121±0.001	0.22±0.01 0.17±0.01 0.26±0.01	90.0±1.2 5.4±0.08	0.05

Table 8. Parameters of the orientation distribution functions defined by the OP and OF methods

It is seen that in the case of the probe in *trans* form, the direction of the main rotational axis ($\theta_{rot} = 10°$, $\varphi_{rot} = 106°$) does not coincide with the direction of the molecular orientation axis ($\theta_t = 12°$, $\varphi_t = 87°$), although the deviation is not large. The order parameter value determined by the OP method is noticeably lower than value obtained by the OF method (0.28 and 0.38, correspondingly). This disagreement becomes dramatic in the case of the probe in *cis* configuration. This comparison demonstrates that the OP method produces understated and unreliable data when the molecular orientation axis of the probe does not coincide with the main rotation axis.

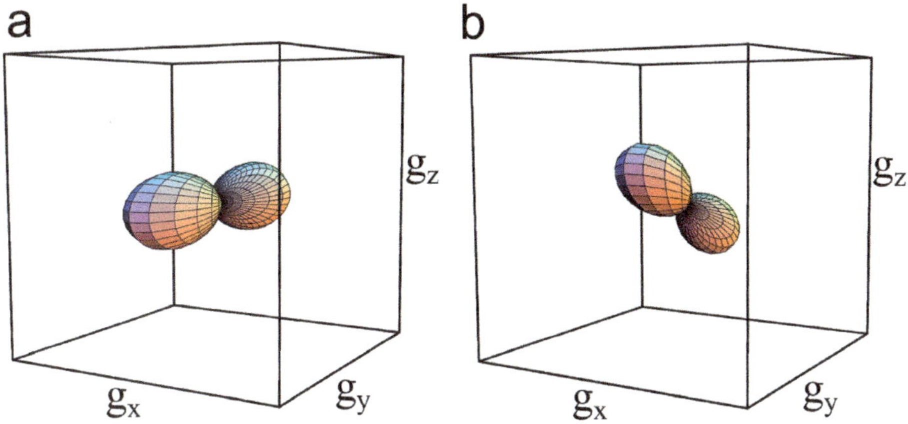

Figure 29. The orientation distribution functions of radical I in *trans* configuration defined by the OF (a) and OP methods (b).

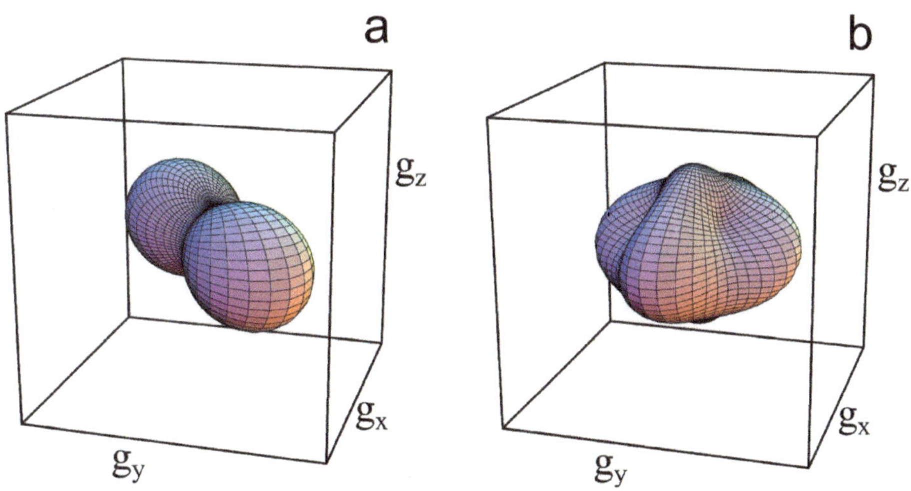

Figure 30. The orientation distribution functions of radical I in *cis* configuration defined by the OF (a) and OP methods (b).

Unfortunately, the OP method has some other disadvantages. One of them is the assumption of Boltzmann distribution of the particles in the field of oriented potential. Boltzmann law is fulfilled locally in most cases. However, each probe molecule is in the surrounding that differs from the surrounding of other molecules by direction and magnitude of potential. As a result, the orientation distribution function as a whole, in general case, cannot be described by simple Boltzmann formula. This drawback becomes a serious obstacle to the application of the model when the local directors in liquid crystal do not coincide with the sample director, for example, when the clear domain structure or cholesteric order exists in the medium. The OP method can be applied for analysis of polydomain samples only in the case of statistical (chaotic) orientation of the domains. Such approach is called microscopic order–macroscopic disorder (MOMD).

The most essential weakness of the OP method is the interdependence of determined parameters. The coefficients of rotation diffusion and the characteristics of ordering potential are not independent when single EPR spectrum is treated. As a result, few different optimal set of parameters produce almost the same calculated spectra. Hence, it is impossible to determine unambiguously the rotation diffusion coefficients and orientation parameters by means of simulation of a single EPR spectrum. This problem is solved to some extent by simultaneous simulation of a series of spectra recorded at various orientations of the sample in the magnetic field. However, such experiment cannot always be performed. For example, recording of the spectrum angular dependence of spin probes in a nematic liquid crystal at the temperature of mesophase is impossible, as the nematic phase is oriented by the magnetic field of the spectrometer. In such case, to record the angular dependence, it is necessary to undertake additional efforts: to align the sample by electric field, to put liquid crystal into the polymer pores, etc. The mutual correlation of different coefficients c_{jm}, which characterize the ordering potential, prevents the reliable determination of the order parameters with a rank more than two even when the set of spectra with different sample orientation is simulated.

5. Conclusions

The data presented in this chapter lead to the following conclusions. Owing to the modern computing technique that allows simultaneous simulation of the set of EPR spectra, some methods for extraction of the quantitative information from the spectra were developed. These methods broaden significantly the area of application of the spin probe technique and make it possible to investigate subtle features of the structure and molecular dynamics of various materials. On the other hand, it is clear that possibilities opened are not used quite effectively.

First of all, it is necessary to note that at present, there is no method for reliable distinction of the influences of rotation and orientation of the radicals on the shape of EPR spectra. One more problem that is close to solution by means of the spin probe method is the determination of the high-rank order parameters. There is no theoretical prohibition for determination of these parameters by the EPR method. However, the existing approach based on the ordering potential is unable to provide reliable values of high-rank order parameters. The direct expansion of the orientation function in a series of generalized spherical harmonics (a model-free technique) overcomes this drawback but is unable to use the spectra recorded at high molecular mobility. One more basic problem in the field is the elaboration of the accurate mathematical criteria for calculation of errors that appear in the course of solving the inverse problem during the spectra simulation.

In our opinion, insufficient attention is paid to consideration of the concentration broadening of the EPR spectra. Whereas the simulation of the exchange-broadened spectra is possible in the program package [4], the description of the spectral line broadening in the case of dipole-dipole interaction of radicals and, moreover, in the cases of dipole-dipole and exchange interactions is impossible. Elaboration of appropriate methods would be useful for determination by means of spin probe technique of not only orientational but also translational order of liquid crystal materials, biological membranes, and films. In such case, one could see the whole spectrum of intermediate states appearing during formation of one or another ordered phase.

We hope that the methods presented in this chapter and the proposed open-code computer programs will serve to develop more complete methods of analysis of the EPR spectra of spin probes.

Author details

Andrey Kh. Vorobiev and Natalia A. Chumakova
Department of Chemistry, M.V. Lomonosov Moscow State University, Leninskie Gory, Moscow, Russia

Acknowledgement

The authors acknowledge the financial support of the Russian Foundation for Basic Research (grant no. 09-03-00900-a). The authors thank Prof. S. Nakatsuji (University of Hyogo) and Prof. R. Tamura (Kyoto University) for the provision of paramagnetic probes and Prof. G. K. Elyashevitch for the provision of stretched porous polyethylene films.

Appendix

Program ODF3

The program ODF3 is elaborated for simulation of the EPR spectra and determination of the spectra parameters by the fitting procedure. The program is a working tool used in our laboratory to test different models, approaches, and algorithms of spectrum simulation. Thus, it is not a finished software product, and it is not optimized by efficiency. Nevertheless, we believe that the program can be useful to reproduce the presented results or treat similar spectral data.

The program is written using Fortran, but as different subprograms were created by different programmers, in different times, and with different aims, the project as a whole does not meet any language standard. Of course, this circumstance produces troubles at compilation and debugging of the program. The program is presented free for any use with the only conditions that the authors are not responsible for any consequences and insist on citation of the source if the results obtained using the program are published.

The program can be loaded from the site
http://www.chem.msu.ru/eng/lab/chemkin/ODF3/

The program allows taking into account the following:

1. Tilt of hfi-tensors relative to g-axes.
2. Forbidden transitions.
3. Convolution of the Gaussian and Lorentzian functions for description of the shape of individual spectral line (Voigt profile).
4. Anisotropy of the line widths described by the second-rank tensors of the Gaussian and Lorentzian line widths.
5. Tilt of the Gaussian and Lorentzian tensors relative to the g-axes.
6. Stochastic rotational oscillations of paramagnetic molecules with limited amplitude and high frequency (quasi-librations).
7. Tilt of the libration axes relative to the g-axes.
8. The orientation distribution functions of the paramagnetic probe.
9. Up to 5 magnetic nuclei.
10. Up to 5 different paramagnetic centers.

The EPR spectra are computed in accordance with explicit formulas presented in [81]. Two types of spectrum calculation can be performed. Both types of calculation use the Hamiltonian within the perturbation theory of the second order. The first type of calculation assumes the coincidence of g-tensor frame with hfi-tensor frame. The second type of calculation takes into account any tilt of hfi-tensors relative to g-axes and forbidden transitions.

The formats for the input-output files are described in the document ODF3.pdf that can be loaded from the same site with the examples of the program applications.

Program esrD

The purpose of the program esrD is the treatment of experimental EPR spectra and preparation of spectra to the fitting procedure. This preparation ordinarily involves removal of the unnecessary tails, subtraction of the base line or base spectrum, double integration, normalizing of area under the spectrum, etc. All these procedures can be carried out using the standard software. On the other hand, when the spectra were recorded in different and possibly irregular points of the magnetic field, the mentioned operations were rather time-consuming. The program esrD allows carrying out necessary operations with the set of such spectra at once. It is useful as well for visual comparison of experimental and calculated spectra.

The program is presented free for any use with only the conditions that the authors are not responsible for any consequences and insist on the citation of the source if the results obtained using the program are published. The program can be loaded from the site http://www.chem.msu.ru/rus/lab/chemkin/esrD/.

6. References

[1] Buchachenko A.L, Wasserman A.M (1973) Stable Radicals, Moscow: Khimiya. 410 p. (in Russian).

[2] Robinson B.H, Tomann H, Beth A.H, Fajer P, Dalton L.R (1985) In EPR and Advanced EPR Studies of Biological Systems. Dalton L.R, editor. Boca Raton: C.R.C.Press, pp. 68-72.

[3] Kuznetsov A.N (1976) Spin Probe Method. Theory and Application. Moscow: Nauka. 210 p. (in Russian).

[4] Schneider D, Freed J (1989) Calculation of slow motion magnetic resonance spectra: a user`s guide in: Biological magnetic resonance. In: Berliner L.J, Reuben J, editors. Spin Labeling: Theory and Application. New York: Plenium Press. pp. 1– 76.

[5] Budil D.E, Lee S, Saxena S, Freed J.H (1996) Nonlinear-Least-Squares Analysis of Slow-Motion EPR Spectra in One and Two Dimensions Using a Modified Levenberg–Marquardt Algorithm. J. Magn. Reson. A 120: 155-189.

[6] Xu D, Budil D.E, Ober C.K, Freed J.H (1996) Rotational diffusion and order parameters of a liquid crystalline polymer studied by ESR: molecular weight dependence. J. Phys. Chem. 100 (39): 15867-15872.

[7] Polimeno A, Freed J (1995) Slow motional ESR in complex fluids: the slowly relaxing local structure model of solvent cage effects. J. Phys. Chem. 99: 10995-11006.

[8] Earle K.A, Moscicki J.K, Polimeno A, Freed J (1997) A 250 GHz ESR study of o-terphenyl: dynamic cage effects above Tc. J. Chem. Phys. 106 (24): 9996-10015.

[9] Barnes J.P, Freed J.H (1998) Dynamics and Ordering in Mixed Model Membranes of Dimyristoylphosphatidylcholine and Dimyristoylphosphatidylserine: A 250-GHz Electron Spin Resonance Study Using Cholestane. Biophysical Journal 75: 2532-2546.

[10] Lou Y, Ge M, Freed J.H (2001) A multifrequency ESR study of the complex dynamics of membranes. J. Phys. Chem. B 105 (45): 11053-11056.

[11] Dzikovski B, Tipikin D, Livshits V, Earle K, Freed J (2009) Multifrequency ESR study of spin-labeled molecules in inclusion compounds with cyclodextrins. Phys. Chem. Chem. Phys. 11: 6676-6688.

[12] Nayeem A, Rananavare S.B, Sastry V.S.S, Freed J.H (1989) Heisenberg spin exchange and molecular diffusion in liquid crystals. J. Chem. Phys. 91: 6887-6905.

[13] Salikhov K.M (2010) Contributions of Exchange and Dipole–Dipole Interactions to the Shape of EPR Spectra of Free Radicals in Diluted Solutions. Appl. Magn. Reson. 38: 237-256.

[14] Freed J.H (1966) On Heisenberg Spin Exchange in Liquids. J.Chem.Phys. 45: 3452

[15] Freed J.H (1967) Theory of saturation and double resonance effects in electron spin resonance spectra. II. Exchange vs. dipolar mechanisms. J.Phys Chem. 71: 38-51.

[16] D.Marsh (1989) Experimental methods in spin-label spectral analysis In: Berliner L.J, Reuben J, editors. Spin Labeling: Theory and Application. New York: Plenium Press. pp. 255-304.

[17] Chumakova N.A, Pergushov V.I, Vorobiev A.Kh, Kokorin A.I (2010) Rotational and Translational Mobility of Nitroxide Spin Probes in Ionic Liquids and Molecular Solvents. Appl. Magn. Res. 39: 409-421.

[18] Molin Yu.N, Salikov K.M, Zamaraev K.I (1980) Spin Exchange. Principles and Applications in Chemistry and Biology. Berlin; New York: Springer-Verlag, 1980. 242 p.
[19] Kokorin A.I in Method of Spin Labels and Probes: Problems and Perspectives (Nauka, Moscow, 1986) pp. 61-78
[20] Kurban M.R, Peric M, Bales B.L (2008) Nitroxide spin exchange due to re-encounter collisions in a series of n-alkanes. J. Chem. Phys. 129: 064501.
[21] Kurban M.R (2009) Experimental correlation of nitroxide recollision spin exchange with free volume and compressibility in alkane and aromatic compounds. J. Chem. Phys. 130: 104502.
[22] Kurban M.R (2011) A study of the relation between translational and rotational diffusion through measurement of molecular recollision. J. Chem. Phys. 134: 034503.
[23] Chumakova N.A, Nikitina V.A, Pergushov V.I (2012) Russ.J.Phys.Chem. In press.
[24] Khairy K, Budil D, Fajer P (2006) Nonlinear-least-squares analysis of slow motional regime EPR spectra. J. Magn. Reson. 183: 152–159.
[25] Dennis J.E, Gay D.M, Welseh R.E (1981) An Adaptive Nonlinear Least-Squares Algorithm. ACM Transactions on Mathematical Software 7(3): 348-383.
[26] Abragam A, Bleaney B. (1970) Electron paramagnetic resonance of transition ions. Oxford: Clarendon press. 911 p.
[27] Abramowitz M, Stegun I.A Eds. (1964) Handbook of Mathematical Functions. National Bureau of Standards 1050 p.
[28] Lebedev Ya.S, Grinberg O.Ya, Dubinsky A.A, Poluektov O.G (1992) Investigation of Spin Labels and Probes by Millimeter Band EPR. In: Zhdanov R.I, editor. Bioactive Spin Labels. Berlin, Heidelberg, New York, London, Paris, Tokyo, Hong Kong, Barcelona, Budapest: Springer-Verlag.
[29] Seber G.A.F, Wild C.J (1989) Nonlinear regression. New York: Wiley. 768 p.
[30] Freed J.H (1976) Theory of slow tumbling ESR spectra for nitroxides. In: Berliner L.J, editor. Spin Labeling: Theory and applications. New York: Academic Press. pp. 53-132.
[31] Chernova D.A, Vorobiev A.Kh (2009) Temperature Dependence of ESR Spectra of Nitroxide Spin Probe in Glassy Polymers. Journal of Polymer Science: Part B: Polymer Physics. 47: 563–575.
[32] Redfield A.G (1966) The theory of relaxation processes. Adv. Magn. Res. 1: 1–32.
[33] Kuznetsov A.N, Wasserman A.M, Volkov A.U, Korst N.N (1971) Determination of rotational correlation time of nitric oxide radicals in a viscous medium. Chem.Phys.Lett. 12(1): 103-105.
[34] Kovarski A.L (1997) Molecular dynamics of Additives in Polymers. Utrecht: VSP. 276 p.
[35] Polanszek C.F, Freed J.H (1975) Electron spin resonance studies of anisotropic ordering, spin relaxation, and slow tumbling in liquid crystalline solvents. J Phys Chem. 79: 2283–2306.
[36] Meirovitch E, Nayaman A, Freed J.H (1984) Analysis of protein-lipid interactions based on model simulations of electron spin resonance spectra. J Phys Chem. 88: 3454–3465.

[37] Meirovitch E, Freed J.H (1984) Analysis of slow-motional electron spin resonance spectra in smectic phases in terms of molecular configuration, intermolecular interactions, and dynamics. J Phys Chem. 88: 4995–5004.
[38] Chernova D.A, Vorobiev A.K (2011) Molecular Mobility of Nitroxide Spin Probes in Glassy Polymers: Models of the Complex Motion of Spin Probes. Journal of Applied Polymer Science.121(1): 102–110.
[39] Bercu V, Martinelli M, Massa C.A, Pardi L.A, Rossler E.A, Leporini D. (2008) Anomaly of the rotational nonergodicity parameter of glass formers probed by high field electron paramagnetic resonance. J Chem Phys.129: 081102.
[40] Maresch G.G, Weber M, Dubinskii A.A, Spiess H.W (1992) 2D-ELDOR detection of magnetization transfer of nitroxides in disordered solid polymers. Chem Phys Lett. 193: 134-140.
[41] Dzuba S.A, Tsvetkov Y.D, Maryasov A.G. (1992) Echo-induced EPR spectra of nitroxides in organic glasses: model of orientational molecular motions near equilibrium position. Chem Phys Lett. 188: 217-222.
[42] Dzuba S.A. (1992) Echo-induced EPR spectra of nitroxides: Study of molecular librations. Pure Appl Chem. 64(6): 825-831.
[43] Dzuba S.A (1996) Libration motion of guest spin probe molecules in glassy media. Phys Lett A. 213(1-2): 77-84.
[44] Dzuba S.A (2000) Libration motion of guest spin probe molecules in organic glasses: CW EPR and electron spin echo study. Spectrochim Acta A. 56(2): 227-234.
[45] Van S. P, Birrell G. B, Griffith O. H (1974) Rapid anisotropic motion of spin labels. Models for motion avaraging of the ESR parameters. J Magn Reson. 15: 444–459.
[46] Griffith O, Jost P. (1976) Instrumental aspects of spin labeling. In: Spin Labeling, Theory and Applications. Berliner L.J, editor. New York: Academic Press. pp 251–272.
[47] Chernova D.A, Vorobiev A.Kh (2009) Molecular Mobility of Nitroxide Spin Probes in Glassy Polymers. Quasi-Libration Model. Journal of Polymer Science: Part B: Polymer Physics. 47: 107–120.
[48] Vorobiev A.K, Gurman V.S, Klimenko T.A (2000) Rotational mobility of guest molecules studied by method of oriented spin probe. Phys Chem Chem Phys. 2(3): 379–385.
[49] Vorob'ev A.K, Gurman V.S, Klimenko T.A (2000) Models of slow rotational mobility of paramagnetic probes in glassy media. Russ Chem Bull. 49: 1059–1067.
[50] Paschenko S.V, Toropov Y.V, Dzuba S.A, Tsvetkov Y.D, Vorobiev A.K (1999) Temperature dependence of amplitudes of libration motion of guest spin-probe molecules in organic glasses. J Chem Phys. 110: 8150–8154.
[51] Tenhu H, Rantala J, Samarinov B, Timofeev V (1998) Phase transitions of poly(octyl isocyanate). Polymer. 39: 4057–4063.
[52] Timofeev V, Samarinov B (1995) Dynamics of macromolucele spin-labelled side-chain groups by electron paramagnetic resonance spectra simulation. J Chem Soc Perkin Trans. 2: 2175–2181.

[53] Veksli Z, Andreis M, Rakvin B (2000) ESR spectroscopy for the study of polymer heterogeneity. Prog. Polymer Sci. 25(7): 949-986.

[54] Dzuba S.A, Salnikov E.S., Kulik L.V (2006) CW EPR, echo-detected EPR, and fieldstep ELDOR study of molecular motions of nitroxides in o-terphenyl glass: Dynamical transition, dynamical heterogeneity and beta-relaxation. Appl. Magn. Reson. 30(3-4): 216-222.

[55] Cameron G.G, Miles I.S, Bullock A.T (1987) Distribution in correlation times for rotational diffusion of spin probes in polymers. Br. Polymer J. 19: 129-134.

[56] Faetti M, Giordano M, Leporini D, Pardi L (1999) Scaling analysis and distribution of the rotational correlation times of a tracer in rubbery and glassy poly(vinyl acetate). Macromolecules. 32(6): 1876 -1882.

[57] Shantarovich V.P, Kevdina I.B, Yampolskii Y.P, Alentiev A.U (2000) Positron Annihilation Lifetime Study of High and Low Free Volume Glassy Polymers: Effects of Free Volume Sizes on the Permeability and Permselectivity. Macromolecules. 33: 7453–7466.

[58] Bercu V, Martinelli M, Massa C.A, Pardi L.A, Leporini D (2005) Signatures of the fast dynamics in glassy polystyrene: First evidence by high-field Electron Paramagnetic Resonance of molecular guests. J Chem Phys. 123: 174906.

[59] Saalmueller J.W, Long H.W, Maresch G.G, Spiess H.W (1995) Two-Dimensional Field-Step ELDOR. A Method for Characterizing the Motion of Spin Probes and Spin Labels in Glassy Solids. J Magn Reson A. 117(2): 193-208.

[60] Saalmueller J.W, Long H.W, Volkmer T, Wiesner U, Maresch G.G, Spiess H.W (1996) Characterization of the motion of spin probes and labels in amorphouspolymers with two-dimensional field-step ELDOR. J Polym Sci Part B Polym Phys. 34(6): 1093-1104.

[61] Bogdanov A.V, Vorobiev A.Kh (2011) Rotational mobility and rate of photoisomerization of spin-labeled azobenzenes in glassy polystyrene. Chemical Physics Letters 506: 46–51.

[62] Nakatsuji S, Fujino M, Hasegawa S, Akutsu H, Yamada J, Gurman V.S, Vorobiev A.Kh (2007) Azobenzene Derivatives Carrying A Nitroxde Radical. J.Org.Chem. 72: 2021-2029.

[63] Friesner R., Nairn J.A, Sauer K (1979) Direct calculation of the orientational distribution function of partially ordered ensembles from the EPR line shape. J. Chem. Phys. 71(1): 358-365.

[64] Friesner R., Nairn J.A (1980) A general theory of the spectroscopic properties of partially ordered ensembles. I. One vector problems J. Chem. Phys. 72 (1): 221-230.

[65] Shimada S., Hori Y., Kashiwabara H (1985) Structure of Peroxy Radicals Trapped in Irradiated Isotactic Polypropylene and Molecular Disorder of the Polymer Chain, Related with Hydrogen Abstraction Reaction of the Radicals. Macromol. 18: 170-176.

[66] Shimada S., Hori Y., Kashiwabara H (1988) Distribution of Molecular Orientation and Stability of Peroxy Radicals in the Noncrystalline Region of Elongated Polypropylene. Macromol. 21: 979-982.

[67] Swartz J.C, Hoffman B.M, Krizek R.J, Atmatzidis D.K (1979) A General Procedure for Simulating EPR Spectra of Partially Oriented Paramagnetic Centers. J. Magn. Res. 36: 259-268.

[68] Hentschel R, Schilitter J, Sillescu H, Spiess H.W (1978) Orientational distributions in partially ordered solids as determined from NMR and ESR line shapes. J. Chem. Phys, 68: 56-66.

[69] Ovchinnikov I.V, Konstantinov V.N (1990) EPR line shape of orientationally ordered solid systems. In: Zaripov M.M, editor. Radio-spectroscopy of the condensed media. Moscow: Nauka. pp. 90-110 (in Russian)

[70] Ajtai K, French A.R, Burghardt Th.P (1989) Myosin cross-bridge orientation in rigor and in the presence of nucleotide studied by electron spin resonance. Biophys. J. 56: 535-541.

[71] Burghardt Th.P, Thompson N.L (1985) Model-independent electron spin resonance for measuring order of immobile components in a biological assembly. Biophys. J. 48: 401-409.

[72] Caldeira J, Figueirinhas J.L, Santos C, Godinho M.H (2004) EPR spectroscopy of protein microcrystals oriented in a liquid crystalline polymer medium. J. Magn. Reson. 170: 213-219.

[73] Vorobiev A.Kh, Chumakova N.A (2005) Determination of orientation distribution function of anisotropic paramagnetic species by analysis of ESR spectra angular dependence. J. Magn. Reson. 175: 146-157.

[74] Vorob'ev A.Kh, Chumakova N.A (2005) Determination of molecular orientation distribution of a stable paramagnetic probe in oriented 4-cyano-4'-n-penthylbphenyl. Russ.Chem.Bull. 54(1): 195-200

[75] Chumakova N.A, Vorobiev A.Kh, Ikuma N, Uchida Y, Tamura R (2008) Magnetic characteristics and orientation of a new nitroxide radical in an ordered matrix, Mendeleev. Commun. 18: 21-23.

[76] Chumakova N.A, Pomogailo D.A, Yankova T.S, Vorobiev A.Kh (2011) The novel stable nitroxide radicals as perspective spin probes for study of orientation order of liquid crystals and polymers. Mol. Cryst. Liq. Cryst. 540: 196-204.

[77] Yankova T.S, Chumakova N.A, Vorobiev A.Kh (2008) Orientation distribution function of HO2· radicals ordered by light irradiation. Appl. Magn. Res. 33: 117-126.

[78] Elyashevich G, Kozlov A, Moneva I (1998) Study of polyethylene orientation in the course of porous structure formation. J. Polym. Sci. series B. 40: 71-74 [Vysokomol. Soyed. 40: 483-486, in Russian].

[79] Michl J, Thulstrup E.W (1986) Spectroscopy with polarized light: Solute Alignment by Photoselection, in Liquid Crystals, Polymers and Membranes. New York: VCH Publishers. 573 p.

[80] Yankova T.S, Chumakova N.A, Vorob'ev A.Kh (2011) Photoinduced Orientational Order of Dichloride Anion Radicals, Russian Journal of Physical Chemistry A 85 (4): 695–701.

[81] Zhidomirov G.M., Lebedev Ia.S., Dobriakov C.A., Shtenshneider N.Ia., Chirkov, A.K., Gubanov V.A. (1975) Interpretation of complex EPR sppectra, Moscow: Nauka, , 216 p. (in Russian).

Section 2

Experiment

Chapter 5

Spin Labels in the Gel Phase and Frozen Lipid Bilayers: Do They Truly Manifest a Polarity Gradient?

Boris Dzikovski and Jack Freed

Additional information is available at the end of the chapter

1. Introduction

Lipid spin labels containing nitroxide groups at different positions in the fatty acid chain, such as 1-palmitoyl-2-stearoyl-(*n*-doxyl)-*sn*-glycero-3-phosphocholines (*n*-PC spin labels) are a useful and proven tool in lipid research. They have provided important insights into the structure of model and biological membranes, reported on the membrane fluidity, polarity, phase state and presence of microscopic domains[1], accessibility of different depth positions in the lipid bilayer for oxygen and other polar and non-polar paramagnetic compounds[2-4] and protein/lipid interactions [5, 6].

It is generally accepted, that, unlike bulky fluorescent labels[7,8], nitroxides are well incorporated into fluid lipid bilayers[9] and not excluded from them. However, it has been shown by NMR that although the most probable location of the nitroxide group for 5-, 10- and 16- PC spin labels in the fluid POPC membrane corresponds to the fully extended conformation, the distribution is relatively broad and other conformations should also be present[10]. Bent conformations were previously found for doxylstearic acids in monomolecular films[11], water/hydrocarbon emulsion particles[12] and micellar systems[13]. In fluid membranes the fluidity, polarity and accessibility parameters reported by ESR using PC spin labels and *n*-doxylstearic acids are, in general, change monotonically with an increase in *n*[3, 14], although there are indications that the spin label groups on the stearates are located nearer to the membrane exterior than the analogous positions of the unlabeled phospholipid chains[15]. However, in the gel phase, which is characterized by denser chain packing and higher order, the preferential location may be different.

In this chapter we focus on the behavior of PC spin labels in the gel phase and frozen membranes. We show how the superior *g*-factor resolution of HF ESR provides new insights in this behavior and a new look at the vast body of experimental data accumulated with PC spin labels in the last 30 years. In particular, we revisit so-called "polarity profiles" determined from the *g*-factor values and hyperfine splittings of PC spin labels in frozen phospholipid membranes with or without cholesterol and show that these values are affected by a number of factors in the membrane composition, chain packing in the lipid phase and folding properties of the *sn-2* spin labeled chain PC labels rather than reflect gradients of polarity or water content present in the membrane[16].

2. Resolution of different hydrogen-bond states by 240GHz ESR in bulk organic solvents

It has been well established that the g-tensor and hyperfine components of nitroxide radicals are very sensitive to the local environment, to polarity and proticity in particular. In general, changing the local environment of the nitroxide moiety from water (polar) to hydrocarbon (non-polar) causes an increase in the *g*-tensor components and a concomitant decrease in the values of the components of the hyperfine tensor. This effect is most pronounced for the tensor components g_{xx} and A_{zz}. However, the separation of the polarity and proticity effects could complicate the analysis of the ESR spectra. Proticity refers to the propensity to donate hydrogen bonds; whereas aprotic refers to solvents which cannot donate a hydrogen bond. At relatively low frequencies, up to 95GHz, the separation of hydrogen-bonded vs. non hydrogen-bonded states of the nitroxide often relied upon different g versus A plots, discovered for these two states[17]. However, as shown in a recent study using TEMPO, if the correlations are indeed different for TEMPO in protic and aprotic solvents, the difference is rather small[18].

On the other hand, superior g-factor resolution of HF ESR allows for observation of two resolved spectral components corresponding to two (Smirnova et al.[19] at 130GHz) or possibly more (Bordignon et al.[20] at 95, 275 and 360GHz) states of hydrogen-bonding. Two hydrogen bonding states coexisting in frozen deuterated alcohols were previously demonstrated for perdeuterated TEMPONE by X-band ESR[21].

Tables 1 and 2 show the values of A_{zz} and a_{iso} hyperfine splitting and the g_{xx} component of the g-tensor for several nitroxide radicals.

As one sees from Fig.1, whereas at X-band one sees a continuous increase in the ^{14}N hyperfine splitting with change of the local environment from non-polar/aprotic to polar/protic, the 240 GHz ESR shows three distinct values of the g_{xx} parameter. Although the presence/ratio of these components strongly depends on the polarity-proticity of the solvent, there is little variation in the g_{xx} measured for each such component. For all four spin labels studied three distinct components could be detected: (1) "non-polar", as in toluene, DBPh or the minor component in alcohols, (2) "polar", the major component for ethanol and major or minor component in TFE and water/glycerol , depending on the nitroxide used, and (3) "very polar" component observable in TFE and water/glycerol. Although components 2 and 3 cannot be separated at 240GHz as two distinct peaks, their presence is quite obvious (compare Figs.1A-D for different nitroxides). We assign these components to different

Solvent, dielectric constant at room temperature[51]	Oxo-TEMPO	TEMPO	4-Hydroxy TEMPO	3-carboxy-2,2,5,5 tetramethylpyrrolidine - 1-oxyl
Isopentane, 1.8	14.27/**	15.29/68.2	15.15/**	13.92/**
MCH, 2.02	14.315/**	15.32/68.2	15.20/**	13.96/**
Toluene, 2.4	14.49/67.6	15.53/69.3	15.43/69.1	14.22/66.7
DBPh, 6.4	14.61/67.6	15.64/69.4	15.55/69.2	14.37/66.9
Ethanol, 24.3	15.06/70.0	16.21/72.9	16.05/71.0	15.02/70.1
TFE, 26.14	15.63/73.5	17.00/79.7	16.73/77.3	15.77/74.4
Water/Glycerol, 80.4	15.97/73.8	17.19/76.5	16.96/75.5	16.12/73.2

** $2A_{zz}$ value at 77K cannot be reliably determined due to the presence of a singlet-like background.

Table 1. Values of isotropic hyperfine splitting constant a_{iso} and $2A_{zz}$ determined at X-band at 295 and 77K respectively for several solvents. These A_{zz} values were also used to obtain the best fits for the corresponding 240 GHz rigid limit spectra.

Solvent, dielectric constant	Oxo-TEMPO	TEMPO	4-Hydroxy TEMPO	3-carboxy-2,2,5,5 tetramethylpyrrolidine - 1-oxyl
Toluene	2.009438	****	*****	****
DBPH	2.009445	2.010103	2.010129	2.009230
Ethanol	2.009485/**2.008838**	sh/2.009411	2.010040/**2.009456**	2.009250/**2.008533**
TFE	**2.008793**/sh	2.008669	2.009460/**2.009089**	2.008487/sh
Water/Glycerol	**2.008784**/~2.00850	2.008805	2.009460/**2.008951**	2.008532/sh

Table 2. g_{xx} component of the g-tensor determined by 240GHz ESR at 80-85K in several glass forming solvents. A common value of 2.00233 was assigned as g_{zz} for all spin labels and g_{xx} value was accurately determined relative to this g_{zz} value from the corresponding spectral splitting[29, 52]. If two components are present in the spectrum, the component with higher fraction is marked bold. "Sh" denotes the presence of a high/low field component, which manifests itself not as a distinct peak but as a shoulder on the main component.

hydrogen-bonding states of nitroxide radicals and speculate that state 2 corresponds to a single hydrogen bond, while state 3 is double-bonded. Existence of multiple hydrogen bonding to a nitroxide has been predicted theoretically[22-24] and later suggested as an explanation for complex ESR lineshapes observed in spin labeled proteins[20]. Interestingly, the g_{xx} value of the non-hydrogen bonded component for all four spin labels studied shows little dependence on the polarity of the frozen glass-forming solvent (Table 3). This contrasts with some theoretical predictions for the g-factor[25], as well as some room temperature measurements for g_{iso} pointing to a higher g-factor for lower dielectric constants ε[17, 18].

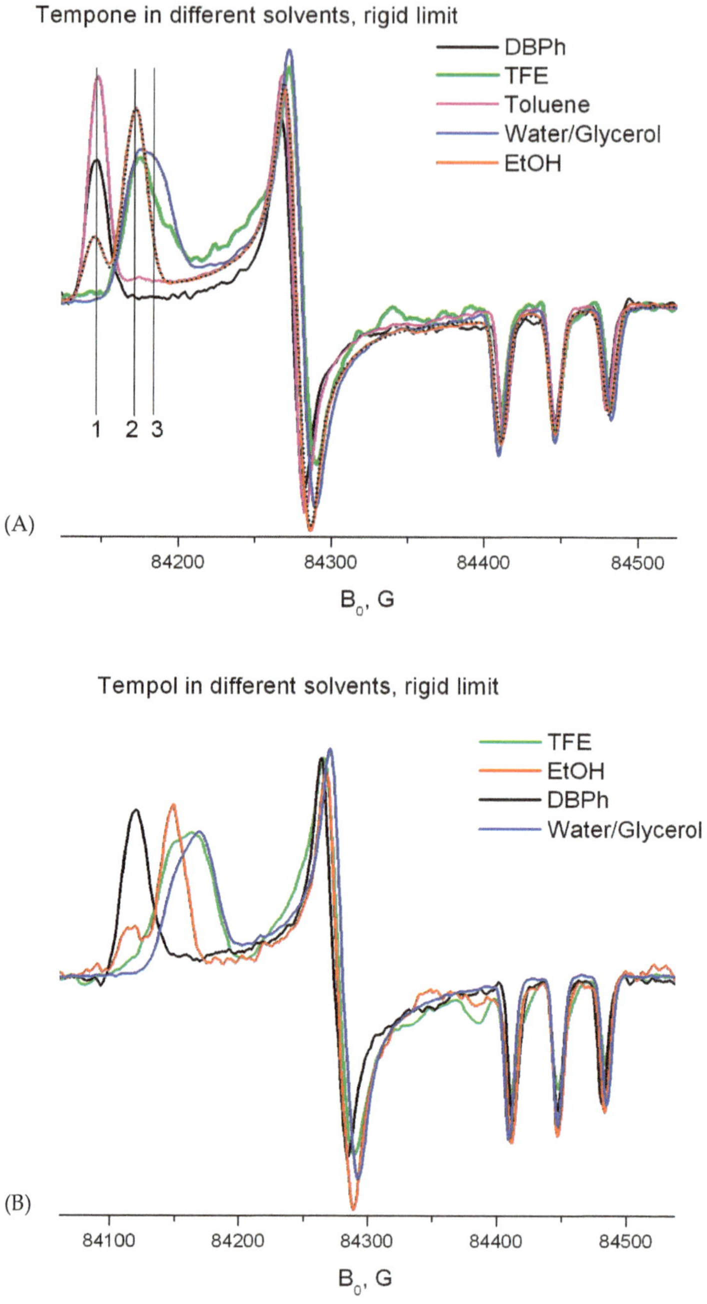
Tempone in different solvents, rigid limit
DBPh
TFE
Toluene
Water/Glycerol
EtOH
1 2 3
(A)
84200
84300
84400
84500
B0, G
Tempol in different solvents, rigid limit
TFE
EtOH
DBPh
Water/Glycerol
(B)
84100
84200
84300
84400
84500
B0, G

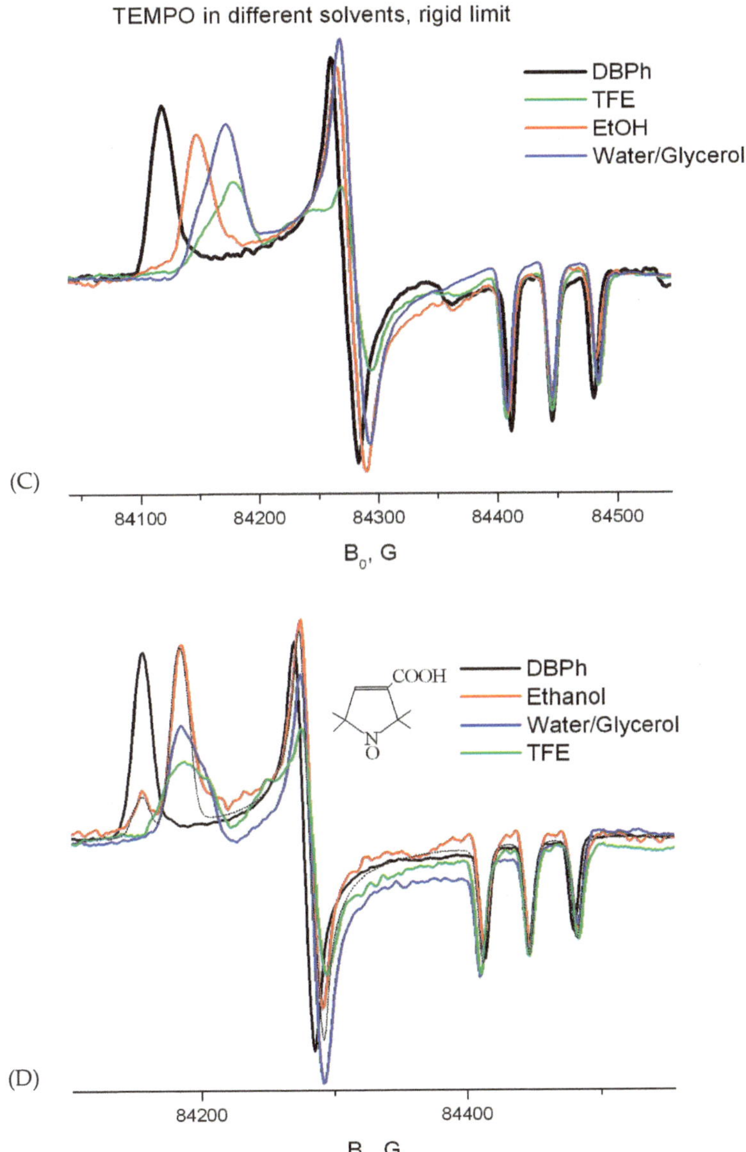

Figure 1. 240 GHz ESR spectra of 4-oxo-TEMPO (TEMPONE) (from [16]) (A), 4-hydroxy-TEMPO (TEMPOL) (B), TEMPO (C) and 2,2,5,5-Tetramethyl-3-pyrrolin-1-oxyl-3-carboxylic acid free radical (D) in a series of glass-forming solvents at 80-85K. Black dotted lines in A and D show two-component rigid simulation of the ethanol spectra. In (A), g values 2.009450, 2.008830 and 2.008500 are noted "1", "2" and "3" respectively.

	DMPC	DMPC/Chol	DPPC	DPPC/Chol
5	68.9	69.7	69.2	70.1
7	69.3	70.1	69.9 (broadening)	69.1
10	69.6	66.6	69.5 (more broadening)	66.5
12	68.4	66.3	67.2	66.4
14	69.3	66.4	66.3	66.8
16	69.3	66.2	66.7	66.5

Table 3. Hyperfine splitting parameter $2A_{zz}$ in DMPC and DPPC membranes with and without cholesterol determined by X-band ESR at 77K. To record the ESR spectra the samples in 1.2 mm ID capillaries after long exposure at 19°C were quickly submerged into liquid nitrogen.

3. Membrane environment

3.1. X-band ESR

The full extent of the 9 GHz spectrum is determined by the largest of the principal values of the ^{14}N hyperfine constant, which is A_{zz}. Hence at X-band, the distance between outer extrema for a well-resolved spectrum in the rigid limit is exactly $2A_{zz}$. This value is known to increase ~ 2G if nitroxide is transferred from a non-polar solvent like hydrocarbons to water and, as mentioned above, can be considered a measure for local polarity. Table.3 shows the $2A_{zz}$ determined by X-band ESR as the outer splitting of the rigid limit spectra at 77K. As seen from the data, DMPC/Cholesterol (Fig. 2B) and DPPC/Cholesterol (Fig. 2D) membranes, consistent with previous observations[3], show abrupt drop in the $2A_{zz}$ value, between *n*-PC positions 7 and 10. However, for DMPC, in the absence of cholesterol, the $2A_{zz}$ value shows little trend throughout the PC spin labels series and remains ~ 69.5G (cf. Fig. 2A), which correspond to a relatively polar environment, more polar than ethanol (see below). DPPC in the absence of cholesterol (Fig. 2C), in general, shows a profile similar to a cholesterol-containing membrane, though the spectra, especially for positions 7-12 show signs of a broad singlet-like component.

The dependencies of relaxation enhancement $\Delta(1/P)$ by 10 mM of $Ni(ClO_4)_2$ on the spin-labeling position for the gel and liquid crystal phase of DMPC are given in Fig. 3. Here P is the relaxation parameter $P=g_e T_1 T_2$, where g_e is the electron gyromagnetic ratio and T_1 and T_2 are corresponding effective relaxation times. In the liquid crystal phase (T=39°C) the interaction with the paramagnetic relaxant decreases with increasing *n*, consistent with an increase in the average immersion depth of the spin label moiety in the membrane. On the contrary, the $\Delta(1/P)$ profile in the gel phase (P_β, T=19°C) is almost flat, with a minor spike at position 10.

3.2. High field/ High frequency ESR

High frequency ESR and PC spin labels were previously used to study polarity profiles in phospholipid membranes[14, 26, 27]. For example, in a detailed ESR study at 250GHz 5,7,10,12,14 and 16 PC were studied in DPPC and DPPC/gramicidin systems[26]. It was found that in pure DPPC most spins are strongly aggregated and the spectrum consists mostly (especially for

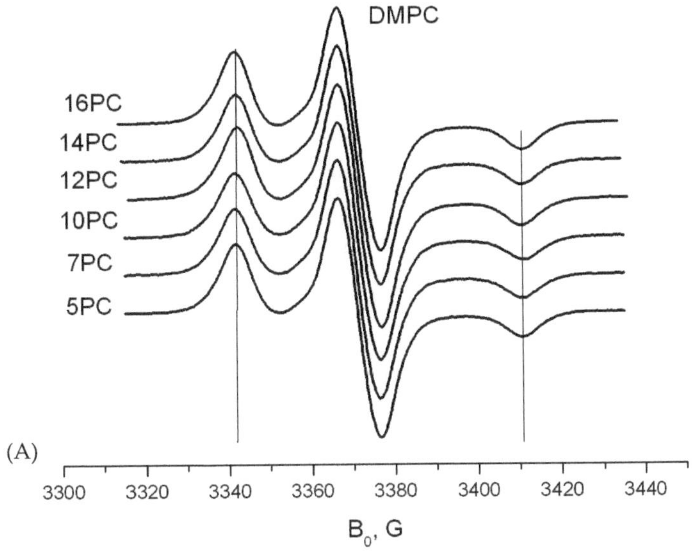
DMPC
16PC
14PC
12PC
10PC
7PC
5PC
(A)
3300
3320
3340
3360
3380
3400
3420
3440
B0, G

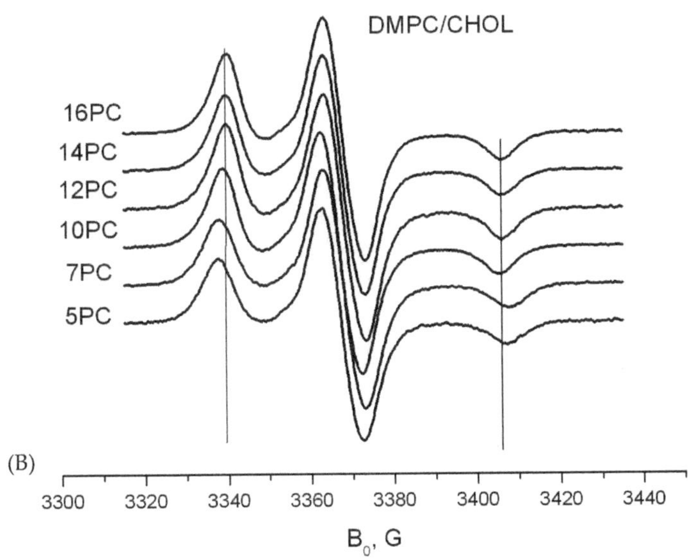
DMPC/CHOL
16PC
14PC
12PC
10PC
7PC
5PC
(B)
3300
3320
3340
3360
3380
3400
3420
3440
B0, G

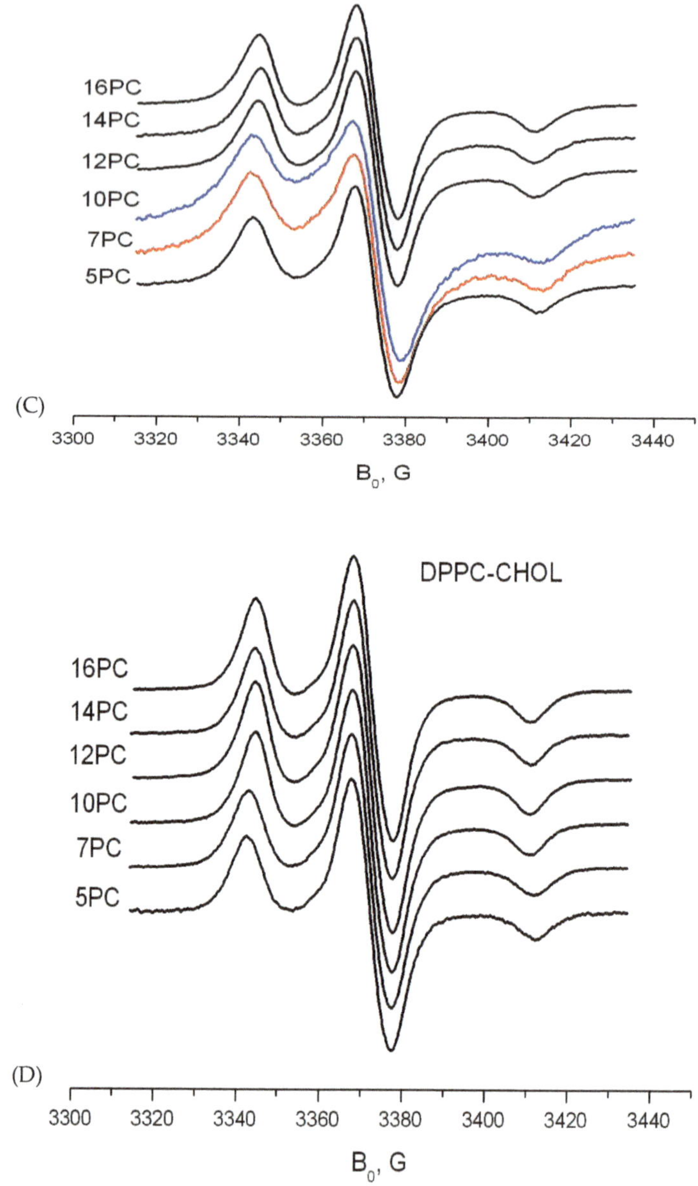

Figure 2. 9 GHz ESR spectra of PC spin labels in DMPC (A) (from [16]), DMPC/30% Cholesterol (B), DPPC (C) and DPPC/30% Cholesterol (D) at 77K. In (C) spectra of 7PC and 10 PC with most broadening are shown in color.

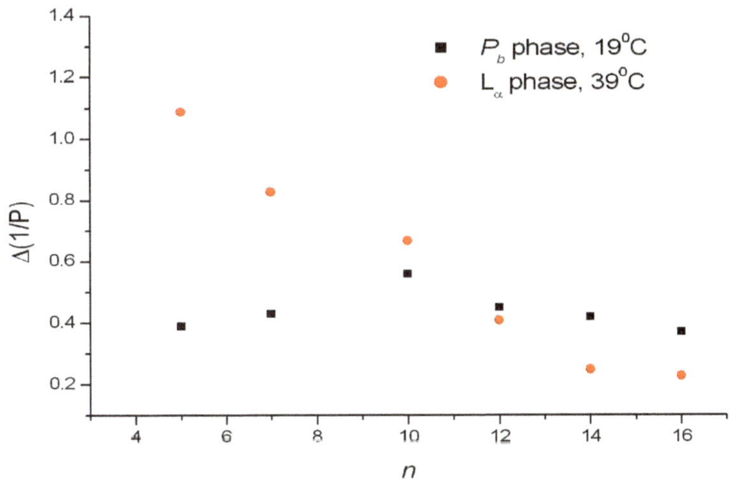

Figure 3. Relaxation enhancement from 10 mM of $Ni(ClO_4)_2$ introduced into the water phase of lipid DMPC dispersions as a function of *n* in the P_β (19°C) and L_α (39°C) lipid phases. An almost flat accessibility profile in the gel might indicate similar average membrane immersion depth for the nitroxide moieties of all studied PC labels (from [16]).

7-12 PCs) of a singlet-like signal. Although the g-factor values for the resolved rigid-limit component seems to indicate increasing immersion of the nitroxide into the hydrophobic core of the membrane with increasing *n*, it is difficult to obtain any information on the location of aggregates which manifest themselves in the broad component. However, addition of cholesterol prevents such aggregation and yields in better resolved spectra[27].

3.2.1. Frozen DMPC/DPPC membranes with cholesterol: Partition-like depth distribution of spin labels

Several W-band (94GHz) studies by Marsh and coworkers[14, 27, 28] on DMPC/Cholesterol membranes utilized all *n*-PC spin labels in the 4-16 range (except 15-PC). Based on the g_{xx} values detailed polarity profiles for these systems were suggested. These polarity profiles appeared to be similar to the polarity profiles previously obtained by rigid limit X-band ESR from the hyperfine splitting[3] values. However, a close inspection of the spectra of [27] shows two partially resolved components, which likely correspond to hydrogen-bonded and non-hydrogen-bonded states of the nitroxide radical. These two components are discernible not only for the area of abrupt "polarity change" (PC 5-10), but also for PC14-16. Higher resolution of 240GHz ESR allows for complete separation and identifying these components. Fig.4 A, B show the spectra for DMPC and DPPC in the presence of cholesterol. The ESR spectra in the two different lipids are very similar, nearly identical. There are two components discernible for n≥7, which are completely resolved for n≥10, with g_{xx}= 2.009435 and g_{xx}= 2.008820 (with g_{zz} taken as 2.00233, see [29]). The two *g* values are nearly the same for all *n*. For 5-PC only the polar component is present.

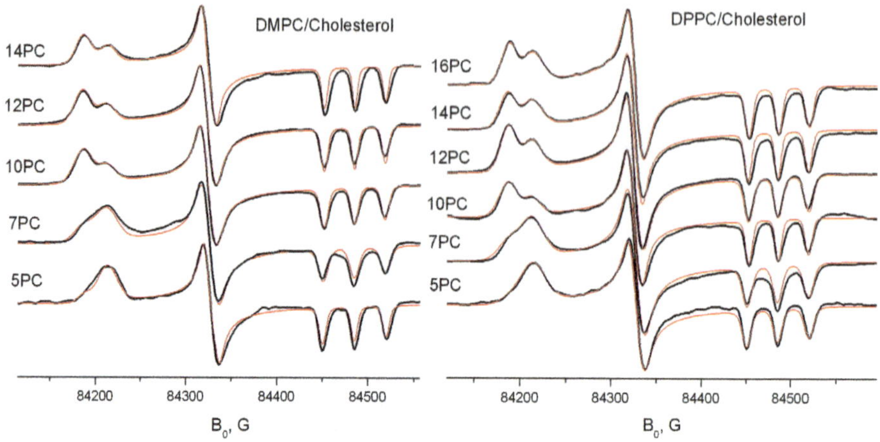

Figure 4. Rigid-limit 240 GHz ESR spectra of *n*-PC spin labels in DMPC and in DPPC containing 30% mol cholesterol. T=80-82K. The samples are gradually frozen in the flow of gaseous nitrogen. Red lines show simulations of these spectra using two components with the g_{xx} = 2.009435±0.000005 and 2.008820±0.000005 for all spectra. g_{yy} and g_{zz} are taken 2.006100 and 2.00233 for both components (from [16]).

Can these two components just reflect a water penetration profile? It has previously been assumed that polarity profile reported by spin labels for fluid and frozen membranes containing cholesterol in general follows the membrane water penetration profile[23, 30, 31]. However, the exact features of the ESR spectrum at low temperature may be defined by interplay of this water penetration and the flexibility of spin-labeled lipid chains, see below. While, due to the short acyl tether, the nitroxide of 5PC is always located in the area with higher water content and shows only hydrogen-bonded component, longer acyl tethers can reach a less polar area with lower water content. Yet, even for 14 and16PC there is a substantial fraction of hydrogen-bonded component. Moreover, the fraction of the hydrogen-bonded component does not decrease monotonically with an increase in the *n* number. As seen in Fig.4 A, B and Table 4, this fraction experiences a dramatic drop between 7 and 10PC and then gradually increases further for positions 12,14 and 16. Based on these observations and what is currently known about flexibility of spin-labeled lipids[10], we believe that the hydrogen-bonded component mainly corresponds to bent conformations of the spin-labeled lipid chain, rather than penetration of water molecules to the middle of the membrane and forming a hydrogen bond with the nitroxide moiety of the PC spin label in a fully extended conformation.

Indeed, theoretical estimates[32] show that the concentration of water in the middle of the DPPC membrane is ~ 1 mM. It should be even lower for membranes containing cholesterol, since cholesterol is known to substantially decrease water permeability across lipid membranes[33, 34]. For the equilibrium constant between hydrogen bonded/unbonded forms of 16-sasl in toluene - trifluoroethanol mixtures Marsh[30] gives a value ~ $1M^{-1}$. This value is determined from the isotropic hyperfine splitting at room temperature. Our estimates for this constant for various nitroxides based on measurements of $2A_{zz}$ values in ethanol –

PC label	Ratio HB/non-HB in DMPC	Ratio HB/non-HB in DPPC
5	~6	>8
7	1.8	2.3
10	0.32	0.36
12	0.45	0.62
14	0.55	0.8
16	---	0.8

Table 4. Ratio of hydrogen bonded, HB (g_{xx}= 2.00882) and non-hydrogen bonded components (g_{xx}=2.009435) determined from rigid-limit simulations of n-PC spin labels in DMPC and DPPC containing 30% mol cholesterol.

DBPH mixtures by 9GHz ESR at 77K or on the ratio of hydrogen-bonded and non-hydrogen bonded components in frozen ethanol (~ 17M hydroxyl concentration) measured by 240GHz ESR give values of the same order, between 0.2M^{-1} and 0.5M^{-1}. A 1:1 ratio of the two spectral components for a nitroxide located in the middle of the bilayer would thus yield a water concentration in the membrane core of ~0.5M, about three orders of magnitude higher than expected from theoretical estimates. Also, consistent with the above estimates of the equilibrium constant, in our test experiments we did not see any appearance of a hydrogen bonded component for nitroxides dissolved in frozen nearly saturated (~16 mM) solutions of water in toluene.

And finally, a strong point in favor of considering bent conformations of *n*-PC spin labels is the non-monotonic dependence of the hydrogen-bonded fraction on *n* (Table 4). In a fluid membrane the membrane depth distribution of nitroxide moieties for each spin-labeled lipid correlates to its *n*-value[10]. Higher *n* show deeper average immersion although the distribution is broad and even high *n* numbers show a substantial fraction of conformations with the nitroxide touching the carbonyl area. In the much denser packed gel phase the situation can be different and the spin labels may prefer defects in the lipid structure[35, 36]. One of the areas with such defects is just above the cholesterol rings and this should correspond to a hydrophobic local environment. It can be reached by the nitroxide of 10PC, but not 7PC. This would explain the jump in the fraction of non-polar component between 7 and 10PC, while further decrease in this fraction could be attributed to U-shaped conformations for higher values of *n*. These conformations put the nitroxide moiety back to the surface region with high water content, while the hydrocarbon chain mostly remains located in the hydrophobic part of the membrane.

Indeed, if we assume that the spin-labeled *sn*-chain takes on mostly the fully extended conformation, we would observe a similar jump after the nitroxide moiety reaches the hydrophobic core of the membrane, but with further increase in the *n* number the hydrogen-bonded fraction would decrease and quickly disappear.

Another limiting case suggests that the acyl-chain hydrocarbon tether connecting the nitroxide to the lipid head group can take all possible coil conformations of the chain. It will put the possible position of the nitroxide of an *n*-PC spin label anywhere within a half sphere with a radius of the all-stretched conformation for the nitroxide tethers. This half sphere would rest on the membrane surface (cf. Fig.5). The ratio of hydrogen bonded/unbonded components would be opposite to the volume ratio of the spherical cap, which is cut off by the border of the hydrophobic core of the membrane, and the rest of the half sphere. If the cap height is half of the sphere radius, the ratio will be 2.2, similar to what we see for 7PC (Table 4). This observation can be then used to set (in a rather arbitrary fashion) the cutoff of the hydrophobic core at "3.5" PC, half of the acyl tether for 7PC in the fully extended conformation. The predictions of the components ratio obtained with this cutoff value by further applying the formula for the spherical cap are given in Table 5.

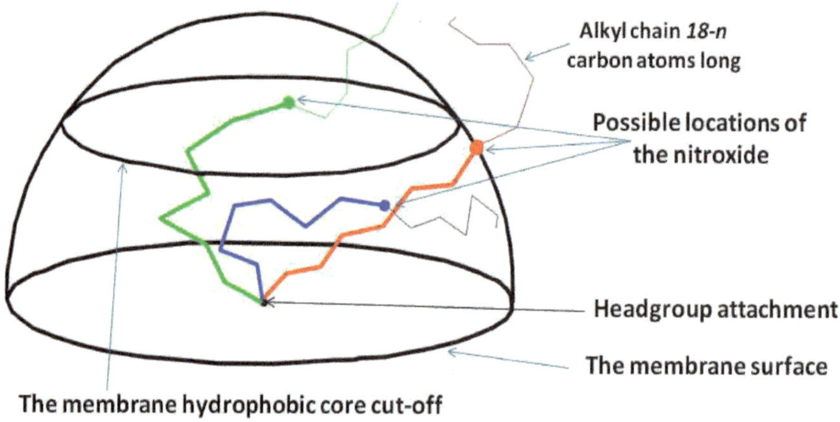

Figure 5. A model of random distribution of conformations of the spin label tethers for 7PC. It puts the spin label in some random position within the half-sphere. The spherical cap is the intersection of the area available for the nitroxide and the hydrophobic core of the membrane.

n-PC	5	7	10	12	14	16
HB/non-HB	7.2	2.2	1.01	0.74	0.58	0.48

Table 5. Estimates for the ratio of hydrogen bonded and non-hydrogen bonded components based on the simple model shown in Fig. 12. The location of the nitroxide moiety is considered in the hydrophobic core of the membrane if the distance to the membrane surface is more than half of the acyl tether for 7PC in the fully extended conformation ("3.5 PC").

Although this model is qualitatively better when compared to Table 4, it does not reproduce the observed abrupt drop between positions 7 and 10. Also, no model assuming some random distribution of nitroxide depth position would reproduce the decrease in the hydrophobic fraction with further increase in *n* beyond 10.

To better explain the observed effects one could assume some set of preferential depth positions in the membrane to be occupied by the nitroxide ring. These positions may be areas of defects in the membrane structure, which can more easily accommodate a structure-disturbing nitroxide moiety than areas with compact alignment of hydrocarbon chains. One of the areas of such defects could start above the end of the rigid fused-ring system of cholesterol[37], which is about the level reachable by *n*-PC labels starting from *n*=9. Once the tether length is sufficient, nitroxides start populating these favorable locations causing the change in the component ratio. However, the preference of the nitroxide ring for the location in the hydrophobic part of a DMPC/Cholesterol or DPPC/Cholesterol membrane apparently does not completely overwhelm its affinity to some sites close to the membrane surface. It gives a partition-like distribution between the two sites which is observable in the spectrum of all *n*-PC labels with *n*>7 as two components.

A gradual increase in the hydrogen-bonded fraction between PC10 and PC16 can be also explained using this partition model. For an *n-PC* spin label, the same carbon atom of the nitroxide ring that is connected to the acyl tether also has a hydrocarbon tail with a length of 18-*n* carbons attached. Bringing the nitroxide label of 10PC to the membrane surface will require, on the average, placing more hydrophobic CH_2 groups to the polar area than for 16PC. This will be associated with some energy penalty preventing the nitroxide of 10PC from leaving the hydrophobic core and affecting the partition. Note also a slight drop in the accessibility of the 10PC position for Ni^{2+} ions for cholesterol-free DMPC membranes, which may have the same origin.

3.2.2. DMPC and DPPC without cholesterol. Nitroxide moiety as a foreign body in the bilayer

It has been previously observed that in the absence of cholesterol, PC spin labels in DPPC[26] and DMPC[27] at rigid limit conditions have a strong broad background signal. This signal points at aggregation of spin labels and is most pronounced for the7-12 PC positions. Since no significant signs of such aggregation are seen in the fluid phase of the membrane or in the gel phase in the presence of cholesterol, this aggregation can be specifically attributed to the gel or crystalline (sub-gel) state of pure-lipid membranes. Also, since diffusion in the gel phase is slow, one can expect hysteresis effects and effects of the sample-treatment procedure, see[38, 39].

Fig. 6 shows 240 GHz ESR spectra of PC spin labels in DMPC membranes without cholesterol. These spectra were recorded after the sample was slowly cooled from 295 to 85K within ~ 2 hours. As in the presence of cholesterol, two components with different g_{xx} are clearly discernible. However, another broad unresolved component, which is present for all spin labels but most pronounced for 7-12PC, follows a previously observed[26] pattern for DPPC at 250GHz . Our 240 GHz data for DPPC obtained under the same conditions as those for DMPC are very similar to the 250 GHz results by Earle et al.[26] and shown in Fig.7. A non-polar (non -hydrogen bonded) component similar to the non-polar component in membranes containing cholesterol can be identified by its characteristic g_{xx} value. However, there is an important and obvious difference between DMPC and DPPC at first inspection. In DPPC there is only one resolved component which shows the g_{xx} value corresponding to the non-hydrogen bonded state of the nitroxide, while in DMPC two components are clearly

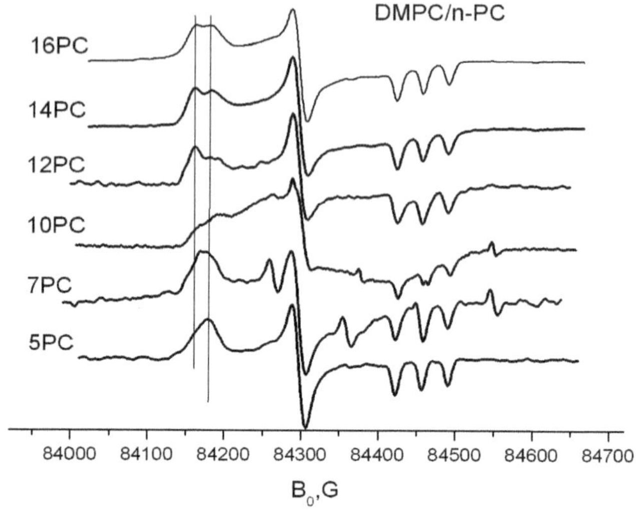

Figure 6. *n*-PC spin labels in DMPC with no cholesterol at 80-82K. The samples are gradually frozen in the flow of gaseous nitrogen. The vertical lines indicate g values of 2.00882 and 2.009435. (The additional superposition signal for 7PC originates from an occasional manganese impurity in the sample holder) (from [16]).

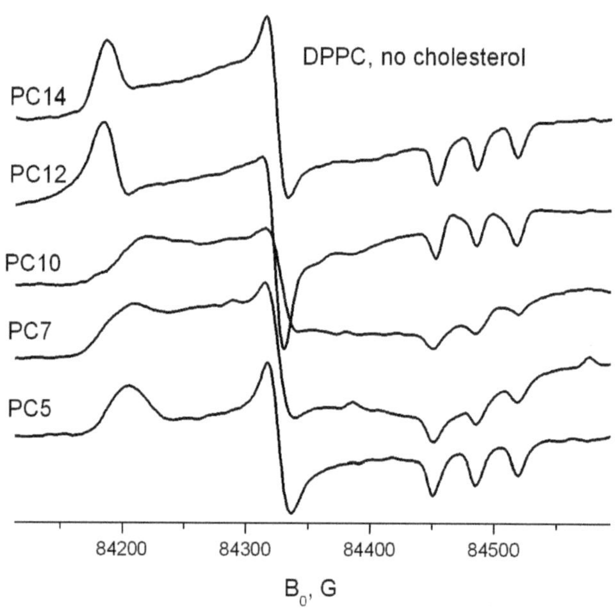

Figure 7. n-PC spin labels in DPPC with no cholesterol at 80-82K. The samples are gradually frozen in the flow of gaseous nitrogen (from [16]).

discernible in the spectrum. This observation is in good accord with the X-band results, which indicate a less polar environment reported by high values of n in DPPC compared to DMPC (cf. Table 3, Figs. 2A, 2C).

It is very unlikely that small structural differences between the gel phase bilayers of DMPC and DPPC[40, 41] cause dramatically different water penetration into these membranes and explain the presence of the hydrogen-bonded component in DMPC, but not DPPC. Moreover, it has been recently shown that water penetration into the membrane depends rather on the surface area of the lipid (which is nearly identical for DMPC and DPPC) than on the membrane thickness[42]. Thus, it makes the explanation of the difference between DMPC and DPPC through dramatically different water penetration very unlikely. Even more intriguing, the average environment of the nitroxide in DPPC looks, at the first glance, less polar in the absence of cholesterol, since no component with a clear peak at smaller g_{xx} can be detected, cf. Fig 8.

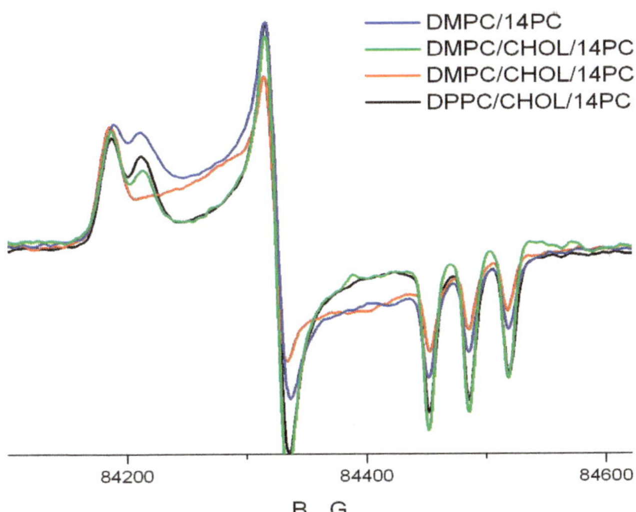

Figure 8. Comparison of 14 PC spin label in DMPC and DPPC with/without cholesterol. All samples are slowly frozen in the flow of gaseous nitrogen. While in the presence of cholesterol the spectra can be described as a superposition of two rigid-limit components, the most salient feature in the absence of cholesterol is a broad singlet-like background signal (from [16]).

However, the most salient feature of all spectra, both DMPC and DPPC, in the absence of cholesterol, is a broad singlet-like background signal which accounts for most of the spins. The location of spins responsible for this background is unclear. In Fig. 8 one can see an apparent broadening of the hydrogen-bonded component in DMPC and complete absence of this component in DPPC, concomitant with an increase in the background. A likely explanation for these observations is that the hydrogen-bonded and non hydrogen- bonded forms of PC spin labels aggregate differently. If the hydrogen bonded form is more prone to aggregation than the non- bonded one, it likely becomes broader due to more interactions between spins. In the superposition, spectral peaks from this broad component are less intense and eventually, with an increase in broadening, not discernible at all. To test this hypothesis we cooled our samples very rapidly in an attempt to trap the hydrogen-bonded component before aggregation could occur.

Slow cooling vs. quick submerging into liquid nitrogen

Quantitative study of aggregation in the gel or subgel phases is difficult because of the slow diffusion rates and various hysteresis effects. For example, even though the gel phase of DPPC does not favor formation of gramicidin channels, it takes between an hour and several days for these channels to dissociate[39]. There are also indications pointing to a relatively slow time scale for aggregation of PC spin labels in the gel-phase. Fig. 9 shows saturation curves for 0.5 mol. % of 7PC in DMPC. At 19⁰C, in the P_β phase, this system show good saturation, with a P parameter of ~ 19. Saturation measurements performed immediately after cooling to 2⁰C give a similarly high P value. However, after longer exposure at this temperature the P value starts to decrease and after several hours drops below 2. This increase in relaxation is very likely due to aggregation, and this aggregation appears to be a relatively slow process at 2⁰C. Also, this increase in relaxation is reversible; return to 19⁰C reverses the aggregation and the P value. Although a temperature of 2⁰C should correspond to the subgel phase, the exact phase state of the lipid at this condition is not obvious due to likely supercooling, see below. It usually takes days at this temperature to form the Lc phase, which then can be characterized by X-Ray diffraction.

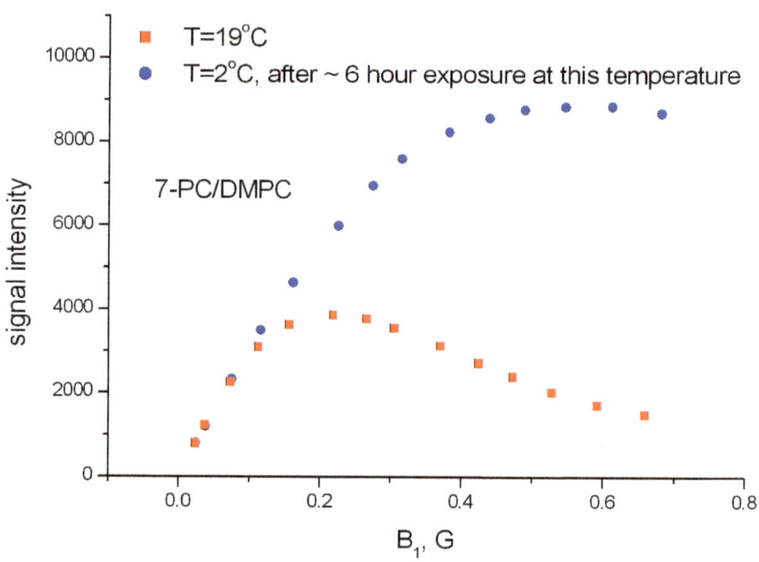

Figure 9. Change in relaxation curve of 7PC in DMPC under exposure at 2⁰C. This increase in relaxation occurs gradually within several hours and it can be completely reversed by a brief exposure at 19⁰C (from [16]).

Previously, similar observations of exclusion of 5PC spin labels from DPPC after exposure at 0°C by using Saturation Transfer ESR[43] were attributed to formation of a sub-gel phase.

We expected that quick freezing (within ~ 0.2s) of the sample by instantly submerging into liquid nitrogen should substantially eliminate these slow lipid rearrangement effects compared to the standard slow cooling procedure using a flow cryostat. Indeed, spectra recorded after this simple quick-freeze procedure were different from the spectra obtained by gradual cooling in a flow of gaseous nitrogen. As seen in Fig. 10A the spectrum of 14PC in DPPC does not have a broad singlet-like component and can be well simulated by a simple superposition of two rigid limit spectra with g_{xx} of 2.00943 and g_{xx}= 2.00882.

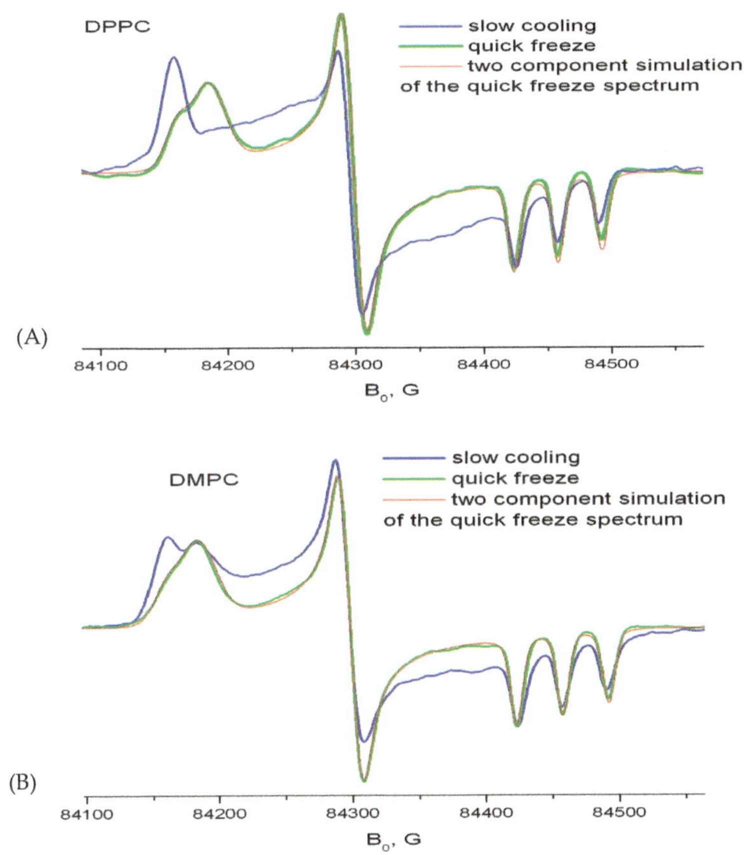

Figure 10. 14PC in DPPC (A) and DMPC (B): gradual cooling vs. quick freeze by submerging into liquid nitrogen. The spectra are recorded at 80-82K. In the case of quick freezing the broad unresolved component disappears and the spectrum can be well described as a superposition of HB/non-HB forms of the nitroxide. The HB/non-HB ratio is 2.2:1 for DPPC and 3.1:1 for DMPC (from [16]).

Importantly, most 14PC spins are present in the form of the hydrogen bonded component, which has a fraction greater than the non-bonded one by a 2.2:1 ratio.

As seen from Fig 11, quick freezing helps to dramatically decrease the amount of the broad background signal and yields well resolved rigid-limit spectra even when a standard cooling procedure results in nearly unresolvable, broad-feature spectra.

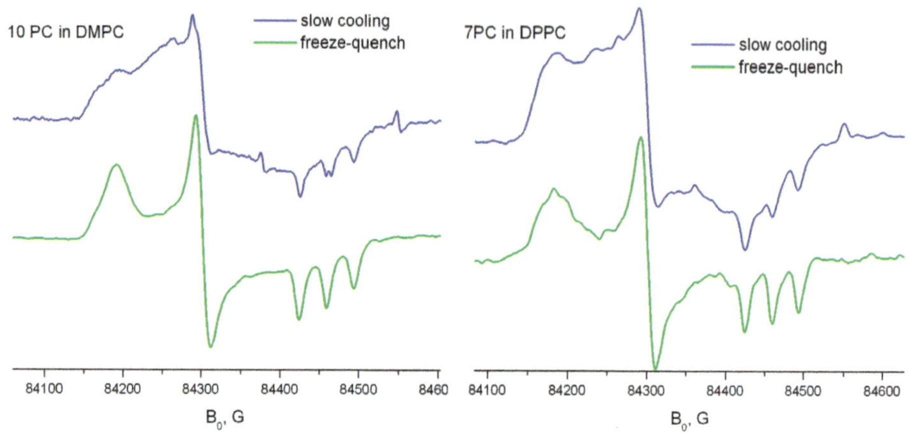

Figure 11. Obtaining better resolved spectra by quick freezing in liquid nitrogen (from [16]).

Nitroxides in the L'_β phase – an analogy to bulk crystallizing solids

One of the possible artifacts of using molecular probes as reporters in biological and model membranes is their exclusion from the membrane interior to the membrane surface. Such exclusion of certain molecules or structural units is a natural feature of membrane biological function. For example, exclusion of tryptophans and their affinity to the membrane interface is an important factor in lipid-protein interactions and stabilization of certain conformations of transmembrane protein/peptides[44]. Exclusion from the bilayer and formation of some U-shaped conformations have been known for a long time for a number of fluorescent labels[7, 9, 11].

Spin labels are usually considered as more adequate molecular probes for lipid bilayers, as evidenced by studies in fluid membranes (see Introduction). However, our current study shows that in the gel phase the situation is different. Several ESR parameters (accessibility for Ni^{2+} ions absorbed on the membrane surface, hyperfine tensor and g-tensor components) unambiguously indicate that in the gel phase a substantial fraction of nitroxide is located in the membrane region with high water content. Pure lipids DMPC and DPPC form the L'_β "tilted gel" phase with densely packed and relatively well ordered hydrocarbon chains. The ability of nitroxide groups on spin labeled stearic acids to be excluded from hydrocarbon environments and form U-shaped conformations has been previously reported[12, 13]. In principle, formation of the L'_β phase can be compared with freezing of a bulk three-dimensional solvent. In this case solutes are excluded from crystallizing solvent and form some regions with high local concentration. If the solute is a nitroxide radical, freezing of its solution in non glass-forming media is well known to yield a broad signal similar to one observed in DMPC or DPPC without cholesterol (Figs 6, 7).

In the case of a lipid in the gel phase the complete exclusion of spin labels into a separate phase with a high spin concentration appears to take two steps. The first one, exclusion of nitroxides from the hydrophobic area of the membrane, occurs once the lipid forms an L_{β}' or even a P_{β} phase (cf. Fig 3 for 19⁰C, the P_{β} phase of DMPC, also the quick-freeze spectra in Fig. 10). The second stage, which is a formation of separate phase by these bent-conformation molecules, can be attributed to lateral aggregation in the supercooled gel, possibly at the gel/subgel phase transition.

Although formation of subgel phases was initially observed after storing a multilamellar suspension of DPPC at ~0⁰C for several days[45], there are reports of quicker gel/subgel transformations in DPPC and DMPC. For example, in one protocol[46] the subgel (SGII) phase in DPPC forms at about 7⁰C upon cooling at 2⁰C/min. These conditions are very similar to cooling conditions observed in our "slow cooling" experiment – and, in general, the cooling conditions usually existing in standard helium/nitrogen cryostats. For DMPC, formation of the subgel phase was reported after incubation at temperatures of -5⁰C or lower for 2h[47]. But, the kinetics of subgel formation is complex[48] and in some cases the final subgel phase contains spectroscopic features characteristic for the L_{β}' phase[49]. It could be possible that aggregation of spin labels does not require a complete transformation into the subgel phase, in the sense of giving a clean X-ray pattern, and can happen sooner during cooling.

Again, we would like to stress that exclusion of nitroxides from the hydrophobic membrane core occurs in all gel phases and is a separate effect from the succeeding lateral aggregation in the supercooled gel phase.

Our quick freeze vs. slow cooling experiments can also help answer an important question about the nature of the two components in the 240GHz spectrum. Does the component with a larger g_{xx} correspond to the location of the nitroxide in the hydrophobic part of the membrane? Or both components arise from equilibrium of the hydrogen-bonded/non-hydrogen bonded forms in the same location, like TEMPONE in ethanol (cf. Fig. 1A)? The different aggregation behavior of spins contributing into these components points to their different location. That is, the major hydrogen-bonded component comes from nitroxides excluded from the hydrophobic part into the area with high water content, while the minor non-polar component may arise from spins somehow trapped in the defects of the hydrophobic membrane core.

3.3. What about biological membranes?

The gel phase rarely exists in biological membranes and the subgel phase is unknown for them. However, polarity measurements in frozen natural membranes using PC spin labels can also be affected by the affinity of spin labels to some structural defects, especially in the presence of proteins. For example, all PC labels in mixtures of DPPC with gramicidin A show about the same g_{xx} value, indicative of the same polarity[26], most likely due to an interaction with the peptide's α-helical structure. Also, natural membranes usually have complex lipid head group composition and variety of lengths and unsaturation in acyl chains. This can affect the free volume available for spin labels both in the hydrophobic core and in the membrane interface. Fig. 12 shows 240 GHz spectra for 14PC in egg-yolk lecithin with and without cholesterol.

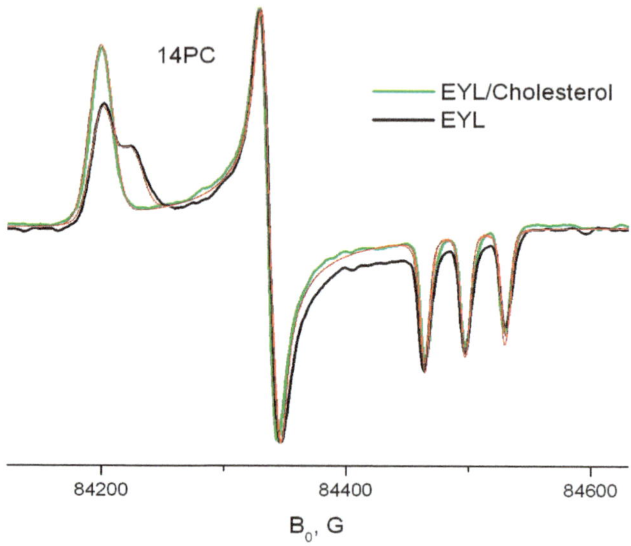

Figure 12. 14 PC in EYL with and without cholesterol at 82K. The samples are gradually frozen in the flow of gaseous nitrogen. The corresponding one and two-component (the ratio of components with g_{xx}=2.009410 and 2.008820 is 0.5) rigid limit simulations are shown in red (from [16]).

One can see that, similar to DMPC and DPPC membranes, the fraction of the hydrogen-bonded component drops with addition of cholesterol (cf. Figs 4 and the quick-freeze spectra in Fig.10). However, compared to DMPC and DPPC membranes this fraction in EYL is lower. Actually, in EYL/Cholesterol only a non-polar component can be observed in the spectrum. It can be explained by the presence of a double bond in the acyl chain of EYL. This unsaturated bond has a kink, which disrupts the packing of lipids. It creates extra free space in the hydrophobic core of the membrane[50], and thus may shift the partition of nitroxides in favor of the membrane interior. Also, a variety of headgroups may improve packing in the polar head area, additionally forcing nitroxides into the hydrophobic core.

3.4. Summary

The ESR parameters of PC spin labels in frozen membranes do not simply represent the membrane polarity or water penetration profile. Instead, they show a distribution between hydrogen-bonded (HB) and non-hydrogen bonded (non-HB) states, which is affected by a number of factors in the membrane composition. In frozen lipid membranes at 240 GHz *n*-

PC labels with $n > 7$ usually show two-component ESR spectra corresponding to HB and non-HB states of the nitroxide. In DMPC and DPPC membranes without cholesterol the ESR spectra recorded after gradual sample cooling show a broad background signal as well. But if the sample is instantly frozen, the broad background signal is absent and one observes mainly the HB component. We attribute the HB component for $n \geq 10$ to bent conformations of the nitroxide tethers. Such conformations at 9 GHz ESR manifest themselves in apparent "polar" A_{zz} values and increased accessibility of the nitroxide to aqueous Ni^{2+} ions. These bent conformations prevail in the absence of cholesterol due to the tendency of the pure gel phase to exclude nitroxides, similar to the exclusion of solutes from crystallizing solvents. The formation of a separate phase of spin-labeled lipid (i.e. precipitate) manifests itself in the broad background ESR signal and occurs slowly in the supercooled gel. In membranes with cholesterol the observed HB/ non-HB ratio can best be described by a partition-like equilibrium between nitroxides located in defects of lipid structure in the hydrophobic core of the membrane and those close to the membrane surface.

Abbreviations

HF ESR, HFHF ESR: High Field ESR, High Field/High Frequency ESR

DMPC: 1,2-dimyristoyl-*sn*-glycero-3-phosphocholine

DPPC: 1,2-dipalmitoyl-*sn*-glycero-3-phosphocholine

EYL: egg yolk lecithin

n-PC spin labels: 1-palmitoyl-2-stearoyl-(n-doxyl)-sn-glycero-3-phosphocholines

TFE: 2,2,2-trifluoroethanol

MCH: methylcyclohexane

TEMPO: 2,2,6,6-tetramethylpiperidine-N-oxyl

TEMPOL, 4-Hydroxy-TEMPO: 4-hydroxy-2,2,6,6-tetramethylpiperidine-N-oxyl

TEMPONE, Oxo-TEMPO: 4-oxy-2,2,6,6-tetramethylpiperidine-N-oxyl

Author details

Boris Dzikovski and Jack Freed
National Biomedical Center for Advanced ESR Technology (ACERT),
Department of Chemistry and Chemical Biology, Baker Laboratory,
Cornell University, Ithaca, NY, USA

Acknowledgement

This work was supported by the National Institute of Health, grants NIH/NIBIB R010EB003150 and NIH/NCRR P41-RR 016292 and NIH/NIGMS P41GM103521.

4. References

[1] Swamy, M.J., L. Ciani, M. Ge, A.K. Smith, D. Holowka, B. Baird, and J.H. Freed, *Coexisting Domains in the Plasma Membranes of Live Cells Characterized by Spin-Label ESR Spectroscopy*. Biophys. J., 2006. 90: p. 4452-4465.

[2] Altenbach, C., D.A. Greenhalgh, H.G. Khorana, and W.L. Hubbell, *A collision gradient method to determine the immersion depth of nitroxides in lipid bilayers: application to spin-labeled mutants of bacteriorhodopsin*. Proc. Natl. Acad. Sci. USA., 1994. 91: p. 1667-1671.

[3] Marsh, D., *Polarity and permeation profiles in lipid membranes*. Proc. Natl. Acad. Sci. USA 2001. 98: p. 7777-7782. .

[4] Dzikovski, B., V.A. Livshits, and D. Marsh, *Oxygen permeation profile in lipid membranes: nonlinear spin-label EPR.* . Biophys. J. , 2003. 85: p. 1005-1012.

[5] Marsh, D., A. Watts, R.D. Pates, R. Uhl, P.F. Knowles, and M. Esmann, *ESR spin-label studies of lipid-protein interactions in membranes*. Biophys. J., 1982. 37: p. 265-274.

[6] Costa-Filho, A.J., P.P. Borbat, R.H. Crepeau, M. Ge, and J.H. Freed, *Lipid-Gramicidin Interactions: Dynamic Structure of the Boundary Lipid by 2D-ELDOR*. Biophys. J., 2003. 84: p. 3364-3378.

[7] Ashcroft, R.G., K.R. Thulborn, J.R. Smith, H.G.L. Coster, and W.H. Sawer, *Perturbation to lipid bilayers by spectroscopic probes as determined by dielectric measurements*. BBA, 1980. 602: p. 299-308.

[8] Krishnan, K.S. and P. Balaram, *Perturbation of lipid structures by fluorescent probes*. FEBS Lett., 1975. 60: p. 419-422.

[9] Chattopadhyay, A. and E. London, *Parallax Method for Direct Measurement of Membrane Penetration Depth Utilizing Fluorescence Quenching by Spin-Labeled Phospholipidst*. Biochemistry, 1987. 26: p. 39-45.

[10] Vogel, A., H.A. Scheidt, and D. Huster, *The distribution of lipid attached spin probes application to membrane protein topology*. Biophysical Journal, 2003. 85: p. 1691-1701.

[11] Cadenhead, D.A., B.M.J. Keller, and F. Muller-Landau, *A comparison of a spin-label and a fluorescent cell membrane probe using pure and mixed monomolecular films*. Biochim. Biophys. Acta., 1975. 382: p. 253-259.

[12] Dzikovski, B. and V. Livshits, *EPR spin probe study of molecular ordering and dynamics in monolayers at oil/water interfaces*. PCCP, 2003. 5: p. 5271-5278

[13] Baglioni, P., L. Dei, E. Rivara-Minten, and L. Kevan, *Mixed Micelles of SDS/$C_{12}E_6$ and DTAC/$C_{12}E_6$, Surfactants*. J. Am. Chem. Soc., 1993. 115: p. 4286-4290.

[14] Marsh, D., *Spin-Label EPR for Determining Polarity and Proticity in Biomolecular Assemblies: Transmembrane Profiles*. Appl. Magn. Reson., 2010. 37: p. 435-454.

[15] Ellena, J.F., S.J. Archer, R.N. Dominey, B.D. Hill, and D.S. Cafiso, *Localizing the nitroxide group of fatty acid and voltage-sensitive spin-labels in phospholipid bilayers*. BBA, 1988. 940: p. 63-70.

[16] Dzikovski, B., D. Tipikin, and J.H. Freed, *Conformational Distributions and Hydrogen Bonding in Gel and Frozen Lipid Bilayers: A High Frequency Spin-Label ESR Study*. J. Phys. Chem. B., 2012. in press: p. DOI: 10.1021/jp211879s.

[17] Kawamura, T., S. Matsunami, and T. Yonezawa, *Solvent effects on the g-value of di-t-butyl nitric oxide*. Bull. Chem. Soc. Japan, 1967. 40: p. 1111-1115.

[18] Smirnov, A.I. and T.I. Smirnova, *Resolving domains of Interdigitated Phospholipid Membranes with 95 GHz Spin labeling EPR.* Appl. Magn. Reson., 2001. 21: p. 453-467.

[19] Smirnova, T.I., A.I. Smirnov, S.V. Pachtchenko, and O.G. Poluektov, *Geometry of Hydrogen Bonds Formed by Lipid Bilayer Nitroxide Probes: A High Frequency Pulsed ENDOR/EPR Study.* JACS, 2007. 129: p. 3476-3477.

[20] Bordignon, E., H. Brutlach, L. Urban, K. Hideg, A. Savitsky, A. Schnegg, P. Gast, M. Engelhard, E.J.J. Groenen, K. Möbius, and H.-J. Steinhoff, *Heterogeneity in the Nitroxide Micro-Environment: Polarity and Proticity Effects in Spin-Labeled Proteins Studied by Multi-Frequency EPR.* Appl. Magn. Reson., 2010. 37: p. 391-403.

[21] Hwang, J.S., R.P. Mason, L.-P. Hwang, and J.H. Freed, *Electron Spin Resonance studies of Anisotropic Rotational Reorientation and Slow Tumbling in Liquid and Frozen Media. III. Perdeuterated 2,2,6,6-Tetramethyl-4-piperidone-N-Oxide and An Analysis of Fuctuating Torques.* J. Phys. Chem., 1975. 79: p. 489-511.

[22] Pavone, M., A. Sillanpää, P. Cimino, O.Crescenzi, and V. Barone, *Evidence of Variable H-Bond Network for Nitroxide Radicals in Protic Solvents.* J. Phys. Chem. B, 2006. 110: p. 6189-16192.

[23] Erilov, D.A., R. Bartucci, R. Guzzi, A.A. Shubin, A.G. Maryasov, D. Marsh, S.A. Dzuba, and L. Sportelli, *Water concentration Profiles in Membranes Measured by ESEEM of Spin-Labeled Lipids.* J. Chem. Phys. B, 2005. 109: p. 12003-12013.

[24] Owenius, R., M. Engström, M. Lindgren, and M. Huber, *Influence of solvent polarity and hydrogen bonding on the EPR parameters of a nitroxide spin label studied by 9-GHz and 95-GHz EPR spectroscopy and DFT calculations.* J. Phys. Chem. A, 2001. 105: p. 10967-10977.

[25] Plato, M., H.-J. Steinhoff, C. Wegener, J.T. Törring, A. Savitsky, and K. Möbius, *Molecular orbital study of polarity and hydrogen bonding effects on the g and hyperfine tensors of site directed NO spin labelled bacteriorhodopsin.* Molecular Physics, 2002. 100: p. 3711-3721.

[26] Earle, K.A., J.K. Moscicki, M. Ge, D.E. Budil, and J.H. Freed, *250-GHz Electron Spin Resonance Studies of Polarity Gradients Along the Aliphatic Chains in Phospholipid Membranes.* Biophys. J., 1994. 66: p. 1213-1221.

[27] Kurad, D., G. Jeschke, and D. Marsh, *Lipid Membrane Polarity Profiles by High-Field EPR.* Biophysical Journal, 2003. 85: p. 1025-1033.

[28] Marsh, D., D. Kurad, and V.A. Livshits, *High-field spin-label EPR of lipid membranes* Magn. Reson. Chem., 2005. 43: p. S20-S25.

[29] Budil, D.E., K. Earle, and J.H. Freed, *Full determination of the rotational diffusion tensor by electron paramagnetic resonance at 250GHz.* . J. Phys. Chem., 1993. 97: p. 1294-1303.

[30] Marsh, D., *Membrane water-penetration profiles from spin labels.* Eur. Biophys. J., 2002. 31: p. 559-562.

[31] Bartucci, R., D.A. Erilov, R. Guzzi, L. Sportelli, S.A. Dzuba, and D. Marsh, *Time-resolved electron spin resonance studies of spin-labeled lipids in membranes.* Chemistry and Physics of Lipids, 2006. 141: p. 142-157.

[32] Berendsen, H.J.C. and S.J. Marrink, *Molecular dynamics of water transport through membranes: Water from solvent to solute.* Pure&Appl.Chem., 1993. 65: p. 2513-2520.

[33] Lande, M.B., J.M. Donovan, and M.L. Zeidel, *The relationship between membrane fluidity and permeabilities to water, solutes, ammonia, and protons.* J. Gen. Physiol., 1995. 106: p. 67-84.

[34] Carruthers, A. and D.L. Melchior, *Studies of the relationship between bilayer water permeability and bilayer physical state.* Biochemistry, 1983. 22: p. 5797-5807.

[35] Rubenstein, J.L., B.A. Smith, and H.M. McConnell, *Lateral diffusion in binary mixtures of cholesterol and phosphatidylcholines.* Proc. Natl. Acad. Sci. U.S.A., 1979. 76: p. 15-18.

[36] Lee, A.G., *Analysis of the Defect Structure of Gel-Phase Lipid.* Biochemistry 1977. 16: p. 835-841.

[37] Yeagle, P.L., *The role of cholesterol in the biology of cells,* in *The structure of biological membranes.* , P.L. Yeagle, Editor. 2005, CRC press: Boca Raton - London - New York - Washington DC. p. 243-254.

[38] Dzikovski, B.G., P.P. Borbat, and J.H. Freed, *Spin-labeled gramicidin A: Channel formation and dissociation.* Biophysical Journal, 2004. 87(5): p. 3504-3517.

[39] Dzikovski, B.G., K.A. Earle, S.V. Pachtchenko, and J.H. Freed, *High-field ESR on aligned membranes. A simple method to record spectra from different membrane orientations in the magnetic field.* Journal of magnetic resonance, 2006. 179: p. 273-277.

[40] Tristram-Nagle, S., Y. Liu, J. Legleiter, and J.F. Nagle, *Structure of gel phase DMPC determined by x-ray diffraction.* Biophysical Journal, 2002. 83(6): p. 3324-3335.

[41] Nagle, J.F. and S. Tristram-Nagle, *Structure of lipid bilayers.* Biochimica et Biophysica Acta, Reviews on Biomembranes, 2000. 1469(3): p. 159-195.

[42] Mathai, J.C., S. Tristram-Nagle, J.F. Nagle, and M.L. Zeidel, *Structural Determinants of Water Permeability through the Lipid Membrane.* J. Gen. Physiol., 2007. 131: p. 69-76.

[43] Páli, T., R. Bartucci, L. Horváth, and D. Marsh, *Kinetics and dynamics of annealing during sub-gel phase formation in phospholipid bilayers. A saturation transfer electron spin resonance study.* Biophys. J., 1993. 64: p. 1781-1788.

[44] Sun, H., D.V. Greathouse, O.S. Andersen, and R.E.I. Koeppe, *The preference of tryptophan for membrane interfaces.* J. Biol. Chem., 2008. 283: p. 22233-22243.

[45] Chen, S.C., J.M. Sturtevant, and B.J. Gaffney, *Scanning calorimetric evidence for a third phase transition in phosphatidylcholine bilayers* Proc. Natl. Acad. Sci. U.S.A., 1980. 77: p. 5060-5063.

[46] Koynova, R., B.G. Tenchov, S. Todinova, and P.J. Quinn, *Rapid Reversible Formation of a Metastable Subgel Phase in Saturated Diacylphosphoatydylcholines.* Biophys. J., 1995. 68: p. 2370-2375.

[47] Chang, H.H., R.K. Bhagat, R. Tran, and P. Dea, *Subgel Studies of Dimyristoyphosphatidylcholine Bilayers.* J. Phys. Chem. B, 2006. 110: p. 22192-22196.

[48] Tristram-Nagle, S., R.M. Suter, W.-J. Sun, and J.F. Nagle, *Kinetic of subgel formation in DPPC: X-ray difraction proves nucleation-growth hypothesis.* BBA, 1994. 1191: p. 14-20.

[49] Lewis, R.N.A.H. and R.N. McElhaney, *Structures of the subgel phases of n-saturated diacyl phosphatidylcholine bilayers:FTIR spectroscopic studies of* ^{13}C*=O and* ^{2}H *labeled lipids.* Biophys. J., 1992. 61: p. 63-77.

[50] Rawicz, W., K.C. Olbrich, T.J. McIntosh, D. Needham, and E. Evans, *Effect of chain length and unsaturation on elasticity of lipid bilayers.* Biophys. J., 2000. 79: p. 328-339.

[51] *CRC Handbook of Chemistry and Physics, 91st Edition,* ed. W.M. Haynes. 2010, Boulder, CO National Institute of Standards and Technology.

[52] Dzikovski, B., D.S. Tipikin, V. Livshits, K.A. Earle, and J.H. Freed, *Multi-frequency ESR Study of Spin-Labeled Molecules in Inclusion Compounds with Cyclodexrins.* PCCP, 2009. 11: p. 6676-6688.

Chapter 6

pH-Sensitive Nitroxide Radicals for Studying Inorganic and Organo-Inorganic Materials and Systems

Elena Kovaleva and Leonid Molochnikov

Additional information is available at the end of the chapter

1. Introduction

Many inorganic materials are widely used as adsorbents and catalysts. For example, silica gels efficiently absorb vapors and gases in chemical reactors and are applied as filters for the purification of mineral oils and water (Buyanov, 1998). Different aluminum oxide modifications show good adsorption and catalytic properties in many organic reactions due to the presence of active sites on their surface (Lisichkin et al., 2003). TiO_2 gels are widely used in heterogeneous catalysis owing to their enhanced chemical stability, the accessibility of active sites on their surface throughout the reaction volume, simplicity of reaction product separation, and the feasibility of repeated regeneration (Petrov et al., 1998).

The efficiency of these materials mainly depends on the degree of surface development, texture and structural characteristics, availability of active centers, and possibly medium acidity near these centers. The specific surface diminishes on drying in any procedure for the preparation of xerogels. It can be maintained constant by a number of methods, including the use of additions (Ur'ev & Potanin, 1992), for example, powder cellulose (PC) (Shishmakov et al., 2007). Cellulose is a linear high molecular polysaccharide, which forms rigid chain structures due to the inter-molecular hydrogen bonding. It functions in nature as an agent that imparts high mechanical stability to plant tissues (Nikitin, 1962). The deposition of SiO_2 and TiO_2 xerogels on the PC surface affords composite materials (CMs) with a high dispersity of particles (Shishmakov et al., 2010).

The use of hybrid organo-inorganic materials as supports is a new area in the development of new metal-containing catalytic materials. It allows to obtain supports with specific surfaces that are capable of retaining the metallic component of a catalytic system more strongly.

Chitosan, poly-D-β-glucosamine, is a commercially available amino polymer that is a perfect complexing agent, due to the strong donor properties of both the amino and hydroxyl groups (Varma et al., 2004). Chitosan is thus widely used in obtaining various catalytic materials, including those containing Au^0 that are used in the hydroamination of alkenes (Corma et al., 2007); Pd^0 used for the reduction of ketones (Yin et al., 1999); the Pd^0–Ni^0 bimetallic system, used for carbonylation (Zhang & Xia, 2003); Os (VIII), used for hydroxylation (Huang et al., 2003); Co^{2+}, used for hydration (Xue et al., 2004); and Cu^{2+}, used for the oxidation of catecholamines (Paradossi et al., 1998).

SiO_2 is usually used as the inorganic component for these systems. The obtained hybrid materials are used to create sorbents of 3d-metal ions (Liu et al., 2002); to immobilize enzymes (Airoldi & Monteiro, 2000); as a solid phase for the liquid chromatography of organic compounds (Budanova et al., 2001), including enantiomers (Senso et al., 1999); and to improve the mechanical properties of other polymers (Yeh et al., 2007). Other oxides in combination with chitosan allow us to obtain biosensors based on ZnO substrate (Khan et al., 2008), selective sorbents of fluoride ions based on Al_2O_3 substrate (Viswanathana & Meenakshib, 2010) and magnetic materials based on Fe_3O_4 substrate (Li et al., 2008). Using an organic polymer (e.g., cellulose) as a substrate also has advantages in the sorption of metal ions (Corma et al., 2007). Metal-containing hybrid organo–inorganic materials can also be used as antibacterial composites (Mei et al., 2009), as sorbents of proteins (Shi et al., 2003), and as pervaporation membranes (Varghese et al., 2010).

Nanostructured metal oxides, which are distinguished by extremely developed surface and porosity of particles, are new promising materials for different fields of science and technology, especially, for heterogeneous catalysis and chemistry of adsorption phenomena (Zakharova et al., 2005).

Many sorption and catalytic processes are pH-dependent. Therefore, the determination of acidity and other acid–base characteristics in pores of inorganic, organo-inorganic materials is of great practical interest, since the catalytic and adsorption properties of solid-phase objects are affected by not only the chemical nature of solutions, but also specific conditions inside pores and on the surface of these materials. The mobility of liquid molecules in pores of inorganic sorbents was investigated by some authors using the spin probe method (Borbat et al., 1990; Martini et al., 1985). Recently, a new method was developed for the determination of medium acidity in pores of solids (pH_{int}) by means of pH-sensitive nitroxide radicals (NRs) as spin probes (Molochnikov et al., 1996 ; Zamaraev et al., 1995). In recent years, this method was used to measure pH_{int} in micropores of various cross-linked organic polyelectrolytes (ion-exchange resins and films) (Molochnikov et al., 1996, 2004) and in pores of some zeolites and kaolin (Zamaraev et al., 1995). We found that pH_{int} inside sorbents differ from the pH of external solutions by 0.8–2.1 units (Molochnikov et al., 1996). The method developed allowed us to study the processes of sorption and hydrolysis in ion-exchange resins and the catalytic properties of Cu^{2+}- containing carboxyl cation exchangers (Kovaleva et al., 2000), to determine ionization constants of functional groups and to give a critical estimation to the regularities previously found for the behavior of adsorbents in aqueous media.

pH-sensitive nitroxide radicals (NR) as labels were also used to determine surface electrical potential (SEP) of different biological objects like phospholipids (SLP) - derivatives of 1,2-dipalmitoyl-sn-glycero-3-phosphothioethanol (PTE) (Voinov et al., 2009) and the mixed bilayers composed of dimyristoylphosphatidylglycerol and dimyristoylphosphatidilcholine (Khramtsov & Weiner, 1988).

This work is aimed to review the applications of pH-sensitive NR as probes and labels for determination of local acidic and electrochemical characteristics of inorganic and organo-inorganic materials and systems such as :

- Al_2O_3 (non-modified and modified with F^- and SO_4^{2-} ions) and TiO_2 as hydrogels and nanopowders doped with Cu(II) (Molochnikov et al., 2007) ;
- pure and Cu^{2+}-containing solid-phase composites based on nanostructured SiO_2 and TiO_2 and powder cellulose (Shishmakov et al., 2010; Parshina et al., 2011) ;
- pure and Co^{2+}-containing hybrid organo-inorganic materials based on the chitosan–SiO_2, chitosan– Al_2O_3, and –chitosan–cellulose systems (Mekhaev et al., 2011a, 2011b).

2. Experimental

2.1. Objects of study

α-Al_2O_3 (basic aluminum oxide), γ-Al_2O_3 and its acid-modified (HF and H_2SO_4) derivatives were supplied by A. M. Volodin. **The IK-02-200 type γ- Al_2O_3** was synthesized by the calcination of aluminum hydroxide at 600°C. γ-Al_2O_3 was modified through the sample impregnation with acids followed by calcination at 600°C that resulted in changes in the acidic properties of its surface (the phase composition and specific surface area of the samples remained unchanged). **α-Al_2O_3** was prepared by the long-term heating of **γ-Al_2O_3** to 1300°C. **γ- Al_2O_3** and **α-Al_2O_3** had specific surface areas of 220 and 145 m^2/g, and an average pore diameter of 6 nm, respectively. The structural characteristics of the matrices were determined from the isotherms of nitrogen adsorption at 77 K measured on a Micromeritics ASAP 1400 volumetric setup and by mercury porosimetry with a Micromeritics Pore Size 9300 setup in the Institute of Catalysis, Siberian Division, Russian Academy of Sciences.

The technique of synthesis of **TiO_2 hydrogel** through the hydrolysis of a tetrabutyl titanate solution in methanol with water at room temperature and under intensive agitation is given in (Shishmakov et al., 2003). The precipitate was washed out with water until no butanol in washing water was observed and heated to 100°C. The resulting product was **TiO_2 xerogel.**

SiO_2 hydrogel was synthesised by dissolving 10 ml of Na_2SiO_3 (TU 6-15-433-92) in 30 ml of H_2O. Then the hydrolysis of Na_2SiO_3 solution in 30 ml of 10% HCl solution was pursued under intensive agitation. During the reaction of condensation a gelatinous SiO_2 gel and NaCl are formed. The precipitate was filtered off until no chlorine ions in washing water was observed, and dried at 100 °C during 24h until it attained a constant weight. The resulting product was **SiO_2 xerogel.**

Powder cellulose (PC) was obtained by hydrolysis of cellulose sulfate (Baikal Cellulose plant, TU OP 13-02794 88-08-91) in 2.5 N hydrochloric acid at 100°C. The hydrolysis was carried out for 2 h. The resulting product was washed on a filter with distilled water to the neutral pH of the washing water and dried at 100°C.

Composite materials (CMs) based on nanostructured TiO_2 and PC called as TiO_2–PC xerogels of 70, 53 and 43 % wt. TiO_2 were prepared by diluting 3 g of tetrabutoxytitanium and 0.5 ; 1 and 1.5 g of PC, respectively, in 3 mL of methanol. The hydrolysis was pursued in 10 mL of water at 20°C under intensive agitation, resulting in the condension of TiO_2 (PC didn't participate in condensation). The TiO_2 particles formed were deposited on a surface of PC.

Composite materials (CMs) based on nanostructured SiO_2 and PC called as SiO_2–PC xerogels of 68, 52 and 35% wt. were prepared .from the solutions of 10 ; 5 ; 5 ml of Na_2SiO_3 and 30 ; 15 ; 15 ml of H_2O which were modified by introducing 2 ; 2 ; 4 g of PC, respectively. The hydrolysis of the first solution were peformed in 30 mL of 10 %HCL, and that of other ones was done in 15 mL of 10% HCL.

The precipitates of the CMs prepared were washed out with hot water, filtered and dried at 100 °C during 24h until they attained a constant weight.

The specific surface (S_{sp}) of the synthesized samples was measured using a SORBIMS instrument (ZAO Meta, Novosibirsk) and calculated by the BET procedure. The data are given in Table 1 (Parshina et al., 2011).

	TiO_2				SiO_2		
PC, %	0	30	47	57	0	32	65
S_{sp}, m²/g	66	177	226.4	261.9	29.5	145	239

Table 1. The specific surface of the pure xerogels of TiO_2 and SiO_2 and the composites with different percentage of PC

Since the specific surface of powder cellulose did not exceed 1 m²/g , the growth of S_{sp} was caused by the fragmentation of TiO_2 and SiO_2 particles deposited on the PC surface during the synthesis of CMs. According to the absolute values of S_{sp}, the procedure used for the preparation of CMs afforded dioxides with a high degree of dispersity.

Powdered samples **nanostructured TiO_2** were prepared through heating a sol for 1 h at 200°C followed by washing with distilled water to remove residual acid used for the sol stabilization and drying at room temperature. The specific surface area of the samples was 240 m²/g, and the average particle diameter ranged from 4 to 5 nm (Poznyak et al., 1999).

Microcrystalline cellulose (MCC) with an ash content of 0.16% and a humidity of 1.1% produced by JSC Polyex; Basic aluminum oxide; BS-50 silica and chitosan produced by JSC Sonat (Moscow) were used to obtain chitosan-containing ahybride organo-inorganic systems. The degree of deacetylation of chitosan (DD) determined by 1H NMR spectroscopy, its molecular weght as determined by viscosimetry and the ash content were found to be 0.84, 250 kDa and 0.19%, respectively (Mechaev et al., 2011a). The BS-50 type silica had a specific surface area of 45 m²/g and an average diameter of pores of 15 nm (Mekhaev et al., 2011a).

The hybrid chitosan–(SiO_2, Al_2O_3, cellulose) systems were obtained by depositing chitosan on the support surface.

0.3 g (1.8 mmol) of citosan was dissolved in 14.5 ml of water containing 0.22 ml (3.84 mmol) of acetic acid with constant stirring. The substrate in quantities of 3 g was then added, and the solution was stirred for 30 min more.

1 M NaOH solution was added to the suspension under stirring until the pH value reached 13. The precipitate was filtered, washed until the pH value was 7, and dried at 60°C until it attained a constant weight.

CHN analysis was performed using an automatic analyzer PerkinElmer, Inc. The data are given in Table 1. IR spectra of diffuse reflection were recorded using the PerkinElmer Spectrum One spectrometer.

System	C	H	N	Cl	Co	Formula
MCC-Chitosan	43.08 (43.11)	6.42 (6.35)	0.37 (0.39)	-	-	$10C_6H_{10}O_5 \cdot 0.5C_6H_{11}NO_4 \cdot 3H_2O$
Al_2O_3-Chitosan	4.07 (4.01)	1.04 (0.61)	0.46 (0.78)	-	-	$16Al_2O_3 \cdot C_6H_{11}NO_4$
SiO_2-Chitosan	4.50 (4.49)	0.94 (0.68)	0.51 (0.87)	-	-	$24SiO_2 \cdot C_6H_{11}NO_4$
MCC-Chitosan-Co^{2+} (I)	43.12 (43.12)	6.15 (6.04)	0.59 (0.40)	1.30 (1.28)	1.38 (1.33)	$10C_6H_{10}O_5 \cdot 0.5C_6H_{11}NO_4 \cdot Co_{0.4}(OH)_{0.17}Cl_{0.65}$
Al_2O_3-Chitosan-Co^{2+} (II)	3.64 (3.82)	0.79 (0.58)	0.47 (0.74)	2.69 (2.63)	2.18 (2.19)	$16Al_2O_3 \cdot C_6H_{11}NO_4 \cdot 0.7CoCl_2$
SiO_2-Chitosan-Co^{2+} (III)	4.02 (4.26)	0.89 (0.65)	0.58 (0.82)	2.03 (2.02)	3.07 (3.07)	$16SiO_2 \cdot C_6H_{11}NO_4 \cdot Co_{0.88}Cl_{0.96}$

Table 2. Composition (%) of hybrid systems (calculated values are shown in brackets)

The surface area of the samples was determined by nitrogen adsorption in accordance with the BET method using a TriStar 3000 V.6.03A instrument. The instrumental error was 0.1 m^2/g. The size of particles was estimated under the assumption that the particles were spherical.

The surface area (S_{sp}) and the diameter of particles (D) were found to be 28.9 m^2/g and 47 nm ; 123.9 m^2/g and 7 nm ; 2.4 m^2/g and 818 nm for chitosan-SiO_2 , chitosan-Al_2O_3 and chitosan-MCC hybride systems, respectively.

2.2. Saturation of samples with Cu^{2+} and Co^{2+} ions

2.2.1. Saturation of samples with Cu^{2+} Ions

A 0.1 M $NaNO_3$ solution (10 ml) was added into weighed samlpes (200 mg) of nanostructured TiO_2, and the samples were kept for one week at a constant solution pH (5.5) held by adding dilute NaOH and HNO_3 solutions. The sorption of Cu^{2+} ions on nanostructured TiO_2 was performed by exposing samples in $Cu(NO_3)_2$ solutions (10 ml) with concentrations of 10^{-4}, 10^{-3}, and 10^{-2} mol/L and ionic strength (μ) of 0.1, which was adjusted using $NaNO_3$. Solution pH equal to 4.3 was maintained by the titration with small volumes of NaOH and HNO_3 solutions. After the equilibrium was established, the residual amount of Cu^{2+} ions the equilibrium solutions was measured to determine the amount of sorbed Cu^{2+}. Then, TiO_2 was separated from the solutions by centrifugation. The samples were washed twice with a 0.1 M $NaNO_3$ solution (pH 4.3) to remove adsorbed Cu^{2+} -ions.

Cu^{2+} ions were sorbed on TiO_2 hydrogel from $CuCl_2$ and $Cu(NO_3)_2$ with subsequent its removal by filtration and drying at 20°C for 3 days upto constant weights of the precipitates. A volume of solution and a mass of hydrogel were changed to vary the content of Cu^{2+} ions in the phase of the studied TiO_2 hydrogel, which was determined by the atomic absorption method on a Perkin Elmer 403 spectrometer. As the ESR spectra of Cu^{2+}-containing hydrated gels are difficult to record, hydrogel samples filtered and dried at room temperature were used. Preliminary experiments were performed to select the sample drying conditions preventing the structural changes of the complexes formed.

Cu^{2+} - containing composites based nanostructured SiO_2, TiO_2, and cellulose powder were prepared by sorption of Cu^{2+} ions on a hydrogel from an aqueous solution of $CuCl_2 \bullet 2H_2O$. The volumes of hydrogels of the samples studied were calculated from the masses of xerogels obtained by hydrogels drying. For preparation of Cu^{2+}-containing TiO_2 and SiO_2 xerogels and the related CMs, a 0.01 g $CuCl_2 \bullet 2H_2O$ containing 0.059 mmol of Cu^{2+} ions and 2 ml of H_2O and the calculated volumes of hydrogels prepared were added into flasks. The samples were kept in the contact with a Cu^{2+}-containing solution about 24 h upto the equilibrium was established. Then, the residual amounts of Cu^{2+} ions in the equilibrium solutions was measured to determine the amount of sorbed Cu^{2+} ions. The initial and residual amounts of Cu^{2+} -ions in a solution were measured using colorimeter KFK-2MP.

2.2.2. Synthesis and characterization of the Cobalt-Containing Chitosan hybrid systems

Cobalt-Containing Chitosan–Supported Systems were synthesised through stirring a mixture containing 0.24 g of $CoCl_2 \bullet 6H_2O$, 2 g of chitosan–supported hybrid system and 20 ml of ethanol under reflux condenser for 24 h. The obtained cake was filtered off, rinsed with ethanol (15 ml × 3 times) and dried at room temperature until the weight became constant. The elemental compositions of the hybrid system surfaces were determined using an analytical setup based on a VEGA II LMH scanning electron microscope and an INCA ENERGY energy dispersive microanalysis system (Mekhaev et al., 2011a, 2011b). The data are shown in Table 2.

2.3. pH probes

The pH values of solutions inside pores and near the surface of the studied inorganic and organo-inorganic materials were determined using spin probes, namely, pH-sensitive NRs of the imidazoline (R1, R2) and imidazolidine (R3) types (Table 3), which were synthesized at the Novosibirsk Institute of Organic Chemistry, Siberian Branch, Russian Academy of Sciences (Volodarskii et al., 1988; Khramtsov et al., 1998; Kirilyuk et al., 2005).

2.4. Recording and processing of the ESR spectra of NR

The ESR spectra were recorded on a PS 100.X ESR spectrometer (ADANI, Belarus) in a three-centimeter (X) wavelength range at room temperature. Quartz sample holders with an internal diameter of 3.5 mm were used for solid samples. Solution spectra were recorded using quartz capillaries.

Figure 1 shows characteristic ESR spectra of the pH-sensitive NR in aqueous solutions. According to the ESR theory, isotropic signals are induced by the fast-motioned NR molecules (correlation times of 10^{-10} s and less) and present the triplet of fine lines. Depending on solution pH, NR can be in protonated (RH^+), deprotonated (R), or intermediate (mixed) forms. Because hyperfine splitting constants a_N for RH^+ and R forms of the radicals are different (Table 3), in their ESR spectra, the distance ***a*** between the low and central-field component of the triplet increases gradually with pH of a solution, from the values characteristic of the RH^+ form to those typical of the R form (Fig. 1). This characteristic is a superposition of hyperfine splitting constants (a_N) characterizing protonated and deprotonated forms of the nitroxide radical. From the results of measuring *a* values during titration, the calibration curves reflecting the dependences *a* vs. pH were plotted for each NR used (for example, see Fig. 2, 3 curve 1). In order to plot the calibration curves, NR solutions (10^{-4} mol/l ; μ = 0.1) were titrated either with dilute HCl and KOH solutions (used for all the samples, excepting hybride systems)(Molochnikov et al.,2007 ; Parshina et al., 2011 ; Shishmakov et al., 2010) or citrate-phoshate (pH 3.5-7.8) and citrate-salt (pH 1.6-4.8) buffer solutions (used for hybride systems) (Mekhaev et al., 2011a, 2011b). to vary the pH within the range of NR sensitivity of 2.5 – 7.5.

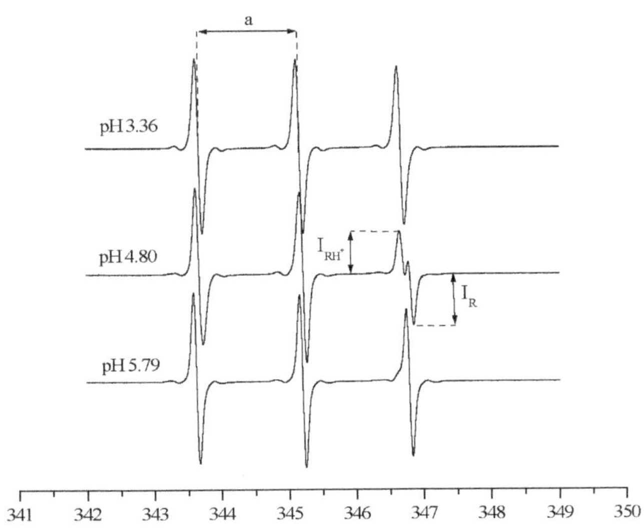

Figure 1. The ESR spectra of the aqueous solution of NR R3 at different pH in the X range of wavelengths at 293K. I_{RH^+} and I_R are the intensities of ESR peaks for RH^+ and R forms of the radical, respectively

Radical	pK_a (±0.1)	g-factor (±0.0001)		a_N (±0.006 mT)	
		R	RH^+	R	RH^+
R1 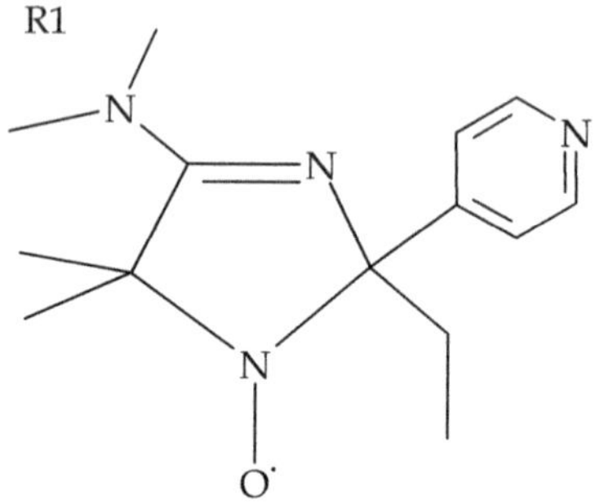	3.15 4.89	2.0048	2.0051	1.520	1.390
R2 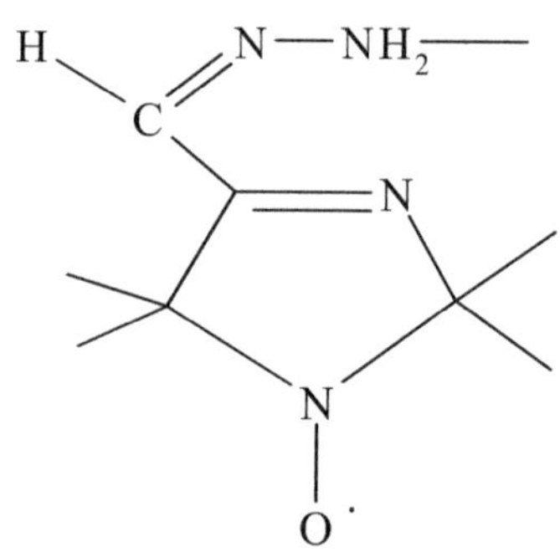	3.55	2.0048	2.0051	1.590	1.515
R3 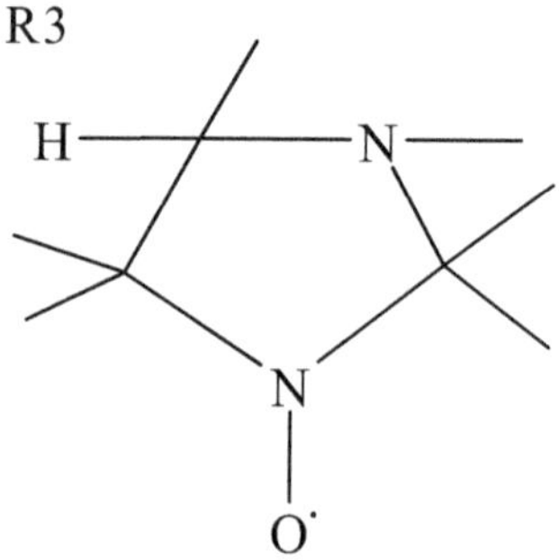	4.70	2.0048	2.0051	1.590	1.485

Table 3. ESR paramters and pK_a values of nitroxide radicals used

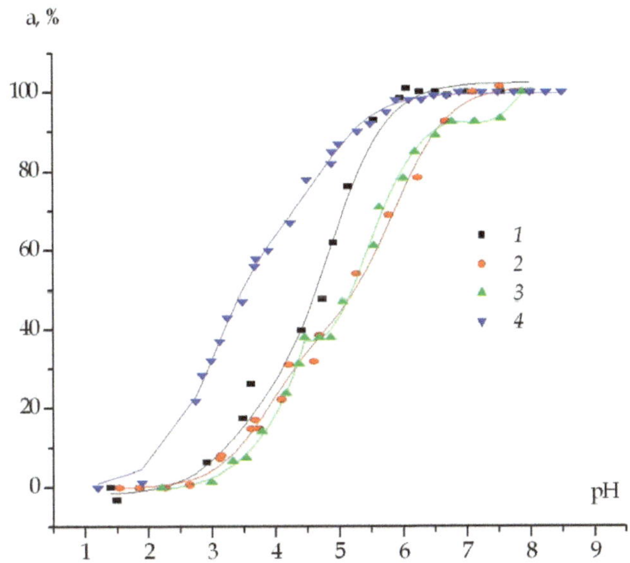

Figure 2. Titration curves for NR R1 in bulk aqueous solution (calibration curve) (1), α-Al_2O_3 (2), the BS-50 SiO_2 (3) and γ- Al_2O_3 (4). a, % = $((a-a_{NRH^+})/(a_{NR}-a_{NRH^+}))\times 100\%$

2.5. Determination of pH in the pore and near the sample surface using pH-sensitive spin probes

An aqueous KCl solution (10 ml) with an ionic strength of 0.1 was added to an oxide sample (200 mg) and the mixture was allowed to stand for a preset time. Then, the solution was thoroughly decanted and an NR solution (10^{-4} mol/L, μ = 0.1) was added to the sample. In some cases, required initial pH values of radical solution were obtained by preliminary mixing of HCl and KOH solutions. After the equilibrium was established, the suspension was titrated with HCl and KOH (HNO_3 and NaOH) solutions to plot the titration curve for the NR present in the sample.

For chitosan cobalt-containing hybrid systems and solid-phase composites based on SiO_2, TiO_2 and cellulose powder the method of multiply batches was used: 0.05 g of sample was kept in 5 ml of buffer aqueous solution containing nitroxide radicals for 2 days (established experimentally). The solution was then decanted.

The pH values of the equilibrium solutions (pH_{ext}) over the samples were measured using a Mettler Toledo pH meter (Switzerland) with an accuracy of 0.01 units. The samples separated from the solutions by centrifugation or filtration were placed into unsealed quartz ampules and their ESR spectra were recorded. After measuring the *a* distances in the ESR spectra of corresponding radicals located in the samples (Fig. 4), the pH_{int} values of the studied materials were determined using the calibration curves (Fig. 2,3).

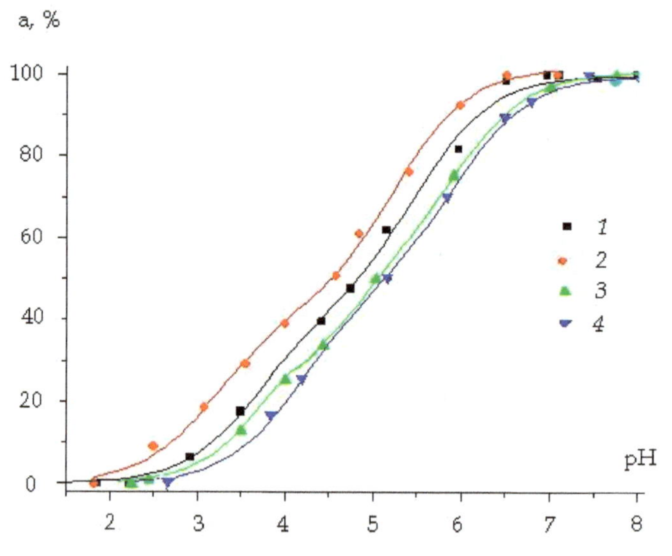

Figure 3. Titration curves for NR R1 in bulk aqueous solution (calibration curve) (1), PC (2), SiO_2 (3) and TiO_2 (4) xerogels. a, % = ((a-aNRH+)/(aNR-aNRH+))×100%

As can be seen from Fig. 4, the ESR spectra of NR in the samples studied represent the superpositions of three components of an isotropic signal of the probes in aqueous solutions inside pores and a spectrum of the probes immobilized on the surface of the objects studied. For determination of pH_{int} values, only the isotropic signals in the ESR spectra were used.

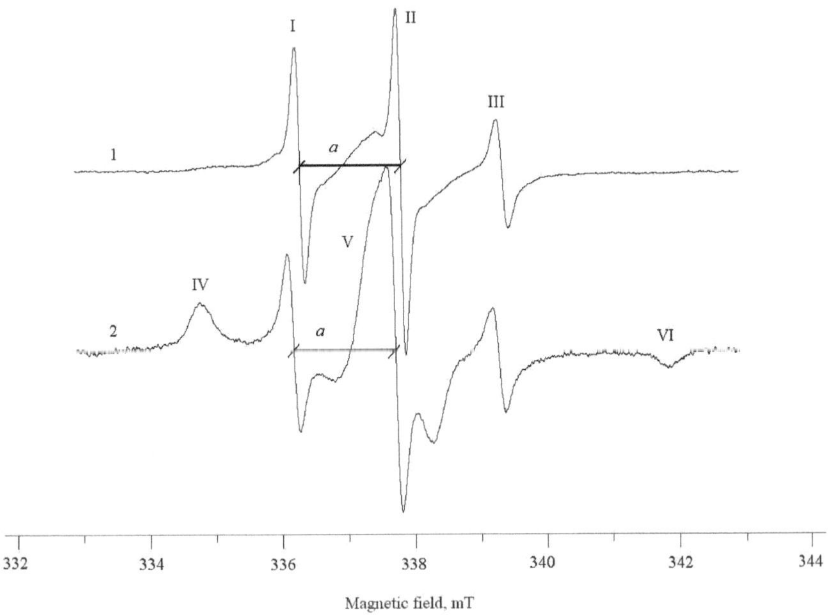

I, II,II – components of an isotropic spectrum ; IV, V, VI – components of the spectrum of the immobilized probes

Figure 4. The ESR spectra of NR R1 in the samples of CMs SiO_2-PC (68% wt.SiO_2) at pH 7.8 (1) and pH 3.7(2)

3. Results and discussion. Acid-base equilibria of inorganic and organo-inorganic materials and systems

3.1. Inorganic oxide materials

3.1.1. Pure oxides

The comparison of the titration curves for NR R1 radical in **γ-Al_2O_3, α-Al_2O_3, the BS-50 type SiO_2 and TiO_2 and SiO_2** xerogels with the calibration curve of this NR (Fig. 2, 3) indicates that for γ-Al_2O_3, the curve is shifted to the left, and for other samples, to the right. As was shown previously for organic sorbents (Molochnikov et al., 2004), the left- side shift of the titration curves for the NRs used is characteristic of anionites in acidic solutions in the pH range, where amino groups bind hydrogen ions, for example,

$$P\text{-}NH_2 + H^+ \leftrightarrows P\text{-}NH_3^+ \quad (1)$$

where *P* is the sorbent matrix. At the same time, the shift in the titration curves to the right is related to the dissociation of acidic functional groups of cationites, for example, a carboxylic one,

$$\text{P-COOH} \leftrightarrows \text{P-COO}^- + \text{H}^+ \quad (2)$$

Thus, the shift of the titration curve of R1 to the left observed for γ-Al_2O_3 indicates the binding of H^+ ions with the surface (pH_{intr} is higher than solution pH) and to a positive charge of γ-Al_2O_3 surface, while the right-hand side shift of the titration curve of this NR in SiO_2 gel suggests the release of H^+ ions and, consequently, a negative surface charge.

According to the data reported in (Hubbard, 2002) the acidic dissociation of hydrated aluminum oxide occurs in two stages, which can be described by the following equations:

$$\text{-AlOH}_2^+ + H_2O \leftrightarrows \text{-AlOH} + H_3O^+, \quad (3)$$

$$\text{-AlOH} + H_2O \leftrightarrows \text{-AlO}^- + H_3O^+, \quad (4)$$

The pK_avalues of these equilibria (pK_{a1}= 5.87 and pK_{a2} = 7.50) are given in (Lidin et al., 1987). Using NR, we can determine only pH values relevant to the ascending part of its S-shaped calibration curve (Fig. 2, 3 curve 1), i.e., the range, in which variations in *a* parameter correspond to variations in pH. The range of pH values appropriate to the ascending part of the calibration curve for a given NR is referred to as the zone of its sensitivity. The sensitivity zone of NR R1 falls in the pH range from 2.5 to 6.5 (Fig. 2, 3). If the above data on pK_a reflect (even if qualitatively) the pK_a of functional group dissociation for γ-Al_2O_3 studied, then the comparison of the sensitivity zone of NR R1 with pK_a values indicates that the use of NR allows one to study only the pH range in which γ-Al_2O_3 either has a positive charge or is electrically neutral. On the other hand, there are data on the point of zero charge (PZC) for Al_2O_3 in the literature.

For example, as was shown previously, the PZC of an Al_2O_3 sample kept in water for a long time was equal to 9.2, and for Al_2O_3 treated at 1400°C, the PZC decreased to 6.7 (Robinson et al., 1964). The first value corresponds to γ-Al_2O_3, and the second one, to α-Al_2O_3. For hydrated Al_2O_3 the PZC was found to be equal to 8, and it decreases as a result of the modification of Al_2O_3 surface with alkyl phosphate-based surfactants (Jeon et al., 1996). Moreover, these data suggest that, in the studied pH range, the surface of γ-Al_2O_3 has a positive charge that is in accordance with the shift of its titration curve to the left.

The pH_{int} of the sorbents differ from the pH_{ext} by 0.5–1.5 units (Table 3). Smaller pH_{intr} values compared to pH of the external solutions for all the studied samples (except for γ-Al_2O_3) are indicative of the shift in titration curves of NR R1 to the right relative to the calibration curve, similarly to that demonstrated in Fig. 2, 3 all the samples excepting γ-Al_2O_3.

It should be noted that the pH_{int} value determined in this study for the hydrated α-Al_2O_3 is in good agreement with the PZC value found in (Robinson et al., 1964). From the data given in Table 3 we notice that in all the studied samples excepting γ-Al_2O_3 acidic centers predominate. However, the acidity of these centers is rather weak; their pK_a of dissociation are presumably close to 6. Only when strongly acidic sulfate residues appear in Al_2O_3, a drastic decrease in pH_{int} is observed as a result of their dissociation. It should be noted that the same decrease in pH_{int} of TiO_2 hydrogel results from the sorption of Cu^{2+} ions by the hydrogel, which leads to the ion-exchange displacement of H^+ ions (Fig. 5).

Sample	pH_{ext} (± 0.01)	pH_{int} (± 0.1)
α - Al_2O_3	7.72	6.6
Al_2O_3 – F^-	6.9	6.4
Al_2O_3 – SO_4^{2-}	4.9	3.8
γ - Al_2O_3	7.16	>7.3
TiO_2-hydrogel	7.0	5.5
BS-50 type SiO_2	7.5	5.7
TiO_2-xerogel	7.0	6.1
SiO_2-xerogel	7.0	6.3

Table 4. pH values of the external bulk solution of NR R1 and in the pores and near a surface of the inorganic materials studied

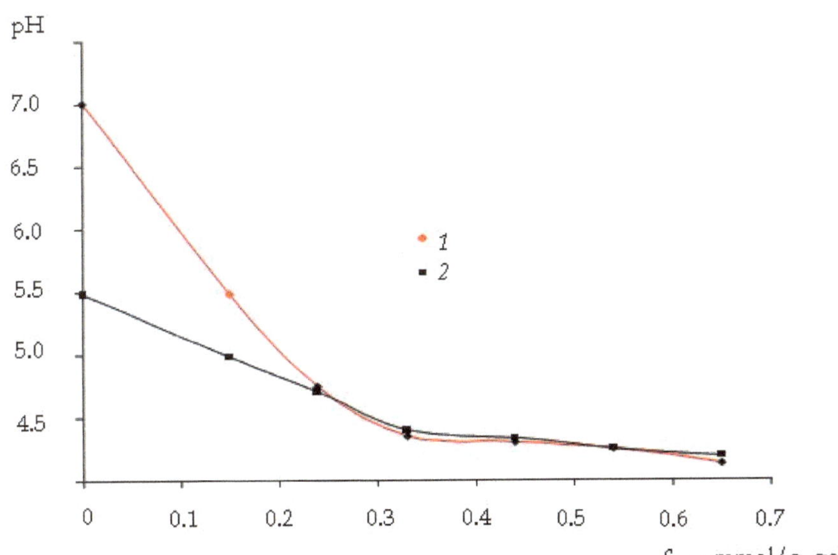

Figure 5. The pH values of a $CuCl_2$ equilibrium solution (pH_{ext}) (1) and a solution inside TiO_2 hydrogel (pH_{int}) (2) vs. the amount of sorbed Cu^{2+} (c_{Cu})

The titration curves of radicals occurring in organic sorbents are characterized by the presence of horizontal plateaus corresponding to constant pH_{int} (parameter *a* remains unchanged), when the pH of the external solution increases (Molochnikov et al., 2004). The presence of the horizontal plateaus on the titration curves is related to the consumption of the titrant solution for the neutralization of functional groups of ionites. There was found no horizontal plateau on the titration curve of NR R1 in γ-Al_2O_3 within its zone of sensitivity (Fig.2). This fact can be explained either by the presence of a small number of acidic functional groups in γ-Al_2O_3 or by noncoincidence of pK_a values of these acid groups with the zone of sensitivity of the radical used. Based on our experimental data, it is difficult to make a conclusion in favor of any of these assumptions.

It should be noted that the titration curve of NR R1 in the BS-50 SiO_2 contains two horizontal plateaus corresponding two constant pH_{int} determined near the SiO_2 surface when the pH of the external solution is changed (pH_{ext}) (Fig.2, curve 3). These plateaus are referred to as the titration process of functional groups on the surface (Golovkina et al., 2008). The pK_a values for these groups were determined from the titration curves, and were pK_{a1} = 4.4–4.7 and pK_{a2} = 6.5–6.8. According to (Long et al., 1999 ; Zhao et al., 1997, 1998) amorphous SiO_2 contains groups of three types: silanol, silandiol, and siloxane groups in a ratio of 59.2, 14.7, and 26.1%, respectively. We can therefore assume that the lower and longer horizontal plateau reflects the titration process of silanol functional groups on the SiO_2 surface, and the upper plateau corresponds to silandiol groups. According to (Méndez et al., 2003), the intervals of changes in the dissociation constants are pK_1 = 3.51–4.65 and pK_2 = 6.17–6.84, respectively. The same values of constants of dissociation for silanol and silandiol groups of silica were measured and published in (Nawrocki, 1997 ; Neue, 2000). The presence of silanol and silandiol groups on the surface of the studied SiO_2 was thus identified.

From two types of the xerogels studied (TiO_2 and SiO_2) electrical potential of TiO_2 xerogel is slightly more than that of SiO_2 xerogel (Fig.3). There was found a small horizontal plateau in the range of pH_{intr} between 3.75 and 4.25 on the titration curve for NR R1 near the SiO_2 surface. From this plateau pK_a values for silanol groups were determined. They were found to be equal to 3.95 ± 0,07 within the zone of constant pH_{int} values (Molochnikov et al., 1996). The measured values of pK_a are in good agreement with those determined before for silicas using chromatography(Méndez et al., 2003) and the method of pH probe in different silica-containing samples (Golovkina et al., 2008). At lesser pH values this curve goes near (but slightly righter) than the calibration curve of the above-mentioned NR. As seen from the figure, the surface of SiO_2 xerogel carries a small negative charge which is characteristic for the sample before titrating silanol groups. The lack of horizontal plateaus on the titration curves of NR R1 for the TiO_2 xerogel indicates the fact of titration of its functional groups at pH above 7. It can be explained by the presence of mainly terminal OH^- -groups with basic properties on a surface of TiO_2 xerogel. The range of their titration lies above the zone of sensitivity of the NR used.

When studying acidity inside organic sorbents or hydrogels, pH probes are placed into water, which penetrates into a solid due to sample swelling. In this case, the probes occur in water environment far away from the chains of organic (synthetic ion-exchange resins) or inorganic (TiO_2 and SiO_2 hydrogel, xerogel) polymers.

The particles of nanostructured oxides in diameter of several nm possess extremely large surface (the ratio of the number of atoms on the surface of a nanoparticle to the number of atoms located inside the particle is equal to 1/7). Therefore, a pH-sensitive NR surrounded by water molecules can approach the hydrated surface of nanoparticles rather closely. It was expected that they as probes and labels would appear to be sensitive to SEP, which would make it possible to estimate its value.

The dependences of NR ionization constants on the surface electric potentials and the polarity of media were reported by Fromherz (Fromherz , 1989) and Khramtsov et al. (Khramtsov et al., 1992) for fluorescent pH indicators and for pH-sensitive NRs, respectively, as follows:

$$\Delta pK_a = \Delta pK_a^{el} + \Delta pK_a^{pol} \tag{5}$$

where ΔpK_a is the total shift in the pK_a of an NR occurring in a sample relative to that of the NR in an aqueous solution and ΔpK_a^{el} and ΔpK_a^{pol} are the shifts in the pK_a of the NR occurring in the sample, which are due to th electric potential arising on the sample surface and to a change in the polarity of the medium relative to the polarity of the NR aqueous solution, respectively.

$$\Delta pK_a^{el} = -e \times \varphi / 2.3k \times T \tag{6}$$

$$\Delta pK_a^{pol} = b \times (\varepsilon - \varepsilon_{HOH}) \tag{7}$$

where φ is the electrostatic potential, e is the electron charge, k is the Boltzmann constant, T is absolute temperature, b is a proportionality coefficient dependent on the radical nature, ε is the medium permittivity, and ε_{HOH} is the permittivity of water.Figures 6 and 7 demonstrate the results of studying nano- structured TiO_2. The measurements were performed using R3 probe at pH ≈ 5.5, which was close to the PZC of TiO_2 (5.7) (Poznyak et al., 1999), and R2 probe at pH ≈ 4.3. Near the surface of TiO_2 nanostructured particles, the titration curves of the radicals are shifted to the left, whereas, for TiO_2 hydrogel and xerogel (Fig.3), a shift to the right was observed. At present, we can not find any explanation for such principal different shift in the titration curves for TiO_2 –based samples. The TiO_2 hydrogel obtained by hydrolysis was washed out with water, removing acidic components of the initial salts. Hereafter, the samples were prepared at different drying temperatures. The samples of xerogels and nanostructured oxides were dried at 100°C and 200°C, respectively.

As was mentioned above, the shift of the titration curve for γ-Al_2O_3 to the left corresponds to the positive charge of the particle surface that, in principle, correlates with the data obtained at pH below PZC. The left-shift (toward more acidic solutions) of the titration curve of the probe located near the surface of TiO_2 nanoparticles indicates that the surface electric potential is also positive. No changes were found in the hyperfine splitting constants a_N for both deprotonated and protonated R3 radicals (Fig. 6). The curves of titration of NR

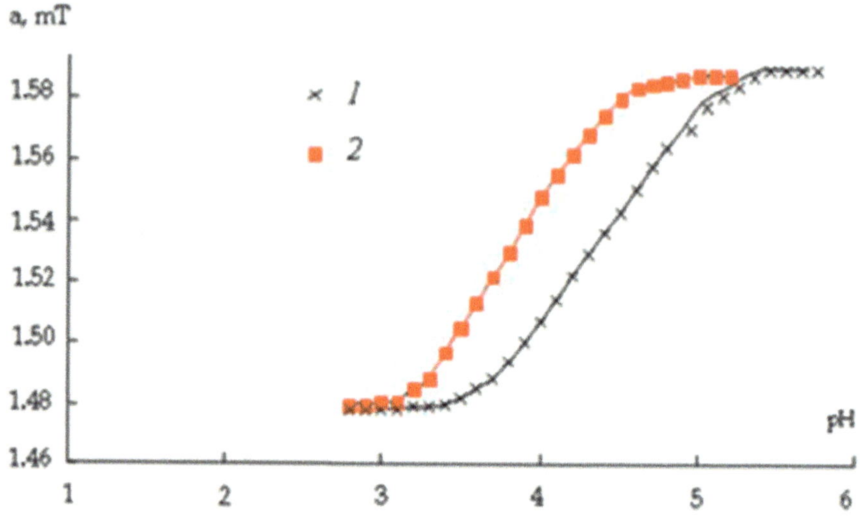

Figure 6. Titration curves for NR R3 in bulk aqueous solution (calibration curve) (1) and near the surface of nanostructured TiO_2 (2) at 25°C and $\mu = 0.1$

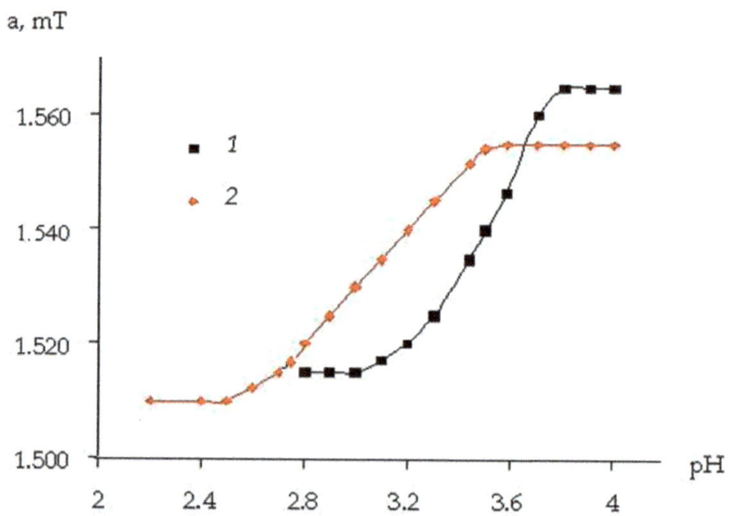

Figure 7. Titration curves for NR R2 (1) in bulk aqueous solution and (2) near the surface of nanostructured TiO_2 at 25°C and $\mu = 0.1$. The amount of Cu^{2+} ions sorbed in the pores of TiO_2 is 0.36 mmol/g of the sample

R3 in the solution (curve 1) and near the surface of the nanostructured TiO_2 (curve 2) begin and end at the same value of parameter *a* characterizing the hyperfine splitting constant a_N for the protonated and deprotonated forms of NR. This fact shows that there are no changes in the polarity of the medium on the surface of nanostructured TiO_2 relative to the external solution, and the term ΔpK_a^{pol} calculated by formula (7) is reduced to zero (Khramtsov et al., 1992). Thus, the shift in pK_a is determined by the electric potential of nanoparticle surface alone ($\Delta pK_a = \Delta pK_a^{el}$). The calculation by formula (6) gives $\varphi = 31.7$ mV at the point of R3 probe location. It should be stressed that the calculated φ value is not the electric potential of TiO_2 nanoparticle surface, but it only characterizes the electric field generated by a nanoparticle at the site where the radical fragment –N–O• of R3 probe is located.

3.1.2. *Cu^{2+}-containing samples*

For TiO_2 hydrogel, pH values of the external solution, which is in contact with the hydrogel (pH_{ext}), and of those inside the hydrogel (pH_{int}) during the sorption of Cu^{2+} ions from copper(II) chloride and nitrate solutions are found to be different (Fig. 5). It was observed that the anions of these copper(II) salts had no effect on the acidity of a medium. In the process of sorption Cu^{2+} ions are involved in the chemical interaction with the active centers of the hydrogel (Kharchuk et al., 2004) that manifests itself both in the parameters of the ESR spectra of Cu^{2+} in the hydrogel and in the impossibility of washing out sorbed Cu^{2+} ions from the hydrogel by Na^+ and Ca^{2+} nitrate solutions. According to the data obtained, initially, pH_{int} is lower by 1.5 units than pH_{ext} (Table 3). As the amount of Cu^{2+} ions in the hydrogel grows, pH_{int} decreases (Fig. 5), and, beginning with an amount of 0.24 mmol/g of the gel, it coincides with pH_{ext}. Thus, the acidity inside the original hydrogel sample is considerably lower than that of the equilibrium external solution. Similar regularities were found for granulated organic ion-exchange resins (Kovaleva et al., 2000). For KB-2 × 4 cationite in a mixed H^+–Na^+ form, it was found that in the process of absorption of copper, the acidity of the external solution diminishes, while the acidity inside the sorbent remains unchanged until a certain amount of Cu^{2+} ions in the ionite is attained, with this content being dependent on the fraction of Na^+ form of the cationite When this value is exceeded, the acidity inside the sorbent begins to decrease and pH_{ext} and pH_{ext} values for the ionite become equal at a certain degree of saturating with Cu^{2+} -ions. For TiO_2 hydrogel, the equalization of pH values is also observed, but the horizontal plateau with a constant pH_{int} value on the titration curve of NR is not observed. The constant pH_{int} values for KB-2 × 4 cationite are explained by the buffer properties of its H^+–Na^+ mixed form. TiO_2 hydrogel has no buffer properties, and the acidity increases smoothly with a rise in the amount of sorbed Cu^{2+} ions.

For nanostructural TiO_2, in the presence of sorbed Cu^{2+} ions, at pH close to 4.3 (Fig. 7), like in the case of pure nano -TiO_2 (Fig.6), a shift of the titration curve toward lower pH values is observed; however, the hyperfine splitting constant a_N decreases (the titration curve is shifted downward). The decrease in a_N is probably related to a lower polarity (the effective permittivity of the medium) near the nanoparticle surface where the probe is located. The theory predicts that the permittivity of a medium must have the same effect on both electroneutral (the deprotonated form of the radical) and charged (the protonated form) species (Griffith et al., 1974). We found that the rise in the Cu^{2+} amount from 0 to 0.36 mmol/g. of the gel results in an increase in the ΔpK_a value. The shift of this curve (ΔpK_a) to the left is significantly smaller that for NR R3, although the change from the PZC to pH 4.3 and the incorporation of Cu^{2+} ions must enhance the positive charge of the nanoparticle surface and, according to formula (6), increase in ΔpK_a. In accordance with formula (7), a reduction in the permittivity near the surface of TiO_2 particles compared to ε_{HOH}, which causes the observed decrease in a_N, must also shift the titration curve to the left.

Hence, the explicit inconsistency between the determined ΔpK_a values for NR R2 and NR R3 radicals and the theoretical predictions is established. We relate this difference to a larger distance of NR R2 from the surface compared to NR R3; this is possible because NR R2 has a substituent in position 4, which is presumably positioned toward the surface due to the tendency of its amino groups to protonation (or complexation with Cu^{2+} ions sorbed on the surface). Thus, the –N–O• fragment of the radical turns out to be removed from the surface of TiO_2 by the chain length of this substituent. The electric potential induced on the –N–O• fragment will decrease with the distance from the surface and will be inversely proportional to the distance from this surface, when it is represented as, for example, a plane. The small NR R3 radical can approach the nanoparticle surface much closer and, therefore, it is affected more strongly by the electrostatic potential of the TiO_2 surface. The sorption of Cu^{2+} ions on the surface of nanoparticles increases the charge of the latter. The observed shift of the titration curve (Fig. 7) and the broadening of the ESR spectra of these radicals attest to the orientation of the substituent in position 4 of NR R2 toward the surface and to the interaction between amino groups of the substituent and the Cu^{2+} ions sorbed on the nanoparticle surface.

3.2. Organic supports of composite and hybride materials

For PC and MCC, the titration curves of NR R1 were found to be shifted to the left of the calibration curve (Fig. 2, 3), indicating a posiive charge on the surface (Kovaleva et al., 2000 ; Molochnikov et al., 2007). Hydroxyl groups of celulose most likely play the role of surface bases whose protonation gives the surface a positive charge.

3.3. Solid-phase composites based on SiO_2, TiO_2 and cellulose powder

3.3.1. *Pure systems*

As shown in Fig. 8 and 9, the titration curves of NR R1 in the CMs TiO_2 – PC with 70 (not shown) and 53 % wt. TiO_2 and in the SiO_2 – PC with 68 and 35 % wt. SiO_2 (unprotonated parts) are shifted to the right relative to the calibration curve, as in the case of TiO_2 and SiO_2 xerogels. For CMs based on SiO_2, TiO_2 and cellulose powder $\Delta pH = pH_{ext} - pH_{int}$ decreased (the curves are shifted to the left relative to those of pure xerogels) as the PC content in the samples increased; this corresponds to a decrease in the negative charge of the CM surface. This is because of the lower acidity of cellulose compared to the acidity of the solution (cellulose has basic alcohol functional groups in its structure) and the positive charge of its surface (Parshina, 2011).

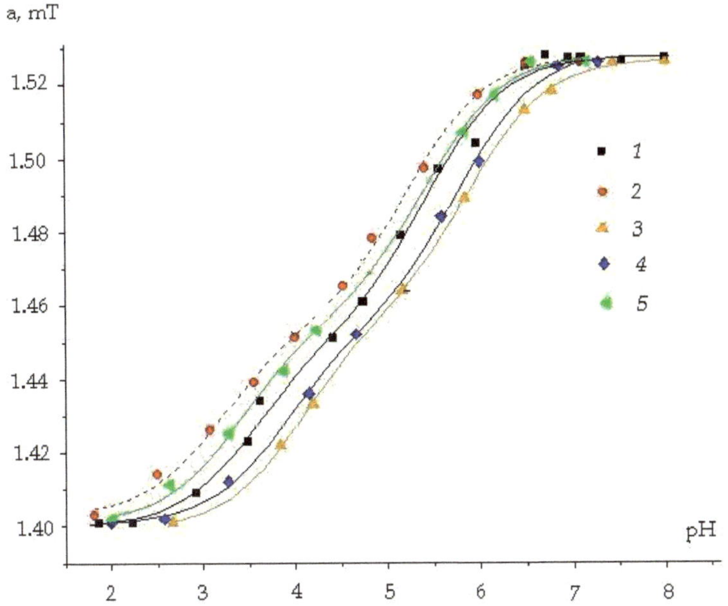

Figure 8. Titration curves for NR R1 in bulk solution (1), PC (2), TiO_2 xerogel (3) и CMs TiO_2 : PC (53% wt.TiO_2)(4) and TiO_2 : PC (43% wt.TiO_2) (5)

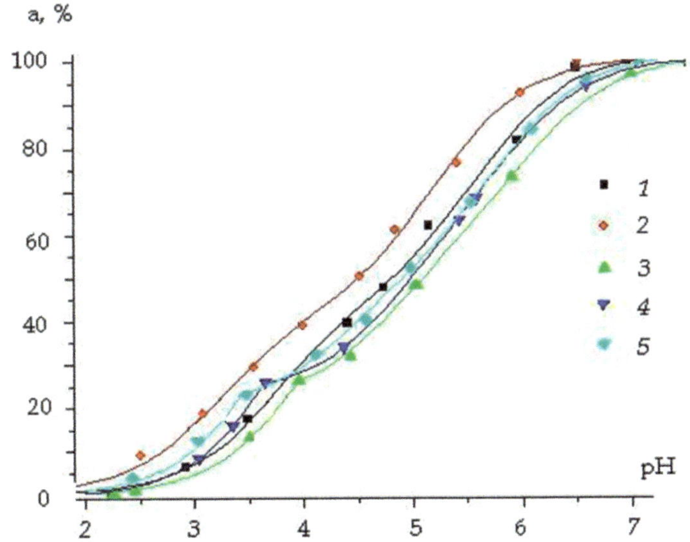

Figure 9. Titration curves for NR R1 in bulk solution (calibration curve) (1), PC (2), SiO_2 xerogel (3) and CMs SiO_2 : PC (58% wt. SiO_2) (4) and SiO_2 : PC (35% wt. SiO_2) (5). a, % = $((a-a_{NRH^+})/(a_{NR}-a_{NRH^+}))\times100\%$

This proves that a surface of the composites studied carries lesser negative charge as compared to that of pure xerogels. The left-shift of the curve of NR R1 in CM TiO_2 – PC with 43% wt.TiO_2 relative to the calibration curve can be explained by initial positive charge of the surface of this CM due to binding H^+ ions. Hence, SEP of CMs based on TiO_2 and PC is varied over a wide range and even changes its sign from negative to positive as an increase in PC content. From the curves plotted, we notice that a decrease in percentage of TiO_2 xerogel in CMs from от 53% до 43% leads to a pH_{intr} greater than a pH_{ext}. This fact can be explained by positive charge of a surface of cellulose due to the presence of alcohol groups in its structure.

For all the synthesized CMs based on SiO_2 xerogel and PC after complete protonating silanol groups, the titration curves of NR R1 were shifted to the left relative to the calibration curve and, hence, a surface of the samples studied remained positively charged (Fig.9). Unlike the composites based on TiO_2 and PC, an increase in a percentage of PC in the SiO_2-PC composites leads to changing a surface charge from positive to negative with increasing pH_{ext} (above the horizontal plateau on the titration curves) due to dissociation of functional groups. Thus, the incorporation of PC into the samples doesn't change acidity of silanol groups and doesn't make a polarizing effect on the SiO-H bond. The length of the horizontal plateau slightly increases in the accordance with the amount of silanol groups and the percentage of PC in the CMs based on SiO_2. This can be caused by the increased dispersity of SiO_2 due to rising in S_{sp} of the CMs (Table 1).

By varying the cellulose percentage in the composites and pH_{ext} values, pH_{int} values and SEP can be selected over a wide range. This information is needed for optimization of the conditions for pH-dependent adsorption and catalytic processes through a choice of CM with a certain pH_{int} and SEP as catalyst substrate and adsorbent.

3.3.2. *Cu^{2+}-containing composites*

The sorption of Cu^{2+} on the TiO_2 and SiO_2 xerogels and on the related CMs is accompanied by a change in the pH in their phases. However, the dependences of $pH_{ext(int)}$ vs c_{Cu} (amount of sorbed Cu^{2+} ions) for the above- mentioned types of xerogels and the CMs based have some differences (Fig.10, 11).

From these figures we notice that :

- pH_{int} is different from pH_{ext}, and with no Cu^{2+} in an external solution (c_{Cu} = 0), pH_{ext} is greater than pH_{int}.
- With an increase in the amount of Cu^{2+} in the CMs, both the pH_{ext} and pH_{int} descrease.
- There is a horizontal plateau on the dependences of pH_{int} on c_{Cu} in the range of sorbed Cu^{2+} from 0.15 up to 0.3 mmol Cu^{2+}/g. TiO_2, SiO_2 xerogel, within which pH_{int} remains constant with increasing in c_{Cu}.
- At greater c_{Cu} pH_{int} and pH_{ext} come closer and become almost equal.

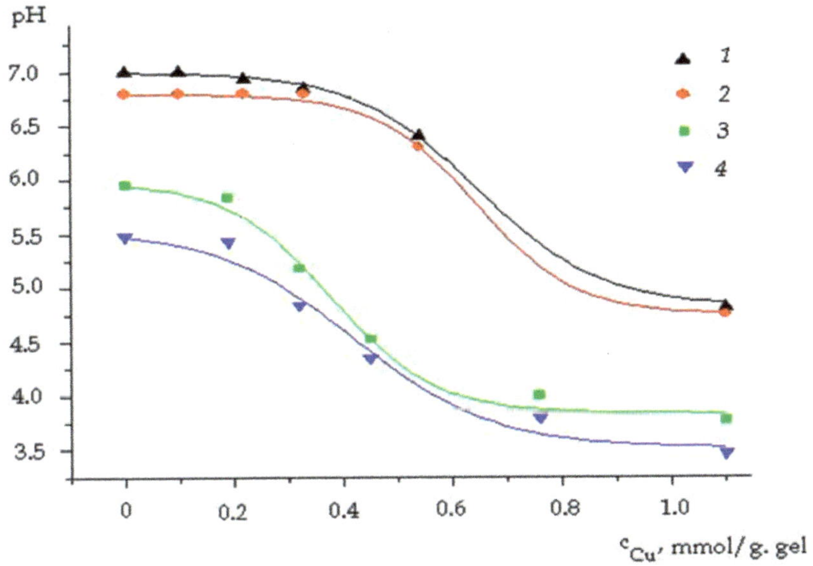

Figure 10. pH of external solution (pH_{ext}) (curves 1,3) and near the surface (pH_{int}) (curves 2,4) of TiO_2 xerogel (curves 1, 2) and the TiO_2 : PC (47% wt. TiO_2) composite (curves 3,4) vs. c_{Cu}

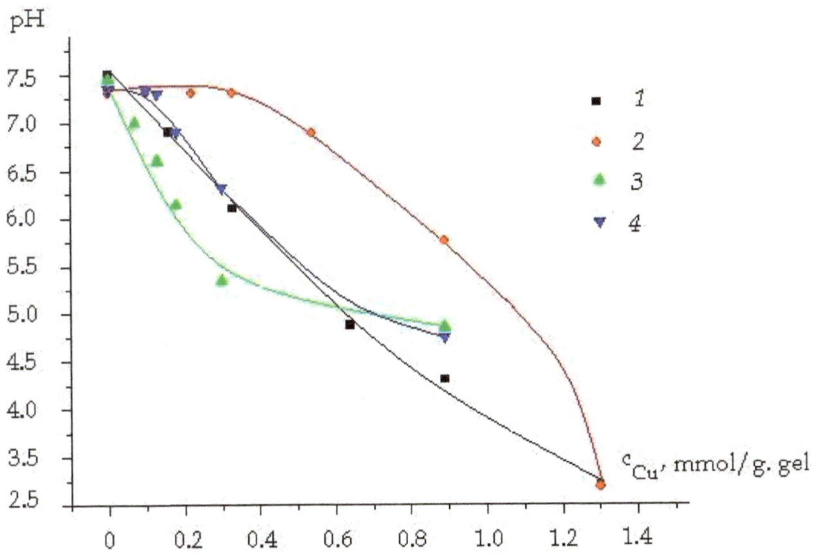

Figure 11. pH of external solution (pH_{ext}) (curves 1,3) and near the surface (pH_{int}) (curves 2,4) of SiO_2 xerogel (curves 1,2) and the SiO_2 : PC (68% wt. SiO_2) composite (curves 3,4) vs. c_{Cu}

The increase in acidity of external solution, first of all, is caused by the hydrolysis of Cu^{2+} at the initial pH_{ext} equal to 6-7 and is accompanied by intense release of H^+ ions. The lesser pH_{int} values as compared to pH_{ext} for both types of xerogels and the related CMs can be explained by a surface negative charge. A surface of the samples studied attracts H^+ ions in its double electrical layer to compensate this charge. Also an increase in acidity of both external and internal (near the surface) solutions is caused by the fact that the sorption of Cu^{2+} ions is accompanied by intense release of H^+ ions into them. The increase in the acidity of a medium due to the competition of H^+ ions almost completely suppresses the ion exchange sorption of Cu^{2+}. An analogous tendency was previously observed for organic ion-exchange resins (as grains) and TiO_2 hydrogel (Kovaleva et al., 2000 ; Molochnikov et al., 2004).

The presence of the horizontal plateaus on the curves pH_{int} vs c_{Cu} (Fig. 9) indicates the buffer properties of the TiO_2 xerogel and the TiO_2 : PC samples by analogy with the same properties of the mixed H^+–Na^+ form of the KB-2 × 4 cationite (Kovaleva et al., 2000). The greater is percentage of PC in a sample, the shorter is the horizontal plateau on the above-mentioned curves. The buffer effect of the TiO_2 xerogel and the related CMs is caused by the existence of some amount of deprotonated functional groups in the samples before the sorption of Cu^{2+}. The complexation of Cu^{2+} ions with these groups occurs first of all and does not lead to the release of H^+ ions, which would reduce both pH_{ext} and pH_{int}. In addition, releasing of hydroxide groups as a result of dissociation of molecules of $Cu(OH)_2$ responsible for complexation with functional groups of a surface of xerogel can retard decreasing pH with increasing c_{Cu}.

Unlike TiO_2 systems, for SiO_2 and the related CMs pH_{int} values were found to be more than pH_{ext} ones (Fig. 11). Apparently, this differance can be explained by the significant differences in pK_a of functional groups of SiO_2 and TiO_2 oxides. pK_a values of silanol groups (3.95 ± 0.07) are significantly lower than pH_{ext} from which sorption of Cu^{2+} has been conducted. Therefore, silanol groups are completely dissociated and H^+ ions don't exchange with Cu^{2+} ions during complexation. Hence, no functional groups can neutralize OH^-groups forming during breakdown of $Cu(OH)_2$ molecules. As a result, pH_{int} values (near a surface of samples) have high values.

As the titratuon curves for TiO_2-based CMs don't contain the horizontal plateaus as well as pure systems, hence, pK_a values of active acidic centers of a surface of these materials are out of zone of sensitivity of the NR used (greater than 7) (Fig. 8). Therefore, these centers are mainly in the protonated form before sorption of Cu^{2+}-ions. During complexation Cu^{2+}-ions exchange with H^+ ions of functional groups releasing them into external solution. Hence, pH_{int} became less than pH_{ext}. It shows up in a general decrease pH (both pH_{int} and pH_{ext}) with increasing c_{Cu} (Fig. 10).

3.4. Hybrid organo-inorganic materials based on the chitosan– Al_2O_3, chitosan–SiO_2, and cellulose –chitosan–systems

Hybrid systems were obtained in accordance with the scheme (Fig.12)

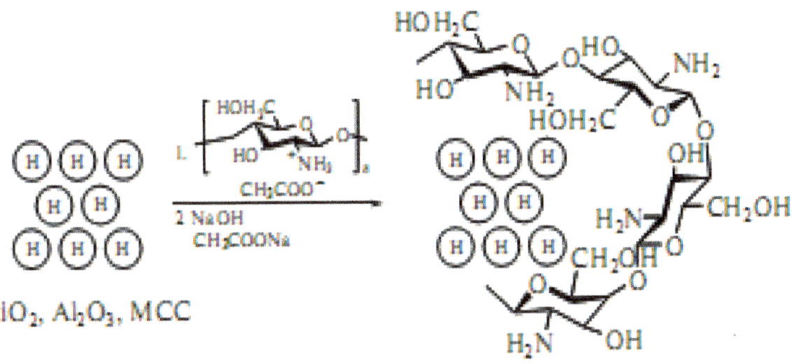

Figure 12. Scheme for obtaining hybrid organo-inorganic materials. s -support

The compositions of the obtained hybrid systems were characterized by elemental analysis (Table 2). According to the obtained data, the reactivity of substrates was different and declined in the order MCC >Al_2O_3 > SiO_2, since the compositions of the obtained hybrids in the case of inorganic oxides do not correspond to the molar ratios of the initial components.

The presence of chitosan molecules on the surface of the obtained systems was confirmed by the IR spectra, which contain characteristic absorption bands at 1652 and 1555 cm^{-1} corresponding to residual acetoamide groups of the polymer.

Since the obtained composites have to be electroneutral, we must assume that in the case of SiO_2, the substrate, being a weak acid, forms an ionic bond with the chitosan's amino group. In case of Al_2O_3, the same situation is possible, but some of the chitosan's amino and hydroxyl groups are involved in complex-forming with aluminum. In the case of MCC, the interaction occurs through hydrogen bonds.

NR were used as pH-probes to obtain a more detailed characterization of the surface structure of hybrid systems, their acid–base properties.

A comparison of the ESR spectra of NR R1 in solution and in the samples studied showed that in all cases, there was an isotropic signal which indicated the lack of covalent bonds between the NR and the sample surface and the presence of this radical near the surfaces of particles (Fig. 4, spectrum 1).

3.4.1. Pure hybride systems

Analysis of the structure of the Al_2O_3–chitosan system according to the adsorbed probe molecules of the nitroxyl radical showed that coating the initial α-Al_2O_3 substrate with chitosan leads to a slight increase in the surface negative charge, as was confirmed by the shift of the NR titration curve to the right (Fig. 13). The titration curves of α-Al_2O_3 and Al_2O_3–chitosan are not parallel to the calibration curve at low pH values due to the gradual dissolving of α-Al_2O_3 particles in weakly acidic media, as has been noted during investigations of samples containing aluminate ions (Iller, R., 1979 ; Golovkina, 2009).

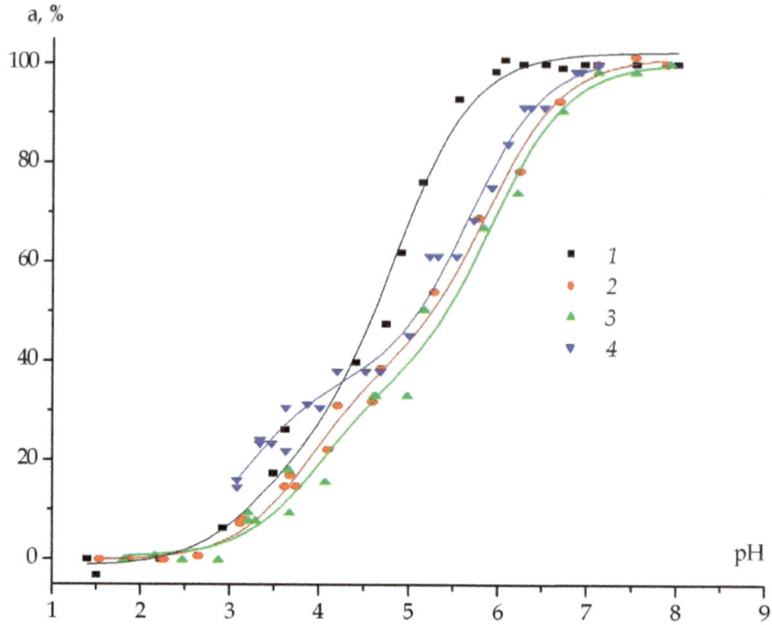

Figure 13. Titration curve for NR R1 in aqueous bulk solution (calibration curve) (1), near the surface of Al_2O_3 (2), hybride Al_2O_3– chitosan (3) and Al_2O_3– chitosan-Co^{2+} (4) systems . a, % = $((a-a_{NRH^+})/(a_{NR}-a_{NRH^+}))\times100\%$

For the SiO_2–chitosan system, the NR titration curve overlaps the corresponding curve for SiO_2 values within the horizontal section of the titration curve for silandiol groups in the range of high pH (Fig. 14). We can correspondingly claim that in this pH region, the surfaces of the samples all have the same charge, demonstrating that chitosan has no influence on the SiO_2 surface. If pH is decreased below 6, the titration curve for the SiO_2–chitosan system shifts to the right of the corresponding curve for SiO_2, indicating the negative charge of the surface. The horizontal pleteau of the titration curve for silanol groups is lower in the case of the SiO_2–chitosan system (pH 4.5–5), and the pK_a value for silanol groups in the presence of chitosan falls slightly. A similar reduction in this parameter has been observed during modification of the surface of mesoporous molecular sieves based on SiO_2 with aluminate and borate ions (Golovkina et al., 2009). Below the titration curve plateau for silanol groups (pH < 4.2), the titration curves of SiO_2 and SiO_2–chitosan overlap again, indicating a similar surface charge.

The deposition of chitosan on MCC causes the titration curve to shift to the right of to both the calibration curve and the NR titration curve of the initial MCC sample (Fig. 15), demonstrating the negative charge of the surface, as was observed for the Al_2O_3–chitosan and SiO_2–chitosan systems. A peculiarity of this substrate is the recharging of the surface, from positive for MCC to negative for the MCC–chitosan system during chitosan deposition. In this case, the type of change in the surface charge occurring upon a reduction in pH is the same for both the MCC and the MCC–chitosan systems. Changing symbatically with the curve typical of the initial MCC sample, the titration curve for the MCC–chitosan system almost overlaps the calibration curve in the acidic region (pH ≤4.5). In this pH range, the potential of the MCC–chitosan system surface is therefore close to zero.

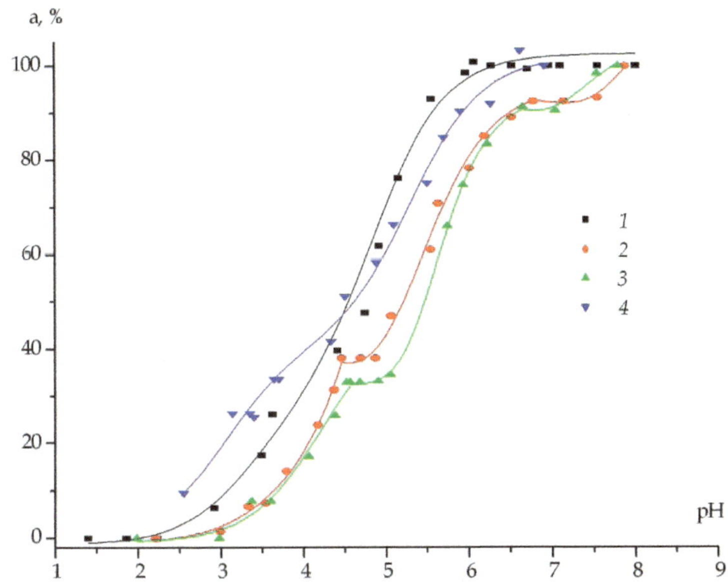

Figure 14. Titration curve for NR R1 in aqueous bulk solution (calibration curve) (1), near the surface of SiO_2 (2), of the hybride SiO_2– chitosan (3), of the SiO_2– chitosan-Co^{2+} (4) systems. a, % = $((a-a_{NRH^+})/(a_{NR}-a_{NRH^+}))\times100\%$

The deposition of chitosan on the substrate always causes the titration curve of the radical near the surface of hybrid material to shift to the right; i.e., it leads to a negative charge on the surface. While the deposition of chitosan leads to relatively slight changes in the surface potential in the case of inorganic substrates, these changes are so great in the case of MCC that they even lead to changes in the surface charge.

Some of differences in the behavior of titration curves for chitosan-containing materials on the inorganic substrates that occur in the range of high pH values could be due to the close pKa values of the silandiol groups of the SiO_2 substrate (pK_a = 6.5–6.8) and of the chitosan (pK_a = 6.42 (Skorik et al., 2003)). When the pH falls, amino groups of chitosan or silandiol groups become protonated and form hydrogen bonds with one another. As a result, the charge of the initial substrate surface and that of the hybrid material become similar. During further protonation of silandiol and amino groups at the same pH range, this interaction becomes impossible, and the surface charge of the SiO_2–chitosan hybrid material becomes more negative than that of the SiO_2.

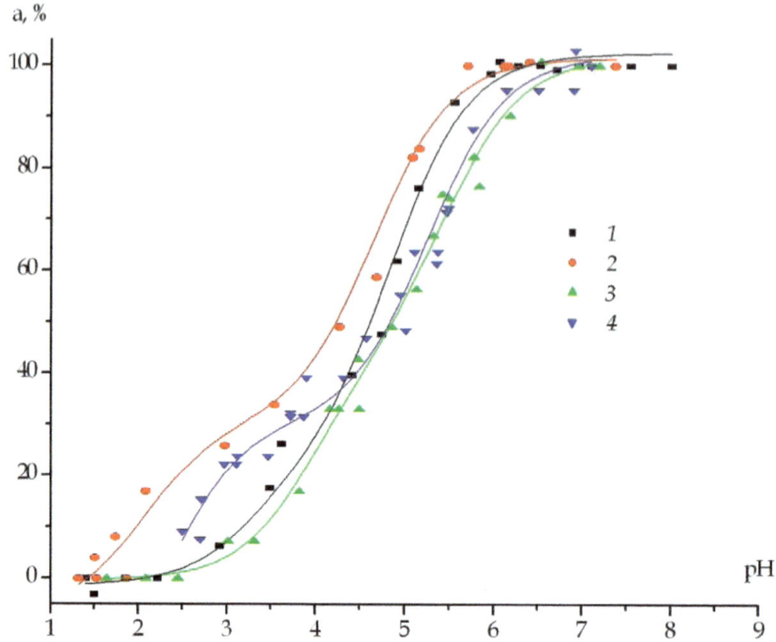

Figure 15. Titration curve for NR R1 in aqueous bulk solution (calibration curve) (1), near the surface of MCC (2), of the hybride MCC– chitosan (3), of the MCC– chitosan-Co^{2+} (4) systems . a, % = ((a-a_{NRH^+})/(a_{NR}-a_{NRH^+}))×100%

This is in complete agreement with the relative positions of the radical titration curves for Al_2O_3 and Al_2O_3–chitosan. It must be emphasized that the pK_a dissociation value of hydroxyl groups of Al_2O_3 is higher, and lies outside the radical sensitivity range.

Deposits of chitosan on the MCC substrate breaks it into smaller particles (Table 2), due most likely to the recharging of the surface from positive to negative. Chitosan molecules are appreciably smaller in size than MCC molecules, that is why a negative charge leads to increased repulsion inside large MCC particles, to their disintegration, and finally to an increase in the total surface area of the material. Chitosan thus plays the role of a disaggregating surfactant. In case of inorganic substrates, the deposition of chitosan leads to an increase in the particle size, and consequently to a decrease in their surface area (Table 1). Chitosan is therefore a weakly aggregating surfactant as to inorganic substrates.

The pK_a value for amino groups of chitosan is 6.42 (Skorik et.al, 2003). In the investigated pH range (pH ≤ 6), polymer molecules must consequently be positively charged due to the formation of – NH_3^+ ammonium groups. If chitosan molecules were adsorbed on the surface of inorganic oxides so that the protonated amino groups were on the material surface, this kind of adsorption would be reflected in the titration curves of the modified samples and would cause the titration curve to shift to the left of the titration curves of the initial oxides. Since this did not happen, we can assume that during the formation of hybrid material, chitosan amino groups form weak hydrogen bonds or Van der Waals bonds with functional groups of substrates (organic or inorganic), leading to an arrangement of chitosan molecules in which glucosamine rings are turned in the direction of the substrate.

3.4.2. Co^{2+} -containing hybride systems

The presence on the support surface of a polymer that is capable of acquiring metal ions of high coordination numbers is needed to fix these ions more firmly on metal-containing hybride materials. This role was played by chitosan on the surfaces of the inorganic (SiO_2, Al_2O_3) and organic (MCC) supports. Co^{2+} ions were sorbed from an aqueous ethanol solution (Fig. 16).

The composition of the studied systems was characterized by elemental analysis. According to the data, the sorption capacity of the hybrid systems relative to Co^{2+} ions is different and declines in the following order: SiO_2 >Al_2O_3 > MCC. We should note that in this case, there is a difference between the mechanisms of binding ions for different supports, since their surfaces have the same coating influencing the process.

As follows from the elemental analysis, chitosan plays the role of a complexing agent. In this case, all amino groups are involved in the coordination by cobalt ions (ratio Co:NH_2 = 1:1-1.5).

In the case of the Al_2O_3–chitosan system, some of the amino groups do not participate in the complexation, but the Co : Cl ratio corresponds to the composition of the initial salt that provides the coordination mechanism for binding cobalt ions (Fig.17). In the case of the SiO_2–chitosan system, the Co : Cl ratio is ~1. The need to obey the law of electroneutrality for the obtained hybrid systems requires assuming that the support, in addition to being a weak acid, also plays a coordinating role with respect to Co^{2+} ions, which accords with the covalent mechanism of binding Co^{2+} ions (Fig. 17). In the case of the MCC–chitosan system, metal ions are sorbed as a basic salt, and hydroxyl groups on the surface provide weaker coordination binding than Al_2O_3.

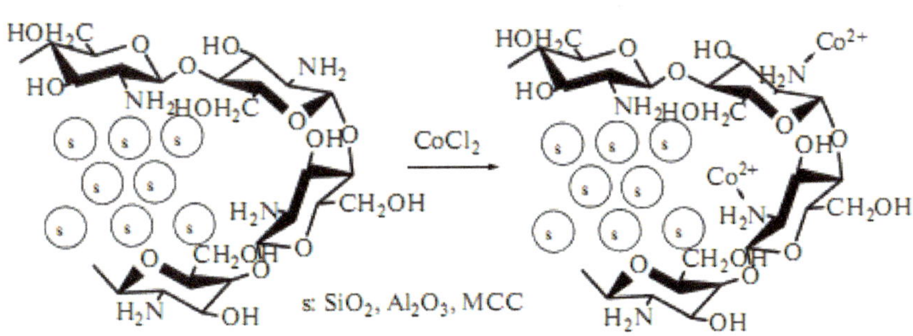

Figure 16. Scheme for synthesis of cobalt-containing hybrid systems. s - support

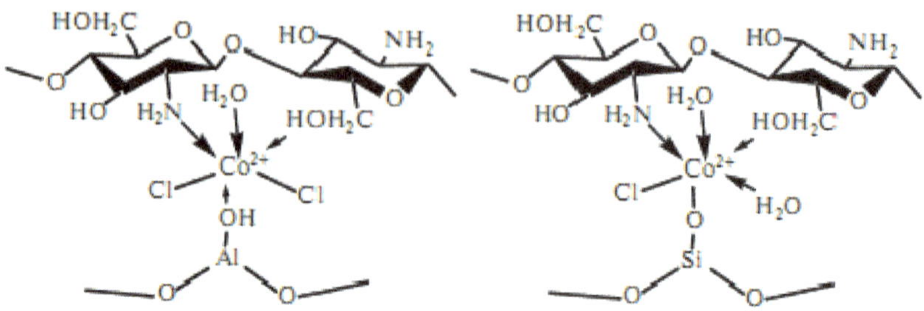

Figure 17. Scheme for the mechanism of binding Co^{2+} by hybrid systems

The elemental composition of the surface shows that chitosan does not cover it completely and some part of the functional groups remains on the support surface in free form. The more complex structure of the surface of hybrid system was characterized by investigating its acid–base properties via ESR spectroscopy of the NR used as pH probes. The titration curves of NR R1 on the surface of the Co^{2+}-containing system are shown in Figs. 13-15. The shift in these curves relative to the calibration curve left or right allows us to determine whether the surface has a positive or negative charge, respectively (Kovaleva et al., 2000 ; Molochnikov et al., 2007).

The technique for introducing Co^{2+} ions into the hybrid materials from the ethanol solutions could lead to the formation of sediments of basic cobalt chloride or chloride–alcoxide micelles on the surface of these materials. In the case of the MCC–chitosan (Fig. 15) and Al_2O_3–chitosan (Fig. 13) systems, the titration curves of the cobalt-containing materials at pH > 5 are shifted slightly to the left relative to the titration curves of the initial samples. Consequently, the occurrence of basic cobalt chloride on the surface of these materials does not lead to a change in the surface charge, but changes its value slightly. This is due to the effective neutralization of a positive charge of Co^{2+} ions by negative chloride and hydroxide ions. In order to confirm our hypothesis, we present the published data on studying the sorption of Co^{2+} ions from aqueous solutions and on the nature of the interaction between the metal center and amino groups of chitosan. According to (Minimisawa et al., 1999) maximum adsorption starts to decrease with increasing pH due to the formation of cobalt hydroxocomplexes. The maximum sorption of cobalt ions by chitosan found at pH 6–8 in (Silva et.al, 2008) is in good agreement with the data from (Minimisawa et al., 1999) and is determined by the formation of $Co(OH)_2$ phase or slightly soluble basic salts. No chemical interaction with amino groups of chitosan occurs in this case (Zhao et al., 1998).

The SiO_2–chitosan system behaves differently. Even at the highest pH values, the titration curve of the sample modified with cobalt (Fig. 17) was appreciably shifted to the left relative to the NR curve of the initial sample, although it remains to the right of the calibration curve. Accordingly, the formation of the chloride hydroxyl cobalt micelles immediately leads to a considerable reduction in the negative surface charge of the SiO_2–chitosan system. The different behavior of the titration curve for system III (Table 2) is most likely associated with the initially lower amount of Cl^- ions in the cobalt micelles (Co : Cl = 1 : 1), i.e., to its higher amount of OH– ions. Acid sites on the SiO_2 surface (silanol groups) also likely interact with the basic cobalt chloride particles precipitating on the surface, thereby replacing the Cl^- ions. In both cases, the titration curve must shift left due to the neutralization of the negative surface charge. In the study of the samples in the aqueous medium at high pH values, the transformation for systems I–III (Table 2) thus occurs; hydration of the surface, followed by hydrolysis leading to the formation of colloidal particles based on chloro-hydroxo complexes of Co^{2+} ions, is observed.

These micelles of Co^{2+} chloro-hydroxocomplexes begin to dissolve at pH ≤ 5 for all investigated systems. The dependence of sorption of cobalt ions on pH (Minimisawa et al., 1999) shows that the maximum adsorption starts to decline at pH < 5, and Co^{2+} ions in the solution are in the form of aqua complexes (Zhao et al., 1998). The latter are sorbed by the chitosan primary amino groups, thereby charging the surface positively; in this case, the titration curves of systems I–III (Table 2) (Figs.13-15) are shifted left relative to the corresponding curves for the initial samples and to the calibration curve. A further reduction in pH leads to the neutralization of OH^- groups, and no further changes in pH near the particle surface are observed upon a change in pH_{ext}.

This indicates the presence of a horizontal plateau on the titration curve. The value of the horizontal section is lower for system III than for systems I and II (Table 2), since some of the OH^- groups are replaced by residues of silicic acid. At high pH values, Co^{2+} ions initially cause a substantial decrease in the negative surface charge for the case of system III; their transformation into the form coordinated by the chitosan - NH_2 groups thus influences the surface charge to a lesser degree upon declining pH. Our results indicate the participation of chitosan amino groups in the complexation with Co^{2+} ions. During the interaction of the hybrid materials with the Co^{2+}-containing solution, at least a part of glucosamine rings are consequently turned outward and are capable of becoming ligands.

4. Conclusions

pH-sensitive NRs gave reliable information on the local acidity of solutions in and the charge of a surface on pure and metal containing inorganic and organo-inorganic materials and systems and allowed to estimate an electric potential near the surface of TiO_2 nanoparticles.

The differences between the acidities of external solutions (pH_{ext}) and inside pores (or near the surface) of all the studied materials and systems (pH_{int}) were found.

The method of spin pH probes allowed to determine the ionization constants of characteristic functional groups of SiO_2-based systems from the horizontal plateaus corresponding to the constant pH_{int} in the samples.

An increase in concentration of H^+ ions (a decrease in pH_{int}) in solutions located inside α-Al_2O_3, TiO_2 hydrogel and near the surface of the BS-50 type SiO_2, TiO_2 and SiO_2 xerogels ; the related CMs and hybrid materials ; metal-containing systems, as compared to those of external solution can be explained by releasing H^+ ions due to dissociation of acidic functional groups, exchange them with metal ions and the partial desruption of hydrogen bonds. It leads to negative charge of a surface of the above-mentioned objects. A decrease in concentration of H^+ ions (an increase in pH_{int}) as compared to those of external solution were characteristic for γ-Al_2O_3 and cellulose matrixes. This resulted from binding H^+-ions by the surface of γ-Al_2O_3 and MCC and PC with basic functional groups such as -AlOH , –AlO $^-$ and OH^-, respectively. As a result, a surface gains a positive charge.

The sorption capacity of Cu^{2+} ions depends on a surface charge of the oxides gels, xerogels and the related CMs studied and decreases as a negative surface charge reduces. The sorption of Cu^{2+} ions on the surface of nanoparticles of nanostructural TiO_2 increases the charge of the latter. An increase in a percentage of PC in the SiO_2 –PC composites leads to an increase in the amount of silanol groups as a result of increasing in dispersivity of SiO_2 particles and specific surface (S_{sp}) of the samples, and to reducing a negative surface charge up to zero, and even its reversing. It led to the formation of $Cu(OH)_2$.

The deposition of chitosan on the substrate always creates a negative charge on the surface. While the deposition of chitosan leads to relatively slight changes in the surface potential in the case of inorganic substrates such as Al_2O_3 and SiO_2, these changes are so great in the case of MCC that they even lead to changes in the surface charge.

The charge of the surface of Co^{2+}-modified organo-inorganic hybrid materials at different pH_{int} was found to effect on the composition and structure of Co^{2+} -containing surface compounds.

The modification of a surface of powder cellulose with nanostructured SiO_2 and TiO_2 xerogels, aluminum oxides, silica and MCC with acidic functional groups and chitosan makes it possible to adjust the local acidity and surface charge over a wide range.

The study of the surface of organo–inorganic composites and hybrid materials and systems using pH-sensitive nitroxyl radicals allows also to reveal regularities in changing their properties during further modification. In addition, this method enables us to describe qualitatively the processes of structure formation in these systems and their effect on catalytic activity in different pH-dependent reactions.

The calculated φ value (31.7 mV) was found not to be the electric potential of TiO_2 nanoparticle surface, but it only characterizes the electric field generated by a nanoparticle at the site where the radical fragment –N–O• of NR is located. Once the anisotropic spectra of NR in nanostructured oxides are simulated, an electical potential of a surface can be determined. When measuring the SEP of solids, the knowledge of the distance between a radical and a surface is of principal importance. The fixation of pH-sensitive NRs on the surface of nanoparticles with linkers of a known length can solve this problem. This will allow one to calculate the potential immediately on the surface of nanoparticles and to compare the calculation results with the experimental data on the electrokinetic potentials.

Author details

Elena Kovaleva*
Ural Federal University,Russia

Leonid Molochnikov
Ural State Forest Engineering University, Russia

Acknowledgement

The authors are very grateful to :

- Prof. A.M. Volodin and Prof. A.I. Kulak for synthesizing and supplying aluminum oxides and nanostructured TiO_2 both original and modified with F– and SO_4^{2-} groups,
- Dr. I.A. Kirilyuk and Prof. I.A. Grigor'ev for synthesis of pH-sensitive nitroxide radicals,
- Dr. A.B. Shishmakov, Mrs. Yu. V. Mikushina and Dr. E. V. Parshina for synthesizing TiO_2 and SiO_2 hydrogels and xerogels, pure and Cu^{2+}-containing solid-phase composites based on nanostructured SiO_2 and TiO_2 and powder cellulose,
- Mr. A.V. Mechaev for synthesis and characteization of pure and Co^{2+}-containing hybrid organo-inorganic materials based on the chitosan–SiO_2, chitosan– Al_2O_3, and –chitosan–cellulose systems,
- Prof. A.I. Kokorin , Dr. A.V. Pestov and Prof. Yu.G. Yatluk for fruitful discussion of the study results.

The studies presented in this manuscript were financially supported by the Ministry of the Education of the Russian Federation (2007-2011, Theme 01.2.007-06425).

5. References

Airoldi C. & Monteiro, O. A. C. (2000). Chitosan-Organosilane Hybrides-Synthesis, Characterization, Copper Adsorption and Enzyme Immobilization. *J. Appl. Polym. Sci.*, Vol. 77, No.4, (July 2000), pp. 797 -804, ISSN 1097-4628

Borbat, P.P., Milov, A.D., Samoilova, R.I. et al. (1990). Vliyanie Razmera Por na Vrashchatel'nuyu Podvizhnost' Spinovogo Zonda v Silikagel'Ach (Effect of Pore Size on Rotational Motion of Spin Probe in Silica Gels). *Kolloidn. Zh. (Russ.Colloid J.)*, Vol. 52, No.2, (February 1990), pp. 341-345, ISSN 1061-933X (in Russian)

Budanova, N. Yu., Shapovalova, E. N. & Lopatin, S. A. (2001). Issledovanie Uderzhivayushchei i Razdelyaushei Sposobnosti Silikagelei, Midifitsirovannych Novym Chiral'nym Selektorom – Nizkomolekulyarnym Khitosanom (Study of Retentivity and Separating Capacity of Silica Gels Modified With a Novel Chiral

* Corresponding Author

Selector – Low Molecular Chitosan) *Vestn. Mosk. Univ., Ser. Khim.* Vol. 42, No.2, pp.112-115, ISSN 0579-9384 (in Russian)

Buyanov, R.A.(Ed.).(1998). *Kataliz i Katalizatory: Fundamental'nye Issledovaniya Instituta kataliza im. G.K. Boreskova* Sib. Otd. Ross. Akad. Nauk *(Catalysis and Catalysts. Fundamental Works of Boreskov Institute of Catalysis),* Siberian Branch of the Russian Academy of Sciences, ISBN 5769201320, Novosibirsk, Russia (in Russian)

Corma A., Concepción, P., Domínguez, I. et al. (2007). Gold Supported on a Biopolymer (Chitosan) Catalyzes the Regioselective Hydroamination of Alkynes. *J. Catal.* ,V.251, No. 1, (1 October, 2007), pp. 39-47, ISSN 0021-9517

Fromherz, P. (1989). Lipid Coumarin Dye as a Probe of Interfacial Electrical Potential in Biomembranes. *Methods Enzymol.,* Vol. 171, pp. 376-387, ISSN 0076-6879

Golovkina, E.L., Kovaleva, E.G., Molochnikov, L.S. et al. (2008). Method of Spin Probe for Study of MMS SBA-15. *Zhurnal Khromatographicheskikh I Sorbtsyonnykh Processov* (*Russ.J. Sorp. Chromat. Proc.),* Vol. 8, No. 6, (June 2008), pp. 971-985, ISSN 1680-0613 (in Russian)

Golovkina E. L. (2009). Electrochemical Properties of Mesoporous Molecular Sieves : Measurements and Study Using Spin pH-probe Method, Ph. D. Dissertation in Chemistry, Ural State Forest Engineering University, Yekaterinburg, Russia (defended at the Chelyabinsk State Pedagogical University, Chelyabinsk) (in Russian)

Griffith, O.H., Dehlinger, P. J. & Van, S. P. (1974). Shape of the Hydrophobic Barrier of Phospholipid Bilayers (Evidence for Water Penetration in Biological Membranes). *J. Membrane Biol.* V.15, No.1, (December 1974), pp. 159-192, ISSN 0022-2631

Hubbard, E.T., (Ed.) (2002). *Encyclopedia of Surface and Colloid Science,* Marcel Dekker, ISBN, New York

Huang, K., Liu, H., Dou, X. et al. (2003). Silica-Supported Chitosan-Osmium Tetroxide Complex Catalysed Vicinal Hydroxylation of Olefins Using Hexacyanoferrate (III) Ion as a Cooxidant. *Polym. Adv. Technol.,* Vol. 14, No.4, (April 2003), pp. 366-370, ISSN 1099-1581

Iller, R. (1979). *The Chemistry of Silica,* Wiley, New York

Jeon, J.S., Sperline, R.P., Raghavan, S. et al. (1996). In Situ Analysis of Alkyl Phosphate Surfactant Adsorption at the Alumina/Aqueous Solution Interface. *Colloids Surf., A,* Vol. 111, No.1, (June 1996) , pp. 29, ISSN 0927-7757

Khan, R., Kaushik A., Solanki, P. R. et al. (2008). Zinc Oxide Nanoparticles-Chitosan Composite Film for Cholesterol Biosensor. *Anal. Chim. Acta* , Vol. 616, No.2, (June 2008), pp. 207-213, ISSN 0003-2670

Kharchuk, V.G., Buldakova, L.Yu., Shishmakov, A.B., et al. (2004). Oxidative Dehydrogenation of 2,3,5-Trimethyl-1,4-Hydroquinine in the Presence of Titanium Dioxide Hydrogel. *Russ.J. General Chem.,* Vol. 74, No.1, (January 2004), pp. 101-104, ISSN 1070-3632

Khramtsov V. & Weiner L. (1988). Proton Transfer Reactions in Free Radicals. Spin pH probes. *Uspekhi Khimii* (*Russian Chem. Rev.*), Vol. 57, No.9, (September 1988), pp. 824-832, ISSN 0036-021X (in Russian)

Khramtsov, V.V., Marsh, D., Weiner, L.M., et al. (1992). The Application of pH-sensitive Spin Labels to Studies of Surface Potential and Polarity of Phospholipid Membranes and Proteins. *Biochim. Biophys. Acta*, Vol. 1104, pp. 317-324, ISSN 0006-3002

Khramtsov, V.V. & Volodarsky, L.B. (1998). Use of Imidazoline Nitroxides in Studies of Chemical Reactions: ESR Measurements of the Concentration and Reactivity of Protons, Thiols and Nitric Oxide. In : *Biological Magnetic Resonance, Vol. 14, Spin Labeling,* Berliner, L., Ed., p. 109-180, Plenum Press, ISBN 0306456443, New York

Kirilyuk, I.A., Bobko, A. A. Khramtsov, V. V. & Grigor'ev, I. A. (2005). Nitroxides with Two pK Values-useful Spin Probes for pH Monitoring Within a Broad Range. *Org. Biomol. Chem.* Vol.3, No. 7, pp. 1269-1274, ISSN 1477-0520

Kovaleva, E.G., Molochnikov, L.S. & Lipunov, I.N. (2000). The Influence of the Mixed H^+-Na^+ Form of KB- 2 Carboxyl Cationite and of pH within Ionite Grains on the State of Cu(II) Ions and the Catalytic Properties of Ionite Catalysts Containing Cu(II). *Rus.J.Phys. Chem.*, Vol. 74, No.8, (August, 2000), pp. 1262-1267, ISSN 0036-0244

Lidin, R.A., Andreeva, L.L. & Molochko, R.A. (1987). *Spravochnik Po Neorganicheskoi Khimii (Handbook on Inorganic Chemistry),* Khimiya (Chemistry) ISBN 5-7245-1163-0, Moscow (in Russian)

Lisichkin, G.V., Fadeev, A.Yu., Serdan, A.A. et al. (2003) *Khimiya Pryvitykh Poverkhnostnykh Soedinenii (Chemistry of Graft Surface Compounds),* FIZMATLIT , ISBN 978-5-9221-0342-8, Moscow, Russia (in Russian)

Li, G., Jiang, Y., Huang, K., Ding, P. et al. (2008). Preparation and Properties of Magnetic Fe_3O_4-chitosan Nanoparticles. *J. Alloys Compd.*, Vol. 466, No.2, (October 2008), pp. 451-456, ISSN 0925-8388

Liu, X. Tokura D. S. & Haruki, M. (2002). Surface Modification of Nonporous Glass Beads with Chitosan and Their Adsorption Property for Transition Metal Ions. *Carbohydr. Polym.*,Vol. 49, No.1, (July 20012), pp. 103-108, ISSN 0144-8617

Long, Y., Xu, T., Sun, Y. et al. (1999). Adsorption Behavior on Defect Structure of Mesoporous Molecular Sieves. *Langmuir,* Vol.15, No.19, (September 1999), pp. 6173-6178, ISSN 0743-7463

Martini, G., Bindi, M., Ottaviani, M.F., et al. (1985). Dipolar and Spin Exchange Effects in the ESR Spectra of Nitroxide Radicals in Solution: Part II. Water Solutions Adsorbed on Porous Silica Gels. *J.Colloid Interface Sci.* , Vol. 108, No.1, (November 1985), pp. 140- 148, ISSN 0021-9797

Mei, N., Xuguang,, L., Jinming, D. et al. (2009). Antibacterial Activity of Chitosan Coated Ag-loaded Nano-SiO_2 Composites. *Carbohydr. Polym.*,Vol. 78, No.1, (August 2009), pp. 54-59, ISSN 0144-8617

Mekhaev, A.V., Pestov, A.V., Moloch nikov, L.S. et al. (2011). Investigation of the Structure of Chitosan Hybrid Systems by pH-sensitive Nitroxide Radical. *Russ. J. of Phys. Chem. A,* Vol. 85, No.6, (June 2011), pp. 987-992, ISSN 0036-0244

Mekhaev, A.V, Pestov, A.V., Molochnikov, L.S. et al. (2011). Structure and Characteristics of Chitosan Cobalt-Containing Hybrid Systems, the Catalysts of Olefine Oxidation. *Russ. J. Phys. Chem. A,* Vol. 85, No.7, (July 2011), pp. 1155-1161., ISSN 0036-0244

Méndez, A.L, Bosch, Roses M. & Neue, U. D. (2003). Comparison of the Acidity of Residual Silanol Groups in Several Liquid Chromatography Columns. *J. Chromatogr. A* , Vol.986, No.1, (January 2003), pp. 33–44, ISSN 0021-9673

Mininisawa, H., Iwanami, H., Azai, N. et al. (1999). Adsorption Behavior of Cobalt(II) on Chitosan and Its Determination by Tungsten Metal Furnace Atomic Absorption Spectrometry. *Anal. Chim. Acta.*, Vol. 378, No.3, (January 1999), pp. 279-285, ISSN 0003-2670

Molochnikov, L.S., Kovalyova, E.G., Grigor'ev, I.A. et al. (1996). Determination of Acidity in the Interior of the Cross-linked Polyelectrolyte Grain by the Use of pH-Sensitive probes, In *Metal-Containing Polymeric Materials*, Pittman, Ch.U. et al., pp. 395-401, Plenum Press, ISBN 10 0306452956, New York

Molochnikov, L. S., Kovalyova, E.G., Golovkina E.L. et al. (2007). Method of Spin Probe for Studying Acidity of Inorganic Materials. *Russ.Colloid J. B,* Vol. 69, No.6, (December 2007), pp. 769-776, ISSN 1061-933X

Molochnikov, L.S., Kovalyova, E.G., Grigor'ev, I.A., et al. (2004). A. Direct Measuring of H^+ Activity Inside Cross-Linked Functional Polymers. *J. Phys. Chem. B,* 2004, Vol. 108, No.4, (December 2004), pp. 1302-1313, ISSN 1520-6106

Nawrocki, J. (1997). The Silanol Group and its Role in Liquid Chromatography *J. Chromatogr. A,* V. 779, No.1, (August 1997), pp. 29-71, ISSN 1520-6106

Neue, U. D. (2000). Silca Gel and Its Derivatization, In : *Encyclopedia of Analytical Chemistry,* Meyers, R. A. Ed.. Wiley, ISBN 9780470027318, Chichester.

Nikitin, N. I. (1962). *Chemistry of Wood and Cellulose,* No ISBN, Nauka, Leningrad, Russia, (in Russian)

Paradossi, G., Chiessi, E., Cavalieri, F. et al. (1998). Networks Based on Chitosan and Oxidized Cyclodextrin-II. Structure and Catalytic Features of a Copper (II)-loaded Network. *Polym. Gels & Networks,* Vol. 5, No.6, (April 1998), pp. 525-54, ISSN 0966-7822

Parshina, E.V., Molochnikov, L.S., Kovaleva, E.G. et al. (2011). Medium Acidity and Catalytic Properties of Composite Materials Based on Silica and Titanium and Powder Cellulose in the Presence of Cu^{2+} Ions. *Russ. J. Phys. Chem. A,*Vol. 85, No.3 , (March 2011), pp. 452-456, ISSN 0036-0244

Parshina, E.V. (2011) *Acid-base, Complexation and Catalytic Properties of Solid Phase Composits Based on Ddioxide Elements Xerogels and Powder Cellulose,* Ph. D. Dissertation in Chemistry, Ural State Forest Engineering University, Yekaterinburg, Russia (defended at the Chelyabinsk State Pedagogical University, Chelyabinsk) (in Russian)

Petrov, L.A., Kharchuk, V.G., Shishmakov, A.B. et al. (1998). Oxidation of 2,3,6-Trimethyl-1,4-Benzenediol by Oxygen in the Presence of Oxide Gels . *Russ. J. Org. Chem.* , Vol. 34, No. 7, (July 1998), pp. 344-346, ISSN 1070-4280

Poznyak, S.K., Pergushov, V.I., Kokorin, A.I., et al. (1999). Structure and Electrochemical Pproperties of Species Formed as a Result of Cu(II) ions Absorption onto TiO_2 Nanoparticles. *J. Phys. Chem., B,* 1999, Vol. 103, No.8, (February 1999), pp. 1308 – 1515, ISSN 1520-6106

Robinson, M., Pask, J.A. & Fuerstenau, D.W. (1964). Surface Charge of Alumina and Magnesia in Aqueous Media. *J. Am.Ceram. Soc.*, Vol. 47, pp. 516-520, No. 10, (October 1964), ISSN 1551-2916

Senso, A., Oliveros, L. & Minguillon, C. (1999). Chitosan Derivatives as Chiral Selectors Bonded on Allyl Silica Gel : Preparation, Characterization, and Study of the Resulting High-PerformanceLiquid Chromatography Chiral Stationary Phases. *J. Chromatogr., A*, Vol. 839, No.1, (April 1999), pp.15-21, ISSN 0021-9673

Silva, R.B., Neto, A.F., Santas, L.S. et al .(2008). Catalysts of Cu(II) and Co(II) Ions Adsorbed in Chitosan Used in Transesterification of Soy Bean and Babassu Oils - A New Route for Biodiesel Syntheses. *Biores. Technol.*, Vol.99, pp.6793-6798, No.15, (October 2008), ISSN 0960-8524

Shi, Q.-H. , Tian, Y., Dong, X.-Y. et al. (2003). Chitosan-Coated Silica Beads as Immobilized Metal Affinity Suppor for Protein Adsorption. *Biochem. Eng.J.* Vol.16, No.3, (December 2003), pp. 317- 322, ISSN 1369-703X

Shishmakov, A.B., Kharchuk, V.G., Kuznetsova, O.V., et al. (2003). Activity of Elements Dioxides in liquid Phase Oxidation of 2,3,5-trimethyl Hydroquinone. *Zh. Fiz. Khim. (Rus. J. Phys. Chem.)*, Vol. 77, pp. 623-628, ISSN 0036-0244 (in Russian)

Shishmakov, A. B., Mikushina, Yu. V., Valova, M. S. et al. (2007). TiO_2 Xerogel Modified with Powder Cellulose in Oxidation Trimethylhydroquinone. *Russ. J. Appl. Chem.*, Vol.80, No.12, (December 2007), pp. 2107-2111, ISSN 1070-4272

Shishmakov, A.B., Kovaleva, E.G.., Mikushina,Yu.V. et al. (2010). EPR study of Cu(II)-Complexes in Matrix of the TiO_2 Gel Modified with Powder Cellulose. *Russ. J. Inorg. Chemistry*, Vol. 55, No.6, pp. 937-942, ISSN 0036-0236

Skorik, Yu. A., Gomes, C. A. R., Teresa,. M. , Vasconcelos, S. D. et al. (2003). N-(2-Carboxyethyl) Chitosans : Regioselective Synthesis, Characterization and Protolytic Equilibruia. *Carbohydr. Res.* , V.338, No.3, (January 2003), pp. 271-276, ISSN 0008-6215

Ur'ev, N. B. & Potanin, A. A. (1992). *Tekuchest' suspenzi' i poroshkov (Fluidity of Suspensions and Powders)* Khimiya (Chemistry), ISBN 5724506572, Moscow (in Russian)

Varghese, J. G., Karuppannan, R. S. & Kariduraganavar, M. Y. (2010). Development of Hybrade Membranes Using Chitosan and Silica Precursors for Pervaporation Separation of Water + Isopropanol Mixtures. *J. Chem. Eng. Data,* Vol. 55, No.6, (June 2010), pp. 2084-2092 , ISSN 0021-9568

Varma, A. J. , Deshpande, S. V.& Kennedy, J. F. (2004). Metal Complexation by Chitosan and Its Derivatives : A Review. *Carbohydr. Polym.* Vol. 55, No.1, (January 2004), pp. 77-93, ISSN 0144-8617

Viswanathana, N. & Meenakshib, S (2010). Enriched Fluoride Sorption Using Alumina-Chitosan Composite. *J. Hazard. Mater.* Vol. 178, No.1, (June 2010), pp. 226-232, ISSN 0304-3894

Voinov, Maxim A., Kirilyuk, Igor A. & Smirnov, Alex I. (2009). Spin-Labeled pH-Sensitive Phospholipids for Interfacial pK Determination: Synthesis and Characterization in Aqueous and Micellar Solutions. *J. Phys. Chem. B*, Vol.113, No. 11, (March 19, 2009), pp.3453–3460, ISSN ISSN 1520-6106

Volodarskii, L.B., Grigor'ev, I.A., Dikanov, S.A. et al. (1988). *Imidazolinovye Nitroksil'nye Radikaly (Imidazoline Nitroxyl Radicals),* Nauka, ISBN 5-0028677-X, Novosibirsk (in Russian)

Xue, L., Zhou, D.-J., Tang, L. et al. (2004). The Asymmetric Hydration of 1-octene to (S)-(+) With a Biopolymer-Metal Complex, Silica-Supported Chitosan-Cobalt Complex. *React. Funct. Polym.* Vol.58, No. 2, (February 2004), pp. 117 - 121 , ISSN 1381-5148

Yeh, J. Chen, C. & Huang, K. (2007). Synthesis and Properties of Chitosan/SiO_2 Hybride Materials. *Mater. Lett.* ,Vol. 61, No.6 , (March 2007), pp.1292- 1295.,ISSN 0167-577X

Zakharova, G.S., Volkov, V.L., Ivanovskaya, V.V. et.a l. (2005). *Nanotrubki i Rodstvennye Nanostruktury Oksidov Metallov (Nanotubes and Related Nano-structures of Metal Oxides),* Ural. Otd. Ross. Akad. Nauk (Ural Branch of the Russian Academy of Sciences), ISBN 5-7691-1559-9, Yekaterinburg (in Russian)

Zamaraev, K.I., Salganik, R.I., Romannikov, V.N. et al. (1995). Modeling of Prebiotic Oligopeptides in the Presence of Zeolites and Kaolin as Catalysts. *Dokl. Akad.* Nauk (Reports of Academy of Sciences), Vol. 340, p. 779-782, ISSN 0869-5652 (in Russian)

Zhang, J. & Xia, C. (2003). Natural Biopolymer-Supported Bimetallic Catalyst Ssystem for the Carbonylation to Esters of Napoxen. *J. Mol. Catal. A: Chem.* Vol.206, No.1, (October 2003), pp. 59-65, ISSN 0021-9517

Zhao, X. S. & Lu, G. Q. (1998). Modification of MCM-41 by Surface Silylation with Trimethylchlorosilane and Adsorption Study. *J. Phys. Chem. B,* Vol. 102, No.9, (February 1998), pp.1556-1561, ISSN 1520-6106

Zhao, X. S., Lu, G. Q., Whittaker, A. K. et al. (1997). Comprehensive Study of Surface Chemistry of MCM-41 Using ^{29}Si CP/MAS NMR, FTIR, Pyridine-TPD and TGA. *J. Phys. Chem. B,* V. 101, No.33, pp. 6525-6531, ISSN 1520-6106

Chapter 7

Magnetic and Electric Properties of Organic Nitroxide Radical Liquid Crystals and Ionic Liquids

Rui Tamura, Yoshiaki Uchida and Katsuaki Suzuki

Additional information is available at the end of the chapter

1. Introduction

Amongst stable organic free radicals such as nitroxides, verdazyls, thioaminyls, a certain hydrazyl, phenoxyls, and carbon-centered radicals, nitroxide radicals (NRs) show outstanding thermodynamic stability ascribed to the delocalization of the unpaired electron over the N—O bond and thereby no dimerization (Aurich, 1989; Hicks, 2010). In fact, sterically protected NRs have found various practical applications in the field of materials science. The landmark is the discovery by Kinoshita et al. in 1991 of the first purely organic ferromagnet (Tc = 0.6 K) with respect to one of several polymorphs of a nitronyl nitroxide, 2-(4-nitrophenyl)-4,4,5,5-tetramethylimidazoline-1-oxy-3-oxide (**1**) (Figure 1) (Tamura et al., 1991). Since then, stable NR structures have been used as the spin source and building block for the elaboration of organic or molecule-based magnetic materials. Up to the late 1990s, more than 20 NR-based organic ferromagnets were prepared (Amabilino & Veciana, 2001); the highest Tc value of 1.48 K was recorded for one of polymorphs of 1,3,5,7-tetramethyl-2,6-diazaadamane-*N,N'*-dioxyl (**2**) prepared by Rassat et al. in 1993 (Figure 1) (Chiarelli et al, 1993). Furthermore, to verify the theoretical prediction for constructing organoferromagnetic conductor, several donors and acceptors carrying NR units were prepared to give the corresponding CT complexes and radical salts (Nakatsuji, 2008). In 2007, Matsushita et al. reported that a radical ion salt of a tetrathiafulvalene (TTF)-based spin-polarized donor NR compound (**3**) successfully exhibited giant negative magnetoresistance, i.e., the decrease in the resistance of the salt by more than 70% under a magnetic field of 9 T at 2K (Figure 1) (Matsushita et al., 2007, 2008). This is the first example detecting the interaction between localized spins and conducting electrons in an organic molecular assembly, i.e., a molecule-based spintronics using not only the charge but also the spin of an electron (Sugawara et al., 2011). Meanwhile, for the last decade the redox properties of NRs have been utilized for the development of environmentally benign organic cathode-active materials for rechargeable batteries with a high energy-density, such as a stable nitroxide polyradical, poly(2,2,6,6-tetramethylpiperidinyloxy methacrylate (**4**) (Figure 1) (Nakahara, 2002; Oyaizu & Nishide, 2010; Suga & Nishide, 2010).

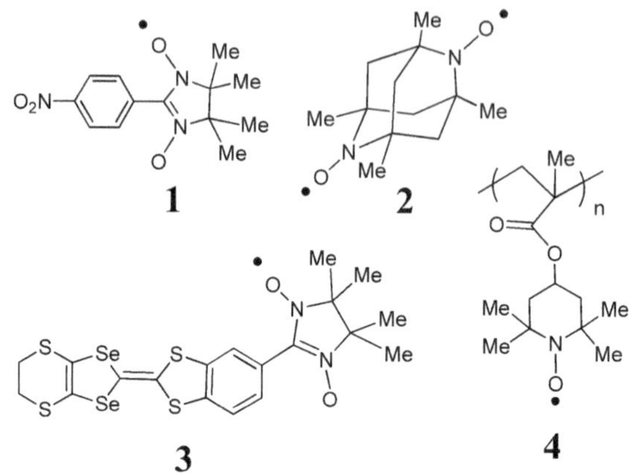

Figure 1. Magnetic materials based on NR chemistry.

Thus, stable NR structures have been used as the spin source or the redox species to develop metal-free solid-state magnetic materials and spintronic devices, or polymer battery devices, respectively. However, the large electric dipole moment (ca. 3 Debye) of a nitroxyl group (N-O •) has never been utilized in these NR-based materials. In this context, with a view to exploiting metal-free magnetic or spintronic soft materials, we have been developing organic liquid-crystalline (LC) and ionic liquid (IL) NRs which can benefit from the unique magnetic and electric properties intrinsic to the NR structure.

Paramagnetic LC compounds have been expected to become novel advanced soft materials that can combine the optical and electrical properties of conventional LCs with the magnetic and electronic properties of paramagnetic compounds (Dunmur & Toriyama, 1999). The magnetic liquid crystals (LCs) are classified into two types; the majority were metal-containing LCs (metallomesogens) with permanent spins originating from transition (d-block) or lanthanide (f-block) metal ions in the mesogen core (Figure 2) (Hudson & Maitlis, 1993; Griesar & Haase, 1999; Binnemans & Gröller-Walrand, 2002; Piguet et al., 2006; Terrazi et al., 2006), while only several all-organic radical LC materials of the first generation were prepared before 2004 (Figure 3), because of the difficulty in the molecular design and synthesis which must satisfy the molecular linearity or planarity necessary for the existence of LC phases (rod-like or disk-like molecules, respectively) as well as the radical stabilization (Kaszynski, 1999; Tamura et al., 2008a, 2012). Moreover, endowing the magnetic LCs with chirality is expected to result in the emergence of unique magneto-electric or magneto-optical properties, intriguing magnetic interactions and so on in the LC state (Tamura et al., 2008b, 2012).

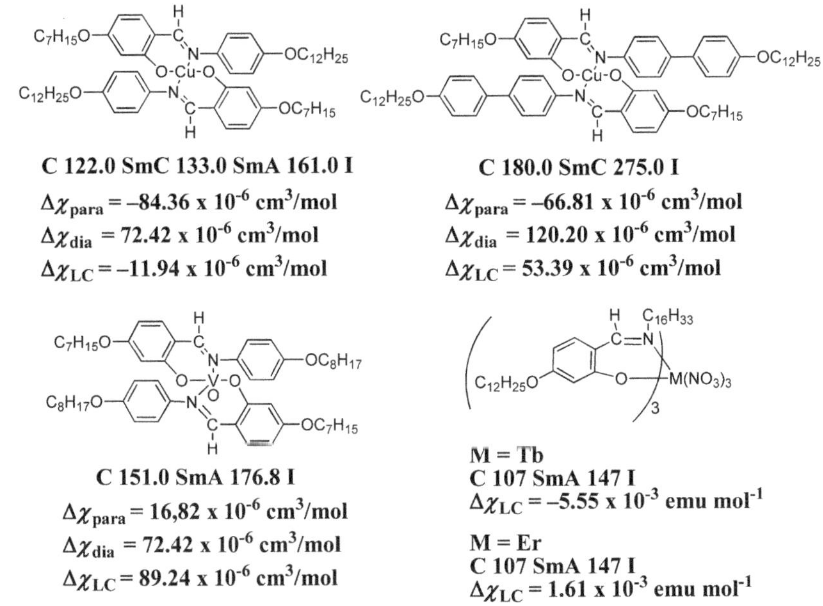

Figure 2. Representative metallomesogens and their transition temperatures and magnetic anisotropies.

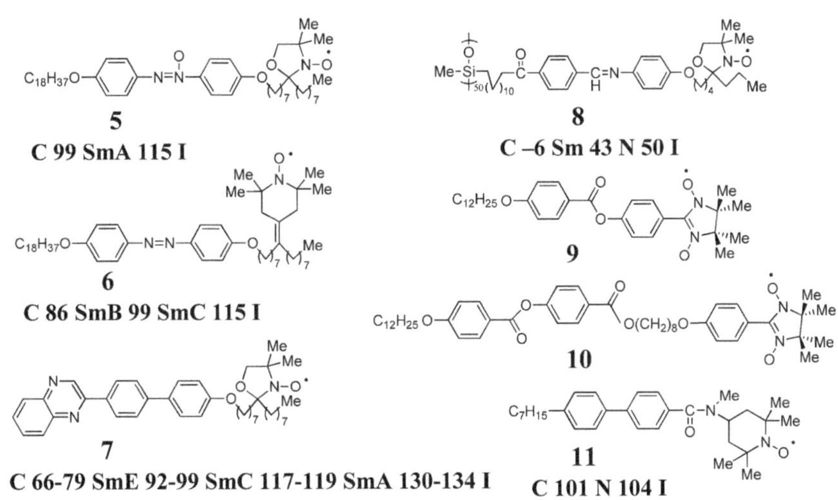

Figure 3. First-generation of rod-like all-organic radical LC compounds **5-11** and their transition temperatures.

In 2004, the present authors reported the preparation and magnetic properties of the prototypic second generation of paramagnetic all-organic rod-like LC compounds **12**, which contain a chiral cyclic NR (PROXYL) unit in the mesogen core and show various chiral and achiral LC phases over a wide temperature range (Figure 4) (Ikuma et al., 2004). In 2006, the chiral smectic C (SmC*) phase of (2*S*,5*S*)-**12** was found to successfully exhibit ferroelectricity in a thin sandwich cell (Ikuma et al., 2006a, 2006b). Furthermore, quite recently, a sort of spin glass (SG)-like inhomogeneous magnetic interactions (the average spin-spin interaction constant $\bar{J} > 0$), which has been referred to as positive 'magneto-LC effects', have been found to be generated in the various chiral and achiral LC phases of compounds **12** and **13** at high temperatures (> 25 °C) under weak magnetic fields (Uchida et al., 2008, 2010; Suzuki et al., 2012). In fact, these LC droplets floating on water were attracted by a permanent magnet and moved freely on water under the influence of an ordinary permanent magnet (Figure 5).

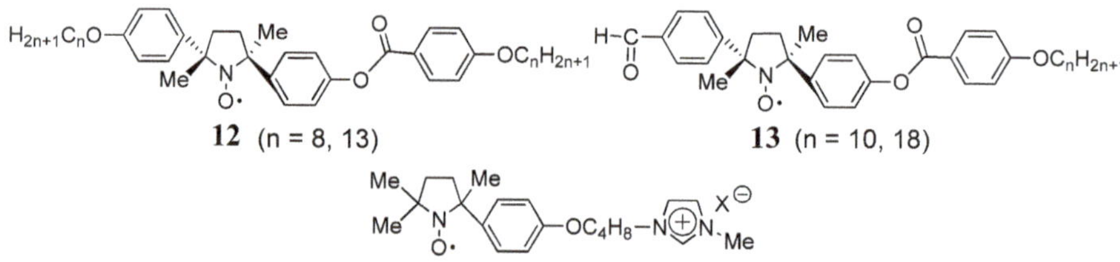

Figure 4. Chiral NR LC compounds **12** and **13**, and IL compounds **14** containing a PROXYL unit.

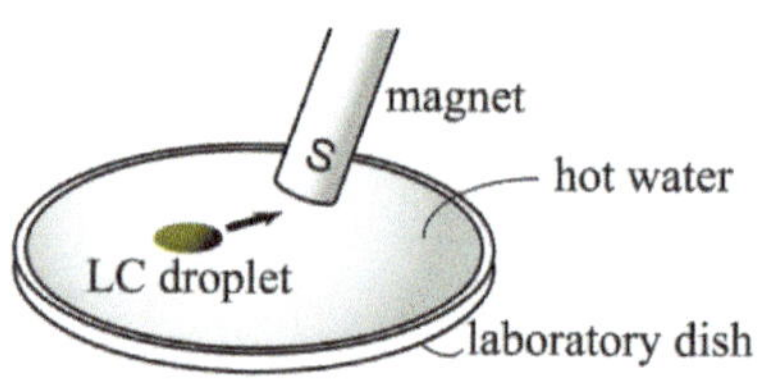

Figure 5. Schematic representation of the attraction by a permanent magnet (maximum 0.5 T) of magnetic LC droplets of **12** and **13** on water in a shallow laboratory dish. Reprinted with permission from ref. (Uchida et al., 2010). Copyright 2012 American chemical Society.

Meanwhile, with the aim of developing room-temperature ionic liquids (ILs) as another type of air-stable, metal-free magnetic soft materials which can act as redox materials or spin probes with molecular shape anisotropy, we designed and synthesized imidazolium compounds **14** containing a chiral PROXYL unit at some distance (Uchida et al., 2009a).

In this chapter, first we briefly introduce the first-generation of all-organic NR LCs, which were prepared before 2004. Then, we report the magnetic and electric properties of the second-generation of NR LCs of compounds **12** and **13**, and the NR IL compounds **14**.

2. First-generation of rod-like all-organic NR LCs

Only a few all-organic radical LC compounds have been prepared, most likely because the geometry and bulkiness of the radical-stabilizing substituents are detrimental to the stability of LCs, which requires molecular linearity or planarity (Kaszynski, 1999). Although several achiral rod-like organic LCs with a stable cyclic NR (DOXYL or TEMPO) unit as the spin source were prepared (Figure 3), their molecular structures were limited to those containing the NR unit within the terminal alkyl chain, away from the rigid core, and hence allowed the free rotation of the NR moiety inside the molecule, resulting in a decrease in the paramagnetic anisotropy ($\Delta\chi_{para}$) as well as the dielectric anisotropy ($\Delta\varepsilon$) of the whole molecule. The molecular structures and magnetic properties of such first-generation of all-organic NR LCs and the objectives of individual studies are briefly summarized below.

Chiral racemic and achiral compounds **5-7** were synthesized by Dvolaitzky et al. to use them as an LC spin-probe for EPR spectroscopic study. Racemic **7** showed stable achiral smectic phases such as SmA, SmC and SmE (Dvolaitzky et al., 1974, 1976a, 1976b). Their temperature dependence of the molar magnetic susceptibility (χ_M) was not measured.

Finkelmann et al. prepared chiral racemic radical polymer **8** which can retain the LC structure in the supercooled glassy phase to measure the magnetic properties of an LC structure at low temperatures (Allgaier & Finkelmann, 1994). The temperature dependence of the χ_M was measured by using a Faraday balance in the temperature range from 6 to 350 K, in which the crystal-to-LC-to-liquid phase transition occurred. As a result, **8** showed neither molecular reorientation nor appreciable change in χ_M at the crystal-to-LC phase transition temperature in the heating run. This is most likely ascribed to the high viscosity of the polymer material.

Greve et al. synthesized the first LC compounds **9** and **10** with an α-nitronyl nitroxide (α-NN) structure as a spin source at a terminal position in the molecule (Greve et al., 2002). They showed a highly viscous monotropic (irreversible) LC phase in the narrow temperature range from 36 to 39 °C in the heating run. The temperature dependence of the χ_M was not measured.

To prepare a supercooled glassy material and crystal polymorphs in the applied magnetic fields and to observe the change in the magnetic behavior accompanying the alteration in the solid-state structure, Nakatsuji et al. synthesized the achiral LC compound **11** (Nakatsuji et al., 2002). Although **11** showed the achiral nematic (N) phase within a narrow temperature range of 3 degree in the heating run, a small but distinct increase in χ_M was observed at the crystal-to-LC transition temperature. The difference in the magnetic behavior between the heating and cooling runs was also observed; **11** showed antiferromagnetic interactions according to a singlet-triplet model at low temperatures before the thermal phase transition in the heating run of the crystals, while the magnetic behavior obeyed the Curie-Weiss law in the cooling run from the isotropic phase (Eq. 1).

$$\chi_{para} = C / (T - \theta) \tag{1}$$

where C is Curie constant, T is temperature, and θ is Weiss temperature.

3. Second-generation of rod-like all-organic NR LCs

3.1. Molecular design and synthesis

The second-generation of chiral NR molecules **12** that could satisfy the following four mandatory requirements were designed and synthesized (Tamura et al., 2008a, 2008b, 2012).

Spin source: A nitroxyl group with a large electric dipole moment (ca. 3 Debye) and known principal g-values (g_{xx}, g_{yy}, g_{zz}) should be the best spin source, because i) the dipole moment is large enough for the source of the spontaneous polarization (*P*s) and ii) the principal g-values are useful to determine the direction of molecular alignment in the LC phase by EPR spectroscopy (Figure 6).

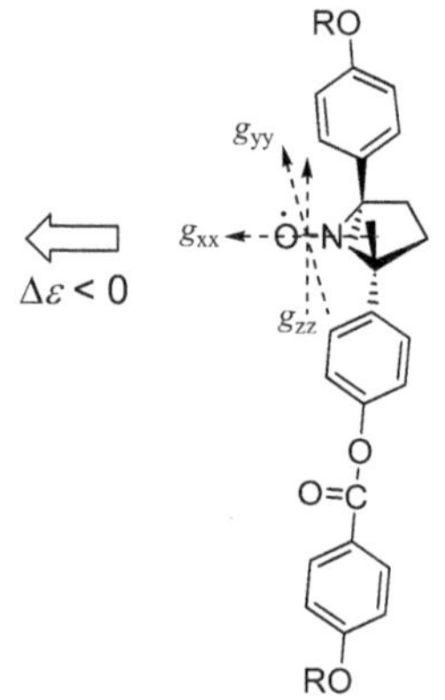

Figure 6. Principal g-values and dielectric anisotropy in the NR molecules **12**.

High thermal stability: A molecule with a 2,2,5,5-tetraalkyl-substituted pyrrolidine-1-oxy (PROXYL) unit is stable enough for repeated heating and cooling cycles below 150 °C in the air.

Molecular structure: (a) To avoid the free rotation of the NR portion inside the molecule so as to maximize the $\Delta\chi_{para}$ and $\Delta\varepsilon$, a geometrically fixed chiral cyclic NR unit should be incorporated into the rigid core of LC molecules. (b) To obtain a slightly zigzag molecular structure and a negative $\Delta\varepsilon$ advantageous for the appearance of a chiral smectic C (SmC*) phase, a *trans*-2,5-dimethyl-2,5-diphenylpyrrolidine-1-oxy skeleton in which the electric dipole moment orients to the molecular short axis is the best choice (Figure 6).

Chirality: Since both chiral and achiral LCs are required for comparison of their optical and magnetic properties in various LC phases, the molecules should be chiral and both enantiomerically-enriched and racemic samples need to be available.

3.2. Magnetic properties

Since the magnetic properties such as $\Delta\chi$–controlled molecular reorientation and magnetic interactions in all-organic magnetic LC phases have been fully characterized for the first time by using the various LC phases of NR compounds **12**, these experimental results are described in detail.

3.2.1. Magnetic anisotropy of LC compounds

Similarly to $\Delta\varepsilon$, $\Delta\chi$ is calculated by subtracting the magnetic susceptibility component ($\chi_{\perp}$) perpendicular to the molecular long axis from the component ($\chi_{//}$) parallel to the same axis (Figure 7 and Eq. 2). Furthermore, the $\Delta\chi$ consists of a paramagnetic component ($\Delta\chi_{para}$) (Eq. 3) and a diamagnetic component ($\Delta\chi_{dia}$) (Eq. 4). Although χ_{para} and χ_{dia} are always positive and negative, respectively, $\Delta\chi_{para}$ and $\Delta\chi_{dia}$ become positive or negative, depending on the magnitude of the respective $\chi_{\perp}$ and $\chi_{//}$ values. Accordingly, the overall molecular magnetic anisotropy ($\Delta\chi_{mol}$) is the sum of $\Delta\chi_{para}$ and $\Delta\chi_{dia}$ (Eq. 5) (Griesar & Haase, 1999; Dunmur & Toriyama, 1999). If $\Delta\chi_{mol}$ is positive (or negative), the molecular long axis becomes parallel (or perpendicular) to the applied magnetic field (H_0), when the applied field is larger than the critical magnetic field (H_c) (Eq. 6 and Figure 7). Such is a driving force for molecular alignment by magnetic fields.

$$\Delta\chi = \chi_{||} - \chi_{\perp} \tag{2}$$

$$\Delta\chi_{para} = \chi_{para||} - \chi_{para\perp} \tag{3}$$

$$\Delta\chi_{dia} = \chi_{dia||} - \chi_{dia\perp} \tag{4}$$

$$\Delta\chi_{mol} = \Delta\chi_{para} + \Delta\chi_{dia} \tag{5}$$

$$H_c = \pi\, d^{-1}\, k^{1/2}\, (\Delta\chi)^{-1/2} \tag{6}$$

where d represents the cell thickness and k is the elastic constant.

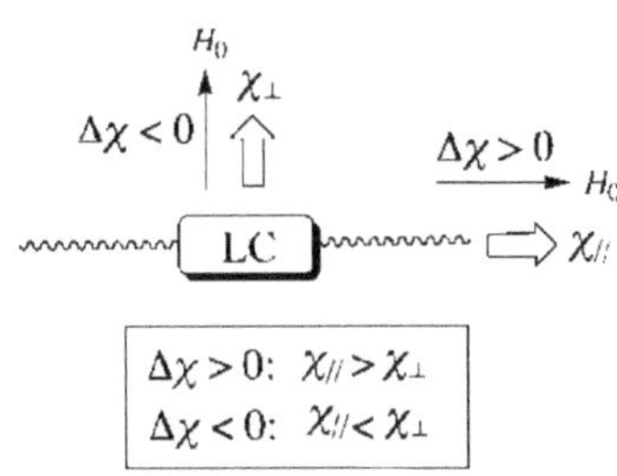

Figure 7. Magnetic anisotropy ($\Delta\chi$) of an LC molecule.

Furthermore, the overall LC magnetic anisotropy ($\Delta\chi_{LC}$) (Figure 2), which is the sum of $\Delta\chi_{para,LC}$ and $\Delta\chi_{dia,LC}$ calculated by EPR spectroscopy and SQUID magnetization measurement, respectively, depends on the orientational order parameter (S) of LCs as shown in Eq. 7 (de Gennes & Prost, 1993) .

$$\Delta\chi_{LC} = N\, \Delta\chi_{mol}\, S \tag{7}$$

where N is the number of molecules.

Diamagnetism resides in all atoms. Particularly aromatic rings show a strong diamagnetic effect in applied magnetic fields. Therefore, the diamagnetic rod-like LC molecules orient themselves such that the axis with the most negative χ_{dia} is perpendicular to the magnetic field. Since $|\chi_{dia\perp}|$ is usually larger than $|\chi_{dia\parallel}|$ with regard to organic LC molecules, the $\Delta\chi_{dia}$ becomes positive and the molecules orient with the director parallel to the magnetic field (Figure 7). For organic LCs, the magnitude of the $\Delta\chi_{dia}$ which is produced by two diamagnetic phenyl groups is approximately +50 x 10^{-6} emu mol^{-1} (Müller & Haase, 1983). Accordingly, a relatively strong magnetic field (> 0.2 T) is necessary to align diamagnetic LCs, depending on the type of LC phases (Boamfa et al., 2003). Meanwhile, the $\Delta\chi_{para}$ of all-organic rod-like LC materials with a stable NR unit in the rigid core is considered to be too small to control the molecular orientation by magnetic fields due to the p-orbital origin.

3.2.2. *Magnetic-field-induced molecular alignment*

It is known that rod-like metallomesogens with high viscosity are not always suited for the investigation on the alignment of LC molecules by magnetic fields. In contrast, LC compounds **12** with low viscosity, low phase transition temperature, and known principal *g*-values of the NR moiety are considered to be a good spin-labeled candidate for the studies on the $\Delta\chi$–controlled molecular orientation by weak magnetic fields. Therefore, to confirm that the magnetic-field-induced molecular alignment in the LC phases of **12** is $\Delta\chi_{dia}$-controlled, the $\Delta\chi_{para}$ and $\Delta\chi_{dia}$ values of **12** and the approximate magnitude of H_c for each LC phase of **12** were evaluated by EPR spectroscopy and SQUID magnetization measurement and by POM observation under variable magnetic fields, respectively (Uchida et al., 2009b).

First, the temperature-dependent $\Delta\chi_{para}$ value of compound **12a** (n=13) was calculated to be –1.7 x 10^{-6} emu mol^{-1} at 300 K from the *g*-value obtained by EPR spectroscopy, while the temperature-independent $\Delta\chi_{dia}$ value was calculated to be +6.5 x 10^{-5} emu mol^{-1} from the experimental molar magnetic susceptibility of (±)-**12a** measured on a SQUID magnetometer. Thus, $|\chi_{dia}|$ has turned out to be 30 times larger than $|\chi_{para}|$; the molecular alignment of **12a** by magnetic fields is definitely $\Delta\chi_{dia}$-controlled.

Next, to identify the direction of molecular alignment in the bulk LC state under a weak magnetic field, the temperature dependence of the experimental *g*-value (g_{exp}) of (±)-**12a** was measured at a magnetic field of 0.33 T by EPR spectroscopy (Figure 8). In the heating run, the g_{exp} of (±)-**12a** was constant at around 2.0065 in the crystalline state, then increased at the crystal-to-SmC phase transition, became constant at around 2.0068 in the SmC phase, then decreased abruptly to 2.0058 at the SmC-to-N phase transition, and finally returned to the level (~ 2.0065) of the crystalline state in the isotropic phase. In the cooling run, the g_{exp} of (±)-**12a** was constant at around 2.0065 in the isotropic phase, then decreased at the Iso-to-N phase transition, became constant at around 2.0055 in the N phase, then increased to 2.0063 at the N-to-SmC phase transition, and finally increased to 2.0067 in the crystalline state.

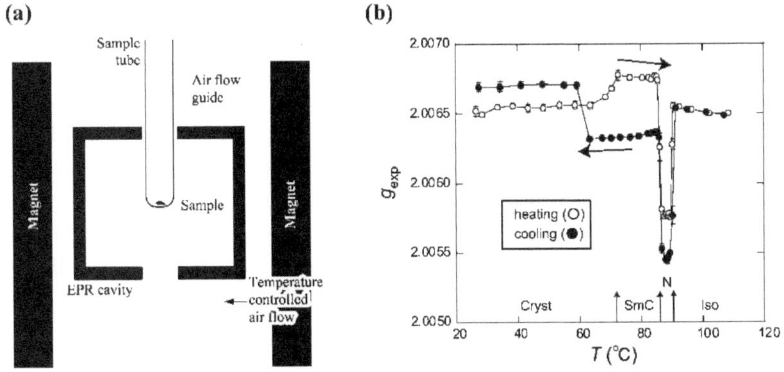

Figure 8. EPR spectroscopy of (±)-**12a** (n=13). (a) Experimental setup and (b) temperature dependence of the g-value measured through the first heating (white circles) and cooling (black circles) processes. Ref. (Uchida et al., 2009b) – Reproduced by permission of The Royal Society of Chemistry.

From these results and the calculated principal g-values (g_{iso}=2.00632, $g_{//}$=2.00540, $g_{\perp}$=2.00678) of **12a**, it is concluded that i) in the N phase the majority of molecules align their long axis along the applied magnetic field of 0.33 T, owing to the $\Delta\chi_{dia}$-controlled molecular reorientation (Figure 9a), whereas in the SmC phase in the heating run the molecular short axis is rather parallel to the field (Figure 9b), most likely due to the viscous layer structure and the natural homeotropic anchoring effect by quartz surface, and ii) that the molecular alignment in each LC phase is influenced by that in the preceding LC phase, although the molecular orientation modes are quite different between the N and SmC phases.

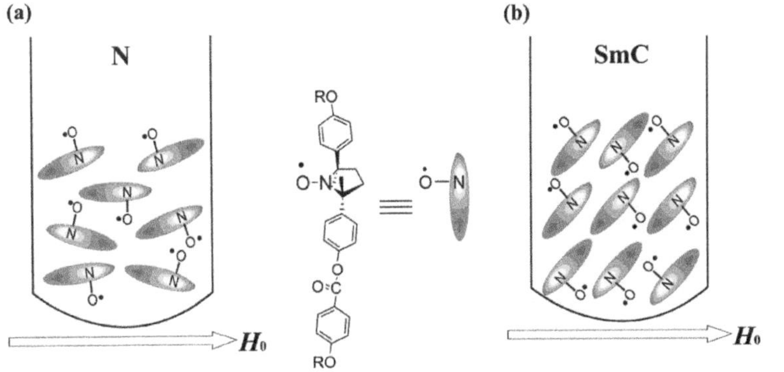

Figure 9. Molecular alignment in the LC phases of (±)-**12a** under a weak magnetic field (0.34 T). (a) The N phase in both heating and cooling runs. (b) The SmC phase in the heating run.

To evaluate the H_c for each LC phase of (±)-**12a**, we observed the texture change by POM observation under variable magnetic fields. Figure 10 shows the experimental setup: the direction of applied magnetic fields is perpendicular to the LC cell surface. The inner glass surface in the sandwich cell with 40 μm thickness was neither chemically nor physically treated. The natural Schlieren texture of the N phase gradually became dark with the increasing magnetic field until 0.5 T, resulting in the complete homeotropic orientation of molecules at 1.0 T, whereas the natural Schlieren texture of SmC phase of (±)-**12a** scarcely changed below 1.0 T, largely changed between 1.0 T and 1.5 T, and finished the change at less than 2.0 T to show another Schlieren texture (Figure 11), which is similar to the SmC Schlieren texture of (±)-**12a** observed under alternative homeotropic boundary conditions (Dierking, 2003). Accordingly, it has been concluded that the smectic layer planes became parallel to the glass plates at 2.0 T. Furthermore, no texture change was noted for N* and SmC*phase of (2*S*,5*S*)-12a below 5 T using the same experimental setup.

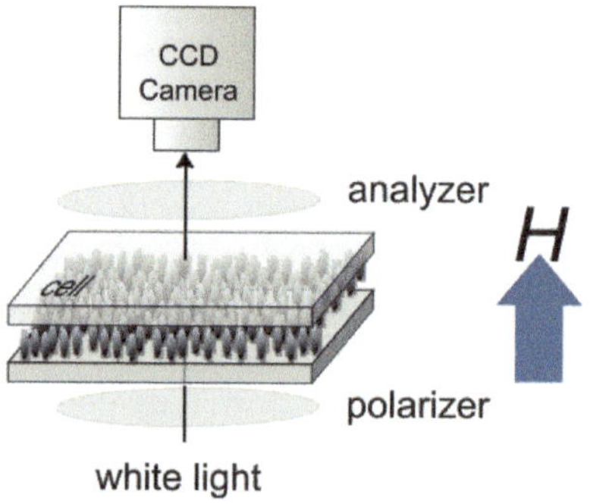

Figure 10. Experimental setup for the polarized optical microscopy observation of (±)-**12a** under variable magnetic fields applied perpendicular to the cell surface. Ref. (Uchida et al., 2009b) – Reproduced by permission of The Royal Society of Chemistry.

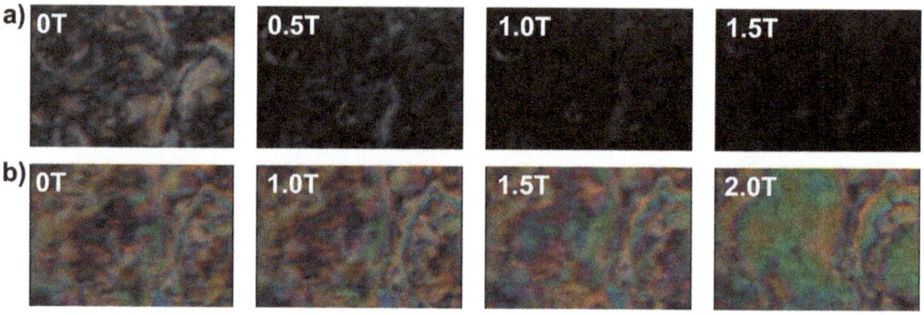

Figure 11. Polarized optical photomicrographs of (±)-**12a**. Texture changes (a) from the natural texture to the pseudo-isotropic texture in the N phase at 88.0 °C and (b) from the natural texture to the Schlieren texture in the SmC phase at 83 °C, with increasing magnetic fields. Ref. (Uchida et al., 2009b) – Reproduced by permission of The Royal Society of Chemistry.

Thus, the H_c of each LC phase turned out to be largely affected by the superstructure; H_c(N) (< 1.0 T) < H_c(SmC) (< 2.0 T) < H_c(N*, SmC*) (> 5.0 T).

3.2.3. *Magneto-LC effects*

The possibility of a ferromagnetic rod-like LC material has been considered unrealistic due to the inaccessibility of long-range spin-spin interactions between rotating molecules in the LC state. However, low viscous all-organic rod-like LC materials with a stable NR unit in the rigid core may show unique intermolecular magnetic interactions owing to the swift coherent collective properties of organic molecules in the LC state.

a. Magnetic LCs with negative dielectric anisotropy ($\Delta\varepsilon < 0$)

Interestingly, the present authors observed a nonlinear relationship (S-curve) between the applied magnetic field (*H*) and the molar magnetization (*M*) in chiral and achiral LC phases of **12** (Figure 12), which implies the generation of an unusual magnetic interaction in the LC phases under applied magnetic fields (Uchida et al., 2008, 2010). Such a nonlinear relationship was not observed in the crystalline phases of the same compounds which showed a usual linear relationship indicating a paramagnetic nature and no contamination of magnetic impurities in the sample. The in-depth investigation on the magnetic properties of LC compounds **12** strongly suggested that the generation of a sort of spin glass (SG)-like inhomogeneous magnetic interactions (the average spin-spin interaction constant $\bar{J} > 0$), which has been referred to as positive 'magneto-LC effects', induced by weak magnetic

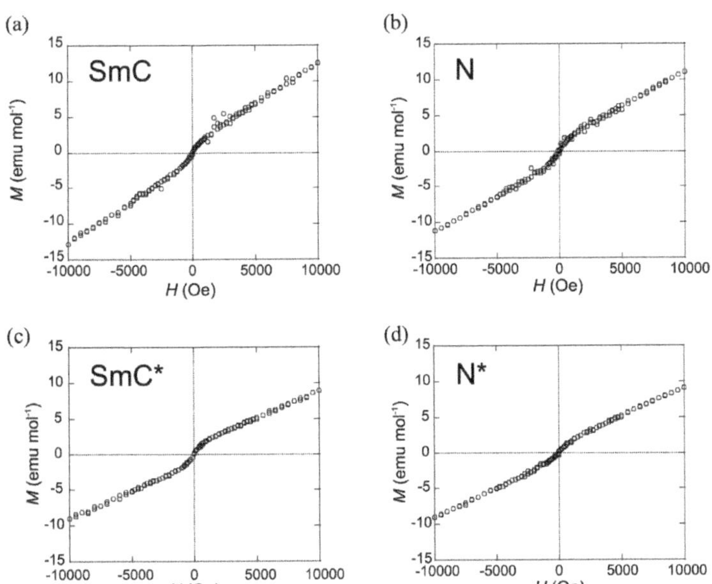

Figure 12. Magnetic field (*H*) dependence of molar magnetization (*M*) at 77 °C for (a) the SmC phase of (±)-**12a**, (b) the N phase of (±)-**12b** (n=8), (c) the SmC* phase of (2*S*,5*S*)-**12a** (88% *ee*), and (d) the N* phase of (2*S*,5*S*)-**12b** (96% *ee*). Reprinted with permission from ref. (Uchida et al., 2010). Copyright 2012 American chemical Society.

fields in the various LC phases of compounds **12** is responsible for the observed nonlinear relationship between the *H* and *M*; the magnitude of magnetic interactions depended on the LC phase type, or the superstructure (Figure 13) (Uchida et a., 2010). Furthermore, it was confirmed that the molecular reorientation effect arising from the simple molecular magnetic anisotropy ($\Delta\chi$) has nothing to do with the positive magneto-LC effects observed in the LC phases of **12**. Thus, it was concluded that the origin of such strong SG-like inhomogeneous magnetic interactions can be interpreted in terms of the anisotropic spin-spin dipole interactions induced by weak magnetic fields in the anisotropic LC superstructure.

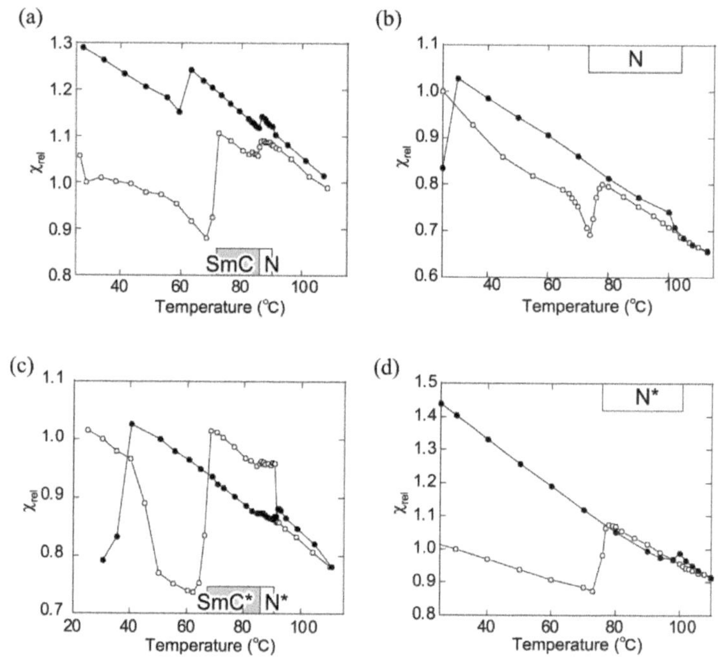

Figure 13. Temperature dependence of relative paramagnetic susceptibility (χ_{rel}) for (a) (±)-**12a**, (b) (±)-**12b**, (c) (2*S*,5*S*)-**12a** (88% *ee*), and (d) (2*S*,5*S*)-**12b** (96% *ee*) at a magnetic field of 0.33 T. Open and filled circles represent the first heating and cooling runs, respectively. The LC temperatures shown in a box refer to the first heating process. Reprinted with permission from ref. (Uchida et al., 2010). Copyright 2012 American chemical Society.

In this study, we could indicate that EPR spectroscopy is the much better means than SQUID magnetization measurement to evaluate the temperature dependence of the χ_{para} for organic NR LC compounds at high temperatures. This is due to the following four reasons: (i) The χ_{para} can be derived from the Bloch equation (Eq. 8) by using the parameters obtained from the EPR differential curves, such as maximum peak height (I'm and $-I'$m), g-value (g), and peak-to-peak line width (ΔH_{PP}).

$$\chi_{para} = \frac{2\mu_B g I'_m \Delta H_{pp}^2}{\sqrt{3h\nu H_1}} \quad (8)$$

where μ_B is the Bohr magneton, h is Planck's constant, ν is the frequency of the absorbed electromagnetic wave, and H_1 is the amplitude of the oscillating magnetic field. Accordingly, the temperature dependence of relative paramagnetic susceptibility (χ_{rel}), which is defined as

$$\chi_{rel} = \frac{\chi_{para}}{\chi_0} \quad (9)$$

where χ_0 is the standard paramagnetic susceptibility, e.g., at 30°C in the heating run (Eq. 9), can be actually used (Figure 12). (ii) Treatment of the χ_{dia} term is totally unnecessary. (iii) The experimental error is very small even at such high temperatures. (iv) The analysis of microscopic magnetic interactions such as spin-spin dipole and exchange interactions is also feasible.

b. Magnetic LCs with positive dielectric anisotropy ($\Delta\varepsilon > 0$)

To examine the effects of $\Delta\varepsilon$ on the magneto-LC effects, compounds **13** with a terminal formyl group (Figure 4) which have a positive $\Delta\varepsilon$ were synthesized. Under weak magnetic fields, positive magneto-LC effects ($\bar{J} > 0$) operated in the chiral nematic (N*) phase of (2*S*,5*S*)-**13a** (n=10) and in the smectic A (SmA*) phase of (2*S*,5*S*)-**13b** (n=18) (Figure 14a,b), whereas negative magneto-LC effects ($\bar{J} < 0$) were observed in the achiral nematic (N) phase of (±)-**13a** (Figure 14c) (Suzuki, et al., 2012). The origin of such negative magneto-LC effects operating in the N phase of (±)-**13a** was interpreted in terms of the generation of antiferromagnetic interactions which is associated with the formation of the *RS* magnetic dipolar interaction due to the strong electric dipole interactions (Figure 15c), while ferromagnetic head-to-tail spin-spin dipole interactions should dominate in the N* and SmA* phases (Figure 15a,b).

c. Attraction of magnetic LC droplet by a permanent magnet

Furthermore, these radical LC droplets floating on water were attracted by a permanent magnet and moved freely on water under the influence of this magnet (Figure 5), whereas the crystallized particles of the same compounds never responded to the same magnet. The response of the LC droplets to the magnet also varied depending on the LC phase type, i.e., the extent of the magnetic interaction ($\bar{J}$). These results indicate that the LC phase domain can help to induce magnetic interactions under applied magnetic fields (Uchida et al., 2008, 2010; Suzuki et al., 2012). This unique magnetic attraction will find use in the development of the metal-free magnetic soft materials usable at ambient temperature, such as a magnetic LC carrier for the magnetically targeted drug-delivery system or an MRI contrast agent (Kumar, 2009).

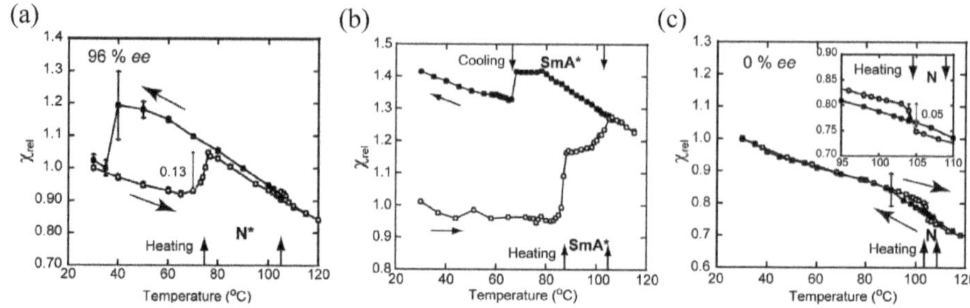

Figure 14. Temperature dependence of χ_{rel} value for (a) (2*S*,5*S*)-**13a**, (b) (2*S*,5*S*)-**13b** and (c) (±)-**13a**. Ref. (Suzuki et al., 2012) – Reproduced by permission of The Royal Society of Chemistry.

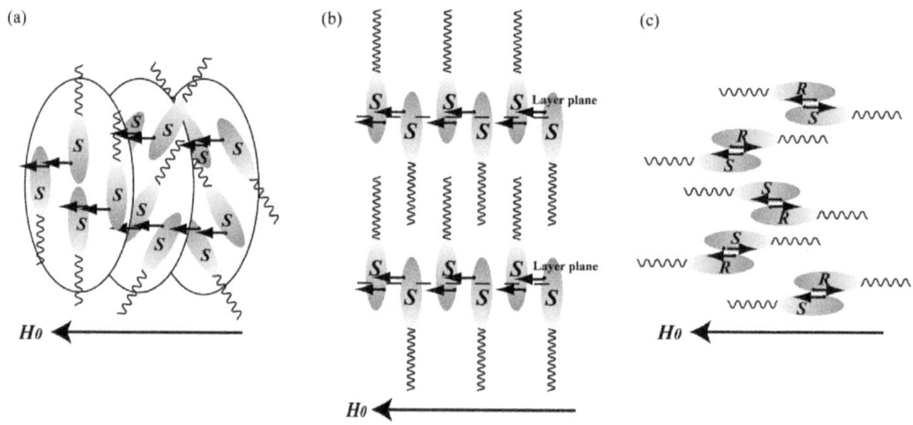

Figure 15. Spin-spin dipole interactions in LC phases. (a) N* phase of (2*S*,5*S*)-**13a**, (b) SmA* phase of (2*S*,5*S*)-**13b** and (c) N phase of (±)-**13a**. Ref. (Suzuki et al., 2012) – Reproduced by permission of The Royal Society of Chemistry.

3.3. Ferroelectric properties

It is known that when an SmC* phase is confined to a thin sandwich cell with a gap smaller than the pitch of the helical superstructure, an unwinding of the helix occurs and a bistable, ferroelectric device is formed (Figure 16) (Goodby et al., 1991; Lagerwall, 1999; Dierking, 2003). Consequently, *Ps* is generated in the sandwich cell in which ferroelectric switching occurs by changing the polarity of the electric field. The ferroelectric properties of the SmC* phase of each sample are characterized by measuring the *Ps*, the optical response time of bistable switching to an applied electric field, and the tilt angle.

The SmC* phase of (2*S*,5*S*)-**12a** indeed exhibited ferroelectricity in a planar anchoring thin sandwich cell (4 μm thickness) (Ikuma et al, 2006a, 2006b); a *Ps* value of 24 nC cm^{-2}, an optical response time ($\tau_{10\text{-}90}$) of 0.213 msec and a layer tilt angle (θ) of 29° were recorded. The ferroelectric LC data of (2*S*,5*S*)-**12a** were superior to those of the typical chiral

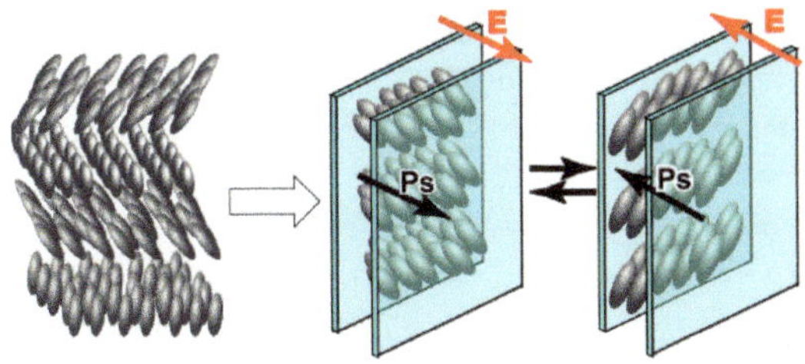

Figure 16. Ferroelectric switching in a thin sandwich cell.

metallomesogen **15** (Figure 17); **15** showed a higher *Ps* value of 38 nC cm^{-2} with a θ of 23° than (2*S*,5*S*)-**12a**, but the optical response was very slow ($\tau_{10\text{-}90}$ = 8.5 msec) due to the high viscosity (Iglesias et al., 1996).

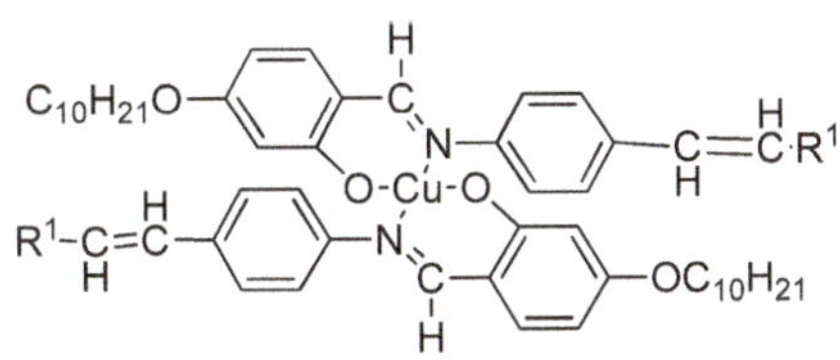

15 $R^1 = COOC^*H(CH_3)COO(CH_2)_3CH_3$

C 109.2 SmC* 120 SmA 160.1 I

Ps(–10°C) = 38 nC cm^{-2}

τ(–10°C) = 8.5 ms (4 V μm^{-1})

Figure 17. Ferroelectric metallomesogen **15**.

Furthermore, second-harmonic generation (SHG) was clearly observed by Kogo and Takezoe et al. under a phase-matching condition in the SmC* phase of (2*S*,5*S*)-**12a** loaded into an LC cell (20 μm thickness), validating the existence of ferroelectricity. The effective second-order nonlinear optical (NLO) constant was evaluated to be a 4.8 x 10^{-2} pm V^{-1}, 3 orders of magnitude smaller than that of quartz known as a standard NLO material (Kogo et al., 2010).

4. NR Ionic Liquids (ILs)

The synthesis and electric, electrochemical and magnetic properties of IL compounds (±)-**14** were reported (Figure 4), coupled with the first use of this type of magnetic IL as an EPR spin probe in typical achiral diamagnetic ILs (Uchida et al, 2009a). Although the chloride (±)-**14a** (X = Cl) was hygroscopic and miscible with water, anhydrous and fairly hydrophobic ILs were obtained for other salts of (±)-**14** (X = BF_4, NTf_2, PF_6) which showed a glass transition between –37 and –22 °C and decomposed between 162 and 170 °C in air.

The temperature dependence of χ_M of the least viscous (±)-**14b** (X = NTf_2) measured on a SQUID magnetometer at a field of 0.5 T showed the high radical purity and antiferromagnetic interactions below 10K. Ionic conductivity and viscosity of (±)-**14b** were determined to be 5.23 x 10^{-5} S cm^{-1} and 1.087 x 10^{-3} cP, respectively, at 25 °C. Electrochemical studies using cyclic voltammetry (CV) were carried out for a CH_3CN solution (1 mM) of (±)-**14b** and the neat IL (ρ = 1.49 g cm^{-3}; the radical concentration is 2.34 M at 25°C), without an additional supporting electrolyte (Figure 18). The voltammogram measured in CH_3CN exhibited a quasireversible wave with half-wave oxidation potential ($E_{1/2}^{ox}$) of +0.363 V (*vs.* Fc/Fc$^+$), while that in the neat IL measured with a micro CV cell showed a wider quasireversible wave with $E_{1/2}^{ox}$ = +0.458 V. The diffusion coefficient of **14b** or the corresponding oxoammonium ion in the neat IL was determined to be 2 x 10^{-10} cm^2 s^{-1}. Thus, paramagnetic (±)-**14b** with high radical purity turned out to be air-stable room temperature IL which can exhibit ionic conductivity and a quasireversible redox behavior in the absence of additional solvent and electrolyte. Notably, (±)-**14b** with molecular shape anisotropy proved to serve as the first IL EPR spin probe that can recognize the local structure or the molecular shape anisotropy of achiral diamagnetic IL solvents.

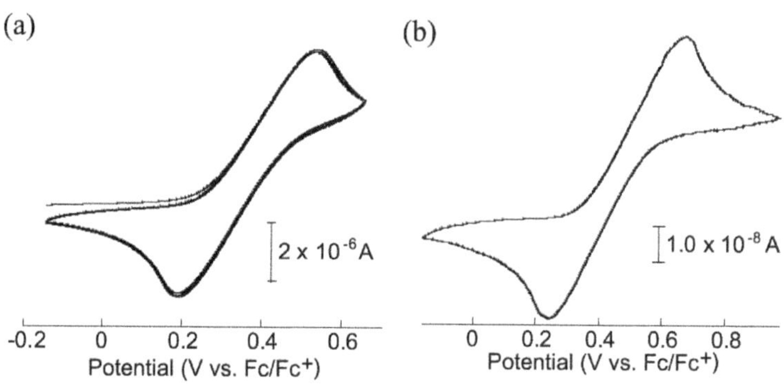

Figure 18. Cyclic voltammograms of (±)-**14b** recorded at 25°C for (a) the MeCN solution (1.0 mM) at a scan rate of 100 mV s^{-1} (five scans) using Pt working and counter electrodes and an Ag reference electrode without an additional supporting electrolyte and (b) the neat ionic liquid sample at a scan rate of 0.5 mV s^{-1} (single scan) using a micro cell with Pt working, counter and reference electrodes without solvent and an additional supporting electrolyte. Ref. (Uchida et al., 2009a) – Reproduced by permission of The Royal Society of Chemistry.

5. Conclusions and prospects

The unique magnetic and electric properties of organic NR LCs and ILs were briefly surveyed. Noteworthy is the first observation of positive magneto-LC effects ($\bar{J}$ > 0) under weak magnetic fields in both chiral and achiral rod-like LC phases of the second-generation of all-organic NR compounds **12** with negative dielectric anisotropy ($\Delta\varepsilon < 0$), while positive and negative magneto-LC effects ($\bar{J}$ > 0 and $\bar{J}$ < 0) were observed in the chiral and achiral

LC phases of **13**, respectively, with positive dielectric anisotropy ($\Delta\varepsilon > 0$). Meanwhile, the ferroelectric properties of the SmC* phase of **12a** were fully characterized in a thin sandwich cell; the NR unit proved to act as the sufficient source of spontaneous polarization (*Ps*). Such chiral organic NR LCs with low viscosity showed faster ferroelectric switching than chiral metallomesogens with high viscosity. The control of magnetic properties of magnetic ferroelectric LCs of **12** by electric fields is the next step. In addition, it is of great advantage to be able to use EPR spectroscopy as the tool for observing the microscopic dynamic behavior of molecules in the NR LC phases, coupled with the use of SQUID magnetization measurement to observe the macroscopic behavior. Furthermore, EPR spectroscopy turned out to be an excellent tool for analyzing the temperature dependence of the χ_{para} for organic NR LC phases at high temperatures, for which SQUID magnetization measurement is not suited. Thus, such second generation of all-organic chiral LC NR compounds would open up a novel research field of metal-free magnetic or spintronic LC materials.

Meanwhile, the advent of IL NR EPR spin probes would make possible the in-depth understanding of the local structure or the molecular shape anisotropy of diamagnetic IL solvents, which cannot be available by using conventional spin probes such as TEMPO derivatives.

The research on metal-free magnetic soft materials is still in its infancy. The development of novel metal-free magnetic soft materials such as LCs, ILs, emulsions, and gels based on the NR chemistry is strongly expected.

Author details

Rui Tamura, Yoshiaki Uchida and Katsuaki Suzuki
Graduate School of Human and Environmental Studies, Kyoto University, Japan

Acknowledgement

We Acknowledge Professor Takeji Takui, Professor Hiroyuki Nohira, Dr. Yoshio Aoki, Professor Hideo Takezoe, Dr. Yoshio Shimbo, Ms. Reiri Kogo, Professor Jun Yamauchi, Dr. Yohei Noda, Dr. Naohiko Ikuma and Dr. Satoshi Shimono for their collaboration.

6. References

Allgaier, J. & Finkelmann, H. (1994). Synthesis and Magnetic Properties of Mesogenic Side-Chain Polymers Containing Stable Radicals, *Macromolecular Chemistry and Physics*, Vol.195, pp. 1017-1030.

Amabilino, D. B & Veciana, J. (2001). Nitroxide-Based Organic Magnets, In: *Magnetism: Molecules to Materials II*, J.S.Miller & M.Drillon, (Eds.), 1-60, Wiley-VCH, ISBN 3-527-30301-4, Weinheim, Germany.

Aurich, H. G. (1989). Nitroxides, In: *Nitrones, Nitronates and Nitroxides*, E.Breuer, H.G.Aurich, A.Nielsen, (Eds.), 313-370, John Wiley & Sons, ISBN 0-471-91709-5, Chichester, UK.

Binnemans, K. & Gröller-Walrand, C. (2002). *Lanthanide-Containing Liquid Crystals and Surfactants*, Chemical Reviews, Vol.102, pp.2303-2345.

Boamfa, M. I.; Kim, M. W.; Maan, J. C. & Rasing, T. (2003). Observation of Surface and Bulk Phase Transitions in Nematic Liquid Crystals, *Nature*, Vol.421, 149-152.

Chiarelli, R.; Novak, M. A.; Rassat, A. & Tholence, J. L. (1993). A Ferromagnetic Transition at 1.48 K in an Organic Nitroxide. *Nature*, Vol. 363, pp. 147-149.

De Gennes, P. G. & Prost, J. (1993). The Physics of Liquid Crystals, Oxford University Press, ISBN 0-19-852024-7, New York, USA.

Dierking, I. (2003). *Textures of Liquid Crystals*, Wiley-VCH, ISBN 3-527-30725-7, Weinheim, Germany.

Dunmur, D. & Toriyama, K. (1999). Magnetic Properties of Liquid Crystals, In: *Physical Properties of Liquid Crystals*, D.Demus, J.Goodby, G.W.Gray, H.-W.Spiess, V.Vill, (Eds.), 102-112, Wiley-VCH, ISBN 3-527-29747-2, Weinheim, Germany.

Dvolaitzky, M.; Billard, J. & Poldy, F. (1974). Notes des Membres et Correspondants et Notes Présentées ou transmises par Leurs Soins, *Comptes Rendus de L'académie des Science*, Vol. 279C, pp.533-535.

Dvolaitzky, M.; Billard, J. & Poldy, F. (1976b). Smectic E, C and A Free Radicals, *Tetrahedron*, Vol. 32, 1835-1838.

Dvolaitzky, M.; Taupin, C. & Poldy, F. (1976a). Nitroxides Piperidiniques – Synthese de Nouvelles Sondes Paramagnetiques, *Tetrahedron Letters*, pp. 1469-1472.

Goodby, J. W.; Blinc, R.; Clark, N. A.; Lagerwall, S. T.; Osipov, M. A.; Pikin, S. A.; Sakurai T.,; Yoshino, K. & Boštjan, Ž. (1991). *Ferroelectric Liquid Crystals*, Gordon and Breach Science Publishers, ISBN 2-88124-282-0, Philadelphia, USA.

Greve, S.; Vill, V. & Friedrichsen, W. (2002). Novel Nitronyl Nitroxides: Synthesis and Properties, *Zeitschrift für Naturforschung*, Vol.57b, pp. 677-684.

Griesar, K. & Haase, W. (1999). Magnetic Properties of Transition-Metal-Containing Liquid Crystals, In: Magnetic Properties of Organic Materials, P.M.Lahti, (Ed.), 325-344, Marcel Dekker, ISBN 0-8247-1976-X, New York, USA.

Hicks, R. G. (2010) (Ed). *Sable Radicals: Fundamentals and Applied Aspects of Odd-Electron Compounds*, John Wiley & Sons, ISBN 978-0-470-77083-2, Chichester, UK.

Hudson, S. A. & Maitlis, P. M. (1993). Calamitic Metallomesogens: Metal-Containing Liquid Crystals with Rodlike Shapes. *Chemical Reviews*, Vol.93, pp. 861-885.

Iglesias, R.; Marcos, M.; Martinez, J.; Serrano, J. L.; Sierra, T. & Perez-Jubindo, M. A. (1996). Ferroelectric Behavior of Chiral Bis(salicylideneaniline) Copper(II), Vanadium(IV), and Palladium(II) Liquid Crystals, *Chemistry of Materials*, Vol.8, pp. 2611-2617.

Ikuma, N.; Tamura, R.; Masaki, K.; Uchida, Y.; Shimono, S.; Yamauchi, J.; Aoki, Y. & Nohira, H. (2006b). Paramagnetic FLCs Containing an Organic Radical Component, *Ferroelectrics*, Vol.343, 119-125.

Ikuma, N.; Tamura, R.; Shimono, S.; Kawame, N.; Tamada, O.; Sakai, N.; Yamauchi, J. & Yamamoto, Y. (2004). Magnetic Properties of All-Organic Liquid Crystals Containing a Chiral Five-Membered Cyclic Nitroxide Unit within the Rigid Core, *Angewandte Chemie International Edition*, Vol.43, 3677-3682.

Ikuma, N.; Tamura, R.; Shimono, S.; Uchida, Y.; Masaki, K.; Yamauchi, J.; Aoki, Y. & Nohira, H. (2006a). Ferroelectric Properties of Paramagnetic, All-Organic, Chiral Nitroxyl Radical Liquid Crystals, *Advanced Materials*, Vol.18, 477-480.

Kaszynski, P. (1999). Liquid Crystalline Radicals: An Emerging Class of Organic Magnetic Materials, In: *Magnetic Properties of Organic Materials*, P.M.Lahti, (Ed.), 325-344, Marcel Dekker, ISBN 0-8247-1976-X, New York, USA.

Kogo, R.; Araoka, F.; Uchida, Y.; Tamura, R.; Ishikawa, K. & Takezoe, H. (2010). Second Harmonic Generation in a Paramagnetic All-Organic Chiral Smectic Liquid Crystal, *Applied Physics Express*, Vol.3, 041701.

Kumar, C. K. (Ed). (2009). *Magnetic Nanomaterials*, Wiley-VCH, ISBN 978-3-527-32154-4, Weinheim, Germany.

Lagerwall, S. T. (1999), *Ferroelectric and Antiferroelectric Liquid Crystals*, Wiley-VCH, ISBN 3-527-29831-2, Weinheim, Germany.

Matsushita, M. M.; Kawakami, H.; Kawada, Y. & Sugawara, T. (2007). Negative Magneto-Resistance Observed on an Ion-Radical Salt of a TTF-Based Spin-Polarized Donor. *Chemistry Letters*, pp. 110-111.

Matsushita, M. M.; Kawakami, H.; Sugawara, T. & Ogata, M. (2008). Molecule-Based System with Coexisting Conductivity and Magnetism and Without Magnetic Inorganic Ions. *Physical Review. B*, Vol.77, 195208.

Müller, H. J. & Haase, W. (1983). Magnetic Susceptibility and the Order Parameter of Some 4,4'-Disubstitued Biphenyl Cyclohexanes, *Journal de Physique (Paris)*, Vol.44, pp. 1209-1213.

Nakahara, K.; Iwasa, S.; Satoh, M.; Morioka, Y.; Iriyama, J.; Suguro, M. & Hasegawa, E. (2002). Rechargeable Batteries with Organic Radical Cathodes. *Chemical Physics Letters*, Vol.359, pp. 351-354.

Nakatsuji, S. (2008). Preparation, Reactions, and Properties of Functional Nitroxide Radicals, In: *Nitroxides: Applications in Chemistry, Biomedicine, and Materials Science*, G.I.Likhtenshtein, J.Yamauchi, S.Nakatsuji, A.I.Smirnov & R.Tamura, (Eds.), 161-204, Wiley-VCH, ISBN 978-3-31889-6, Weinheim, Germany.

Nakatsuji, S.; Mizumoto, M.; Ikemoto, H.; Akutsu, H. & Yamada, J. (2002). Preparation and Properties of Organic Radical Compounds with Mesogenic Cores, *European Journal of Organic Chemistry*, pp. 1912-1918.

Oyaizu, K. & Nishide, H. (2010). Mesoscale Radical Polymers: Bottom-Up Fabrication of Electrodes in Organic Polymer Batteries, In: *Advanced Nanomaterials*, Vol.2, K.E.Geckeler, H.Nishide, (Eds.), 319-332, Wiley-VCH, ISBN 978-3-527-31794-3, Weinheim, Germany.

Piguet, C.; Bünzli, J-C. G.; Donnio, B. & Guillon, D. (2006). Thermotropic Lanthanidomesogens, Chemical Communications, pp. 3755-3768.

Suga, T. & Nishide, H. (2010). Rechargeable Batteries Using Robust but Redox Active Organic Radicals, In: *Sable Radicals: Fundamentals and Applied Aspects of Odd-Electron Compounds*, R.G.Hicks, (Ed.), 507-520, John Wiley & Sons, ISBN 978-0-470-77083-2, Chichester, UK.

Sugawara, T.; Komatsu, H. & Suzuki, K. (2011). Interplay between Magnetism and Conductivity Derived from Spin-Polarized Donor Radicals. *Chemical Society Reviews*, Vol.40, pp. 3105-3118.

Suzuki, K.; Uchida, Y.; Tamura, R.; Shimono, S. & Yamauchi, J. (2012). Observation of Positive and Negative Magneto-LC Effects in All-Organic Nitroxide Radical Liquid Crystals by EPR Spectroscopy, *Journal of Materials Chemistry*, DOI: 10.1039/c2jm16278d.

Tamura, M.; Nakazawa, Y.; Shiomi, D.; Nozawa, K.; Hosokoshi, Y.; Ishikawa, M.; Takahashi, M. & Kinoshita, M. (1991). Bulk Ferromagnetism in the β-Phase Crystal of the *p*-Nitrophenyl Nitronyl Nitroxide Radical. *Chemical Physics Letters*, Vol.186, No.4 & 5, pp. 401-404.

Tamura, R. (2008a). Organic Functional Materials Containing Nitroxide Radical Units, In: *Nitroxides: Applications in Chemistry, Biomedicine, and Materials Science*, G.I.Likhtenshtein, J.Yamauchi, S.Nakatsuji, A.I.Smirnov & R.Tamura, (Eds.), 303-329, Wiley-VCH, ISBN 978-3-31889-6, Weinheim, Germany.

Tamura, R.; Uchida, Y. & Ikuma, N. (2008b). Paramagnetic All-Organic Chiral Liquid Crystals, *Journal of Materials Chemistry*, Vol.18, pp. 2872-2876.

Tamura, R.; Uchida, Y. & Suzuki, K. (2012). Magnetic Liquid Crystals, In: *Liquid Crystals Beyond Displays: Chemistry, Physics, and Applications*, Q, Li (Ed.), Chap. 3, John Wiley & Sons, ISBN 978-1-118-07861-7, Chichester, UK.

Terazzi, E.; Suarez, S.; Torelli, S.; Nozary, H.; Imbert, D.; Mamula, O.; Rivera, J-P.; Guillet, E.; Bénech, J-M.; Bernardinelli, G.; Scopelliti, R.; Donnio, B.; Guillon, D.; Bünzli, J-C. G. & Piguet, C. (2006). Introducing Bulky Functional Lanthanide Cores into Thermotropic Metallomesogens: A Bottom-Up Approach. Advanced Functional Materials, Vol.16, pp. 157-168.

Uchida, Y.; Ikuma, N.; Tamura, R.; Shimono, S.; Noda, Y.; Yamauchi, J.; Aoki, Y. & Nohira, H. (2008). Unusual Intermolecular Magnetic Interaction Observed in an All-Organic Radical Liquid Crystal, *Journal of Materials Chemistry*, Vol.18, pp. 2950-2952.

Uchida, Y.; Oki, S.; Tamura, R.; Sakaguchi, T.; Suzuki, K.; Ishibashi, K. & Yamauchi, J. (2009a). Electric, Electrochemical and Magnetic Properties of Novel Ionic Liquid Nitroxides, and Their Use as an EPR Spin Probe. *Journal of Materials Chemistry*, Vol.19, pp. 6877-6881.

Uchida, Y.; Suzuki, K.; Tamura, R.; Ikuma, N.; Shimono, S.; Noda, Y. & Yamauchi, J. (2010). Anisotropic and Inhomogeneous Magnetic Interactions Observed in All-Organic Nitroxide Radical Liquid Crystals, *Journal of the American Chemical Society*, Vol.132, 9746-9752.

Uchida, Y.; Tamura, R.; Ikuma, N.; Shimono, S.; Yamauchi, J.; Shimbo, Y.; Takezoe, H.; Aoki, Y. & Nohira, H. (2009b). Magnetic-Field-Induced Molecular Alignment in an Achiral Liquid Crystal Spin-Labeled by a Nitroxyl Group in the Mesogen Core, *Journal of Materials Chemistry*, Vol.19, pp.415-418.

Chapter 8

Synthesis and Utilization of α-Substituted Nitroxides

Toshihide Yamasaki, Fumiya Mito, Yuta Matsuoka, Mayumi Yamato and Ken-ichi Yamada

Additional information is available at the end of the chapter

1. Introduction

Nitroxides are stable free radicals which have the >N-O moiety. In most cases, nitroxides have a ring structure. For example, imidazoline, isoindoline, piperidine and pyrrolidine ring nitroxides (Fig. 1) have been used as agents for spin labeling, imaging, and as antioxidants. These nitroxides have four substituents at the α-position; two substituents on each α-carbon. All four substituents are needed for avoiding the disproportionation reaction of nitroxides except for the case of a bridgehead at the α-position. Methyl groups have been chosen as simple and inert substituents. However, it has been reported that other types of substituents, especially ethyl groups, showed unique characteristics that were unlike those of the conventional methyl group.

In this chapter, we will introduce the conventional as well as the latest synthetic methods used to introduce the various substituents to the α-position. Also, we will describe the structure–reactivity relationships of α-substituted nitroxides.

2. Synthetic methods of α-substituted nitroxides

2.1. Imidazoline

Imidazoline nitroxides have been synthesized from α-hydroxyaminoketone with carbonyl compounds (Scheme 1) (Volodarsky and Igor A, 1988). The R_1 groups of α-hydroxyaminoketone and R_2 groups of carbonyl compounds correspond to the α-position of the imidazoline ring. α-Hydroxyaminoketones are synthesized from appropriate olefines *via* nitrosylation, treatment with hydroxylamine, and hydrolysis (Kirilyuk et al., 2004). This is a practical method because this key compound can be synthesized on the gram scale. For R_2 groups, chain ketones or cyclic alkyl ketones are used as carbonyl compounds (Bobko et al., 2007; Kirilyuk et al., 2004; Yan'shole et al., 2010; Zubenko et al., 2006). These give the 2,2,5,5-tetrasubstituted 2,5-dihydro-1*H*-imidazole-1-ol. After oxidation, they give the desired imidazole nitroxides. Imidazolidine nitroxides can be synthesized from imidazoline nitroxides by reduction (Zubenko et al., 2006). Therefore, α-substituted imidazoline nitroxides are also synthesized from α-substituted imidazolidine nitroxides.

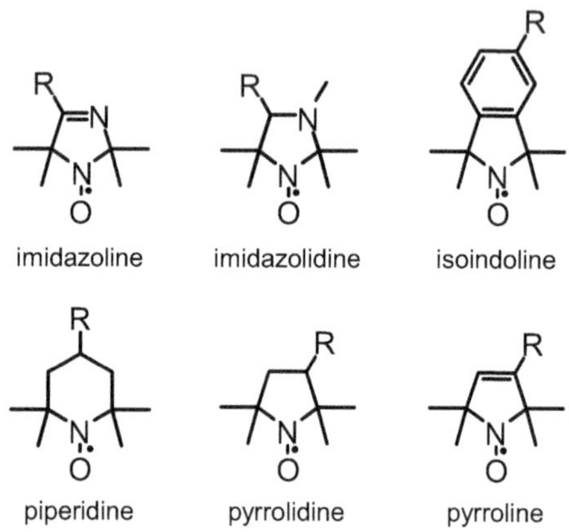

Figure 1. Basic structure of nitroxides

Scheme 1. α-Substitution synthesis of imidazoline and imidazolidine nitroxides

R_1 = Me; R_2 = Me, Et, n-Bu, $(CH_2)_4$, $(CH_2)_5$

R_1 = Et; R_2 = Me, Et, $(CH_2)_5$, $(CH_2)_2COONa$

2.2. Isoindoline

Isoindoline nitroxides have been prepared by the addition of a greater than fourfold excess of a single Grignard reagent to *N*-benzylphthalimide (Scheme 2) (Griffiths et al., 1983). Therefore, according to the type of Grignard reagent, various substituents can be introduced to the α-position of the nitroxide moiety (Chan et al., 2010). Corresponding 2-benzyl-1,1,3,3-tetrasubstituted isoindoline derivatives are then deprotected and subsequent oxidation gives the α-substituted isoindoline nitroxides. The isoindoline skeleton has the potential to have low solubility in water; longer alkyl chains cause a decrease in water solubility. Hence, many studies have been conducted with ethyl groups with respect to biological applications (Fairfull-Smith et al., 2009; Marx et al., 2000).

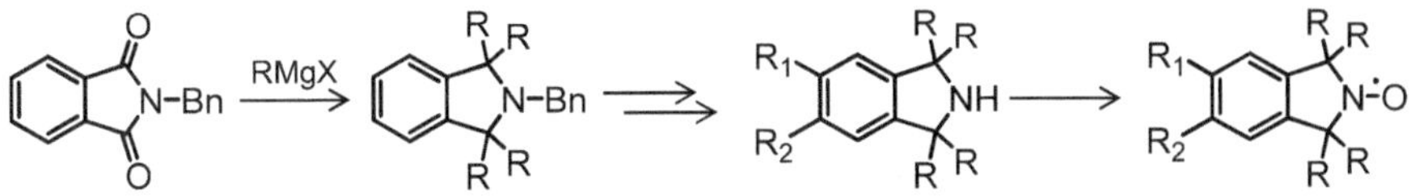

Scheme 2. α-Substitution synthesis of isoindoline nitroxides

R = Me, Et, n-Pr, n-Bu, Ph

2.3. Piperidine

Piperidine nitroxides have been synthesized by two main approaches. One is the synthesis from acetonin (2,2,4,4,6-pentamethyl-2,3,4,5-tetrahydropyrimidine) with carbonyl compounds (Schemes 3a and b). Murayama *et al.* reported the isolation of 1,9-diaza-2,2,8,8,10,10-hexamethylspiro[5.5]undecan-4-one as a byproduct in the method development for 2,2,6,6-tetramethyl-4-piperidone (Murayama et al., 1969). Subsequently, using acetonin, various α-substituted piperidine derivatives were reported by Ma *et al.* (Ma et al., 1993) and Miura *et al.* (Scheme 3b) (Mannan et al., 2007; Miura et al., 2003; Miura et al., 2001; Okazaki et al., 2007). These methods have the advantage of synthesis in a few steps, but the key compound used at the start of the synthesis, acetonin, lacks stability.

The other main approach is a stepwise synthesis from the appropriate starting material (Schemes 3c, d and e). For example, Yoshioka *et al.* reported on the synthesis for piperidine nitroxides having a spirocyclohexyl group *via* α, β-unsaturated ketone derivatives (Yoshioka et al., 1972). In addition, focusing on substitution of the tetraethyl group, Studer *et al.* reported a stepwise synthesis *via* bisphosphonates (Wetter et al., 2004), and also β-lactams for obtaining piperidine nitroxides on a large scale (Schulte et al., 2005) which involved several synthetic steps and a high-pressure reactor.

Recently, an alternative synthetic method has been developed (Scheme 3f) (Sakai et al., 2010). This method involves 2,2,6,6-tetramethyl-4-piperidone as a starting compound; this compound is more stable than acetonin and is available commercially. This compound with cyclohexanone directly gave the piperidone derivative having spirocyclohexyl groups at the 2,6-position under a mild reaction condition. Moreover, the reaction yield was increased by using 1,2,2,6,6-pentamethyl-4-piperidone and a base. With this starting compound, various substituents have been introduced to the α-position (Yamasaki et al., 2011; Yamasaki et al., 2010). From the investigation of the reaction mechanism, the nitrogen derived from ammonium chloride was introduced to the piperidone ring. Therefore, using ^{15}N-labeled NH_4Cl instead of $^{14}NH_4Cl$, ^{15}N-labeled 2,2,6,6-tetrasubstituted piperidin-4-one-1-oxyls can be produced with high (>98%) ^{15}N content. Thus, the external NH_4X compound seems to be the source of nitrogen during this reaction.

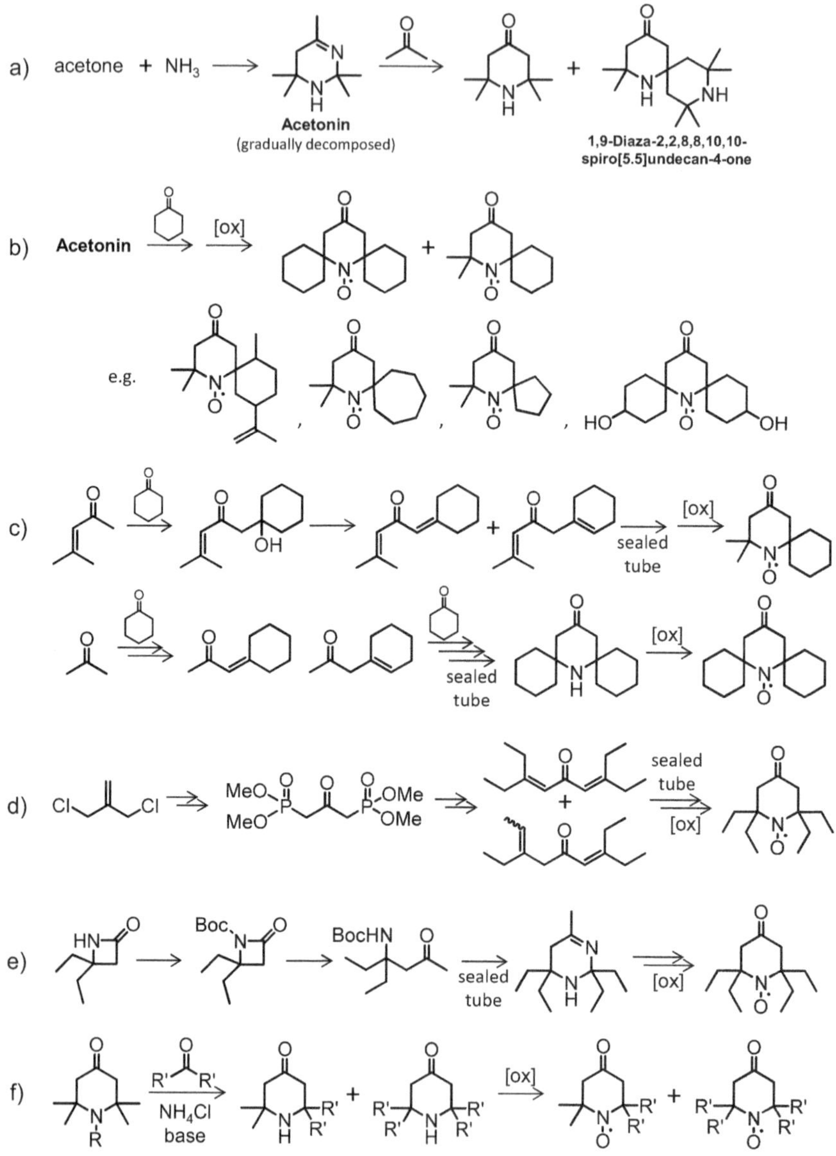

Scheme 3. α-Substitution synthesis of piperidine nitroxides. a), b) from acetonin, c), d), e) stepwise synthesis, f) from piperidone

2.4. Pyrrolidine

α-Substituted pyrrolidine nitroxides has been synthesized *via* the nucleophilic addition of organometallic reagents such as 2,2,5-trimethyl-3,4-dihydro-2*H*-pyrrole-1-oxide or 2,5-dimethyl-3,4-dihydro-2*H*-pyrrole-1-oxide to nitrone (Ikuma et al., 2004; Keana et al., 1983). Grignard reagents are also used as nucleophiles. These moieties are introduced to the α-position of the nitroxide. Therefore, various substituents could be introduced depending on the organometallic reagent. Moreover, substitution of the starting 3,4-dihydro-2*H*-pyrrole-1-oxide at the 2 and 5 positions has been carried out (Black et al., 2000). This also led to α-substituted pyrrolidine nitroxides. These derivatives have been synthesized *via* intramolecular reductive amination from 5-nitrohexan-2-one derivatives. Additionally, the stepwise synthesis from 5-methylhex-5-en-2-one to introduce the phosphonate group has been reported (Le Moigne et al., 1991).

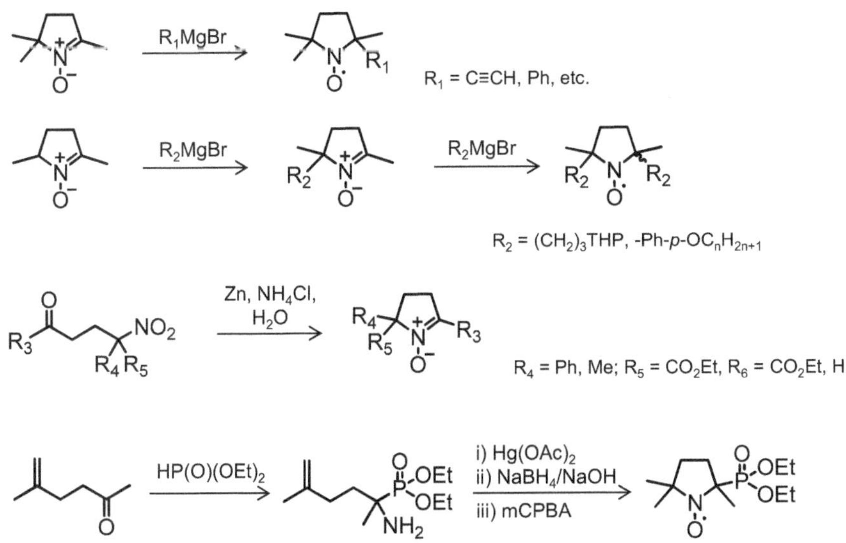

Scheme 4. α-Substitution synthesis of isoindoline nitroxides

R = Me, Et, n-Pr, n-Bu, Ph

3. Evaluation of α-substituted nitroxides

3.1. Common reactivity of nitroxides

Nitroxides have several potential advantages as spin probes (Kuppusamy et al., 2002; Yamada et al., 2006), spin labels (Borbat et al., 2001), contrast agents (Soule et al., 2007) and antioxidants (Wilcox and Pearlman, 2008). These applications are based on the complementary nature of the radical moieties in nitroxides; paramagnetism allows them to react with free radicals and interact with nuclear spin. For instance, these properties allow nitroxides to be used as contrast agents for magnetic resonance imaging (MRI) to give images of the morphological nature and redox imbalance in animal models of oxidative stress.

In biological systems, understanding of biophysical properties is helpful to promote effective utilization and control of the reactivity of nitroxides. Nitroxides are readily oxidized to oxoammonium cations or reduced to hydroxylamines by various *in vivo* oxidants or reductants (Fig. 2) (Kocherginsky and Swartz, 1995). Upon reaction with superoxides, nitroxides undergo one-electron oxidation and subsequent two-electron reduction with reductants, which is driven forward by the redox potential of the nitroxide redox couple (Krishna et al., 1992). Conversely, nitroxides are reduced by ascorbic acid and converted into hydroxylamines (Saphier et al., 2003). This effect is one of the limitations of using nitroxides for monitoring *in vivo* redox status because the lifetime *in vivo* is shortened. The reduction rate by ascorbic acid is dependent mainly upon their ring structures and substituent groups. The rate constants decrease in the order oxazolidines > piperidines > pyrrolines > pyrrolidines, and the increasing inductive effects by the substituent group at the β- or γ-positions of pyrrolidine and piperidine nitroxide also affect the rate (Kocherginsky and Swartz, 1995). In this section, the effect of α-substituted nitroxides (especially piperidine nitroxide) on reduction by ascorbic acid was summarized and its *in vivo* application discussed.

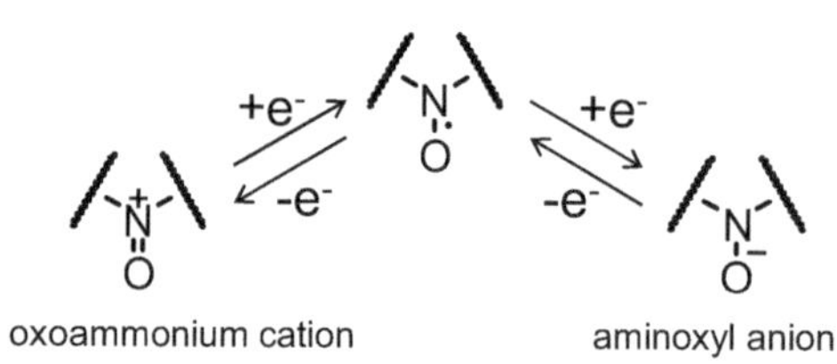

Figure 2. Redox couples of nitroxide

3.2. Reduction stability

As well as at β- or γ-positions, substituent groups at α-positions in a nitroxide ring can change their reactivity. For instance, phosphorylated pyrrolidinyl nitroxide showed moderate increase toward ascorbate reduction compared with the tetramethyl pyrrolidine nitroxide (Mathieu et al., 1997). On the other hands, the tetraethyl-substituted isoindoline (Marx et al., 2000), imidazoline (Kirilyuk et al., 2004), and imidazolidine (Kirilyuk et al., 2004) nitroxides showed high resistivity to ascorbate reduction than the corresponding tetramethyl compounds. Furthermore, Kirilyuk *et al.* reported that tetraethyl-substituted imidazoline nitroxides had significantly longer half-lives in rat blood samples as compared with the corresponding tetramethyl-nitroxides (Kirilyuk et al., 2004). Although piperidine nitroxides with spirocyclohexyl groups at α-positions showed no difference in reactivity with ascorbic acid, the enzymatic reaction was efficiently protected (Okazaki et al., 2007). Recently, tetraethyl-substituted piperidine nitroxide was also reported to exhibit resistance to ascorbate reduction (Kinoshita et al., 2009). The electron spin resonance (ESR) or electron paramagnetic resonance (EPR) signal intensities of tetraethyl-substituted piperidine nitroxide remained 97% after ascorbate addition (Fig. 3). These reports indicated that the reduction rates of nitroxides vary significantly depending on their structure and microenvironment resulting from the α-substitution of nitroxides. Tetraethyl-type nitroxides could be potentially useful for monitoring *in vivo* free-radical reactions, pH changes, and redox status without undesirable reduction by ascorbic acid.

Tetraethyl-nitroxides, having higher lipophilicity than tetramethyl compounds, have been reported to be less toxic to cells (Kinoshita et al., 2010) although the toxicity is reported to be correlated with the structure and lipophilicity of nitroxides (Ankel et al., 1987). Furthermore, single-dose administration of tetraethyl piperidine nitroxide has been shown to have lower blood pressure-lowering effects compared with that of Tempol (Kinoshita et al., 2010).

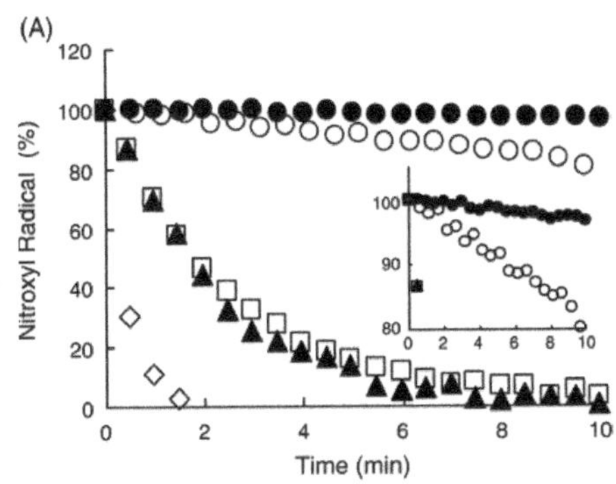

Figure 3. Effect of AsA on the decay of ESR signals of nitroxides. Nitroxides (100 μM) were mixed with AsA (1 mM) in phosphate buffered saline and their ESR spectra measured as a function of time. •: 4-oxo-2,2,6,6-tetraethyl-piperidine-1-oxyl; ○: carbamoyl-PROXYL; ▲: hydroxy-TEMPO (TEMPOL); □: oxo-TEMPO (TEMPONE); ◊: 7-Aza-3,11-dioxa-15-oxodispiro[5.1.5.3]hexadec-7-yl-7-oxyl. (Kinoshita et al., *Free Radic Res* 2009;43: 565, copyright© 2012, Informa Healthcare. Reproduced with permission of Informa Healthcare)

3.3. Electrochemical behaviours

The change in nitroxide reactivity due to the presence of tetraethyl substituents suggests that introduction of bulky alkyl groups at α-positions in a nitroxide ring are responsible for their reduction stability. Steric hindrance around the radical moiety is one of the most important factors inhibiting access to reductants. However, the ESR signal intensities of 7-Aza-3,11-dioxa-15-oxodispiro[5.1.5.3]hexadec-7-yl-7-oxyl (which has also bulky spirocyclohexyl rings at α-positions) decrease rapidly in the presence of ascorbate

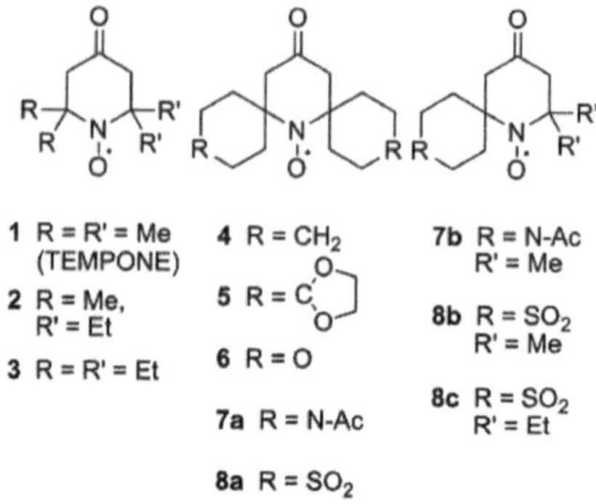

Figure 4. Structure of α-substituted piperidine nitroxides

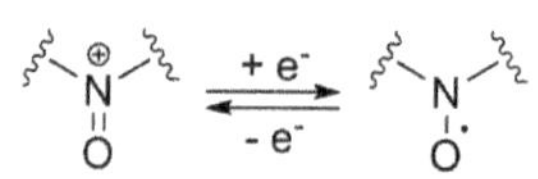

Nitroxide	E_{pa} [b]	E_{pc} [c]	$E^{1/2}$ [d]	ΔE [e]	i_{pa}/i_{pc} [f]
1	0.939	0.869	0.904	0.070	2.30
2	0.926	0.860	0.893	0.065	1.56
3	0.806	0.747	0.776	0.058	2.64
4	0.878	0.811	0.845	0.067	2.97
5	1.013	0.874	0.944	0.139	3.29
6	1.053	0.940	0.997	0.113	3.06
7a	1.113	-[g]	-	-	-
7b	1.019	0.902	0.961	0.117	3.34
8a	1.303	-[g]	-	-	-
8b	1.122	1.012	1.067	0.110	3.51
8c	1.116	1.020	1.068	0.096	3.65

[a] glassy carbon electrode, Ag/AgCl, Pt, sweep rate: 0.1 V s^{-1}. Potentials were shown as vs. SHE. [b] anodic peak potential. [c] cathodic peak potential. [d] $E^{1/2} = (E_{pa} + E_{pc})/2$. [e] $\Delta E = E_{pa} - E_{pc}$. [f] The peak currents (i_{pa} and i_{pc}) were measured from the respective baseline currents. [g] The current of the cathodic peak was too low to determine the potential value. Adapted with permission from (Yamasaki *et al.*, *J Org Chem* 2011;76:435). Copyright (2012) American Chemical Society.

Table 1. Experimental redox potentials of oxoammonium cation / nitroxide redox couple[a]

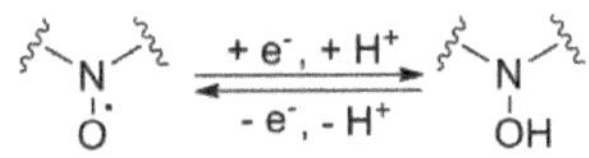

Nitroxide	E_{pa} [b]	E_{pc} [c]	$E^{1/2}$ [d]	ΔE [e]	i_{pa}/i_{pc} [f]
1	0.524	-0.371	0.076	0.895	0.20
2	0.453	-0.401	0.026	0.854	0.20
3	0.226	-0.298	-0.036	0.524	2.12
4	0.244	-0.141	0.051	0.385	0.99
5	0.515	-0.364	0.076	0.879	0.09
6	0.644	-0.386	0.129	1.029	0.19
7a	0.613	-0.247	0.183	0.860	0.27
7b	0.585	-0.381	0.102	0.967	0.31
8a	0.653	-0.357	0.148	1.010	2.18
8b	0.694	-0.494	0.100	1.189	0.16
8c	0.416	-0.427	-0.006	0.844	0.22

[a] glassy carbon electrode, Ag/AgCl, Pt, sweep rate: 0.1 V s^{-1}. Potentials were shown as vs. SHE. [b] anodic peak potential. [c] cathodic peak potential. [d] $E^{1/2} = (E_{pa} + E_{pc})/2$. [e] $\Delta E = E_{pa} - E_{pc}$. [f] The peak currents (i_{pa} and i_{pc}) were measured from the respective baseline currents. Adapted with permission from (Yamasaki *et al.*, *J Org Chem* 2011;76:435). Copyright (2012) American Chemical Society.

Table 2. Experimental redox potentials of nitroxide / deprotonated hydroxylamine redox couple[a]

(Kinoshita et al., 2009). This suggests that the electronic environment around the N-O moiety also influences its reduction stability. In fact, the rate of reduction of β- or γ-substituted nitroxides by ascorbate has been reported to be primarily dependent upon their structure and correlation with $E_{1/2}$ (Blinco et al., 2008; Kocherginsky and Swartz, 1995). The reactivity of α-substituted nitronyl nitroxides is also dependent upon the electronic properties of the substituent groups (Wu et al., 2006). The α-substitution of piperidine nitroxide has been reported to change dramatically their redox potentials for one-electron oxidation and reduction (Yamasaki et al., 2011). In the oxidation step, electron-donating substituents are likely to stabilize oxoammonium cations, and substituents with heteroatoms destabilize them because of the electron-withdrawing inductive effect (Fig. 4, Table 1). The redox potentials for one-electron reduction are listed in Table 2. The electron-withdrawing groups at the α-positions of the piperidine ring destabilize the reduced form of nitroxides, whereas electron-donating substituents stabilize them.

3.4. Structure–reactivity relationships

As described above, ascorbate can readily convert nitroxides into the corresponding hydroxylamines. The reduction rate is correlated with the inductive effects from the β-position in the piperidine ring and the γ-position in the pyrrolidine ring (Morris et al., 1991).

Also, nitroxides with heteroatoms in their ring are unstable for the reduction (Couet et al., 1985).

Imidazole, isoindoline and piperidine nitroxides have a common feature: tetraethyl-nitroxides at α-positions adjacent to the radical moiety have high resistance to reduction by ascorbate compared with the widely used tetramethyl-nitroxides (see above). The rate of decay of the ESR signals of nitroxides seems to be inversely proportional to the number of ethyl groups (Yamasaki et al., 2010). Nitroxides containing four ethyl groups are more resistant to the reduction than those with two ethyl groups. The reduction rates of nitroxides which have heteroatoms in their spirocyclohexyl ring have been found to be higher than tetramethyl nitroxides. Electron-withdrawing groups at spirocyclohexyl rings decrease the electron density around the N-O moiety, thereby favoring the reduction reaction. The trend of redox potentials for nitroxide reduction from electrochemical experiments is likely to be exactly the same as that of the nitroxide reduction rate by ascorbate. The ESR signal decay rate and the electromotive force between nitroxide and ascorbate (ΔE_{N-A}) or the change in Gibbs free energy (ΔG) demonstrates very good correlations with ΔG in the negative ΔG region (r^2 = 0.988) (Fig. 5) (Yamasaki et al., 2010). This indicates that reduction of the nitroxide by ascorbate occurred spontaneously if the ΔG value is negative, and that the reduction is not spontaneous if the ΔG value is positive. The factors influencing the reduction process of the nitroxide are dependent not only upon steric hindrances but also on redox potentials. The α-substitutions of piperidine nitroxides would be an effective approach to control the reactivity of nitroxides as a function of their applications.

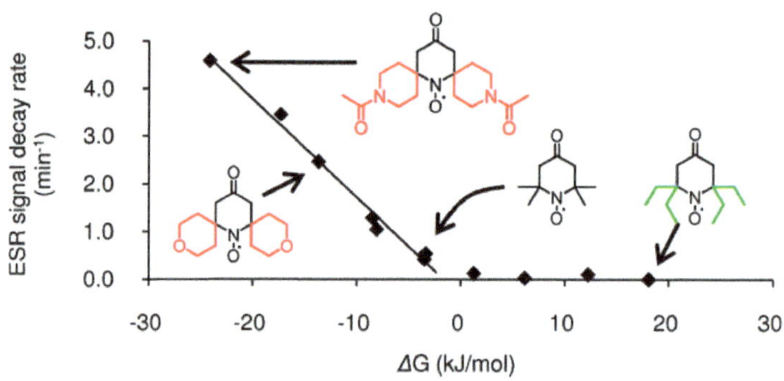

Figure 5. Relationship between the rates of decay of ESR signals and ΔG. ΔG was calculated from ΔE_{N-A} with use of the standard expression $\Delta G = -nF\Delta E_{N-A}$, where ΔE_{N-A} is a subtraction of the redox potential of ascorbate from that of nitroxides, n is the number of electrons per mole of product, and F is the Faraday constant. The correlation coefficient when ΔG was negative was 0.988. Reprinted with permission from (Yamasaki *et al.*, *J Org Chem* 2011;76:435). Copyright (2012) American Chemical Society)

3.5. *In vivo* evaluation and imaging

Nitroxides are reduced to mainly the hydroxylamine form *in vivo* by non-enzymatic processes involving gluthathione, the reduced form of nicotinamide adenine dinucleotide phosphate (NADPH), and ascorbic acid. Tetramethyl-nitroxides are widely used as biological reporters such as superoxide-dismutase mimics (Krishna et al., 1996), antioxidants (Wilcox and Pearlman, 2008) and radiation-protecting agents (Metz et al., 2004). However, non-specific reactions limit their applications to monitor changes in redox status. To increase their general versatility, improvement of the stability of nitroxides *in vivo* is important.

In general, the stability of nitroxide is reflected by their type of ring, substituent groups, and lipophilicity. Piperidine-nitroxides show a short half-life compared with that of pyrroridine-nitroxides. A typical tetramethyl-piperidine nitroxide, Tempone (oxo-TEMPO), has a short life-time (≈2 min) in blood due to rapid reduction (Ishida et al., 1989; Schimmack et al., 1976). However, piperidine nitroxides with spirocyclohexyl groups show resistance to enzymatic reduction in mouse liver homogenates (Okazaki et al., 2007). Conversely, tetraethyl nitroxides show resistance to reduction by ascorbic acid and seem to be stable *in vivo*. In fact, the half-lives of 4-oxo- and 4-hydroxy-2,2,6,6-tetraethyl-piperidine-1-oxyl in the mouse tail were 16.4±1.3 min and 20.0±3.0 min, respectively, although that of carbamoyl-PROXYL was 8.5±2.7 min (Kinoshita et al., 2010). The factor influencing the reactivity of nitroxides to reductants is thought to be their redox potential. Also, higher inductive substitution constants at the 4-position of piperidine nitroxides were found to confer higher rate constants for ascorbic acid reduction, emphasizing the importance of the electronic effects of the substituents of nitroxides.

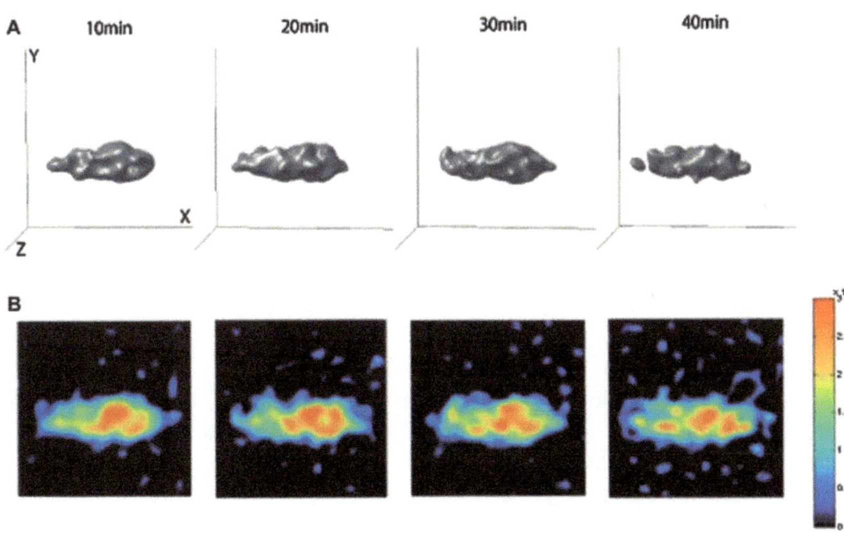

Figure 6. Temporal changes in three-dimensional surface-rendered EPR images (A) and two-dimensional slice images (B) of TEEPONE in mouse brains. EPR images were obtained 10, 20, 30, and 40 min after intravenous injection of TEEPONE. (Emoto *et al., Free Radic Res* 2011;45: 1325, copyright© 2012, Informa Healthcare. Reproduced with permission of Informa Healthcare)

A nitroxide with a long half-life can also be a candidate *in vivo* imaging agent. *In vivo* evaluations have revealed that tetraethyl-nitroxides have several advantages for application as contrast agents for monitoring redox status: non-toxicity, lack of blood pressure-lowering effects, and stability *in vivo*. Recently, it has been reported that the half-life of 4-oxo-2,2,6,6-tetraethyl-piperidine-1-oxyl (TEEPONE) obtained in mice brains is 81.1±0.7 min, indicating that TEEPONE is more stable compared with other nitroxides that are readily converted into hydroxylamines in tissues (Fig. 6) (Emoto et al., 2011). The decay rate of TEEPONE has no relation with ascorbic acid reduction: this compound can effectively scavenge free radicals such as carbon- and oxygen-centered radicals.

4. Conclusions

Recently, various types of α-substituted nitroxides have been synthesized and their *in vivo* and *in-vitro* characteristics evaluated. Introducing α-substituent groups can modify and control their nature according to usage as spin probes, antioxidants, and contrast agents. Although tetramethyl-nitroxides have been widely used in physiological and biological systems, the unique properties of α-substituted nitroxides will extend the possibilities of their applications. Development of α-substituted nitroxides creates the possibility of expansion of their use in a wide range of fields.

Author details

Toshihide Yamasaki, Fumiya Mito, Yuta Matsuoka, Mayumi Yamato and Ken-ichi Yamada
Faculty of Pharmaceutical Sciences, Kyushu University, Japan

5. References

Ankel, E.G.;Lai, C.S.;Hopwood, L.E. & Zivkovic, Z. (1987). Cytotoxicity of commonly used nitroxide radical spin probes. *Life Sci* 40, 495-498.

Black, D.S.;Edwards, G.L.;Evans, R.H.;Keller, P.A. & Laaman, S.M. (2000). Synthesis and Reactivity of 1-Pyrroline-5-carboxylate Ester 1-Oxides. *Tetrahedron* 56, 1889-1897.

Blinco, J.P.;Hodgson, J.L.;Morrow, B.J.;Walker, J.R.;Will, G.D.;Coote, M.L. & Bottle, S.E. (2008). Experimental and theoretical studies of the redox potentials of cyclic nitroxides. *J Org Chem* 73, 6763-6771.

Bobko, A.A.;Kirilyuk, I.A.;Grigor'ev, I.A.;Zweier, J.L. & Khramtsov, V.V. (2007). Reversible reduction of nitroxides to hydroxylamines: Roles for ascorbate and glutathione. *Free Radic Biol Med* 42, 404-412.

Borbat, P.P.;Costa-Filho, A.J.;Earle, K.A.;Moscicki, J.K. & Freed, J.H. (2001). Electron spin resonance in studies of membranes and proteins. *Science* 291, 266-269.

Chan, K.S.;Li, X.Z. & Lee, S.Y. (2010). Ligand-Enhanced Aliphatic Carbon–Carbon Bond Activation of Nitroxides by Rhodium(II) Porphyrin. *Organometallics* 29, 2850-2856.

Couet, W.R.;Brasch, R.C.;Sosnovsky, G. & Tozer, T.N. (1985). Factors affecting nitroxide reduction in ascorbate solution and tissue homogenates. *Magn Reson Imaging* 3, 83-88.

Emoto, M.;Mito, F.;Yamasaki, T.;Yamada, K.I.;Sato-Akaba, H.;Hirata, H. & Fujii, H. (2011). A novel ascorbic acid-resistant nitroxide in fat emulsion is an efficient brain imaging probe for in vivo EPR imaging of mouse. *Free Radic Res* 45, 1325-1332.

Fairfull-Smith, K.E.;Brackmann, F. & Bottle, S.E. (2009). The Synthesis of Novel Isoindoline Nitroxides Bearing Water-Solubilising Functionality. *Eur J Org Chem* 2009, 1902-1915.

Griffiths, P.;Moad, G. & Rizzardo, E. (1983). Synthesis of the radical scavenger 1,1,3,3-Tetramethylisoindolin-2-yloxyl. *Aust J Chem* 36, 397-401.

Ikuma, N.;Tamura, R.;Shimono, S.;Kawame, N.;Tamada, O.;Sakai, N.;Yamauchi, J. & Yamamoto, Y. (2004). Magnetic Properties of All-Organic Liquid Crystals Containing a Chiral Five-Membered Cyclic Nitroxide Unit within the Rigid Core. *Angew Chem Int Ed* 43, 3677-3682.

Ishida, S.;Kumashiro, H.;Tsuchihashi, N.;Ogata, T.;Ono, M.;Kamada, H. & Yoshida, E. (1989). In vivo analysis of nitroxide radicals injected into small animals by L-band ESR technique. *Phys Med Biol* 34, 1317-1323.

Keana, J.F.W.;Seyedrezai, S.E. & Gaughan, G. (1983). Difunctionalized trans-2,5-disubstituted pyrrolidine (azethoxyl) nitroxide spin-labels. *J Org Chem* 48, 2644-2647.

Kinoshita, Y.;Yamada, K.;Yamasaki, T.;Mito, F.;Yamato, M.;Kosem, N.;Deguchi, H.;Shirahama, C.;Ito, Y.;Kitagawa, K.;Okukado, N.;Sakai, K. & Utsumi, H. (2010). In vivo evaluation of novel nitroxyl radicals with reduction stability. *Free Radic Biol Med* 49, 1703-1709.

Kinoshita, Y.;Yamada, K.;Yamasaki, T.;Sadasue, H.;Sakai, K. & Utsumi, H. (2009). Development of novel nitroxyl radicals for controlling reactivity with ascorbic acid. *Free Radic Res* 43, 565-571.

Kirilyuk, I.A.;Bobko, A.A.;Grigor'ev, I.A. & Khramtsov, V.V. (2004). Synthesis of the tetraethyl substituted pH-sensitive nitroxides of imidazole series with enhanced stability towards reduction. *Org Biomol Chem* 2, 1025-1030.

Kocherginsky, N. & Swartz, H.M. (1995). Nitroxide Spin Labels: Reactions in Biology and Chemistry.

Krishna, M.C.;Grahame, D.A.;Samuni, A.;Mitchell, J.B. & Russo, A. (1992). Oxoammonium cation intermediate in the nitroxide-catalyzed dismutation of superoxide. *Proc Natl Acad Sci U S A* 89, 5537-5541.

Krishna, M.C.;Russo, A.;Mitchell, J.B.;Goldstein, S.;Dafni, H. & Samuni, A. (1996). Do nitroxide antioxidants act as scavengers of O2-. or as SOD mimics? *J Biol Chem* 271, 26026-26031.

Kuppusamy, P.;Li, H.;Ilangovan, G.;Cardounel, A.J.;Zweier, J.L.;Yamada, K.;Krishna, M.C. & Mitchell, J.B. (2002). Noninvasive imaging of tumor redox status and its modification by tissue glutathione levels. *Cancer Res* 62, 307-312.

Le Moigne, F.;Mercier, A. & Tordo, P. (1991). β-Phosphorylated cyclic nitroxides. 2. Synthesis of pyrrolidin- and piperidin-2-yl phosphonates and the corresponding stable nitroxides. *Tetrahedron Lett* 32, 3841-3844.

Ma, Z.;Huang, Q. & Bobbitt, J.M. (1993). Oxoammonium salts. 5. A new synthesis of hindered piperidines leading to unsymmetrical TEMPO-type nitroxides. Synthesis and enantioselective oxidations with chiral nitroxides and chiral oxoammonium salts. *J Org Chem* 58, 4837-4843.

Mannan, M.A.;Ichikawa, A. & Miura, Y. (2007). Living radical polymerization of styrene mediated by a piperidinyl-N-oxyl radical having very bulky substituents. *Polymer* 48, 743-749.

Marx, L.;Chiarelli, R.;Guiberteau, T. & Rassat, A. (2000). A comparative study of the reduction by ascorbate of 1,1,3,3-tetraethylisoindolin-2-yloxyl and of 1,1,3,3-tetramethylisoindolin-2-yloxyl. *J Chem Soc, Perkin Trans 1*, 1181-1182.

Mathieu, C.;Mercier, A.;Witt, D.;Debmkowski, L. & Tordo, P. (1997). beta-phosphorylated nitroxides in the pyrrolidine series: reduction by ascorbate. *Free Radic Biol Med* 22, 803-806.

Metz, J.M.;Smith, D.;Mick, R.;Lustig, R.;Mitchell, J.;Cherakuri, M.;Glatstein, E. & Hahn, S.M. (2004). A phase I study of topical Tempol for the prevention of alopecia induced by whole brain radiotherapy. *Clin Cancer Res* 10, 6411-6417.

Miura, Y.;Ichikawa, A. & Taniguchi, I. (2003). 'Living' radical polymerization of styrene mediated by spiro ring-substituted piperidinyl-N-oxyl radicals. The effect of the spiro rings on the control of polymerization. *Polymer* 44, 5187-5194.

Miura, Y.;Nakamura, N. & Taniguchi, I. (2001). Low-Temperature "Living" Radical Polymerization of Styrene in the Presence of Nitroxides with Spiro Structures. *Macromolecules* 34, 447-455.

Morris, S.;Sosnovsky, G.;Hui, B.;Huber, C.O.;Rao, N.U.M. & Swartz, H.M. (1991). Chemical and electrochemical reduction rates of cyclic nitroxides (nitroxyls). *J Pharm Sci* 80, 149-152.

Murayama, K.;Morimura, S.;Amakasu, O.;Toda, T. & Yamao, E. (1969). Convenient Synthetic Methods of 2,2,6,6-Tetramethyl-4-piperidone. *Nippon kagaku zassi* 90, 296-299.

Okazaki, S.;Mannan, M.A.;Sawai, K.;Masumizu, T.;Miura, Y. & Takeshita, K. (2007). Enzymatic reduction-resistant nitroxyl spin probes with spirocyclohexyl rings. *Free Radic Res* 41, 1069-1077.

Sakai, K.;Yamada, K.-i.;Yamasaki, T.;Kinoshita, Y.;Mito, F. & Utsumi, H. (2010). Effective 2,6-substitution of piperidine nitroxyl radical by carbonyl compound. *Tetrahedron* 66, 2311-2315.

Saphier, O.;Silberstein, T.;Shames, A.I.;Likhtenshtein, G.I.;Maimon, E.;Mankuta, D.;Mazor, M.;Katz, M.;Meyerstein, D. & Meyerstein, N. (2003). The reduction of a nitroxide spin label as a probe of human blood antioxidant properties. *Free Radic Res* 37, 301-308.

Schimmack, W.;Deffner, U. & Michailov, M.C. (1976). E.s.r study of the decay of the nitroxyl free radical TAN in whole rats and rat-tissue homogenates. *Int J Radiat Biol Relat Stud Phys Chem Med* 30, 393-397.

Schulte, T.;Siegenthaler, K.O.;Luftmann, H.;Letzel, M. & Studer, A. (2005). Nitroxide-Mediated Polymerization of N-Isopropylacrylamide: Electrospray Ionization Mass Spectrometry, Matrix-Assisted Laser Desorption Ionization Mass Spectrometry, and Multiple-Angle Laser Light Scattering Studies on Nitroxide-Terminated Poly-N-isopropylacrylamides. *Macromolecules* 38, 6833-6840.

Soule, B.P.;Hyodo, F.;Matsumoto, K.;Simone, N.L.;Cook, J.A.;Krishna, M.C. & Mitchell, J.B. (2007). The chemistry and biology of nitroxide compounds. *Free Radic Biol Med* 42, 1632-1650.

Volodarsky, L.B. & Igor A, G.e. (1988). Synthesis of Heterocyclic Nitroxides. In *Imidazoline Nitroxides*, L.B. Volodarsky, ed. (Boca Raton: CRC Press), 5-28.

Wetter, C.;Gierlich, J.;Knoop, C.A.;Müller, C.;Schulte, T. & Studer, A. (2004). Steric and Electronic Effects in Cyclic Alkoxyamines—Synthesis and Applications as Regulators for Controlled/Living Radical Polymerization. *Chem Eur J* 10, 1156-1166.

Wilcox, C.S. & Pearlman, A. (2008). Chemistry and antihypertensive effects of tempol and other nitroxides. *Pharmacol Rev* 60, 418-469.

Wu, Y.;Bi, L.;Bi, W.;Li, Z.;Zhao, M.;Wang, C.;Ju, J. & Peng, S. (2006). Novel 2-substituted nitronyl nitroxides as free radical scavengers: synthesis, biological evaluation and structure-activity relationship. *Bioorg Med Chem* 14, 5711-5720.

Yamada, K.;Yamamiya, I. & Utsumi, H. (2006). In vivo detection of free radicals induced by diethylnitrosamine in rat liver tissue. *Free Radic Biol Med* 40, 2040-2046.

Yamasaki, T.;Ito, Y.;Mito, F.;Kitagawa, K.;Matsuoka, Y.;Yamato, M. & Yamada, K.-i. (2011). Structural Concept of Nitroxide As a Lipid Peroxidation Inhibitor. *J Org Chem* 76, 4144-4148.

Yamasaki, T.;Mito, F.;Ito, Y.;Pandian, S.;Kinoshita, Y.;Nakano, K.;Murugesan, R.;Sakai, K.;Utsumi, H. & Yamada, K.-i. (2010). Structure–Reactivity Relationship of Piperidine Nitroxide: Electrochemical, ESR and Computational Studies. *J Org Chem* 76, 435-440.

Yan'shole, V.;Kirilyuk, I.;Grigor'ev, I.;Morozov, S. & Tsentalovich, Y. (2010). Antioxidative properties of nitroxyl radicals and hydroxyamines in reactions with triplet and deaminated kynurenine. *Russ Chem Bull* 59, 66-74.

Yoshioka, T.;Higashide.S & Murayama, K. (1972). Studies on Stable Free-Radicals .8. Synthesis and Oxidation of Hindered 4-Oxopiperidine Derivatives. *Bull Chem Soc Japan* 45, 636-638.

Zubenko, D.;Tsentalovich, Y.;Lebedeva, N.;Kirilyuk, I.;Roshchupkina, G.;Zhurko, I.;Reznikov, V.;Marque, S.R.A. & Bagryanskaya, E. (2006). Laser Flash Photolysis and CIDNP Studies of Steric Effects on Coupling Rate Constants of Imidazolidine Nitroxide with Carbon-Centered Radicals, Methyl Isobutyrate-2-yl and tert-Butyl Propionate-2-yl. *J Org Chem* 71, 6044-6052.

Chapter 9

Kinetics and Mechanism of Reactions of Aliphatic Stable Nitroxide Radicals in Chemical and Biological Chain Processes

Eugene M. Pliss, Ivan V. Tikhonov and Alexander I. Rusakov

Additional information is available at the end of the chapter

1. Introduction

Stable nitroxide radicals ($>NO^\bullet$) play very important role in experimental studies and in theoretical analysis of chemical and biochemical processes' mechanism for over 50 years.

For a long time it was believed that cyclic stable nitroxide radicals (piperidine, pyrroline, and imidazoline $>NO^\bullet$) inhibit oxidation of hydrocarbons (RH) and 1- and 1,1- ethylene substituted monomers (M) only via reaction with substrate's alkyl radicals in accordance with Scheme 1 (Buchachenko, 1965; Denisov & Afanas'ev, 2005; Mogilevich & Pliss, 1990). If substrate is M then chains are propagated by polyperoxide radicals ($MO_2^\bullet$) through its addition to monomer's double bond in accordance with the reaction: $MO_2^\bullet + M \rightarrow M^\bullet$ (Mogilevich & Pliss, 1990).

(i)	Initiator	$\xrightarrow{O_2,RH}$	$R^\bullet$	W_i
(1)	$R^\bullet + O_2$	$\longrightarrow$	$RO_2^\bullet$	k_1
(2)	$RO_2^\bullet + RH$	$\longrightarrow$	$ROOH + R^\bullet$	k_2
(3)	$RO_2^\bullet + RO_2^\bullet$	$\longrightarrow$	molecular products	k_3
(4)	$R^\bullet + >NO^\bullet$	$\longrightarrow$	$>NOR$	k_4

Scheme 1. Classical mechanism of organic compounds oxidation inhibited by nitroxide radicals

At the same time a probability of direct reaction of nitroxide with peroxide radicals is considered in a number of works (Barton, 1998; Goldstein & Samuni, 2007; Offer & Samuni, 2002; Pliss et al., 2010a, 2010b, 2012). Therefore it is necessary to review the most significant results in the field of kinetics and mechanism of elementary reactions of piperidine, pyrroline, and imidazoline >NO• with active particles of chemical and biochemical oxidation processes. Such an attempt has been made in this review.

Structures of nitroxide radicals and corresponding hydroxylamines which reactions' mechanisms were analyzed in this work are presented in Figure 1.

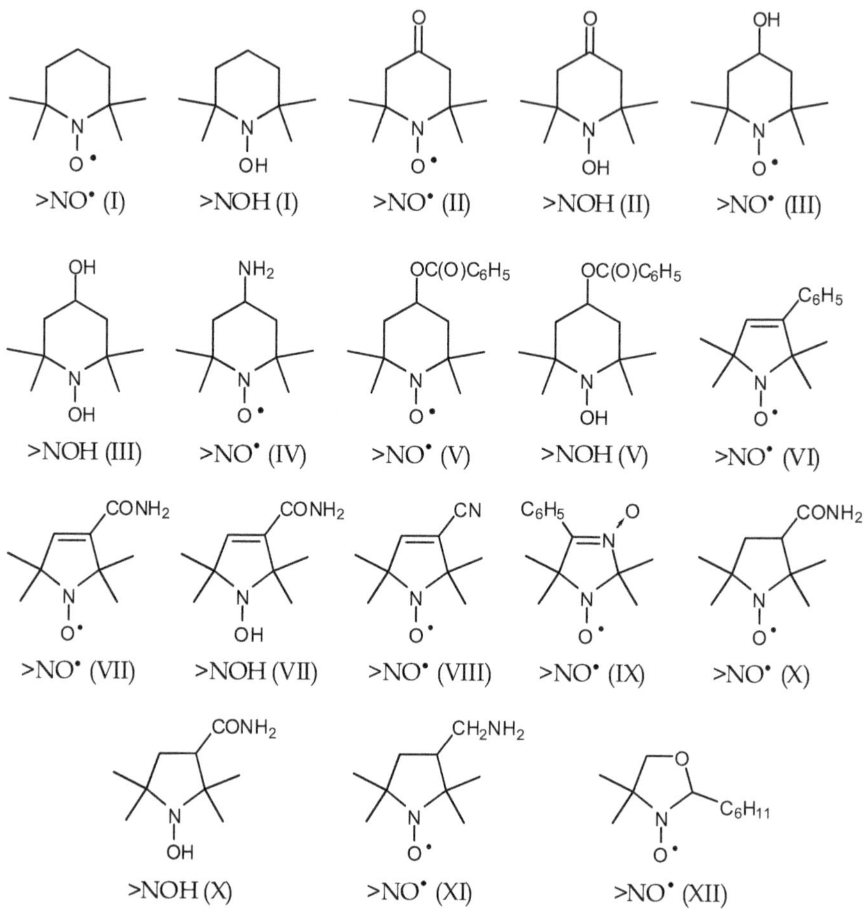

Figure 1. Structures of nitroxide radicals and corresponding hydroxylamines

2. Classical mechanism of organic compounds liquid phase oxidation inhibited by stable nitroxide radicals

For the first time Scheme 1 was proposed by A. Buchachenko with staff to describe the oxidation of ethylbenzene in presence of >NO• (II) and (III) (Buchachenko, 1965). k_4/k_1 Values (Table 1) were obtained from kinetic data of some organic compounds' initiated oxidation in presence of $>NO^•$. It can be seen that nitroxide radicals inhibit oxidation of methacrylic and acrylic ethers more effectively rather than oxidation of alkylaromatic compounds: k_4/k_1 values on average an order of magnitude greater for methacrylates and acrylates as compared with styrene, benzene, and cumene (Table 1).

At the same time k_4/k_1 value for methyl methacrylate (0.36) is close to that for cyclohexyl methyl ether (0.33) (Kovtun et al., 1974). Such results prove an important role of polar effects in reactions of $>NO^•$ with alkyl radicals during vinyl monomers oxidation.

So it should be noted that inhibiting activity of $>NO^•$ increases along with length of methacrylic ethers' alkyl substitute: k_4/k_1 value increases more than 3 times from methyl ether to ionylic ether. The similar trend also takes place for acrylic ethers (Table 1).

$M^•$ ($R^•$)	k_4/k_1 Value		Reference
	$>NO^•$ (I)	$>NO^•$ (II)	
$\sim CH_2C^•HC_6H_5$	—	0.10	(Browlie & Ingold, 1967)
$\sim CH_2C^•(CH_3)COOCH_3$	0.36	0.37	(Pliss & Aleksandrov, 1977)
$\sim CH_2C^•(CH_3)COO$–n-C_4H_9	0.72	0.72	(Pliss & Aleksandrov, 1977)
$\sim CH_2C^•(CH_3)COO$–iso-C_4H_9	0.72	0.72	(Pliss & Aleksandrov, 1977)
$\sim CH_2C^•(CH_3)COO$–n-C_9H_{19}	1.20	1.20	(Pliss & Aleksandrov, 1977)
$\sim CH_2C^•HCOOCH_3$	2.27	2.27	(Pliss & Aleksandrov, 1977)
$\sim CH_2C^•HCOO$–n-C_4H_9	5.20	5.20	(Pliss & Aleksandrov, 1977)
$CH_3C^•HC_6H_5$	0.04	—	(Browlie & Ingold, 1967)
$(CH_3)_2C^•C_6H_5$	0.05	—	(Kovtun et al., 1974)

Table 1. k_4/k_1 Values in oxidizing monomers and hydrocarbons at 323 K

Absolute k_4 values were measured in (Aleksandrov et al., 1979) by ESR spectroscopy method (Table 2). We note that $>NO^•$ is one of the strongest acceptors for alkyl radicals. k_4 Values are close to ones for molecular oxygen addition to alkyl radicals ($k_1 \geq 1 \cdot 10^7$ $M^{-1} \cdot s^{-1}$ (Aleksandrov et al., 1979)). k_4 Values for reactions of $>NO^•$ (I) – (VII) with alkyl radicals of methyl methacrylate at 323 K are within a limits of $(0.8 - 2.0) \cdot 10^7$ $M^{-1} \cdot s^{-1}$ (Table 2). Even higher k_4 values for the reaction of $>NO^•$ (I) with $R^•$ of low molecular weight at 291 K were obtained in (Bowry & Ingold, 1992) by laser flash photolysis method: these values are within a limits of $1 \cdot 10^6 - 2 \cdot 10^9$ $M^{-1} \cdot s^{-1}$.

The probability of $>NO^•$ participation in chain initiation process via its reaction with monomer's π-bond was estimated in (Ruban et al., 1967) using the reaction of $>NO^•$ (I) and (III) with styrene and α-methyl styrene as example. Reaction

$$>NO^• + CH_2{=}C(X)C_6H_5 \longrightarrow >NO{-}CH_2{-}C^•(X)C_6H_5 \text{ (where X = H or } CH_3)$$

at 393 K proceeds with constant rate value which equals to 4.6 $M^{-1} \cdot s^{-1}$ in styrene, but in α-methyl styrene this reaction doesn't proceed at all even at 453 K (Ruban et al., 1967). It's clear that $>NO^•$ initiating function completely suppresses by its participation in reaction (4).

Further we will see that reaction (4) is not the only one in oxidation inhibition by nitroxide radicals.

$M^•$	$>NO^•$	k_4, $M^{-1} \cdot s^{-1}$
$\sim CH_2C^•HC_6H_5$	I	$8.0 \cdot 10^6$
$\sim CH_2C^•(CH_3)COOCH_3$	I	$1.2 \cdot 10^7$
$\sim CH_2C^•(CH_3)COOCH_3$	II	$0.8 \cdot 10^7$
$\sim CH_2C^•(CH_3)COOCH_3$	III	$0.8 \cdot 10^7$
$\sim CH_2C^•(CH_3)COOCH_3$	V	$1.4 \cdot 10^7$
$\sim CH_2C^•(CH_3)COOCH_3$	VI	$1.6 \cdot 10^7$
$\sim CH_2C^•(CH_3)COOCH_3$	VII	$1.2 \cdot 10^7$
$\sim CH_2C^•(CH_3)COOCH_3$	VIII	$2.0 \cdot 10^7$
$\sim CH_2C^•(CH_3)COO$–*n*-C_4H_9	I	$1.2 \cdot 10^7$
$\sim CH_2C^•(CH_3)COO$–*iso*-C_4H_9	I	$0.9 \cdot 10^7$
$\sim CH_2C^•HCOOCH_3$	I	$3.0 \cdot 10^7$
$\sim CH_2C^•HCOO$–*iso*-C_4H_9	I	$1.8 \cdot 10^7$
cyclo-$C^•_6H_{11}$	I	$3.0 \cdot 10^7$
$(CH_3)_2C^•CN$ (in benzene)	I	$8.6 \cdot 10^7$

Table 2. Rate constants for the reaction $M^•$ ($R^•$) + $>NO^•$ at 323 K (Aleksandrov et al., 1979)

3. Multiple chain-breaking by stable nitroxide radicals

It was proved on oxidation of a number of compounds that oxidation chains propagate by peroxide radicals which possess redox properties. These are $HO_2^•$ radicals (cyclohexadiene (Howard & Ingold, 1967), 1,2-ethylene substituted and 1,4-butadiene substituted monomers (Mogilevich & Pliss, 1990)), $>C(OH)O_2^•$ (alcohols (Kharitonov & Denisov, 1967)), and $>CH$–$CH(OO^•)N<$ (aliphatic amines (Aleksandrov, 1987)). Dual reactivity of these radicals results in multiple $>NO^•$ participation in chain termination processes (Denisov, 1996). So for hydroperoxide radical this process can be described with the following reactions (Denisov, 1996):

$$HO_2^• + >NO^• \longrightarrow >NOH + O_2 \tag{5.1}$$

$$HO_2^• + >NOH \longrightarrow >NO^• + H_2O_2 \tag{5.2}$$

Let's perform the analysis of oxidation mechanism with $>NO^•$ regeneration and one without it.

3.1. Analysis of oxidation mechanism without nitroxide radical regeneration

In accordance with Scheme 1 initial rate process (W) without $>NO^•$ regeneration would be described by the following equation ($W = W_0$ when $[>NO^•]_0 = 0$):

$$W_i\left(\frac{W_0}{W} - \frac{W}{W_0}\right) = \frac{k_4[>NO^•]_0 W_0}{k_1[O_2]} \tag{1}$$

If $k_4[>NO^\bullet]_0 >> k_3[RO_2^\bullet]$ then

$$W = \frac{k_1[O_2]}{k_4[>NO^\bullet]_0} W_i \qquad (2)$$

So the rate process in linear termination mode (high $[>NO^\bullet]_0$) is directly proportional to the partial oxygen pressure (P_{O_2}). Therefore if oxygen is substituted by air the oxidation rate is to decrease 5 times. Such results were gained in (Browlie & Ingold, 1967; Pliss & Aleksandrov, 1977). But here is one important circumstance. From literature data (Aleksandrov, 1987) it's known that the reduction of nitroxide to corresponding hydroxylamine (>NOH) occurs via the reaction of aminoalkyl radical $>N–C^\bullet H–CH<$ with nitroxide radical as $>NO^\bullet$ attack to β-C–H bond of alkyl radical.

$$>N–C^\bullet H–CH< + NO^\bullet \longrightarrow >NOH + >N–CH=C<$$

The hydroxylamines being formed are thermally stable under experimental conditions (Aleksandrov, 1987). If we assume that >NOH is able to react with $R^\bullet$

$$>N–C^\bullet H–CH< + >NOH \longrightarrow >N–CH_2–CH< + >NO^\bullet,$$

then $>NO^\bullet$ stoichiometric coefficient must be more than 1.

Let's consider the probability of $>NO^\bullet$ cross-dispropotionation with other alkyl radicals. Such a consideration is quite useful cause at physiological P_{O_2} values in body tissues of higher animals and humans (5 – 50 torus) oxygen concentration is less than $1{\cdot}10^{-4}$ M (Porter & Wujek, 1984). In this case inequality $[RO_2^\bullet] >> [R^\bullet]$ is not satisfied and it's necessary to take into account alkyl radicals' participation contribution. This can be done by modifying Scheme 1 for vinyl monomers' oxidation (Scheme 2).

(i)	Initiator	$\xrightarrow{O_2,M}$	$M^\bullet$	W_i
(1)	$M^\bullet + O_2$	$\longrightarrow$	$MO_2^\bullet$	k_1
(2)	$MO_2^\bullet + M$	$\longrightarrow$	$M^\bullet$	k_2
(3)	$MO_2^\bullet + MO_2^\bullet$	$\longrightarrow$	molecular products	k_3
(4.1)	$M^\bullet + >NO^\bullet$	$\longrightarrow$	>NOR	$k_{4.1}$
(4.2)	$M^\bullet + >NO^\bullet$	$\longrightarrow$	M_{-H} + >NOH	$k_{4.2}$
(4.3)	$M^\bullet$ + >NOH	$\longrightarrow$	>NOR + $>NO^\bullet$	$k_{4.3}$

Scheme 2. Mechanism of organic compounds oxidation inhibited by nitroxide radicals taking into account $M^\bullet$ with $>NO^\bullet$ disproportionation

Let's estimate $>NO^\bullet$ recombination and disproportionation shares ratio according to reactions (4.1) and (4.2). There's almost no any experimental data for such estimation, so we have to use the results of quantum-chemical calculations.

Table 3 represents the values of quantum energies of $>NO^{\bullet}$ reactions[1] with peroxyalkyl radicals ~OOM$^{\bullet}$ (DFT B3LYP/6-31G* calculation similar to one in (Becke, 1993)). As a structural unit we've used $-OOCH_3$ fragment. As can be seen from the table, such operation is quite acceptable: substitution of $-OOCH_3$ to $-CH_2CH_3$ or to $-CH_2CH_2CH_3$ doesn't result in significant changes in energy values calculated.

It should be noted that in accordance with calculated results recombination's probability is significantly greater as compared with disproportionation's one: mean difference in energies is greater or equal of 30 kJ/mol. However, cross-disproportionation in liquid phase also can not be excluded: polar effects may have a significant effect especially if there are polar groups in conjugation with β-C–H bond (Roginskii, 1987). Recombination and disproportionation energies lowering during methyl group addition to α-position of radical center ~OOM$^{\bullet}$ also seem to be a logical cause as a steric effects appear in this case.

$M^{\bullet}$	Reaction type	
	$M^{\bullet} + >NO^{\bullet} \rightarrow MON<$	$M^{\bullet} + >NO^{\bullet} \rightarrow M_{-H} + >NOH$
$C_6H_5CH^{\bullet}CH_2OOCH_3$	–121	–77
$C_6H_5CH^{\bullet}CH_2CH_3$	–136	–69
$C_6H_5CH^{\bullet}CH_2CH_2CH_3$	–125	–70
$C_6H_5C^{\bullet}(CH_3)CH_2OOCH_3$	–79	–70
$C_6H_5C^{\bullet}(CH_3)CH_2CH_3$	–94	–68
$C_6H_5C^{\bullet}(CH_3)CH_2CH_2CH_3$	–69	–67
$CH_3OC(=O)CH^{\bullet}CH_2OOCH_3$	–127	–111
$CH_3OC(=O)CH^{\bullet}CH_2CH_3$	–172	–145
$CH_3OC(=O)CH^{\bullet}CH_2CH_2CH_3$	–144	–93
$CH_3OC(=O)C^{\bullet}(CH_3)CH_2OOCH_3$	–79	–61
$CH_3OC(=O)C^{\bullet}(CH_3)CH_2CH_3$	–93	–75
$CH_3OC(=O)C^{\bullet}(CH_3)CH_2CH_2CH_3$	–88	–79
Mean value	**–111**	**–82**

Table 3. Reaction energies of alkyl radicals with $>NO^{\bullet}$ (I) (kJ/mol)

Reactions (4.2) and (4.3) rates ratio can be estimated on the basis of experimental kinetic data of $>NO^{\bullet}$ (I) consumption in cumene, styrene, or methyl methacrylate in inert atmosphere. Experiment conditions: 323 K, atmosphere of argon, $[>NO^{\bullet}(I)]_0 = 5\cdot10^{-3}$ M, initiator – azobisisobutyronitrile, $W_i = 1\cdot10^{-7}$ M·s^{-1}. The dynamic equilibrium is to set over time in case of reactions (4.2) and (4.3) and observed residual ESR signal value is to increase along with $[>NO^{\bullet}]_0$ growth since the following equalities are valid during that equilibrium:

[1] In chemical thermodynamics quantum-chemically calculated reaction energy is used. It is a difference between full energies of reaction's products and reagents. This value often correlates with experimental value – enthalpy of reaction (http://cccbdb.nist.gov/).

$$k_{4.2}[R^\bullet]\overline{[>NO^\bullet]} = k_{4.3}[R^\bullet]\overline{[>NOH]}$$

$$\overline{[>NOH]} = [>NO^\bullet]_0 - \overline{[>NO^\bullet]}$$

$$\overline{[>NO^\bullet]} = \frac{k_{4.3}[>NO^\bullet]_0}{k_{4.2}+k_{4.3}}, \text{ whence } \frac{k_{4.2}}{k_{4.3}} = \frac{[>NO^\bullet]_0}{\overline{[>NO^\bullet]}} - 1$$

After completely >NO$^\bullet$ consumption the residual ESR signal amplitude doesn't exceed a noise level under experimental conditions. This corresponds to potential stationary >NO$^\bullet$ concentration of less than 10^{-7} M (ESR spectrometer Adani CMS 8400). In this case [>NOH] ≈ $[>NO^\bullet]_0$ and at $[>NO^\bullet]_0 = 5\cdot10^{-4}$ M we have the following value of $k_{4.2}/k_{4.3}$ ratio:

$$k_{4.2}/k_{4.3} = (5\cdot10^{-4}/1\cdot10^{-7}) - 1 \approx 5\cdot10^{4}$$

And now the ratio of (4.3) to (4.2) reaction rates can be estimated:

$$\frac{W_{4.3}}{W_{4.2}} = \frac{k_{4.3}\overline{[>NOH]}}{k_{4.2}\overline{[>NO^\bullet]}} \approx 2\cdot10^{-5}\frac{\overline{[>NOH]}}{\overline{[>NO^\bullet]}}.$$

Therefore in range of up to 99% of >NO$^\bullet$ consumption reaction's (4.3) share is less than 1% of reaction (4.2), so practically there's no any >NO$^\bullet$ regeneration at all. That is $f = 1$ in inert atmosphere and in these substrates' medium. It's obvious that reaction (4.2) would be completely suppressed by reaction of >NOH with $RO_2^\bullet$ at $[O_2] > 1\cdot10^{-4}$ M.

3.2. Analysis of oxidation mechanism with nitroxide radical regeneration

In case of >NO$^\bullet$ regeneration the oxidation's mechanism describes by Scheme 3.

(i)	Initiator	$\xrightarrow{O_2,RH}$	$R^\bullet$	W_i
(1)	$R^\bullet + O_2$	⟶	$RO_2^\bullet$	k_1
(2)	$RO_2^\bullet$ + RH	⟶	ROOH + $R^\bullet$	k_2
(3)	$RO_2^\bullet + RO_2^\bullet$	⟶	molecular products	k_3
(4)	$R^\bullet$ + >NO$^\bullet$	⟶	>NOR	k_4
(5.1)	$RO_2^\bullet$ + >NO$^\bullet$	⟶	product + >NOH	$k_{5.1}$
(5.2)	$RO_2^\bullet$ + >NOH	⟶	ROOH + >NO$^\bullet$	$k_{5.2}$

where $RO_2^\bullet$ = $HO_2^\bullet$, >C(OH)$O_2^\bullet$, or >CH–CH(OO$^\bullet$)N<

Scheme 3. Mechanism of organic compounds oxidation inhibited by nitroxide radicals taking into account >NO$^\bullet$ regeneration

The following equation is valid for this scheme:

$$W_i\left(\frac{W_0}{W}-\frac{W}{W_0}\right)=\frac{k_4[>NO^\bullet]_0 W_0}{k_1[O_2]}+\frac{2k_5[>NO^\bullet]_0 W_i^{0.5}}{k_3^{0.5}} \quad (3)$$

where $k_5 = (k_{5.1}[>NO^\bullet] + k_{5.2}[>NOH])/2[>NO^\bullet]_0$.

Kinetic analysis shows that at $[O_2] \sim 1\cdot10^{-2}$ M and $[>NO^\bullet]_0 < 10^{-4}$ M the contribution of reaction (4) to chain termination process is negligible, and then

$$\frac{W_0}{W}-\frac{W}{W_0}=\frac{2k_5[>NO^\bullet]_0}{(W_i k_3)^{0.5}} \quad (4)$$

With the drop of P_{O_2} and small share of quadratic chain termination the oxidation rate will decrease not linearly, but slower. Such facts were found for instance in (Pliss & Aleksandrov, 1977; Ruban et al., 1967).

Reaction (5.1) proceeds as disproportionation of nitroxide and peroxide radicals (Denisov, 1996):

$$HO_2^\bullet + >NO^\bullet \longrightarrow >NOH + O_2$$

$$>C(OH)O_2^\bullet + >NO^\bullet \longrightarrow >NOH + >C{=}O + O_2$$

$$>CH{-}CH(OO^\bullet)N< + >NO^\bullet \longrightarrow >NOH + >C{=}CH{-}N< + O_2$$

$>NO^\bullet$ regeneration and multiple chain termination processes are caused just by subsequent reaction (5.2). Measured kinetic inhibiting factors[2] for different nitroxide radicals and substrates presented for example in review (Denisov, 1996). The most of f values greater than ten and reflects just a lower bound of this value.

4. Specific features of nitroxide and peroxide radicals reactions in biological systems and in liquid phase organic substrates

Like transition metals, nitroxide radicals can easily transform both to oxidized (oxoammonium cation) and to reduced (hydroxylamine) forms (Berliner, 1998; Sen' & Golubev, 2009; Zhdanov, 1992). This fact, along with $>NO^\bullet$ ability to penetrate through cell membranes and with its paramagnetic properties, suggests that nitroxide radicals poses a number of unique features unlike the compounds of any other class. Like antioxidants and mimetics of superoxide dismutase enzyme, $>NO^\bullet$ are able to supply an effective protection for cells and tissues (Denisov & Afanas'ev, 2005; Soule et al., 2007; Wilcox, 2010). Nitroxide radicals inhibit oxidation of lipids in fatty acids' micelles (Damiani et al., 2002; Noguchi et al., 1999), liposome membranes (Damiani et al., 2002; Wilcox, 2010), lipoproteins (Damiani et al., 1994), and in low density microsomes (Antosiewicz et al., 1995; Nilsson et al., 1990). At the moment there's not enough information concerning $>NO^\bullet$ reactions with redox peroxide

[2] Inhibiting factor is ratio of real induction period (t) to theoretical period of inhibitor conversion (φ), i.e. $f_{inh} = t/\varphi$, where $\varphi = f[>NO^\bullet]/W_i$ (f – stoichiometric inhibiting factor).

radicals to understand such reactions' mechanisms. In fact, there is no any certain agreement on this reaction, so some authors consider it as unlikely one (Blough, 1988; Browlie & Ingold, 1967; Damiani et al., 2002; Denisov, 1996).

Our recent results concerning reactions of >NO•(I) – (VII) with $MO_2^•$ radicals of 1,1-substituted ethylenes allows suggestion that such reaction's probability is quite high.

Let's review some experimental data in this field to understand the situation.

4.1. Nitroxide and peroxide radicals reaction under conditions modeling biological systems

Barton et al. assumed that in reaction of >NO• (I) with $(CH_3)_3COO^•$ the formation of quite stable intermediate should take place (Barton et al., 1998):

$$>NO^• + (CH_3)_3COO^• \longrightarrow (CH_3)_3COOON<$$

Further this intermediate should decompose to a number of products with concomitant formation of molecular oxygen and >NO• regeneration according to the following very speculative scheme (Scheme 4) with no any kinetic evidence (Barton et al., 1998).

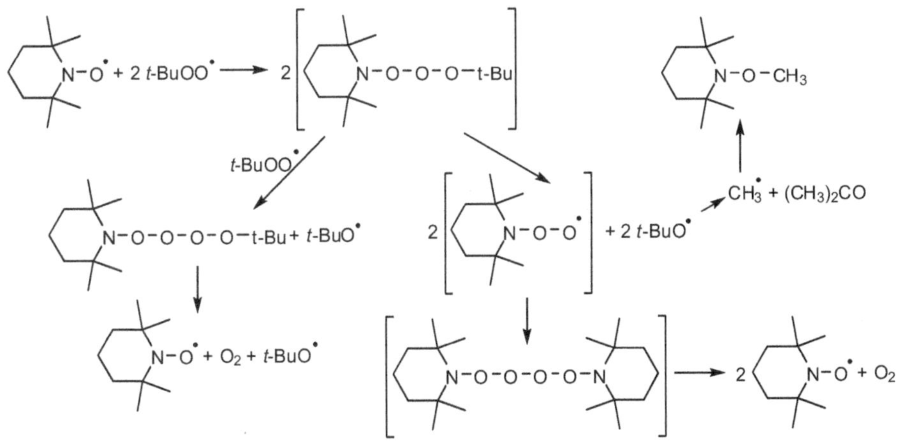

Scheme 4. Mechanism of nitroxide with tert-butyl peroxide radicals reaction according to (Barton et al, 1998)

Stipa attempted to prove this scheme with *ab initio* quantum-chemical calculations (Stipa, 2001). The calculations were performed using Gaussian 98 Hartree-Fock method and complete basis set CBS-QB3. $H_2NO^•$ Radical was used as a model of >NO• (I) in view of available computer resources limitations (work was submitted in April 2001). The calculations' results indicate the possibility of Scheme 4.

This conclusion seems to be quite controversial since radical $H_2NO^•$ can not provide an adequate >NO• (I) model. For example, standard enthalpy calculated for the first reaction stage was –1.794 kcal/mol (tabelle 2 in (Stipa, 2001)), i.e. this value is not very different from zero within the quantum-chemical calculations accuracy. This suggests that the studied trioxide apparently must be thermodynamically unstable structure. At the same time, radical $H_2NO^•$ is much more active than simulated $>NO^•$ (I). It's easy to show with Density Functional Theory (DFT) calculations (Becke, 1993). Quantum energy values of radicals $H_2NO^•$ and $>NO^•$ (I) recombination with $CH_3O^•$ and $CH_3OO^•$ were calculated by DFT B3LYP/6-31G* (similar calculation is shown in Table 3). The results are shown below.

Radical	Recombination energy, kJ/mol	
	$>NO^•$(I)	$H_2NO^•$
$HO^•$	–28.3	–98.4
$HOO^•$	48.1	–20.4
$CH_3^•$	–150.6	–232.8
$CH_3O^•$	11.5	–58.2
$CH_3OO^•$	60.3	9.92

Table 4.

Even this simple example shows that stable nitroxide radical $>NO^•$ (I) is significantly less active than $H_2NO^•$.

Offer and Samuni also suggest that stable trioxide formation proceeds in reaction of tert-amidinopropyl radicals with $>NO^•$ (I) and (III) in phosphate buffer at pH 7.4 and 310 K (Offer & Samuni, 2002):

$$(H_2N)_2^+CC(CH_3)OO^• + >NO^• \longrightarrow (H_2N)_2^+CC(CH_3)OOON<$$

This reaction was studied with combination of ESR-spectroscopy method and cyclic voltammetry, but no any kinetic evidences of product formation and its subsequent transformation were provided.

Brede et al. studied the reaction of n-$C_{17}N_{35}OO^•$ with $>NO^•$ (I) with pulse radiolysis at room temperature (Brede et al., 1998). The rate constant value $k < 10^5$ $M^{-1}·s^{-1}$ was specified but the reaction's mechanism wasn't discussed.

Goldstein S. and Samuni also determined the rate constants of reactions of $CH_3OO^•$, $(CH_3)_3COO^•$, $CH_2(OH)OO^•$, and $HO_2^•$ with $>NO^•$ (I) – (IV), (X) – (XII) with pulse radiolysis at room temperature (Goldstein & Samuni, 2007). The values obtained are shown in Table 4. The mechanism of $RO_2^• + >NO^•$ reaction was discussed; therefore let's consider this paper in more detail.

It was shown that nitroxide with peroxide radicals' reaction mainly results in formation of corresponding oxoammonium cations. In case of $(CH_3)_3COO^•$ the formation of stable cation radical $>NO^•$ (I) and decay products of relatively unstable cation radical $>NO^•$ (II) was spectrophotometrically detected. In case of $HOO^•$ the reaction with piperidine nitroxide radicals is catalyzed by anion $H_2PO_4^-$ as noted in (Goldstein & Samuni, 2007). Therefore under physiological conditions (pH 7.4 and $5·10^{-2}$ M phosphate) observed rate constants for piperidine nitroxide radicals are slightly greater than those shown in Table 4. Moreover, catalysis by $H_2PO_4^-$ anions implies that nitroxide with peroxide radicals' reaction proceeds according to inner-sphere electron transfer mechanism when >NOOOR adduct's decomposition can undergo general acid catalysis:

$$>NO^{\bullet} + ROO^{\bullet} \rightleftarrows >NOOOR$$

$$>NOOOR \rightleftarrows >N^{+}=O + ROO^{-}$$

$$>NOOOR + H_2PO_4^{-} \rightleftarrows >N^{+}=O + ROOH + HPO_4^{2-}$$

It's obvious that in absence of catalysis the adduct's formation may also occur, but inner-sphere electron transfer mechanism still can not be excluded:

$$>NO^{\bullet} + ROO^{\bullet} \rightleftarrows >N^{+}=O + ROO^{-}$$

Known rate constants of reaction $HO_2^{\bullet}$ ($>C(OH)O_2^{\bullet}$) + $>NO^{\bullet}$ (I) and (III) in organic solvents are in the range of $1.1\cdot10^4 - 2.1\cdot10^5$ $M^{-1}\cdot s^{-1}$ at 323 K (Aleksandrov, 1987; Wilcox, 2010), which is two-three orders of magnitude lower than values shown in Table 5. Such a huge difference may hardly be explained only by the reaction's specificity in phosphate buffer. Further reaction of radicals $HO_2^{\bullet}$ and $CH_2(OH)OO^{\bullet}$ with $>NO^{\bullet}$ proceeds, as is well known, as disproportionation rather than recombination (Denisov, 1996; Denisov & Afanas'ev, 2005; Mogilevich & Pliss, 1990).

$>NO^{\bullet}$	$CH_3OO^{\bullet}$	$(CH_3)_3COO^{\bullet}$	$CH_2(OH)OO^{\bullet}$	*) $HO_2^{\bullet}$
I	$2.8\cdot10^7$	$5.1\cdot10^7$	$1.0\cdot10^8$	$1.1\cdot10^8$
II	$2.8\cdot10^5$	$5.4\cdot10^5$	—	$<6\cdot10^6$
III	$3.3\cdot10^6$	$5.4\cdot10^6$	$4.4\cdot10^7$	$2.7\cdot10^7$
IV	$1.0\cdot10^6$	$1.5\cdot10^6$	—	$<1\cdot10^7$
X	$8.1\cdot10^5$	$1.1\cdot10^5$	$9.0\cdot10^6$	$1.1\cdot10^6$
XI	$9.6\cdot10^5$	$1.5\cdot10^6$	—	$1.1\cdot10^6$
XII	$5.0\cdot10^4$	$6.7\cdot10^4$	—	$1.6\cdot10^5$

*) Calculated on the base of (Goldstein et al., 2003; Goldstein & Samuni, 2007).

Table 5. Rate constants ($M^{-1}\cdot s^{-1}$) of $ROO^{\bullet}$ + $>NO^{\bullet}$ reaction (Goldstein & Samuni, 2007).

It's important that authors (Goldstein & Samuni, 2007) on the base of experimental data suggested that rate constant of reaction $RO_2^{\bullet}$ + >NOH doesn't exceed $1\cdot10^5$ $M^{-1}\cdot s^{-1}$, i.e. this reaction is much slower as compared with $RO_2^{\bullet}$ + $>NO^{\bullet}$ (Table 4). Thereby we note that, as it well known (Denisov, 1996; Denisov & Afanas'ev, 2005; Mogilevich & Pliss, 1990), upon a competition of reactions (5.1) and (5.2) (see Scheme 2) the rate-limiting reaction is just the first one.

As can be seen from the results above, the mechanism of nitroxide with peroxide radicals' reactions in biological systems isn't elucidated at all. The authors of works analyzed are inclined to possibility of $>NO^{\bullet}$ reaction not only with peroxide radicals poses redox properties: $HO_2^{\bullet}$, $>C(OH)O_2^{\bullet}$, but with $RO_2^{\bullet}$ which are oxidants only: $(CH_3)_3COO^{\bullet}$, n-$C_{17}H_{35}OO^{\bullet}$ (Barton et al., 1998; Brede et al., 1998; Goldstein et al., 2003; Goldstein & Samuni, 2007; Offer & Samuni, 2002 ; Stipa, 2001). It is assumed that such a reaction occurs as recombination with formation of trioxide. According to the authors of papers cited, nitroxide radicals are able to provide protection against oxidation at extremely low concentrations due to the regeneration process resulting from the reaction of corresponding oxoammonium cations with common biological reducing agents.

Hereby it's interesting to consider the reaction of $>NO^{\bullet}$ with $RO_2^{\bullet}$ which are only oxidants under conditions when oxoammonium cations' formation is improbable, i.e. during an oxidation in organic phase.

4.2. Nitroxide radicals reactions with peroxide radicals of 1- and 1,1- ethylene substituted monomers

As it has already been mentioned above, reactions of $>NO^{\bullet}$ with peroxide radicals of 1- and 1,1- ethylene substituted monomers were discovered recently (Pliss et al., 2010a, 2010b, 2012). Authors (Pliss et al., 2010a, 2010b) on the basis of kinetic data of styrene's and (meth)acrylates' oxidation in presence of piperidine, pyrroline, and imidazoline $>NO^{\bullet}$ suggested that nitroxide radicals inhibit the oxidation process via $>NO^{\bullet}$ reaction with substrate's both alkyl and peroxide radicals, and moreover $>NO^{\bullet}$ regeneration occurs during chain termination process in accordance with reactions:

$$MO_2^{\bullet} + >NO^{\bullet} \longrightarrow \text{product} + >NOH \qquad (5.1)$$

$$MO_2^{\bullet} + >NOH \longrightarrow MOOH + >NO^{\bullet} \qquad (5.2)$$

The following results served grounds for these assumptions: oxygen consumes linearly in presence of $>NO^{\bullet}$ for all monomers for a long period of time. That period is greater than the theoretical induction period, and if P_{O2} reduces to five times (from $1{\cdot}10^5$ to $0.2{\cdot}10^5$ Pa) the oxidation rate decreases to less than two times (Pliss et al., 2010a, 2010b). But of course it is unacceptable to make any definite conclusions about detailed mechanism just on the basis of the oxygen consumption kinetic data only.

In (Pliss et al., 2012) $>NO^{\bullet}$ antioxidant activity was studied during styrene's oxidation using a complex of kinetic methods in combination with quantum-chemical calculations and kinetic modeling. The choice of styrene is caused by the following circumstances. Firstly, the reaction of styrene's inhibited oxidation is not complicated by complexation process as it is in case of many other vinyl compounds (acrylic monomers for example). Secondly, high reactivity of styrene's double bond makes it possible to study this process under long chains conditions even if oxidation is quite strongly inhibited. Thirdly, in case of styrene the rate constants of elementary stages are known for many key reactions, and this knowledge makes kinetic modeling significantly easer to carry out.

Let's consider the results of this work and some of our new data.

Oxidation kinetics was studied in area of initial O_2 consumption rates in temperature range of 310 – 343 K with highly sensitive capillary microvolumometer according to technique (Loshadkin et al., 2002). Initial >NO$^\bullet$ concentrations were in range of $10^{-7} - 10^{-3}$ M. Experiments were carried out at $P_{O2} = 20$ or 100 kPa. In special cases oxygen-argon mixes were prepared to obtain the oxidation rates dependences on [O_2]. Initiation rate W_i was determined with inhibition method by detection of induction period ending time τ_{ind} and application of known equation $W_i = 2[InH]_0/\tau_{ind}$. 6-Hydroxy-2,2,5,7,8-pentamethylbenzochroman was used as inhibitor (InH). Kinetic modeling was performed as described in (Loshadkin et al., 2002).

Values of styrene's oxidation rates inhibited by different >NO$^\bullet$ under oxygen and air saturation conditions and when quadratic termination share is no more than 25% are presented in Table 6. As seen from the table, the oxidation's rate for different >NO$^\bullet$ from oxygen to air decreases substantially less than five times. This fact according to (Aleksandrov, 1987; Denisov & Afanas'ev, 2005; Kharitonov & Denisov, 1967; Kovtun et al., 1974; Mogilevich & Pliss, 1990; Pliss & Aleksandrov, 1977; Denisov, 1996) suggests that nitroxide radicals react with both M$^\bullet$ and $MO_2^\bullet$.

Kinetics of >NO$^\bullet$(III) consumption at its different initial concentrations are presented in Figure 2. From this data it follows that according to inhibition by reaction (4) >NO$^\bullet$ should consume much faster than it happens in fact. With special experiments it was shown that value of inhibiting factor e.g. for >NO$^\bullet$ (III) and >NO$^\bullet$ (IV) is more than 7 (Pliss et al., 2012), so it suggests that >NO$^\bullet$ regeneration process occurs upon chain termination. We note that this effect doesn't depend on styrene's concentration.

[>NO$^\bullet$]$_0 \cdot 10^5$, M	$W \cdot 10^6$, M·s^{-1}		W_{O2} / W_{air}
	$P_{O2} = 1 \cdot 10^5$ Pa (oxygen)	$P_{O2} = 0.2 \cdot 10^5$ Pa (air)	
0	11.0	11.0	1.00
	>NO$^\bullet$ (I)		
2.8	4.75	2.58	1.84
4.1	4.37	2.84	1.54
	>NO$^\bullet$ (II)		
4.2	3.76	2.12	1.78
14	2.88	1.17	2.45
	>NO$^\bullet$ (III)		
5.7	4.34	2.19	1.98
18	3.13	1.30	2.40
	>NO$^\bullet$ (IV)		
4.0	5.67	4.09	1.38
8.0	5.07	2.20	2.30
	>NO$^\bullet$ (V)		
5.9	4.90	3.21	1.53
17	2.38	0.78	3.06

Table 6. Kinetic parameters of styrene's oxidation inhibited with >NO$^\bullet$; $W_i = 1.0 \cdot 10^{-8}$ M·s^{-1} (Pliss et al., 2012)

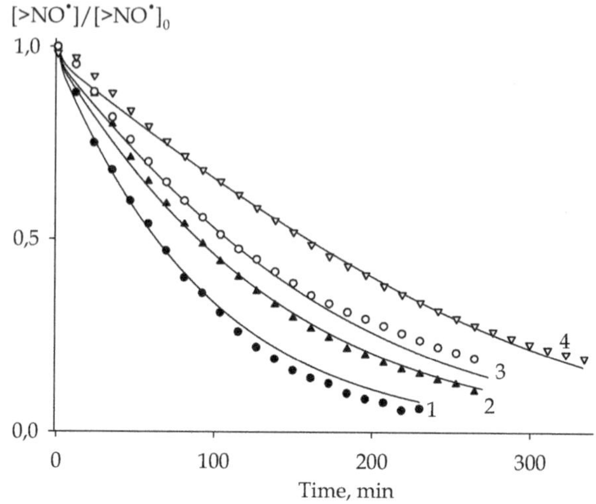

Figure 2. Kinetics of >NO• (III) consumption during styrene oxidation (in relative coordinates); points: experimental data, curves: a result of simulation. P_{O2} = 20 kPa; $W_i = 1.0{\cdot}10^{-8}$ M·s^{-1}; [>NO•]$_0$, M: 1 – $2.3{\cdot}10^{-6}$, 2 – $5.9{\cdot}10^{-6}$, 3 – $1.2{\cdot}10^{-5}$, 4 – $2.8{\cdot}10^{-5}$ (Pliss et al., 2012).

Styrene's oxidation can be effectively inhibited by hydroxylamines. This assertion confirms with distinct induction period on oxygen consumption's kinetic curve (typical one represented at Figure 3 (Pliss et al., 2012)). Rate constants of such antioxidants' reactions with peroxide radicals can be determined from inhibited oxidation's rate dependence on time according to the following equation (Loshadkin et al., 2002):

$$F = \ln\frac{1+W/W_0}{1-W/W_0} - \frac{W_0}{W} = \frac{k_{5.2}W_0}{k_2[\mathrm{M}]}t + const \tag{5}$$

But the problem is that oxidation rate remains significantly lower than W_0 after the inhibiting period (see Figure 3) since >NO• being formed is an inhibitor of oxidation itself. However the kinetic modeling shows that in this case in equation (5) we can substitute W_0 value to the value of oxidation's rate at after-induction period. Herewith $k_{5.2}$ determination error is less than 10%. Calculated $k_{5.2}$ mean value at 323 K for >NOH (III), equal to $4{\cdot}10^6$ M^{-1}·s^{-1} (Pliss et al., 2012), was later used in mechanism's kinetic modeling.

Dependences of >NO• consumption and its accumulation from corresponding >NOH on time during styrene's oxidation at P_{O2} = 20 kPa presented at Fig. 4 (ESR-spectroscopy method). It's seen that hydroxylamine being injected quickly transforms to >NO• and its maximum concentration differs from [>NO•]$_0$ by less than 10%. It's important that hereafter consumption rates of injected >NO• and one being formed from hydroxylamine are almost the same (see curves 1 and 2 at Fig. 4).

As already been mentioned (see Section 3), if chains propagates by $HO_2^•$, >C(OH)$O_2^•$, or >CH-CH(OO•)N< radicals then inhibition proceeds as disproportionation of these radicals with >NO•. Herewith hydroxylamine being injected quickly transforms to >NO• which almost doesn't consume but effectively inhibits an oxidation of corresponding substrates. So in this case the inhibiting factor values are more than 100 (Kovtun et al., 1974). Therefore it should be considered that in case of styrene's oxidation the way of irreversible >NO• consumption is reaction (4).

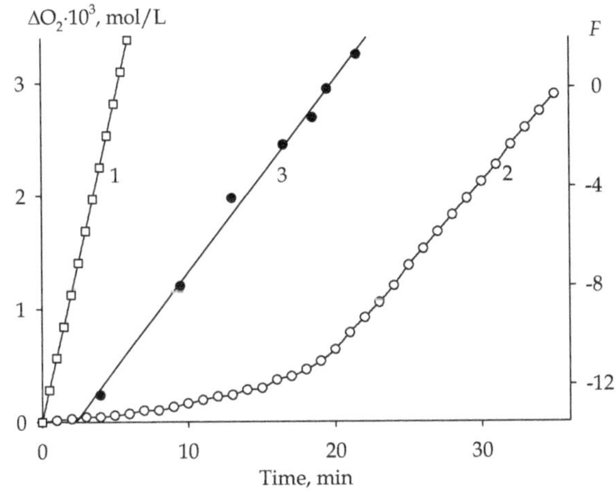

Figure 3. Kinetics of O_2 consumption during styrene's oxidation: 1 – without inhibitor; 2 – [>NOH (III)] = $2 \cdot 10^{-5}$ M; 3 – anamorphous of curve 2 in coordinates of equation (5); P_{O_2} = 20 kPa; $W_i = 1.0 \cdot 10^{-8}$ M·s^{-1} (Pliss et al., 2012)

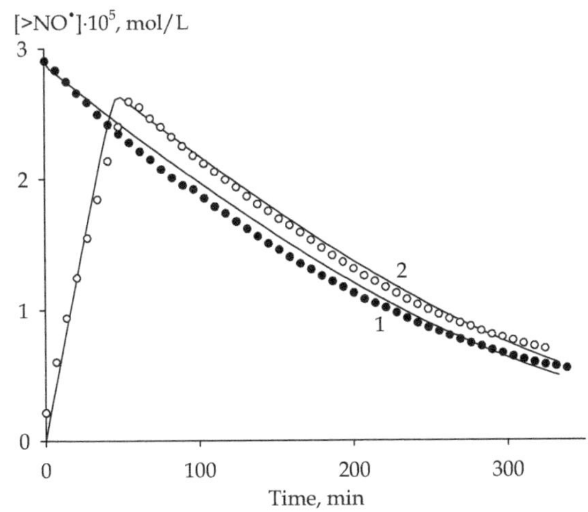

Figure 4. Kinetics of >NO• (III) consumption (1) and its accumulation from corresponding hydroxylamine (2) during styrene's oxidation: dots – experimental data, curves – modeling results; [>NO• (III)]$_0$ = [>NOH (III)]$_0$ = $2.9 \cdot 10^{-5}$ M; P_{O_2} = 20 kPa; $W_i = 1 \cdot 10^{-8}$ M·s^{-1} (Pliss et al., 2012)

As it was mentioned above (see section 3.1), reaction (4) may proceed both as recombination and as disproportionation. Target experiment's results presented at Figure 5. Rate of $>NO^{\bullet}$ (III) consumption in styrene (atmosphere of argon) during initiated oxidation was equal to initiation rate, and this $>NO^{\bullet}$ consumption was proceeded up to detection limit of ESR-spectrometer ($\leq 1 \cdot 10^{-7}$ M). After that argon was substituted to oxygen and, as can be seen at Figure 5, $>NO^{\bullet}$ signal was appeared again.

These cycles were repeatedly detected several times until we reached the spectrometer's detection limit. Each time $>NO^{\bullet}$ was recrudesced in share about 25–30% of its initial concentration. This fact confirms the assumption that reaction (4) also may proceed as disproportionation due to β-C–H bond of styrene's alkyl radical ($\sim CH_2C^{\bullet}HC_6H_5$) with olefin's formation $\sim CH{=}CHC_6H_5$ (M_{-H}):

$$M^{\bullet} + {>NO^{\bullet}} \longrightarrow {>NOR} \quad (4.1)$$

$$M^{\bullet} + {>NO^{\bullet}} \longrightarrow M_{-H} + {>NOH} \quad (4.2)$$

Thus $>NO^{\bullet}$ signal may appear due to the reaction (5.2) upon oxygen blow (see Scheme 5 below).

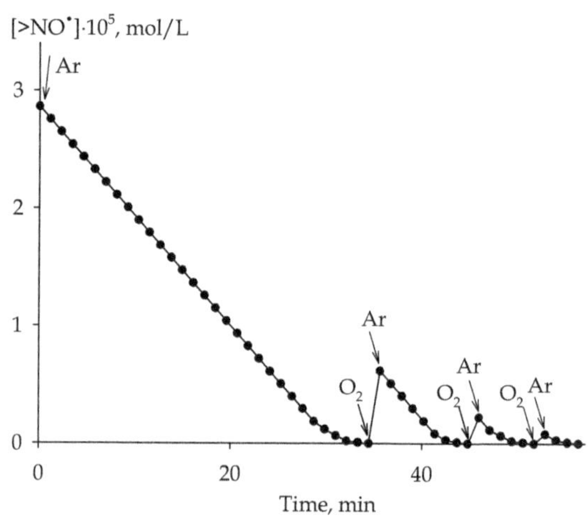

Figure 5. Kinetics of $>NO^{\bullet}$ (III) consumption in styrene: $[>NO^{\bullet}\text{ (III)}]_0 = 2.8 \cdot 10^{-5}$ M; $W_i = 1 \cdot 10^{-8}$ $M \cdot s^{-1}$ (Pliss et al., 2012)

Another one probable reason of this effect (Figure 5) is reaction $MO_2^{\bullet} + {>NOR} \rightarrow$ product $+ {>NO^{\bullet}}$ (5.3) that was first proposed in (Denisov, 1982). As $>NO^{\bullet}$ regeneration source, this reaction was studied in detail for reactions of some alkoxyamines $>NO^{\bullet}$ (I) with peroxide radicals of cumene and cyclohexylmethyl ether at 338 K (Kovtun et al., 1974). Obtained $k_{5.3}$ values (1 – 26 $M^{-1} \cdot s^{-1}$) suggest that >NOR being studied in (Kovtun et al., 1974) are weak inhibitors. This conclusion also is confirmed with the dependences of styrene's oxidation rates on >NOR (I) and $>NO^{\bullet}$ (III) concentrations (Pliss et al., 2012).

All of the experimental data and our previous results (Pliss et al., 2010a, 2010b, 2012) allow us to provide the following formal kinetic scheme of styrene's oxidation inhibited by aliphatic stable nitroxide radicals:

(i)	Initiator	$\xrightarrow{O_2,RH}$	$M^\bullet$	$W_i = 1\cdot10^{-8}$ M·s^{-1}	
(1)	$M^\bullet + O_2$	$\longrightarrow$	$MO_2^\bullet$	$k_1 = 3\cdot10^8$	(Aleksandrov, 1979)
(2)	$MO_2^\bullet + M$	$\longrightarrow$	$MOOM^\bullet$ ($\equiv M^\bullet$)	$k_2 = 100$	(Mogilevich & Pliss, 1990)
(3.1)	$M^\bullet + M^\bullet$	$\longrightarrow$	products	$2k_{3.1} = 7\cdot10^7$	(Mogilevich & Pliss, 1990)
(3.2)	$M^\bullet + MO_2^\bullet$	$\longrightarrow$	products	$k_{3.2} = 1\cdot10^8$	(Mogilevich & Pliss, 1990)
(3.3)	$MO_2^\bullet + MO_2^\bullet$	$\longrightarrow$	products	$2k_{3.3} = 5.4\cdot10^7$	(Mogilevich & Pliss, 1990)
(4.1)	$M^\bullet$ + >$NO^\bullet$	$\longrightarrow$	>NOR	$k_{4.1} = 6\cdot10^6$	(Aleksandrov, 1979)
(4.2)	$M^\bullet$ + >$NO^\bullet$	$\longrightarrow$	M_{-H} + >NOH	$k_{4.2} = 2\cdot10^6$	
(4.3)	$M^\bullet$ + >NOH	$\longrightarrow$	MH + >$NO^\bullet$	$k_{4.3} < 10$	(He et al., 2000)
(5.1)	$MO_2^\bullet$ + >$NO^\bullet$	$\longrightarrow$	product + >NOH	$k_{5.1} = 2.5\cdot10^4$	
(5.2)	$MO_2^\bullet$ + >NOH	$\longrightarrow$	MOOH + >$NO^\bullet$	$k_{5.2} = 4\cdot10^6$	(estimated above)
(5.3)	$MO_2^\bullet$ + >NOR	$\longrightarrow$	product + >$NO^\bullet$	$k_{5.3} = 20$	(Kovtun et al., 1974)

Scheme 5. Detailed mechanism of vinyl monomers oxidation inhibited by nitroxide radicals

We've used this scheme for kinetic modeling (Pliss et al., 2012). Values of $k_1 - k_{4.1}$, $k_{4.3}$, $k_{5.3}$ (M^{-1}·s^{-1}) were taken from the literature and values of $k_{4.2}$ and $k_{5.1}$ were obtained from modeling. Figures 2 and 4 shows that calculated curves are of satisfactorily consistent with experimental data. This indirectly confirms the reliability of Scheme 5.

Kinetic analysis shows that $[M^\bullet] \ll [MO_2^\bullet]$ when $[O_2] \sim 1\cdot10^{-2}$ M. In this case reactions (3.1), (3.2), (4.2), (4.3), and (5.3) can be neglected. Then the scheme including reactions (i), (1), (2), (3.3), (4.1), (5.1), and (5.2) can be described by equation:

$$W_i\left(\frac{W_0}{W}-\frac{W}{W_0}\right)=\frac{k_4[\text{>NO}^\bullet]_0 W_0}{k_1[O_2]}+\frac{2k_5[\text{>NO}^\bullet]_0 W_i^{0.5}}{k_{3.3}^{0.5}},$$

where $k_5 = (k_{5.1}[\text{>NO}^\bullet] + k_{5.2}[\text{>NOH}])/2[\text{>NO}^\bullet]_0$. If $[\text{>NO}^\bullet] < 10^{-4}$ M then chain termination by reaction (4) can be neglected, therefore that scheme can be described by equation:

$$\frac{W_0}{W}-\frac{W}{W_0}=\frac{2k_5[\text{>NO}^\bullet]_0}{(W_i k_{3.3})^{0.5}}. \tag{6}$$

Previously (Pliss et al., 2010a, 2010b) we've analyzed a simplified scheme including reaction (i), (1), (2), (3.3), (5.1), and (5.2). We've calculated the values of $k_5 = (5 \pm 3)\cdot10^4$ M^{-1}·s^{-1} for >$NO^\bullet$ (I) – (V) in oxidizing vinyl monomers at 323 K and $P_{O_2} = 1\cdot10^5$ Pa. These values are close enough to estimated in this present work value $k_{5.1} = 2.5\cdot10^4$ M^{-1}·s^{-1}.

A fundamental question about the detailed mechanism of the reaction (5.1) remains open. By analogy with the oxidation of 1,2-substituted ethylenes and 1,4-substituted butadienes (Mogilevich & Pliss, 1990) we can assume that hydroxylamine's formation is facilitated by conjugation of β-C–H bond with peroxide bridge of styrene's polyperoxide radical:

$$\sim OO{-}CH_2{-}CH(C_6H_5){-}OO^{\bullet} + {>}NO^{\bullet} \longrightarrow \sim OO{-}CH_2{=}CH{-}C_6H_5 + {>}NOH + O_2$$

Peroxide bridge is an important structural unit of $\sim MO_2^{\bullet}$ radical. It alters the reaction center's electronic characteristics and increases the electrostatic term's contribution to the transition state's total energy (Denisov, 1996; Denisov & Afanas'ev, 2005; Mogilevich & Pliss, 1990). Probable reason of this effect is the difference in the triplet repulsion, which is close to zero in transition state of disproportionation reaction of $MO_2^{\bullet}$ with $>NO^{\bullet}$ and is sufficiently large for the reaction of $>NO^{\bullet}$ with nonconjugated C–H bond of hydrocarbon (Denisov, 1996). The latter probably explains the fact that aliphatic nitroxide radicals inhibit the hydrocarbon's oxidation via reaction with alkyl radicals only.

The results obtained in the present study draws attention to the results gained for biological systems where it is assumed that reaction of aliphatic $>NO^{\bullet}$ with peroxide radicals proceeds via >NOOOR adduct formation decomposing to corresponding oxoammonium cations (Barton et al., 1998; Goldstein & Samuni, 2007; Offer & Samuni, 2002). The probability of such intermediate's existence is also considered in quantum-chemical analysis (Hodgson & Coote, 2010; Stipa, 2001). Further regeneration of nitroxide radicals may be due to reaction of oxoammonium cations with common biological reducing agents (Goldstein & Samuni, 2007; Offer & Samuni, 2002).

Direct reaction $MO_2^{\bullet} + {>}NO^{\bullet} \longrightarrow MOOON{<}$ that results to stable trioxide's formation is seems quite doubtful for aliphatic $>NO^{\bullet}$ in organic phase at moderate temperatures (≤ 373 K). First, it's easy to reject on the base of kinetic reasons cause in this case the kinetics of $>NO^{\bullet}$ consumption and stoichiometry of chain termination would have a different nature than those observed in numerous studies (Browlie & Ingold, 1967; Kovtun et al., 1974; Pliss et al., 2010a, 2010b, 2012; Pliss & Aleksandrov, 1977). Second, it's easy to refute by direct quantum-chemical calculations (DFT B3LYP/6-31G*, Table 7). It's easy to see that peroxide radicals' addition to $>NO^{\bullet}$ is thermodynamically unfavorable.

Radical	$H^{\bullet}$	$HO^{\bullet}$	$HOO^{\bullet}$	$^{\bullet}CH_3$	$CH_3O^{\bullet}$	$CH_3OO^{\bullet}$
Energy	–243.4	–28.3	48.1	–150.6	11.5	60.3

Table 7. Energy of some radicals' addition to $>NO^{\bullet}$ (I), kJ/mol

5. Conclusions

Thus, we must conclude that reaction of nitroxide with peroxide radicals plays an important role during styrene's oxidation in presence of aliphatic stable $>NO^{\bullet}$. This reaction proceeds probably as disproportionation and results to a partial $>NO^{\bullet}$ regeneration.

At the same time we emphasize that detailed mechanism of chemical and biological oxidation processes inhibited by stable nitroxide radicals is still far from being established. Therefore kinetic experiments on the key reactions involving nitroxide radicals and its conversion products (hydroxylamines, alkoxyamines, oxoammonium cations) in solutions of organic substrates and in biological systems must be carried out to solve this problem.

Author details

Eugene M. Pliss and Alexander I. Rusakov
P.G. Demidov Yaroslavl State University, Russia

Ivan V. Tikhonov
Yaroslavl Branch of the Institute of Physics and Technology, Russian Academy of Sciences, Russia

Acknowledgement

The authors gratefully acknowledge Vasily Sen' for providing of nitroxide radicals and hydroxylamines that were applied in this work. The experimental part was carried out on equipment of the Scientific and educational center «Physical organic chemistry» and the Center for collective usage of scientific equipment for diagnostics of micro and nanostructures of Yaroslavl State Univercity. The work was financially supported by Ministry of Education and Science of the Russian Federation (contracts N. 02.740.11.0636, 29.03.2010 and N. 16.552.11.7006, 29.04.2011).

6. References

Aleksandrov, A. L. (1987). The Specificity of Liquid-Phase Oxidation of Aliphatic, Alkylaromatic, and N-Alkylamides by Molecular Oxygen. *Kinetika i Kataliz (in Russian)*, Vol. 28, No. 3, (March 1987), pp. 536-549, ISSN 0453-8811

Aleksandrov, A. L.; Pliss, E. M. & Shuvalov, V. F. (1979). Rate Constants of the Reaction between Alkyl Radicals and Oxygen and Stable Nitroxyl Radicals. *Russian Chemical Bulletin*, Vol. 28, No. 11, (November 1979), pp. 2262-2267, ISSN 1066-5285

Antosiewicz, J.; Popinigis, J.; Wozniak, M.; Damiani, E.; Carloni, P. & Greci, L. (1995). Effects of Indolinic and Quinolinic Aminoxyls on Protein and Lipid Peroxidation of Rat Liver Microsomes. *Free Radical Biology & Medicine*, Vol. 18, No. 5, (May 1995), pp. 913-917, ISSN 0891-5849

Barton, D. H. R.; Le Gloahec, V. N. & Smith, J. (1998). Study of a New Reaction: Trapping of Peroxyl Radicals by TEMPO. *Tetrahedron Letters*, Vol. 39, No. 41, (October 1998), pp. 7483-7486, ISSN 0040-4039

Becke, A. D. (1993). Density-Functional Thermochemistry. III. The Role of Exact Exchange. *Journal of Chemical Physics*, Vol. 98, N. 7, (April 1993), pp. 5648-5652, ISSN 0021-9606

Berliner, L. J. (Ed.). (1998). *Spin Labeling: The Next Millennium*, Plenum Press, ISBN 978-0-306-45644-2, New York, USA

Blough, N.V. (1988). Electron paramagnetic resonance measurements of photochemical radical production in humic substances. 1. Effects of oxygen and charge on radical scavenging by nitroxides. *Environmental Science & Technology*, Vol. 22., No. 1, (January 1988), pp. 77-82, ISSN 0013-936X

Bowry, V. W. & Ingold, K. U. (1992). Kinetics of Nitroxide Radical Trapping. 2. Structural Effects. *Journal of the American Chemical Society*, Vol. 114, No. 13, (June 1992), pp. 4992-4996, ISSN 0002-7863

Brede, O.; Beckert, D.; Windolph, C. & Gottinger, H. A. (1998). One-Electron Oxidation of Sterically Hindered Amines to Nitroxyl Radicals: Intermediate Amine Radical Cations, Aminyl, α-Aminoalkyl, and Aminylperoxyl Radicals. *Journal of Physical Chemistry A*, Vol. 102, No. 9, (February 1998), pp. 1457-1464, ISSN 1089-5639

Browlie, J. P. & Ingold, K. U. (1967). The Inhibited Autoxidation of Styrene. Part VII. Inhibition by Nitroxides and Hydroxylamines. *Canadian Journal of Chemistry*, Vol. 45, No. 20, (October 1967), pp. 2427-2432, ISSN 0008-4042

Buchachenko, A. L. (1965). *Stable Radicals*, Consultants Bureau, ISBN 978-0-608-30037-5, New York, USA

Damiani, E.; Paganga, G.; Greci, L. & Riceevans, C. (1994). Inhibition of copper-mediated low density lipoprotein peroxidation by quinoline and indolinone nitroxide radicals. *Biochemical Pharmacology*, Vol. 48, No. 6, (September 1994), pp. 1155-1161, ISSN 0006-2952

Damiani, E.; Castagna, R. & Greci, L. (2002). The Effects of Derivatives of the Nitroxide Tempol on UVA-mediated in Vitro Lipid and Protein Oxidation. *Free Radical Biology & Medicine*, Vol. 33, No. 1, (July 2002), pp. 128-136, ISSN 0891-5849

Denisov, E. T. (1982). Role of alkyl radical reactions in polymer oxidation and stabilisation, In: *Developments in Polymer Stabilisation*, G. Scott, (Ed.), pp. 23-40, Applied Science Publishers, ISBN 978-0-85334-967-9, London, UK

Denisov, E. T. (1996). Cyclic Mechanisms of Chain Termination in the Oxidation of Organic Compounds. *Russian Chemical Reviews*, Vol. 65, No. 6, (June 1996), pp. 505–520, ISSN 0036-021X.

Denisov, E. T. & Afanas'ev, I. B. (2005). *Oxidation and Antioxidants in Organic Chemistry and Biology*, CRC Press, ISBN 978-0-8247-5356-6, Boca Raton, Florida, USA

Goldstein, S.; Merenyi, G.; Russo, A. & Samuni, A. (2003). The Role of Oxoammonium Cation in the SOD-Mimic Activity of Cyclic Nitroxides. *Journal of the American Chemical Society*, Vol. 125, No. 3, (January 2003), pp. 789-795, ISSN 0002-7863

Goldstein, S. & Samuni, A. (2007). Kinetics and Mechanism of Peroxyl Radical Reactions with Nitroxides. *Journal of Physical Chemistry A*, Vol. 111, No. 6, (February 2007), pp. 1066-1072, ISSN 1089-5639

He, J.; Li, L. & Yang, Y. (2000). Effect of Hydrogen Transfer Reaction on Kinetics of Nitroxide-Mediated Free-Radical Polymerization. *Macromolecules*, Vol. 33, No. 6, (March 2000), pp. 2286-2289, ISSN 0024-9297

Hodgson, J.L. & Coote M.L. (2010). Clarifying the Mechanism of the Denisov Cycle: How do Hindered Amine Light Stabilizers Protect Polymer Coatings from Photo-oxidative

Degradation? *Macromolecules*, Vol. 43, No. 10, (May 2010), pp. 4573-4583, ISSN 0024-9297

Howard, J. A. & Ingold, K. U. (1967). Absolute Rate Constants for Hydrocarbon Autoxidation. V. The Hydroperoxy Radical in Chain Propagation and Termination. *Canadian Journal of Chemistry*, Vol. 45, No. 8, (April 1967), pp. 785-792, ISSN 0008-4042

Kharitonov, V. V. & Denisov, E. T. (1967). Ambiguous Reactivity of Hydroxyperoxide Radicals in Reactions With Aromatic Amines. *Russian Chemical Bulletin*, Vol. 16, No. 12, (December 1967), pp. 2634-2636, ISSN 1066-5285

Kovtun, G. A.; Aleksandrov, A. L. & Golubev V. A. (1974). Interaction of Peroxide Radicals with Esters of Hydroxylamines. *Russian Chemical Bulletin*, Vol. 23, No. 10, (October 1974), pp. 2115-2121, ISSN 1066-5285

Loshadkin, D.; Roginsky, V. & Pliss E. (2002). Substituted p-Hydroquinones as a Chain-breaking Antioxidant during the Oxidation of Styrene. *International Journal of Chemical Kinetics*, Vol. 34, No. 3, (March 2002), pp. 162-171, ISSN 0538-8066

Mogilevich, M. M. & Pliss, E. M. (1990). *Oxidation and Oxidative Polymerization of Unsaturated Compounds*, Khimiya, ISBN 5-7245-0564-9, Moscow (in Russian)

Nilsson, U. A.; Carlin, G. & Bylundfellenius, A. C. (1990). The Hydroxylamine OXANOH and its Reaction Product, the Nitroxide OXANO$^•$, Act as Complementary Inhibitors of Lipid Peroxidation. *Chemico-Biological Interactions*, Vol. 74, No. 3, (December 1990), pp. 325-342, ISSN 0009-2797

Noguchi, N.; Damiani, E.; Greci, L. & Niki, E. (1999). Action of quinolinic and indolinonic aminoxyls as radical-scavenging antioxidants. *Chemistry and Physics of Lipids*, Vol. 99, No. 1, (May 1999), pp. 11-19, ISSN 0009-3084

Offer, T. & Samuni, A. (2002). Nitroxides Inhibit Peroxyl Radical-Mediated DNA Scission and Enzyme Inactivation. *Free Radical Biology & Medicine*, Vol. 32, No. 9, (May 2002), pp. 872-881, ISSN 0891-5849

Porter, N. A. & Wujek, D. G. (1984). Autoxidation of Polyunsaturated Fatty Acids, an Expanded Mechanistic Study. *Journal of the American Chemical Society*, Vol. 106, No. 9, (May 1984), pp. 2626-2629, ISSN 0002-7863

Pliss, E. M. & Aleksandrov, A. L. (1977). Relative Rate Constants of Reactions of Alkyl Radicals of Methacrylates and Acrylates with Oxygen and Stable Nitroxyl Radicals. *Russian Chemical Bulletin*, Vol. 26, No. 4, (April 1977), pp. 684-687, ISSN 1066-5285

Pliss, E. M.; Grobov, A. M.; Postnov, M. G.; Loshadkin, D. V.; Tikhonov, I. V. & Rusakov, A. I. (2010). Kinetic Features of Styrene Oxidation in The Presence of Stable Nitroxyl Radicals. *Bashkirskii Khimicheskii Zhurnal (in Russian)*, Vol. 17, No. 2, (June 2010), pp. 14-16, ISSN 0869-8406

Pliss, E. M.; Grobov, A. M.; Postnov, M. G.; Loshadkin, D. V.; Tikhonov, I. V. & Rusakov, A. I. (2010). Multiple Oxidation Chain Breaking of Vinyl Monomers by Aliphatic Stable Nitroxyl Radicals. *Bashkirskii Khimicheskii Zhurnal (in Russian)*, Vol. 17, No. 2, (June 2010), pp. 25-27, ISSN 0869-8406

Pliss, E. M.; Tikhonov, I. V. & Rusakov, A. I. (2012). The Kinetics and Mechanism of Reactions of Aliphatic Stable Nitroxyl Radicals with Alkyl and Peroxide Radicals of

Oxidizing Styrene. *Russian Journal of Physical Chemistry B. Focus on Physics*, Vol. 6, No. 3, (June 2012), pp. 424-431, ISSN 1990-7931

Roginskii, V. A. (1987). Thermochemistry, Mechanism, and Kinetics of Disproportionation of Organic Free Radicals, *Zhurnal Organicheskoi Khimii (in Russian)*, Vol. 25, No. 3, (March 1987), pp. 449-461, ISSN 0514-7492

Ruban, L. V.; Buchachenko, A. L. & Neiman, M. B. (1967). Kinetics of Addition of Nitroxide Mono- and Biradicals to Olefins. *Vysokomolekulyarnye soedineniya A (in Russian)*, Vol. 9, No. 7, (July 1967), pp. 1559-1564, ISSN 0507-5475

Sen', V. D. & Golubev, V. A. (2009). Kinetics and Mechanism for Acid-catalyzed Disproportionation of 2,2,6,6-Tetramethylpiperidine-1-oxyl. *Journal of Physical Organic Chemistry*, Vol. 22, No. 2, (February 2009), pp. 138-143, ISSN 0894-3230

Soule, B. P.; Hyodo, F.; Matsumoto, K. & Mitchell, J. (2007). The Chemistry and Biology of Nitroxide Compounds. *Free Radical Biology & Medicine*, Vol. 42, No. 11, (June 2007), pp. 1632-1650, ISSN 0891-5849

Stipa P. (2001). The Reactivity of Aminoxyls towards Peroxyl Radicals: an Ab Initio Thermochemical Study. *Journal of the Chemical Society, Perkin Transactions 2*, No. 9, (September 2001), pp. 1793-1797, ISSN 0300-9580

Wilcox, C. S. (2010). Effects of Tempol and Redox-cycling Nitroxides in Models of Oxidative Stress. *Pharmacology & Therapeutics*, Vol. 126, No. 2, (May 2010), pp. 119-145, ISSN 0163-7258

Zhdanov, R. I. (Ed.). (1992). *Bioactive Spin Labels*, Springer, ISBN 978-3-540-53811-0, Berlin, Germany

Chapter 10

Uniform EPR Spectra Analysis of Spin-Labeled Macromolecules by Temperature and Viscosity Dependences

Yaroslav Tkachev and Vladimir Timofeev

Additional information is available at the end of the chapter

1. Introduction

Spin labeling method crossed its 47-th anniversary. It was invented in 1965 by Harden McConnel. He was the first who reported EPR spectra of bovine serum albumin and poly-(L-lysine) with stable nitroxide radicals chemically bound, and attempted to interpret them (Stone et al., 1965). As it stands today, we must conclude that there is little consensus between different scientists on the problem of such EPR spectra interpretation. When stepped into area of studying much more sophisticated biological objects comparing to these McConnel used, spin labeling methods this problem appeared to be very difficult. Its difficulty arises not only from the objects complexity, but also from method limitations, which relies on the solution of EPR reverse problem. Even with all of extensive developments done since 1965, such as availability of new microwave frequencies and pulsed EPR techniques, which, in fact, grew into separate method, the problem remains. It originates from large discrepancy between the information content of EPR spectrum and underlying object's behavior. Systems studies modern life science is interested in displays such a large degree of versatility that simply does not fit into single spectrum. This work summarizes our attempts to address this issue during the past 30 years.

The key for interpreting EPR spectra uniquely is to use as much of experimental data as possible. The most of informational content is found in spectra containing broad outer peaks (BOPs), as otherwise strongly narrowed spectra are highly degenerate, and attempting to solve reverse problem from them with underlying model which is slightly more complex than trivial is prone to non-uniqueness. Our method is primarily based on spectra with BOPs, to extract the most of useful information. In spectra of liquid solutions, their position is strongly affected by the mobility of nitroxide, and it is modulated by nitroxide environment and macromolecule Brownian motion. These are two main contributions into EPR lineshape the method is capable to resolve with most objects being studied.

We have previously suggested that monitoring the position of the broad outer peaks (BOPs) in EPR spectra of spin-labeled macromolecules unravels the system dynamics to the maximum extent conventional EPR can deliver. This may be achieved by studying of so-called temperature and viscosity dependencies, or TVDs, for short. In many cases, they provide a key to the interpretation of EPR spectra of spin-labeled macromolecules, by eliminating the large degree of non-uniqueness arising from different kinds of molecular motion. Similar approach is applicable to the non-covalent spin probe EPR studies of membranes. From these dependencies one can experimentally determine two very important quantities: the rotational correlation time (τ) of the protein molecule, and McConnell's order parameter (S) of rapid spin-label motion. Possibility of immediate use of these parameters for computational simulation of the EPR spectra significantly contributes to unique interpretation.

McConnell pioneered the use of the BOPs position in EPR spectra depending on the value of correlation time of a nitroxide radical motion. He conducted theoretical (McCalley et al., 1972), using Bloch equation, and experimental (Shimshick & McConnell, 1972) studies of the spin labeled alpha-chymotrypsin. Kuznetsov (Kuznetsov et al., 1971) theoretically showed the dependence of low-field BOP shift on the value of nitroxide molecular motion correlation time. J. Freed (Goldman et al., 1972) theoretically derived the expression for separation between BOPs depending on the value of τ for nitroxide radical.

We in parallel worked with the same problem, but for some reasons our first work has been published only in 1977 (Dudich et al., 1977). In this work it was emphasized for the first time that it is necessary to consider partial averaging of nitroxide magnetic tensors due to the fast reorientation of spin label relative to the macromolecule. This fast motion effectively changes the values of nitroxide magnetic tensors components. Most of the studies mentioned above were focusing on influence of polarity of an immediate $NO^{\bullet}$ moiety environment. Although this influence is not challenged, the polarity effects from nearby amino acid residues are very object-specific, as the surface charge of protein is. It is also pH-dependent, making it very difficult to follow. Additionally, in many practical cases the nitroxide moiety of attached label is significantly exposed to solvent, reducing polarity change effects. The fast motion is still ubiquitous, as neither label-to-protein linkage, nor the local protein structure at labeling site is perfectly rigid. In J. Freed's work (Mason and Freed, 1974) this fast motion relative to the carrier (molecular unit label is bound to) was considered as very rapid motion with respect to one of the molecular axes, while setting the diffusion tensor components in perpendicular plane to very slow motion, with an artificial introduction of tilt between the diffusion tensor principal axes relative to magnetic tensors ones.

The method based on averaging and is similar to known Model Free (MF) approach widely accepted in NMR (Lipari and Szabo, 1982). In this method, fast motion is considered to be independent of overall tumbling. The complicated Slowly Relaxing Local Structure (SRLS) model developed by Freed (Meirovitch et al., 2007) is claimed to benefit from including the coupling between these motions. It is achieved by substitution of composite diffusion operator into stochastic Liouville equation. This point was recently shown to be challengeable (Halle B., 2009). With simplified two-dimensional SRLS model Halle showed that attempting to exploit an advantage over MF it takes from motions coupling, lead to rather unphysical conditions. The contention was that complication introduced in SRLS probably does not worth its theoretical benefits.

Here we will present a clear algorithm suitable for correct and unequivocal interpretation of EPR spectra according to two independent motions (TwIM) model. As an example we will present an EPR study of spin-labeled Barstar in solution as well as attached to sepharose adsorbent to eliminate slow tumbling. Parameters measured from temperature and viscosity dependencies will be fed into the EPR spectra simulation program, and resulting spectra compared to experimental ones. We also show how the given approach may be naturally extended to membrane structures. Joint use of this method, together with Molecular Dynamics simulations, allows deeper investigation of the object's nature.

2. The temperature and viscosity dependence of EPR spectra of spin-labeled macromolecules

As it was noted in Introduction, an informational content of single EPR spectrum is rather limited. A case when both slow tumbling and fast label motions are indistinguishable from single lineshape is frequently encountered. The method of temperature and viscosity dependences (TVD) is found extremely helpful here, although other approaches also exist (e.g., mutifrequency EPR, where the relative sensitivity to motions on different time scales at various radiofrequency bands is utilized to resolve them). The TVD method developed throughout many years in a close connection with progress in computer hardware and software technologies. At present time, it can be utilized for unambiguous simulation of EPR spectra of spin-labeled macromolecules (Dudich et al., 1977; Timofeev, 1986, 1993, 1995) with easily available equipment (X-band spectrometer and personal computer) and minimum effort. The main idea this method is based on is pretty straightforward, once the TwIM model is assumed. In this model, the overall motion of spin label is composed of two independent components, one related to rapid motion of nitroxide-containing molecular fragment, and another to entire macromolecule tumbling. The former usually have correlation times in order of 100 ps or less, and, therefore, is *fast* on timescale of EPR method at X-Band. If this motion would be perfectly isotropic, its main effect is limited to some Lorentzian line broadening. In spin labeling applications, however, this is not usually the case, as the object (typically protein) spin label is attached to, significantly hinders nitroxide mobility. This is regarded as *motion anisotropy* in TwIM model, and it is quantified by order parameters. Its effect is much stronger: it changes the position of resonances, and this shift has severe impact on EPR spectrum. Effective spin Hamiltonian partial averaging is an efficient method to account for these changes. In fact, the only change in simple nitroxide spin Hamiltonian is different values of magnetic tensor g and hyperfine tensor A. This partial tensor averaging technique was utilized for studying anisotropic phases since early history of the spin labeling. Macromolecule tumbling correlation times lie typically in nanosecond range. The anisotropy of this motion is generally determined by the shape of macromolecule in liquid solution, and for most globular proteins the single isotropic "effective" correlation time (τ) is sufficient to describe the lineshape change induced by it. At X-Band, presumed here if no otherwise stated, and nitroxide spin labels, the τ values in range of 1-100 ns leads to significant changes in lineshape, which may be calculated by means of stochastic Liouville equation (SLE).

Henceforth we assume that the main EPR-observable effect of fast nitroxide fragment motion is changes in its *anisotropy*, quantified by order parameter S. For the slow tumbling, the corresponding quantity is its correlation time τ. This anisotropy-correlation time splitting presumed in TwIM model allows to eliminate much of the ambiguity from EPR spectra interpretation. But even in this case it may be very difficult to distinguishingly quantify the effects of both contributions from single lineshape. This is where temperature-viscosity dependence comes into play. Varying sample temperature, one can modulate both kinds of motion: τ decreases with temperature, and order parameter S typically do so due to increasing of molecular motion amplitudes. The viscosity of solution (which is also dependent on temperature), can be independently varied by addition of sucrose, glycerol or some polymer media. It mainly affects the correlation time τ, while interference with small-scale fast molecular motions is typically observed only at very high viscosity, where it is, strictly speaking, not "fast" anymore. In the TVD experiment, the set of EPR spectra is recorded, at different temperatures and solution viscosities. The processing of resulting dependence starts with analysis of separation between broad outer peaks in spectrum, according to the procedure described below.

2.1. The theory of temperature-viscosity dependences

The method of temperature and viscosity dependence was originally proposed (Timofeev & Samarianov, 1995) to determine the correlation time τ of slow isotropic rotational diffusion of a macromolecule, and McConnell's order parameter S in axially symmetric case. This method was further developed and modified (Tkachev, 2010) to be suitable for smooth joining into spectra interpretation and simulation framework based on TwIM model. The primary quantity measured experimentally in this method is the separation between the broad outer peaks (BOPs) in X-Band conventional EPR spectra (absorption spectra derivative). Independence of slow Brownian diffusion of entire macromolecule treated as rigid entity, and fast anisotropic reorientations of spin label, as the basic concept of TwIM, is the fundamental proposition in TVD method. Both motion types narrow the lineshape, shrinking the separation between BOPs (referenced as $2A'$ here). If one considers fast motion by anisotropic (partial) averaging of effective spin Hamiltonian, the amount $2A'$ decreases will depend on motion ordering, or, speaking other way, its degree of anisotropy. The slow tumbling, which in many cases of globular proteins can be thought isotropic, lead to the same effect but now it is dependent on the correlation time. This means one can express an effective shift of BOPs as the sum of the shifts originating from these two motion components:

$$2A_{ZZ} - 2A' = \left(2A_{ZZ} - 2\bar{A}\right) + \left(2\bar{A} - 2A'\right) = \Delta_1 + \Delta_2 \tag{1}$$

where $2A_{ZZ}$ – the separation between BOPs in rigid-limit EPR ("powder") spectrum. It was originally found from experiment, that plotting the $2A'$ value versus the temperature and viscosity ratio ., yields linear dependencies at constant temperature. The point where the extrapolated line crosses the ordinate axis yields the $2\bar{A}$ value. In approximation of axial symmetry of the fast motion, it is possible to derive dependence of the first term Δ_1 , on McConell's order parameter S:

$$S = \frac{2\bar{A}_{||} - 2\bar{A}}{2A_{zz} - 2A} = \frac{2\bar{A} - 2a_0}{2A_{zz} - 2a_0} \tag{2}$$

where $a_0 = \left(A_{XX} + A_{YY} + A_{ZZ}\right)/3$ is the isotropic hyperfine splitting constant of nitroxide radical. Therefore, the value of Δ_1 is represented as:

$$\Delta_1 = 2A_{ZZ} - 2\bar{A} = \left(2A_{ZZ} - 2a_0\right)\left(1 - S\right) \tag{3}$$

The second term Δ_2 gives the shift of BOPs in relative to the $2\bar{A}$ value. The value of $\Delta_2 = 2\bar{A} - 2A'$ can be evaluated in the following way. Spectral lineshape narrowing due to an exchange (Slichter, 1981), caused by slow rotation of a macromolecule with correlation time τ, is defined by the following expression (γ is a magnetogyric ratio):

$$\Delta H' \approx \Delta H\left[1 - \left(\frac{2}{\gamma\tau\Delta H}\right)^2\right] \tag{4}$$

In the model of an exchange between two states we assume them to be parallel and perpendicular orientations relative to magnetic field (Z axis). Corresponding values of ΔH and $\Delta H'$ will be

$$\Delta H = 2\left(\bar{A}_{||} - \bar{A}\right) = 3\left(\bar{A}_{||} - a_0\right) = 3S\left(A_{ZZ} - a_0\right) \tag{5}$$

$$\Delta H' = 2\left(A'_{||} - A'\right) = 3\left(A'_{||} - a_0\right) \tag{6}$$

Using expressions (2, 5, 6), we write $\Delta H'$ as follows:

$$\Delta H' = 2\left(A'_{||} - A'\right) = \Delta H - \frac{3}{2}\Delta_2 \tag{7}$$

Now, combining (7) with the expressions (4, 5) we finally obtain

$$\tau S = \frac{1}{3\gamma\Delta_0}\left(\frac{\Delta_2}{\Delta_0 S}\right)^{-\frac{1}{2}}, \tag{8}$$

where $\Delta_0 = 2\left(A_{ZZ} - a_0\right)$.

Hence, the dependence between the BOP shift, correlation time τ of a macromolecule tumbling, and order parameter S of fast spin label motion, should be found in the following form:

$$\tau S = a\left(\frac{\Delta_2}{\Delta_0 S}\right)^{-b} \tag{9}$$

Empirical parameters a and b can be found from a set of simulated EPR spectra at various values of S and τ. The formula (9) is deduced in the dimensionless form, therefore parameter a has the dimension of nanoseconds. Parameters a and b are dependent on changes of initial magnetic tensors and individual line width. But their mutual dependence (they always derived simultaneously, and are not independent) makes splitting (1) weakly affected by these changes. For the most of experiments with spin-labeled samples the value of a is close to 1.1 ns, and parameter b ≈ 1.3.

To establish a link with experimentally measurable values – temperature T and viscosity of a solution η, the straightforward way is to call for Stokes-Einstein relation (describing rotational motion of a sphere in a viscous media):

$$\tau = \frac{V\eta}{kT} \tag{10}$$

In this case expression (9) takes a form:

$$\Delta_2 = \Delta_0 S\left(\frac{akT}{SV\eta}\right)^{\frac{1}{b}} \Leftrightarrow 2A' = 2\bar{A} - \Delta_0 S\left(\frac{akT}{SV\eta}\right)^{\frac{1}{b}} \tag{11}$$

The overall result here is that the separation between BOPs is a linear function of $(T/\eta)^{\frac{1}{b}}$. Experimental measurement of this dependence by varying the temperature and the viscosity, it is possible to find $2\bar{A}$, and corresponding order parameter S, and, according to the formula (9), rotational correlation time of a macromolecule tumbling in solution. It's important to remember, however, that derived expressions are valid only when motions differ significantly on time scale. This may not be the case at very high viscosities where shear forces may interfere with the fast motion, which is believed to be affected solely by temperature changes. At these conditions motion of nitroxide cannot be regarded as 'fast' anymore.

2.2. The uniform method for resolving an EPR spectroscopy reverse problem

Combining the EPR spectra simulation using TwIM model for motion of spin label, and experimental approach of TVD, the following scheme can be used for uniform spectra interpretation. It allows unequivocally treat the line shape modulated by different dynamic effects. It calls for stochastic Liouville equation approach for calculation of slow motional EPR spectra. The flowchart is shown on Fig. 1, and individual steps are described in the following sub-sections. It contains two feed-back optimization cycles: one on the right relies on TVD data (experimental), but it corresponds to axially symmetrical case. The one on the left is more general, and deals with rhombic components of magnetic tensors (more than one order parameter is possible to describe fast motion). Right cycle **1** is responsible for locking the BOP separation in simulated spectra to experimental values, and for distinguishing between fast and slow motional contribution to BOP shift (ca. expression (1)). The generalized cycle **2** allows obtaining finer details of the motion (additional order parameters) from fitting the XY manifold (central region of EPR spectrum). The procedure is considered complete when both cycles are consistent.

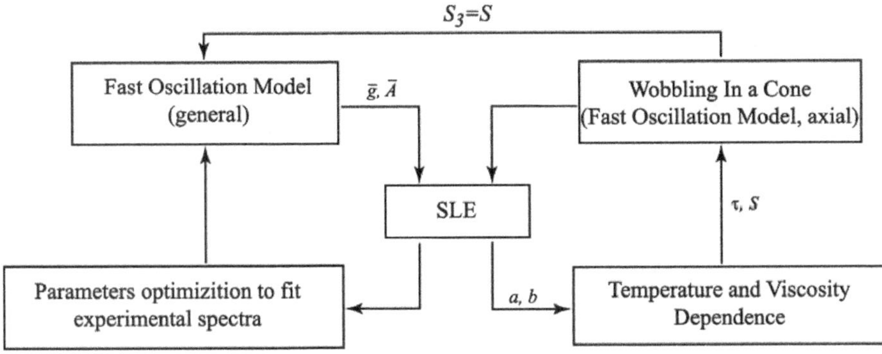

Figure 1. The scheme of uniform EPR spectra interpretation based on TwIM model and TVD experiment. SLE states for stochastic Liouville equation-based procedure for calculation of slow motional spectra.

2.2.1. Determination of magnetic tensors components

The entire procedure was designed to eliminate as many degrees of freedom as possible while retaining physical sense. The lineshape changes induced by molecular motions are controlled via τ and S. To set up spectra simulation, magnetic parameters have to be determined in some way. Blind fitting of them would scramble all the averaging effects, and therefore is not suitable here. The most correct way is using diluted crystal, but is production in many cases is not feasible. The compromise is to use values of $2a_0$ from free label solution EPR spectrum at room temperature and $2A_{ZZ}$ from spectrum of frozen labeled sample at 77K. The rhombic component still remains unknown, and has to be fitted. For hyperfine A tensor, it is small, and the values of A_X and A_Y are almost equal. This mostly eliminates the A-tensor problem, which is mostly significant as it determines the distance between BOPs used to quantify molecular motion from in TVD experiment. The components of g tensor are still subject to fitting, at least its rhombic component which is not negligible as it is in case of hyperfine tensor.

2.2.2. Empirical parameters in temperature and viscosity dependences

Once the magnetic tensor components were determined, it is possible to calculate EPR spectra. The spectra simulation procedure based on tensor averaging for accounting of fast motion, and SLE for slow rotational diffusion is used for calculation of empirical parameters **a** and **b** in the equation (9). For this purpose, a set of EPR spectra is calculated axially symmetrical case (that is, no rhombic component in hyperfine tensor). The shift of separation between BOPs in theoretical EPR spectra measured with different values of correlation time τ and order parameter S, if plotted in logarithmic scale versus τS according to the formula (9), allows to calculate the parameter b from the slope of linear approximation, and parameter a from its intersection with ordinate axis (Tkachev, 2010).

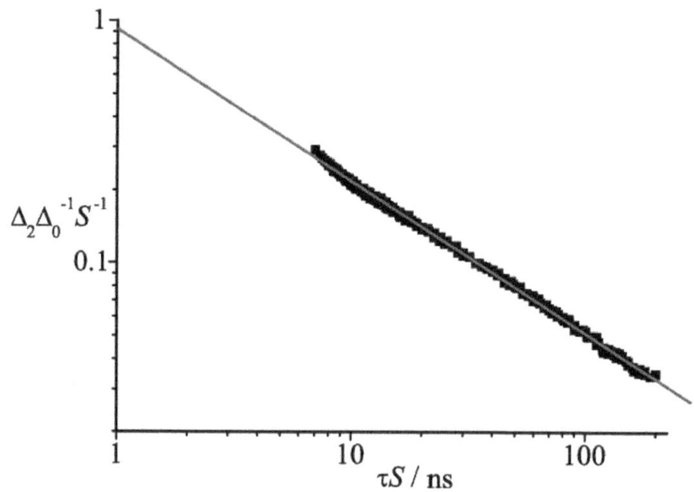

Figure 2. Theoretical dependence of BOP shift vs. global tumbling correlation time τ and fast label motion order parameter S. It is used to find a and b parameters in eqation (9). This particular dependence was calculated for g_{xx} = 2.0088, g_{yy} = 2.0058, g_{zz} = 2.0023, A_{xx} = A_{yy} = 6,0 G A_{zz} = 35.3 G, and individual line width of 1.1 G. Every black point corresponds to simulated spectrum, and the gray line is least-square approximation.

Values of empirical TVD parameters determined from simulation of set of spectra according to the method described above are dependent on magnetic tensors and line widths used for calculation. High performance in spectra simulation which may be routinely achieved on modern workstations and personal computers makes possible to obtain these parameters interactively, upon changing line width or magnetic tensor components. This is important part of the EPR spectra interpretation method described here, as it may require multiple processing of experimental TVD data with different a and b values to achieve good coincidence of simulated spectra with experimental ones.

2.2.3. Experimental determination of τ and S values

When empirical parameters are known, it is possible to use equation (11) for experimental data processing. According to equation (11), the dependence of $2A'$ on $\left(T/\eta\right)^{\frac{1}{b}}$ should be linear. Therefore, an intersection of linear approximation of experimental results will give the value of $2\bar{A}$ which encodes the order parameter according to expression (2). Then, from equation (9) it is possible to find the correlation time of slow global tumbling of the macromolecule.

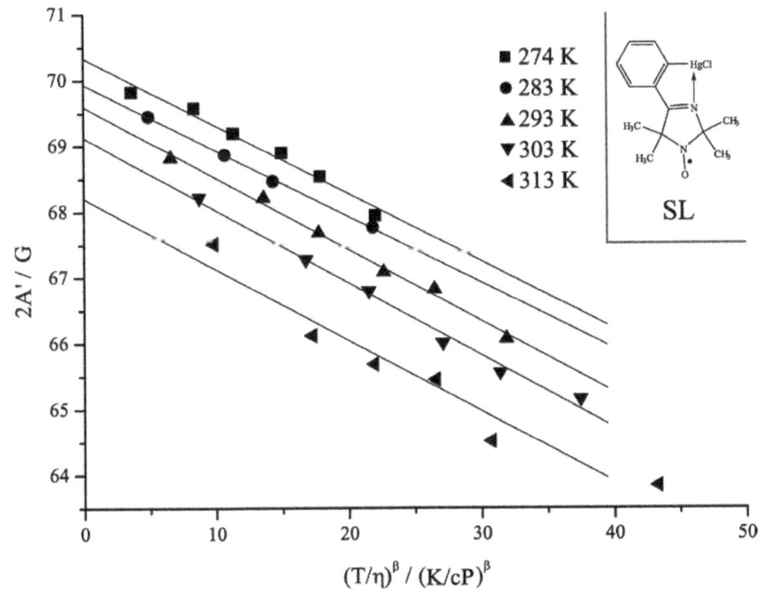

Figure 3. Typical plot of temperature and viscosity dependences of BOP separation. This data corresponds to bovine serum albumin (BSA) labeled with 4-(2-chloromercurophenyl)-2,2,5,5-tetramethyl-3-imidazoline- Δ^3 -1-oxyl (on inset). It was used as model system for procedure test purposes. For every temperature the linear data approximation crosses ordinate axis on $2\bar{A}$, and order parameter value may be derived from it.

2.2.4. Simulation of EPR spectra of spin-labeled macromolecules

The simulation of EPR spectra of spin-labeled macromolecules used here is also based on the two independent motions (TwIM) model (Dudich et al., 1977; Timofeev 1986; Timofeev & Samaryanov 1993, 1995; Tkachev 2009, 2010). Therefore, fast motion of the spin label relative to the macromolecule is taken into account by effective spin Hamiltonian partial averaging, and the slow motion of the macromolecule is determined by the rotational correlation time of isotropic diffusion.

By means of the generalized model of fast nitroxide oscillations the partially averaged magnetic tensor components can be calculated (Tkachev, 2009, 2010) in such a way that order parameter obtained from TVD experiment can be employed. This is possible because axially symmetrical model presumed in TVD is a special case of the generalized model mentioned. Substitution of averaged tensors into the SLE-based procedure for slow motional nitroxide spectra simulations (Schneider & Freed, 1989) gives the final EPR lineshape with contributions of both fast and slow motions.

On the above basis the simulation program (S_imult6) was developed. The model of fast limited oscillations (Timofeev and Samaryanov, 1995) gives formulas in analytical form by which components of partially averaged magnetic tensors can be calculated. According to model nitroxide fluctuates around some fixed axis $\vec{n}$ defined in polar coordinates by θ, φ angles. Fluctuations are limited by an angular amplitude of oscillation α, and all orientations are equally probable in range from $-\alpha$ to α. However, it was shown (Tkachev 2009, 2010), that partially averaged magnetic tensors aren't diagonal in general case. This implies that tensors have to be diagonalized after averaging, resulting in tilt between principal axes of both partially averaged magnetic tensors to appear. Two parameters defining an axis of oscillation $\vec{n}$, $S=\left(3\cos^2\theta-1\right)/2$, $\kappa=\left(1-2\cos^2\varphi\right)$ can be shown to play role of order parameters. The meaning of S defined this way is exactly the same as in axially-symmetrical case which arises if one let $\alpha=180^0$. In this case it is equal to experimental McConnell's order parameter. This allows using axially symmetrical case with experimentally determined order parameter as the starting point in scheme shown on Fig. 1, when calculating theoretical EPR spectra.

The general scheme for the resolution of the EPR spectroscopy inverse problem for spin-labeled macromolecules presented as follows. Experimentally determined magnetic tensor components, the correlation time τ together with the order parameter S, found from temperature-viscosity dependences, are fed into the simulation procedure based on SLE (cycle 1 on Fig. 1). In axially symmetrical case one obtains a theoretical spectrum similar to experimental one only on its wings (position of BOPs). In order to fit calculated spectra to experiment in the central region of magnetic fields, it is necessary to deviate from initial θ, φ and α values, breaking axial symmetry (switch to cycle 2 on Fig.1). Additionally, the linewidth and initial tensor rhombic parts may be adjusted as well, to achieve the best fit of a theoretical spectrum to the experimental one. If these are modified, it may be necessary to recalculate empirical TVD parameters (a and b), and repeat the procedure from cycle 1. Once it is possible to obtain consistent results from both cycles, process is complete. It utilizes the maximum of information from experimental TVD dependence, and yields the set of simulated spectra with the only difference in order parameters and macromolecular tumbling correlation time. The starting conditions are provided by axially symmetrical data obtained directly from experiment. This eliminates most of non-uniqueness which the usual multidimensional fitting procedures is prone to.

An ensemble may be inhomogeneous in sense of averaging parameters. One of the reasons for this is existence of macromolecule's structure fluctuations, or set of different structural conformations of the labelling site. This leads to concept of sub-ensembles of nitroxides, called clusters (Timofeev & Samarianov, 1995). They are clustered according to averaging order parameters (all kinds of equivalent dynamical behaviors, producing identical averaged tensors), and these may be assumed to be normally distributed. The spectrum depends on $\alpha(\alpha_0,\sigma)$, where α_0 is an average, and σ is a standard deviation of the Gaussian distribution (Tkachev, 2010, Timofeev & Samarianov, 1995). The resulting spectrum is calculated as Gaussian-weighted sum of spectra of individual clusters. This adds single fitting parameter, but in many real-world cases helps to dramatically improve fits quality.

3. The temperature and viscosity dependences for spin-labeled Barstar in solution

Spin labeling, EPR and all the method described above, based on the temperature-viscosity dependence experiment, was used to study the protein-protein interaction between the enzyme barnase (Bn) and its inhibitor barstar (Bs) (Timofeev et al., 2008). A mutant of barstar (C40A), containing only one cysteine residue, C82 was selectively modified by spin label (SL) 4-(2-chloromercurophenyl)-2,2,5,5-tetramethyl-3-imidazoline- Δ^3 -1-oxyl (Shapiro et. al., 1979). We used spin label which ensured higher ability of his reporter group to access different protein microenvironments. To estimate the mobility of the spin labeled C82 side group and the whole globular protein quantitatively, the temperature-viscosity approach was used.

To extract unambiguous information from experimental EPR spectra of spin-labeled macromolecules, it is necessary to have maximum possible control over the degrees of freedom (DOFs) an EPR line shape depends on. This effectively reduces linehape 'degeneracy' with respect to the set of parameters used to interpret these spectra. Generally, for complex objects like mutually interacting proteins, unequivocal interpretation of EPR spectra is possible only when BOPs are present. For small molecules undergoing relatively rapid rotations narrowing the spectrum at ambient temperatures, decreasing the temperature and increasing solvent viscosity increase order parameters, as well as macromolecule tumbling correlation time. This leads to broad outer peaks (BOPs) of increasing amplitude to appear in low and high field regions of EPR spectra. By processing a series of EPR spectra of spin-labeled samples and measuring the separation between the BOPs, one can estimate the effective rotational correlation time τ of the protein molecule and the order parameter S of the spin label attached to the side chain. EPR spectra of Bs-SL in solution with changing viscosity are shown on Fig. 1 and Fig. 2. Viscosity was controlled by addition of sucrose.

As can be seen from Fig. 4 and Fig. 5, BOPs shifts considerably towards field range end points with increasing viscosity, and this is an effect one expects to be able of studying temperature-viscosity dependence. According to formulas (11) for parameter $2A'$, this dependence is shown on Fig. 6.

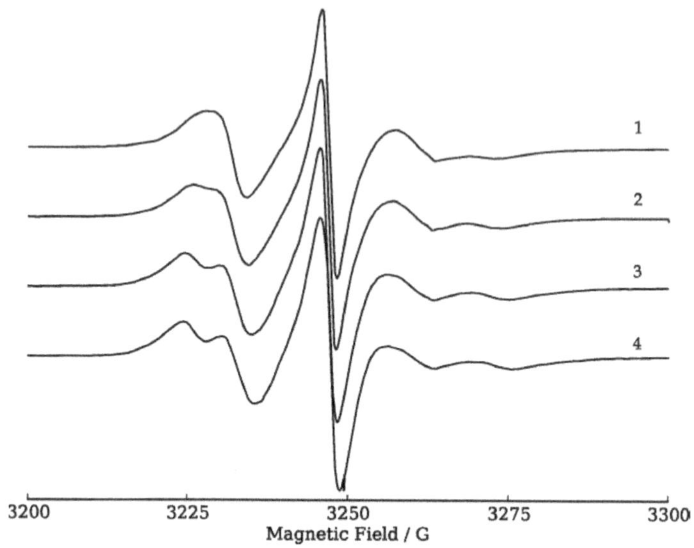

Figure 4. EPR spectra of Bs-SL at 1º C, with viscosity altered by addition of (1) 0%, (2) 6%, (3) 16%, and (4) 19.5 % sucrose (w/w).

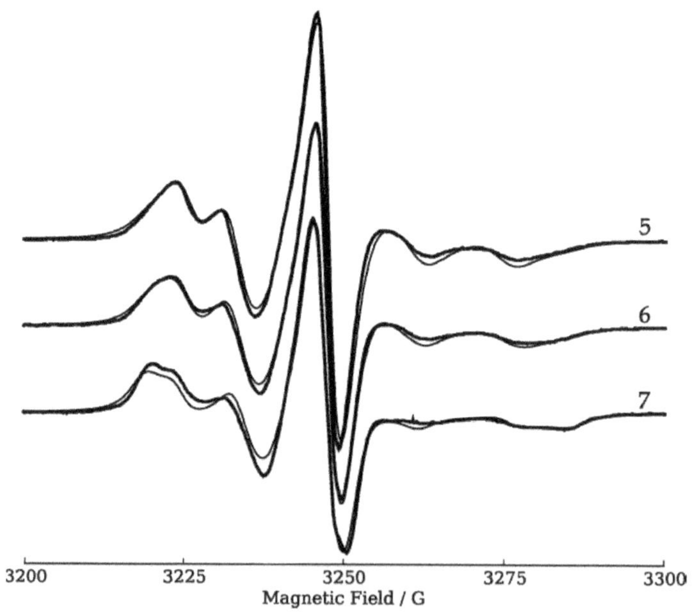

Figure 5. EPR spectra of Bs-SL at 1º C, with viscosity controlled by addition of (5) 27 %, (6) 33 %, (7) 40 % sucrose (w/w). The experimental spectrum is plotted with heavy line and the simulation - with fine line. The theoretical calculation of spectra shown is discussed in the text.

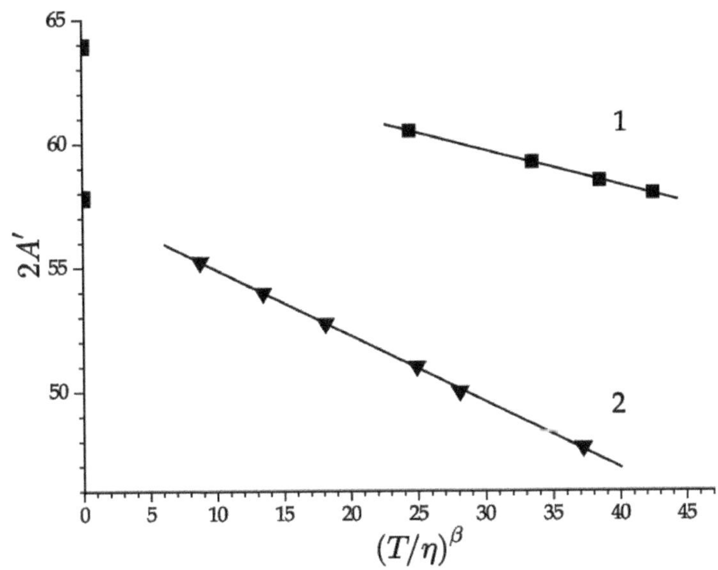

Figure 6. Temperature-viscosity dependences of separations between BOPs ($2A^{'}$) in the EPR spectra of Bs-SL (line **2**), and complex Bs-SL with Bn (line **1**) at 1º C. The data were fitted using the linear least squares method. Line 1 crosses the ordinate axis yielding the extrapolated value $2\bar{A} = 63.9$ G, whereas line 2 yields $2\bar{A} = 57.8$ G. Units are T for K, and cP (centipoises) for η . The value of TVD parameter $\beta = 1/b = 0.74$.

It is clearly observable from EPR spectra presented in Fig. 4 and Fig. 5, that at lower viscosity values (6%, 16%, and 19.5% w/w sucrose in solution) BOPs are considerably asymmetrical. In fact, there are two poorly resolved BOP pairs, what is clearly seen in spectra at higher viscosities (27%, 33%, and 40% w/w sucrose in solution). Therefore, it is clear that spectra are composed of two components, first broad, and second, narrower. The separations between BOPs from narrow component (three points) $2A^{'}$ are shown on the Fig. 6 (line 2). The first spectrum on Fig. 4 lacks of low-field BOP, thus there is no point for 0% sucrose on the line 2 (Fig. 6). At higher viscosity values (27%, 33%, and 40% sucrose in solution) BOPs from broad and narrow spectrum components were separated. Hence we define the separations between BOPs near to the center of the spectrum as $2A^{'}$ (narrow) and farther from the center of the spectrum as $2A^{'}$ (broad). As for the three points to $2A^{'}$ (broad) they are visible on the line 2 (Fig. 7). In Fig. 7 for clarity, a line 1 is repeated as line 1 in Fig. 6.

For these double-component spectra, two explanations are possible. This is either due to the Bs-Bs interaction of two macromolecules (Timofeev, et al., 2008), or a spin label SL have two dynamic states with strongly different order parameters. To clarify this alternative, an experiment on the binding of macromolecules Bs-SL with adsorbent has been undertaken.

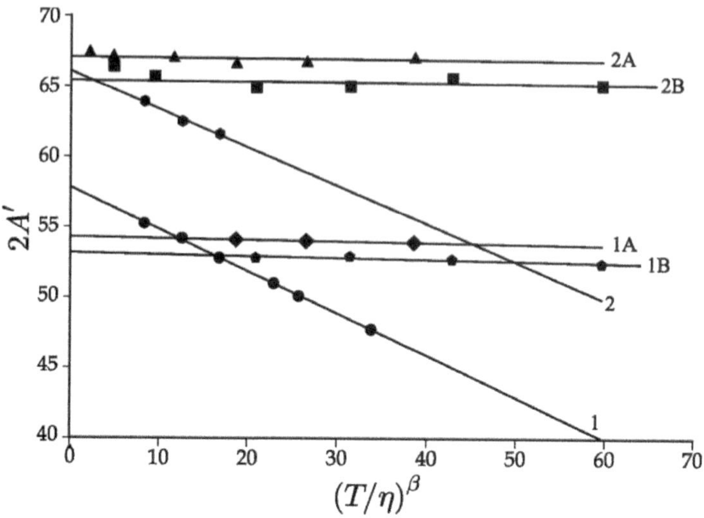

Figure 7. Temperature and viscosity dependences for BOPs separation ($2A^{'}$ /G) in EPR spectra of spin-labeled Barstar in solution, and on adsorbent. Solution: 6 points for BOPs from narrower component (line **1**): (1) 6%, (2) 16%, (3) 19.5 %, (4) 27%, (5) 33%, and (6) 40 % sucrose (w/w), and 3 points for broad component BOPs (line **2**): (1) 27%, (2) 33%, and (3) 40 % sucrose (w/w), at 1ºC. With adsorbent: **1A** line at 1ºC and **1B** line at 20ºC for narrow component BOPs; **2A** line at 1ºC and **2B** line at 20ºC for broad component BOPs. The data were fitted using the linear least squares method. Line 1 crosses the ordinate axis yielding the extrapolated value $2\bar{A} = 57.8$ G, and line 2 yields $2\bar{A} = 66.1$ G. The value of $\beta = 1/b = 0.74$.

4. The temperature and viscosity dependences for immobilized Barstar

If macromolecule with spin label attached to it can be immobilized, rendering effective correlation time to be large enough to mimic ridgid-limit spectrum. In case with spin labeled Barstar we used QFF sepharose to achieve this goal. The sample was prepared by addition of SL to Barstar solution in HEPES buffer, then added 6 mg SL corresponding concentration to have molar excess of protein. Then 200 ml of QFF sepharose suspension was added and sample washed to remove unbound protein by spinning on microcentrifuge. Spin-labeled protein-charged sepharose was used for recording of EPR spectra. Temperature dependence of these spectra (1, 10, 20, 30, 40C) without sucrose is shown on Fig. 8.

It is clear that all five EPR spectra display "quasi-powder" pattern (broad due to strong immobilization) as spin-labeled macromolecules are now attached to an adsorbent (cf. Fig.1). Over entire temperature range (1-40ºC) two components are clearly observable, with corresponding BOP separations of $2A^{'}$ (narrow) and $2A^{'}$ (broad).

Rotational Brownian motion of medium-sized macromolecules in solution is on the nanosecond range. Linking macromolecules to an adsorbent shifts it to the microsecond range, which, in fact, is not distinguishable from rigid-limit ("powder spectrum") at

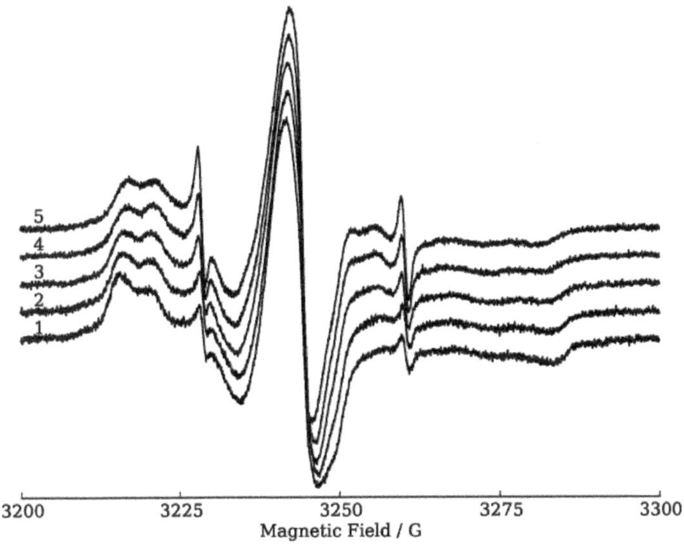

Figure 8. EPR spectra of the QFF-Bs-SL1 complex at 1ºC (1), 10ºC (2), 20ºC (3), 30ºC (4), and 40ºC (5). The spectral width is 100 G.

experimental conditions used. Therefore, the shape of "quasi-powder" EPR spectrum is exclusively determined by the fast reorientation of the spin label in a limited configuration/conformation space. This makes it easy to measure the order parameter of nitroxide (see expression (2)).

On the other hand, once the protein is adsorbed on the sepharose, the arising "quasi-powder" EPR spectrum should not depend on oligomeric state of the protein, to an extent local structure of the protein close to labeling site is disturbed by weak inter-macromolecule interactions. Strong Barstar interaction with an heavily charged adsorbent is expected to significantly impair its ability to form dimer, and is expected to shift equilibrium to monomeric form. Therefore, we conclude that the order parameter S in the dimeric form of the protein (if any) and its monomeric form should be the same. Consequently, there are two conformational states of the spin label attached to the thiol group of the protein, corresponding to two spectral components with BOP separations of $2A'$ (narrow), and $2A'$ (broad).

In further experiments EPR spectra of Bs-SL on the adsorbent at different viscosities have been recorded, for the full temperature-viscosity dependence to be obtained. Corresponding spectra, for different temperatures, are shown on Figure 9 (A, B). Two components found (two BOP pairs) were processed independently to get order parameters for both spin label conformational states. Fig. 7 displays TVD for sepharose-bound Barstar overlaid on these for free protein in solution at 1ºC (lines 1A and 2A) and 20ºC (lines 1B and 2B). As seen in Fig. 7, four lines (1A, 1B, 2A, 2B) are parallel to the x-axis. This behavior is expected, as both types of BOPs (from narrow and broad spectrum components) do not shift with increasing viscosity of the medium in the "quasi-powder" EPR spectra, given the value of τ for protein is virtually infinite. Consequently, the separations between BOPs not change and, thus, in the case with adsorbent: $2A^{'}$ (narrow) = $2\bar{A}$ (narrow) and $2A^{'}$ (broad) = $2\bar{A}$ (broad). It also means that the second term in equation (1), for both spectrum components is zero.

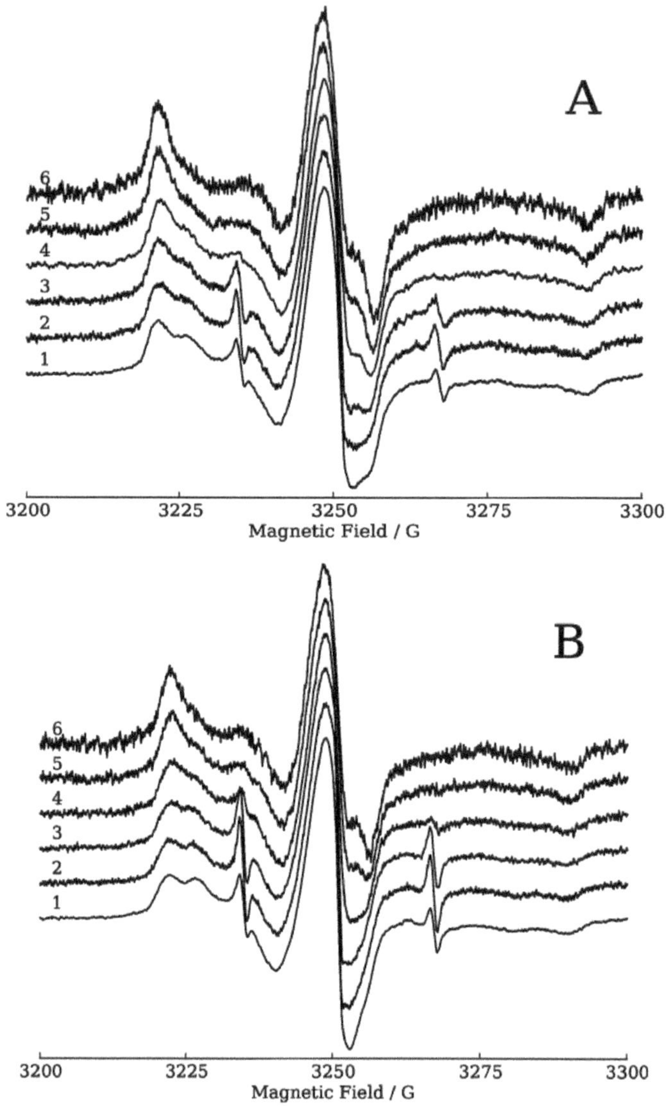

Figure 9. EPR spectra of spin labeled Barstar immobilized on QFF sepharose (QFF•Bs-SL system) at 1ºC (A) and 20ºC (B) with viscosity altered by addition of (1) 0%, (2) 15%, (3) 30%, (4) 37%, (5) 48%, and (6) 56% sucrose (w/w).

It is now straightforward, knowing $2A_{ZZ}$ and $2a_0$, to calculate order parameters S_1 (for narrow component) and S_2 (for broad component) according to the formula (2) for both states of the spin label. Dependences of the order parameters S_1 and S_2 on the temperature are shown in Fig. 7, in this case $S_2 > S_1$.

Thus the ensemble of spin-labeled molecules barstar Bs-SL is divided into two sub-ensembles. In one sub-ensemble spin label is found in 1st state with corresponding order parameter S_1, and another – in 2nd state with order parameter S_2.

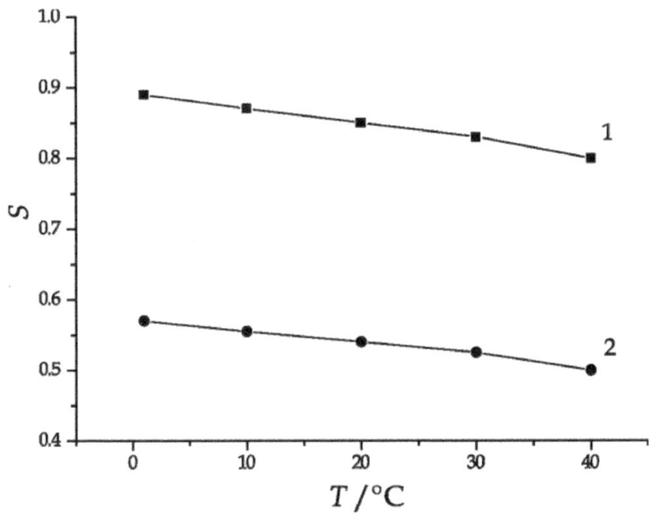

Figure 10. Values of the order parameters S2 (to 2nd state of SL, farther BOPs) – line 1; and S1 (to 1st state of SL, near BOPs) – line 2; in depend on the ambient temperatures.

Interestingly, the fast motion activation by temperature in each of the SL states in a local protein site (Cys82) is negligible. As seen in Fig. 7, parameters S_1 or S_2 drift only slightly over entire temperature range from 1 to 40ºC. This suggests that the rigid spin label samples almost the same amount of configuration space provided by rigid protein frame in this range of temperatures.

Addition of the Barnase (Bn) to Bs-SL solution results in a temperature-viscosity dependence, as shown in Fig. 3 by line 1. In this case, the spin label in the complex of BnBs-SL has a single state with a value of the order parameter (S = 0.86). Line 1 slope is two times less than that of line 2 (Fig. 3). Very strong affinity of Barstar to Barnase explains this, since the molecular weight of BnBs complex (22.7 kD) is about twice larger than the molecular weight of Barstar (10.3 kD).

When introduced to the system QFF•Bs-SL, Barnase had no significant effect on EPR spectrum (see Fig. 8). In case of stable Bn•Bs-SL complex formation, the EPR spectrum was expected to be substantially changed, which is the case in solution. The capacity of the QFF sepharose which is abundant in the sample mixture and its charge make Bn easier to bind to

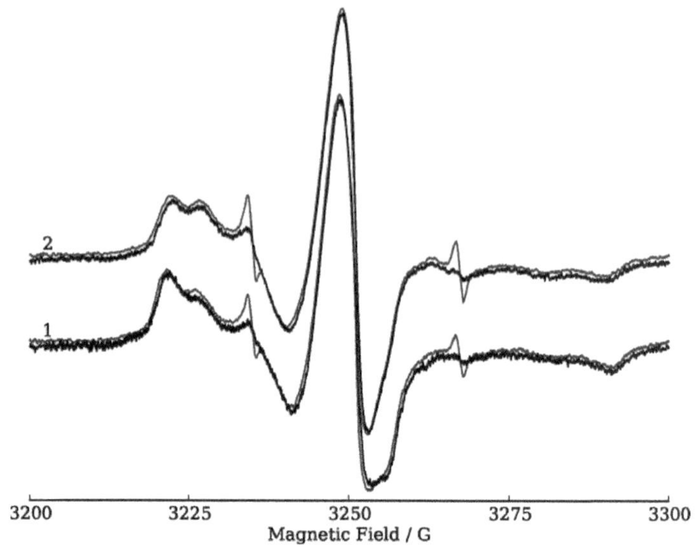

Figure 11. The spectra EPR of QFF-Bs-SL (light line with sharp peaks of free SL) and the same at adding Bn in a bit large excess (dark line without sharp peaks) at 1ºC (1) and 20ºC (2).

the adsorbent, while bound Barstar may have its Bn binding interface unreachable. This leads to disruption of Bn•Bs-SL complex formation equilibrium. No change in EPR spectra at two temperatures, 1ºC and 20ºC indicate that there is apparently no significant amount of complex present. Smaller amount of free label (three narrow lines on Fig. 8) after Bn addition was due to centrifugation, and, therefore, better washing.

5. Spin-labeled Barstar EPR spectra simulation

The procedure for simulating EPR spectra using the proposed model of the two motional nitroxide radical was described in section **2.3**. Here we want to give some examples of EPR spectra simulation, paying an attention to the similarity of theoretical spectra to experimental ones.

5.1. Barstar and formation of complex with Barnase

On the Fig. 5 (A, B, C) fitted simulation EPR spectra superimposed onto experimental ones. As it appears, they can be described by two dynamical states of spin label rapid motion relative to Barstar molecule. If one assumed these states uncoupled, each of them corresponds to its own EPR spectrum, and experimental one corresponds to their weighted sum. On Figures 12 - 14 individual simulated spectra are shown for each state of spin label (1st and 2nd); their superposition results a final spectrum. Using Gaussian distribution on α averaging parameter helps to obtain better fit due to states dynamical coupling (exchange), and the spectra shown on these figures actually account for it this way (see legends).

We used following initial components of nitroxide magnetic tensors: g(X,Y,Z)= 2.00732; 2.0063; 2.0022; A(X,Y,Z)=6.55 G; 5.00 G; 35.40 G. In legends to figures, the rotational correlation times and linewidth, parameters used for calculation of partially averaged tensors components along with themselves, are given.

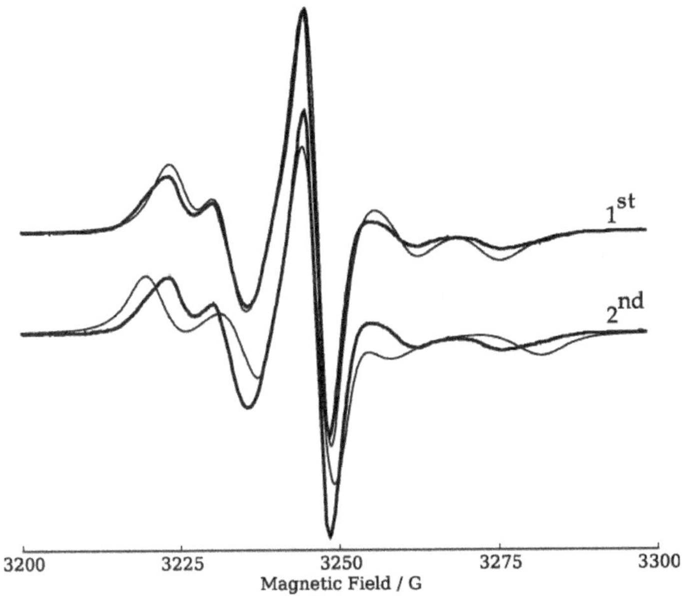

Figure 12. Simulated EPR spectra (fine line) for of 1st state SL and 2nd state of SL in comparison with an experimental spectrum of Bs-SL at 1º C and 27 % sucrose in solution (heavy line).

Spectrum for of 1st state contains BOPs close to the center (narrow). It is described by three parameters which, according to section 2.3, determine partial averaging of tensors in 1st state by the values of $\alpha = 77(15)^{\circ}$, $\theta = 46^{\circ}$, $\phi = 5^{\circ}$. Each spectrum was calculated as the average by α with Gaussian distribution, as 77(15), where the value in brackets is the standard deviation. Magnetic tensors diagonal components: $\overline{g}$ (X,Y,Z) = 2.00696, 2.00562, 2.00323; $\overline{A}$ (X,Y,Z) = 6.66246 G, 11.9003 G, 28.3872 G. BOPs in simulated EPR spectrum (fine line) for 2nd state located farther from the center. This second state is described by parameters $\alpha = 30(44)^{\circ}$, $\theta = 30^{\circ}$, $\phi = 5^{\circ}$. Magnetic tensors diagonal components: $\overline{g}$ (X,Y,Z) = 2.00723, 2.00628, 2.0029; $\overline{A}$ (X,Y,Z) = 6.46487 G, 5.74586 G, 34.7363 G. Rotational correlation time for macromolecule was $\tau = 25$ ns. Linewidth used was 1.7 G. The simulated spectra in both states were summed with the ratio of 28:72, which gives the resulting spectrum shown in Fig. 5(5).

All partial averaging parameters are the same as in the legend to Fig. 12. The protein rotational correlation time is $\tau = 38$ ns. The simulated spectra in both states were summed with the ratio of 45:55, which gives the resulting spectrum shown in Fig. 5(6).

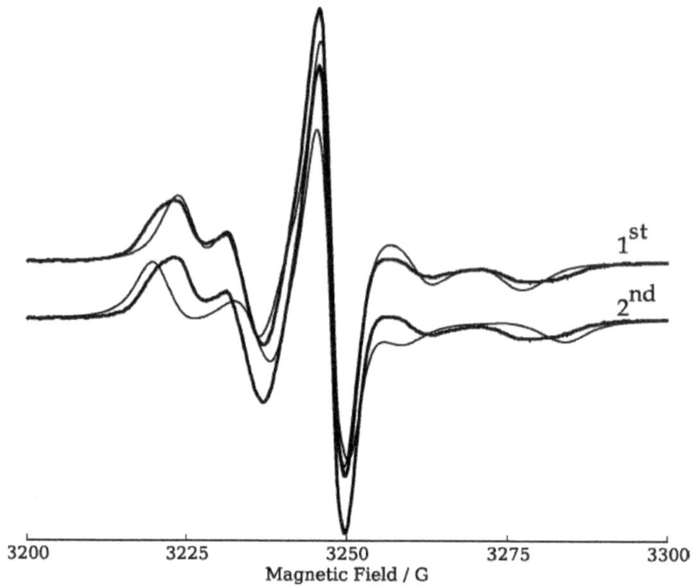

Figure 13. Simulated EPR spectra (fine line) for 1st and 2nd states of SL and 2 state SL in comparison to experimental one of Bs-SL at 1º C and 33 % sucrose in solution (heavy line).

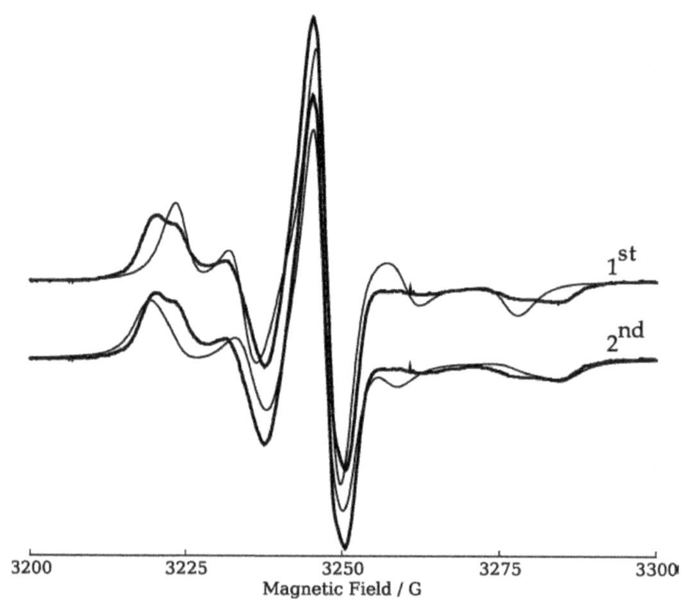

Figure 14. Simulated EPR spectra (fine line) for 1st and 2nd states of SL and 2 state SL in comparison to experimental one of Bs-SL at 1º C and 40 % sucrose in solution (heavy line).

All partial averaging parameters are the same as in the legend to Fig. 12. The protein rotational correlation time is $\tau = 60$ ns. The simulated spectra in both states were summed with the ratio of 40:60, which gives the resulting spectrum shown on Fig. 5 (7).

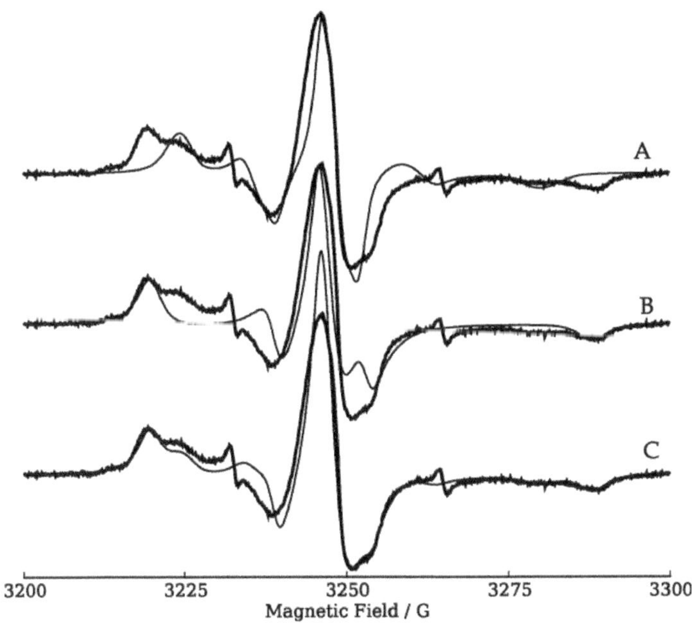

Figure 15. EPR spectrum of QFF-sepharose-Bs-SL at 1º C and 0 % sucrose (heavy line) and the simulated spectrum of 1st or 2nd states (thin line). (A) The 1st state is described (see Fig. 9A) is described by three parameters which, according to section 2.3, determine partial averaging of tensors in 1st state by the values of $\alpha = 77(14)^{\circ}$, $\theta = 46^{\circ}$, $\phi = 5^{\circ}$. Components of averaged tensors: $\bar{g}$ (X,Y,Z) = 2.00696, 2.00562, 2.00323; $\bar{A}$ (X,Y,Z) = 6.66246 G, 11.9003 G, 28.3872 G. Each spectrum was calculated as the average by α with Gaussian distribution, as 77(15), where the value in brackets is the standard deviation. The rotational diffusion tensor components (expressed as correlation times) for macromolecule are $\tau_X = 100$ ns, $\tau_Y = 1000$ ns, $\tau_Z = 1000$ ns. Anisotropic linewidth components are $\delta_X = 1.3$ G, $\delta_Y = 1.5$ G, $\delta_Z = 1.5$ G. (B) The 2nd state is described (see Fig. 9) by three parameter values of $\alpha = 27(32)^{\circ}$, $\theta = 27^{\circ}$, $\phi = 5^{\circ}$. Components of averaged tensors: $\bar{g}$ (X,Y,Z) = 2.00724, 2.00629, 2.00226; $\bar{A}$ (X,Y,Z) = 6.66246 G, 11.9003 G, 28.3872 G. The rotational diffusion tensor components (expressed as correlation times) for macromolecule are $\tau_X = 100$ ns, $\tau_Y = 1000$ ns, $\tau_Z = 1000$ ns. Anisotropic linewidth components are $\delta_X = 1.4$ G, $\delta_Y = 1.9$ G, $\delta_Z = 1.9$ G.
(C) Spectrum of QFF-sepharose-Bs-SL at 1º C and 0 % sucrose in solution, (heavy line) and the simulated spectrum (fine line) obtained by summation of ones for 1st and 2nd states (Fig. 15 A, B), with ratio of 30:70.

The results obtained by fitting of simulated spectra to experimental ones for spin-labeled Bs-SL, are listed in legends to Figures 12, 13, 14. Thus, three points (27, 33 and 40% sucrose) on temperature-viscosity dependence (Fig. 6 line2; or Fig. 7 line1) correspond to simulated EPR spectra on Fig. 5 and Fig. 12, 13, 14. It is important to note that simulated EPR spectra differ solely in the rotational correlation time of macromolecule! This is highly consistent with proposed TVD advantage of separating BOPs shift contributions regarding to slow rotational dynamics of macromolecule, and rapid motion of spin label. The partial averaging of magnetic tensors conserve when temperature is held at constant level.

5.2. Sepharose-immobilized spin-labeled Barstar EPR spectra

The EPR spectra of Bs-SL in solution in the presence of QFF sepharose on Fig. 8 and Fig. 9 are shown. This entire set of spectra can be simulated using the same principle described here. However, for an example we have selected only two experimental spectra to show procedure of simulation (see section **2.2.4**). We will discuss already shown simulated EPR spectra of Bs-SL in solution.

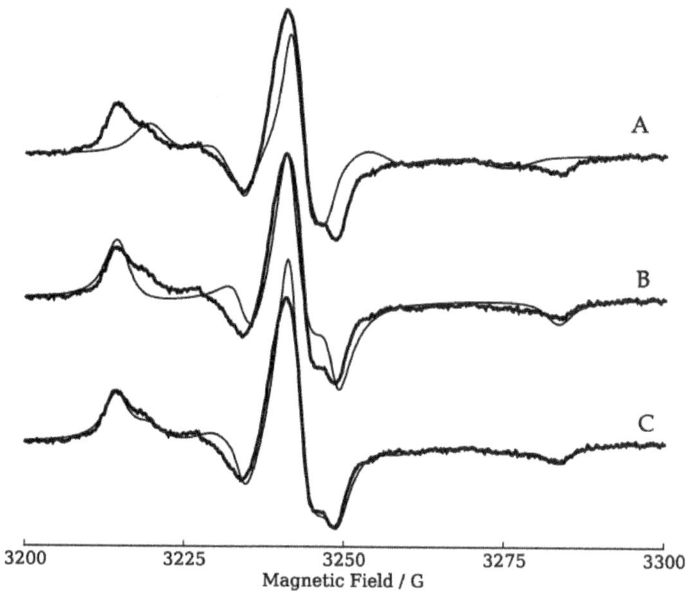

Figure 16. A. EPR spectrum of QFF-sepharose-Bs-SL at 1º C and 35% sucrose in solution (heavy line), and the simulated spectrum of 1st (fine line). Partial averaging parameters are the same as given in legend to Fig. 15A. The rotational diffusion tensor components (expressed as correlation times) for macromolecule are τ_X =200 ns, τ_Y =1000 ns, τ_Z =1000 ns. Anisotropic linewidth components are δ_X =1.5 G, δ_Y =2.4 G, δ_Z =2.4 G.

B. Spectrum of QFF-sepharose-Bs-SL at 1º C and 35% sucrose in solution (heavy line) and the simulated spectrum of 2nd state (fine line). Partial averaging parameters are the same as given in legend to Fig. 15A. The rotational diffusion tensor components (expressed as correlation times) for macromolecule are τ_X = 200 ns, τ_Y = 1000 ns, τ_Z = 1000 ns. Anisotropic linewidth components are δ_X =2.0 G, δ_Y =3.0 G, δ_Z =2.0 G.

C. EPR spectrum of QFF-sepharose-Bs-SL at 1º C and 35 % sucrose in solution, (heavy line) and the simulated spectrum calculated as weighted sum of the above with ratio of 30:70.

On the Fig. 16, experimental spectra of QFF-sepharose-Bs-SL at 1º C and 35 % sucrose in solution, (heavy line), are shown along with simulated ones, demonstrating individual contributions from two SL states. The final spectrum (weighted sum of two states) is shown on Fig.15

5.3. Fast motion analysis by Molecular Dynamics

Method of Molecular Dynamics (MD) has become a very powerful and versatile tool for research connected with protein dynamics. Significant growth of hardware computational capabilities recently makes it possible to incorporate MD simulations into many analysis workflows involving experimental data collected using modern physical methods. It is successfully used in conjunction with nuclear magnetic resonance (NMR) for elucidation of protein structure and dynamics. The spin labelling and EPR may provide similar type of information, in form of order parameters (although on different time scale). Conventional MD method itself is completely theoretical (that is, pure calculation with no use of experimental data regarding particular system), but it is attractive as it provides detailed information on the system (macromolecule) dynamics. As a result of MD run, one obtains a trajectory with the time course of all atomic coordinates of the system. Modern software and visualization hardware allows viewing the resulted trajectory, and one can see a visual representation of the molecular system in time, which makes this technique a powerful tool for analyzing the dynamic properties of macromolecules. The trajectory obtained from MD enables one to calculate virtually any property of the system. We suggest it is very attractive to join the MD method with EPR techniques, once the latter provide experimental data to be guidance for designing MD runs, and/or verifying its results. On the other side, MD calculations may help in interpretation of EPR spectra.

MD method was used here to study the internal structural and dynamical properties of spin-labeled Barstar and its complex with Barnase. It was already mentioned in section 2, that EPR spectrum gives 'digest' type of information about the system dynamics. Most efforts to bridge MD and EPR techniques are now focused on attempt to simulate EPR spectrum from MD trajectory. We adopt different approach, similar to one used in NMR, based on order parameters. As a quantitative estimate for comparing the results of the two methods we used parameter S, which characterizes the angular reorientations of the attached spin label and the spatial constraints of the immediate protein environment. This order parameter, easily and unambiguously determined from the temperature and viscosity dependencies, can be calculated from the MD trajectory. The order parameter was calculated in two ways from the MD trajectories: the McConnell's method (axially symmetrical case), and the method of Model Free approach.

The initial structure of barstar molecules with a resolution of 2.8 A (PDB id 1A19) was obtained from the database Protein Data Bank. The structure was prepared for MD, Cys82 residue was mutated to spin-labeled cystein. New residue SLHG was added to parameter files used for simulation; the parameters were combined from (Stendardo E *et al*, 2010), CHARMM-cgenff and UFF, and optimized to reproduce average geometry of spin label known from X-Ray (Shapiro A. B. et al., 1979)). Explicit solvent model was used, with TIP3 water. After equilibration (NPT ensemble with Langevin method for temperature control), a series of annealing steps has been performed. Annealing ended up in clearly observable two kinds of distinct dynamical behavior of spin label, more and less ordered. For them, the production runs of lengths of 20 ns at 330 K, and 10 ns at 300 K, were performed.

The initial structure of the complex barnase-barstar (PDB id 1AY7) was also obtained from the Protein Data Bank. Spin-labeled complex was reconstructed by Targeted MD using already built model of labeled Barstar. Both dynamical states of spin label attached to Barstar, discovered by annealing, upon virtual Barnase 'binding', yielded trajectory with disordered motional state absent. Trajectories of length of 20 ns at 330 K and 10 ns at 300 K were calculated, and used for determination of order parameters.

All calculations were performed using the software packages NAMD & VMD (Phillips J.C., *et al*, 2005; Humphrey W., *et al*, 1996).

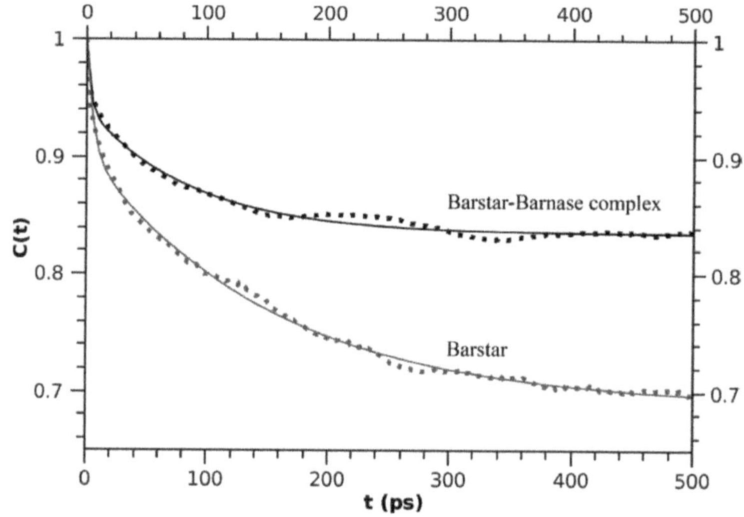

Figure 17. Autocorrelation function calculated from MD trajectories (squares) of Barstar (gray) and its complex with Barnase (black), and corresponding two-exponential model fits (line).

It is straightforward to compute coordinate autocorrelation function from MD trajectory. Its long time asymptotic limit is referred as generalized order parameter, as it does not rely on azimuthal motion symmetry, which is the case of McConnel's parameter *S*. This method of order parameter definition is commonly used in NMR and referred to as Model Free approach (Lipari, G.; Szabo, 1982; K.K. Frederick, K. A. Sharp., 2008). Two-exponential decomposition of autocorrelation function obtained for free Barstar yielded fast-component order parameter (apparent correlation time 3.2 ns for Bs, or 5.4 for complex) to be significantly different from one for Barstar-Barnase complex (0.89 *vs* 0.97). McConnel's azimuthal order parameters calculated along entire trajectories were lower, $S = 0.67$ for the Bs, and $S = 0.93$ for the BnBs complex, which is consistent with motion anisotropy (azimuthal and generalized order parameters coincide when motion is axially symmetrical). The exact calculation by averaging along the trajectory shown $S = 0.67$ for the free Barstar, and $S = 0.93$ for the complex. This is expected, as all the EPR spectra for Barstar and its complex with Barnase presented above were simulated using non-axial model (but with several order parameters). On the other hand, the value of azimuthal order parameter S is in good agreement, as expected, with the experimental value obtained from the temperature-viscosity dependence of EPR spectra. The detailed trajectory analysis for free Barstar showed that it is non-uniform in sense of spin label motion. Namely, some parts of trajectory demonstrated much less disorder in this motion, and order parameter computed from them appeared to be 0.91, which is in good agreement with McConnel's order parameter determined experimentally by TVD for 2nd state EPR spectra component.

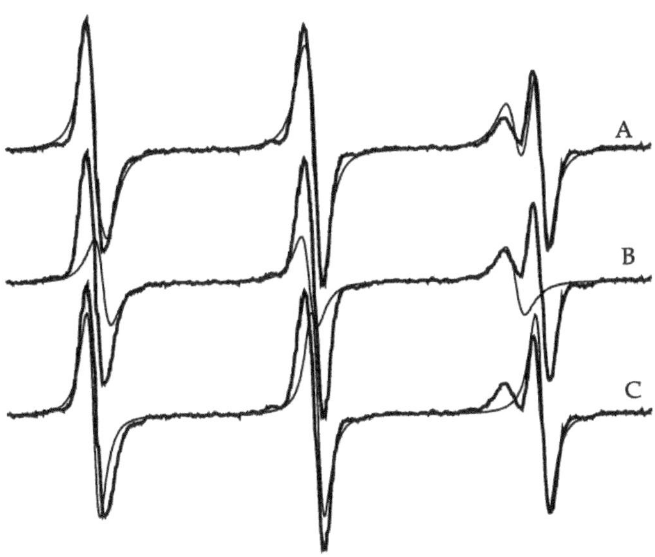

Figure 18. A. Experimental EPR spectrum of TEMPO radical in DPPC at 37.6°C (heavy line) and simulated spectrum (fine line) at X band (microwave frequency 9.15 GHz). All necessary parameters for simulation are summarized in Table 1. Center field $B_0 = 3260$ G, and scan range is 49 G.
B. The simulated EPR spectrum (fine line) of TEMPO radical in lipid phase. Parameters are in Table 1.
C. The simulated EPR spectrum (fine line) of TEMPO radical in aqueous phase. Parameters are in Table 1.

6. Two motion model for EPR spectra simulation of TEMPO radical in membranes

The given approach can naturally be extended to membrane structures. On the Fig. 18, 19 calculated EPR spectra of TEMPO radical in DPPC in X- and W-bands, correspondingly, are shown. The experimental spectrum is taken from the excellent Smirnov's work (Smirnov et. al., 1995). The experimental part of this work is highly valuable, but the interpretation of EPR spectra, from our point of view, should be conducted differently. We propose model, according to which, TEMPO radical, being in lipid phase, subjects to fast anisotropic reorientation, simultaneously experiencing the slow dynamic process motion due to lateral heterogeneity of domain structure in lipid bilayer.

Figures 18 and 19 represent the first example when simulated EPR spectra precisely reproduce experimental EPR spectra both in X and W bands, with exactly the same magnetic and dynamic parameters. The only simulation parameter changed was the frequency of EPR device. This consistency of our approach with multifrequency EPR data strongly argues for approach described here for uniform interpretation of dynamical effects in EPR spectra of spin-labeled samples.

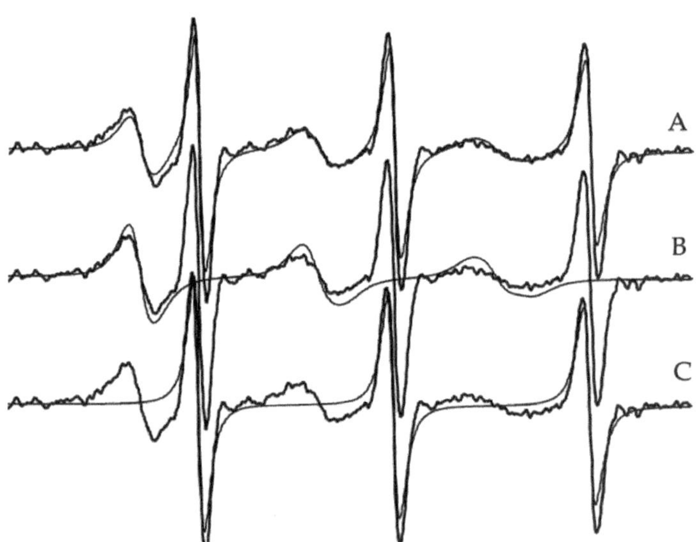

Figure 19. A. Experimental EPR spectrum of TEMPO radical in DPPC at 37.6°C (heavy line) and simulated spectrum (fine line) at W-band (microwave frequency 94.3 GHz), B0 = 33584 G, scan range is 60 G.
B. The simulated EPR spectrum of TEMPO radical in lipid phase. Parameters are in Table 1.
C. The simulated EPR spectrum of TEMPO radical in aqueous phase. Parameters are in Table 1.

Parametrs	Water phase	Lipid phase
g_x	2.0093	2.0098
g_y	2.0063	2.0068
g_z	2.0022	2.00245
A_x	8.0 G	6.5 G
A_y	6.0 G	5.0 G
A_z	37.3 G	35.6 G
α	1	173°
θ	0	53°
ϕ	0	45°
$\bar{g}_X$		2.006477
$\bar{g}_Y$		2.006307
$\bar{g}_Z$		2.006066
$\bar{A}_X$		15.650 G
$\bar{A}_Y$		14.707 G
$\bar{A}_Z$		16.743 G
τ	0.007 ns	30 ns
Line Width	0.8 G	1.0 G
Fraction	0.55	0.45

Table 1. Parameters for simulation of EPR spectra (Fig. 18, 19) of TEMPO radical in DPPC to X and W bands.

7. Conclusion

In the conclusion, we presented a uniform approach to EPR spectra analysis based on two-motion model and temperature and viscosity dependence experiment. It was shown to be consistent with molecular dynamics simulations and multifrequency EPR. The present approach is, in principle, applicable to all kinds of spin-labeled macromolecules and polymers. Although TVD is not limited to X-band, this type of spectrometers is most common and especially easy to maintain. At X-band this method may be utilized to full power, as it allows separating rapid motion of spin label and slow Brownian tumbling of macromolecule in solution by simply changing the temperature and viscosity of solvent by adding sucrose. Where it is impossible to carry out such experiment, for example, in membrane structures with embedded spin probe, the two-motion model still remains applicable, and it is possible to follow the same spectra simulation approach as used in conjunction with temperature and viscosity dependence. The two-motion model for nitroxide spin label dynamics can applied successfully to interpretation EPR spectra of most systems involving covalently bound as well as non-bound spin probes (*eg,* TEMPO radical embedded in membrane). It should be noted, there is one weakness in the proposed method – relatively small (1-4 G) shifts of rather broad peaks has to be measured with sufficient degree of accuracy. Sometimes series of EPR spectra obtained at different conditions may appear very similar, but still they are distinguished by the position of broad outer peaks. With proposed method, it is possible to carry out purposefully the interpretation of EPR spectra of any spin-labeled biological object. Elucidation microscale dynamical features from its EPR spectra represents extremely difficult problem. This inverse problem of EPR spectroscopy still has no unequivocal solution. This is partially due to method limitations, but they may be partially overcome by using TVD approach, which extracts additional information from set of EPR spectra at different conditions. The detailed system dynamics may be obtained from MD simulations and verified against experimental EPR data by using order parameters, as it was shown here on example of Barstar-Barnase complex formation. The uniform method as described here is, therefore, considered to be a very big step forward for the resolution of inverse EPR problem for spin-labeled macromolecules and other biological nanoscale objects.

Author details

Yaroslav Tkachev and Vladimir Timofeev
V.A. Engelhardt Institute of Molecular Biology/RAS, Moscow, Russia

8. References

Dudich I.V., Timofeev V.P., Volkenstein M.V., and Misharin A.Yu. (1977) Measurement of rotational correlation time of macromolecules by the ESR method in the case of a covalently bound spin label. *Molecular Biology (translated from Russian).* 11, 531-538.

Goldman S.A., Bruno G.V., Freed J.H. (1972). Estimating slow motional rotational correlation times for nitroxides by electron spin resonance. *J. Phys. Chem.* 76, 1858-1860.

Halle B. (2009). The physical basis of model-free analysis of NMR relaxation data from proteins and complex fluids. *J. Chem. Phys.* 131, 224507-224521.

Humphrey, W., Dalke, A. and Schulten, K. (1996). "VMD - Visual Molecular Dynamics", *J. Molec. Graphics*, , vol. 14, pp. 33-38.

Frederick K. K., Sharp K. A., Warischalk N., and Wand A.J. (2008). Re-evaluation of the model-free analysis of fast internal motion in proteins using NMR relaxation, *J. Phys. Chem. B* , 112, 12095–12103.

Kuznetsov A.N., Wasserman A.M., Volkov A.Yu., Korst N.N. (1971). Determination of rotational correlation time of nitric oxide radicals in a viscous medium. *Chem. Phys. Letters.* 12, 103-106.

Lipari G., and Szabo A. (1982). Model-free approach to the interpretation of nuclear magnetic resonance relaxation in macromolecules. 1. Theory and range of validity. *J. Am. Chem. Soc.* 104, 4546-4559.

Mason R., and Freed J.H. (1974). Estimating microsecond rotational correlation times from lifetime broadening of nitroxide ESR spectra near the rigid limit. *J. Phys. Chem.* 78, 1321-1323.

McCalley R.C., Shimshick E.J., and McConnell H.M. (1972). The effect of slow rotational motion on paramagnetic resonance spectra. *Chem. Phys. Lett.* 13, 115-119.

Phillips J.C., Rosemary Braun, Wei Wang, James Gumbart, Emad Tajkhorshid, Elizabeth Villa, Christophe Chipot, Robert D. Skeel, Laxmikant Kale, and Klaus Schulten. (2005). Scalable molecular dynamics with NAMD. *Journal of Computational Chemistry*, 26:1781-1802.

Smirnov A.I., Smirnova T.I., and Morse II P.D. (1995). Very High Frequency Electron Paramagnetic Resonance of 2,2,6,6-Tetramethyl-1-Piperidinyloxy in 1,2-Dipalmitoyl-*sn*-Glycero-3-Phosphatidylcholine Liposomes: Partitioning and Molecular Dynamics. *Biophysical J.* 68, 2350-2360.

Shimshick E.J., McConnell H.M. (1972). Rotational correlation time of spin-labeled α-chymotrypsin. *Biochem. Biophys. Res. Commun.* 46, 321-326.

Shapiro A. B., Volodarskii L. B., Krasochka O. N., Avtomyan L. O., and Rozantsev E. G. (1979). *Dokl. Akad. Nauk. SSSR.* 248, 1135-1139.

Slichter C.P., (1980). Principles of Magnetic Resonance. Springer-Verlag. Berlin, Heiddelberg. New York.

Stendardo E, Pedone A., Barone V. (2010). Extension of the AMBER force-field for the study of large nitroxides in condensed phases. *Phys Chem Chem Phys.* 12, 11697-709.

Stone T.J., Buckman T., Nordio P.L., and McConnell H.M. (1965). Spin-labeled biomolecules. *Proc. Nat. Acad. Sci. U.S.* 54, 1010-1017.

Timofeev V. P. (1986). Segmental mobility of poly(U) and the spin label method. *Molecular Biology (translated from Russian).* 20, 557-568.

Timofeev V. P., and Samarianov B. A. (1993). About new universal approach to the EPR spectra simulation of the spin-labeled macromolecules. *Appl. Magn. Reson.* 4, 523-539.

Timofeev V. P., and Samarianov B. A. (1995). Dynamics of macromolecule spin-labeled side chain groups by electron spin resonance spectra simulation. *J. Chem. Soc. PERKIN TRANS.* 2, 2175-2181.

Timofeev V.P., Novikov V.V., Tkachev Y.V., Balandin T.G., Makarov A.A., Deyev S.M. (2008). Spin label method reveals barnase-barstar interaction: a temperature and viscosity dependence approach. *J. Biomol. Struct. Dyn.* 25, 525-534.

Tkachev Ya.V., and Timofeev V.P. (2008). Fast Motion Models Help Separating Dynamic Effects on EPR Spectra of Spin-Labeled Macromolecules. *5th International Conference of Nitroxide Radicals*, Ancona , Italy, September 7-11.

Tkachev Ya.V. (2009). The general form of fast oscillation model and its consequences for nitroxide EPR spectra interpretation in biological systems. *EUROMAR Magnetic Resonance Conference*, Gothenburg, Sweden, July 5-9.

Tkachev Ya.V. (2010). Selective modelling of effects of a fast rotational motion nitroxide in EPR spectra of spin-labeled macromolecules. *PhD Thesis Dissertation*. Moscow.

Section 3

Applications

Chapter 11

In vivo Spectroscopy and Imaging of Nitroxide Probes

Valery V. Khramtsov

Additional information is available at the end of the chapter

1. Introduction

Magnetic resonance spectroscopic and imaging techniques are methods of choice for *in vivo* applications in animals and humans due to sufficient depth of microwave penetration in living tissues. Nuclear magnetic resonance, NMR, and its imaging modality, MRI, have found numerous biomedical and clinical applications but still suffer from limited functional resolution. Low intrinsic NMR sensitivity and overlap of the various endogenous NMR signals limit functional *in vivo* NMR and MRI applications beyond anatomical resolution. Electron paramagnetic resonance, EPR, has the advantage over NMR in functional specificity due to the absence of endogenous EPR signals but requires stable *in vivo* exogenous paramagnetic probes. Nitroxyl radicals, NRs, represent the most diverse class of stable organic radicals varying in stability, spectral properties and functionality which have been successfully used in numerous EPR spectroscopic and imaging applications. Synthesis of stable organic NRs (Lebedev & Kayanovskii, 1959; Neiman et al., 1962), with an unpaired electron localized at a sterically protected NO group ($\rho_0^\pi \approx 0.6$; $\rho_N^\pi \approx 0.4$), revolutionized numerous areas of EPR applications. Half a century of continuous progress in nitroxide chemistry has resulted in the design of specific NRs for spin labeling (Berliner, 1998), site-directed spin labeling (Hubbell et al., 2000), EPR oximetry (Halpern et al., 1994; Swartz, 2004), pH (Khramtsov et al., 1982, 2005), thiols (Khramtsov et al., 1989, 1997), redox (Swartz et al., 2007) and NO measurements (Akaike et al., 1993; Joseph et al., 1993; Woldman et al., 1994). Over the past decade low-field L-band (1.2 GHz) EPR spectrometers and imagers have become commercially available allowing for functional EPR measurements in isolated organs (Komarov et al., 2012) and small animals such as mice (Kuppusamy et al., 2002; Bobko et al., 2012). Moreover, continuous waves (CW) EPR (Halpern et al., 1994; He et al., 2002) and pulsed EPR (Yasui et al., 2010) instruments with lower radio frequencies (RF) down to 250 MHz have been constructed allowing for spectral-spatial functional imaging in larger animals, and potentially in humans. Recently, a functional proton-electron double-resonance imaging (PEDRI) approach (Khramtsov et al., 2010; Efimova et al., 2011) based on MRI detection with EPR excitation of paramagnetic probes at pre-selected EPR fields/frequencies, has been developed. The latter approach inherits high resolution, fast image acquisition and the availability of anatomical information from MRI techniques. This chapter overviews the recent applications of nitroxide probes for functional spectroscopy and imaging of living tissues.

2. Nitroxides as functional EPR probes

2.1. EPR-based oximetric applications of the nitroxides

EPR oximetry is one of the most promising and rapidly developing techniques for measurement of oxygen in living tissues (Swartz, 2004). An advantage of EPR oximetry is that it is based on pure physical interaction, Heisenberg spin exchange, between paramagnetic molecules of probe and oxygen, and does not interfere with oxygen metabolism, therefore providing a basis for noninvasive oxygen measurements in biological systems. Initially, the paramagnetic probes for EPR oximetry were almost exclusively nitroxides (Backer et al., 1977; Lai et al., 1982; Froncisz et al., 1985; Bacic et al., 1989; Chan et al., 1989; Hyde & Subszynski, 1989). Both the longitudinal (T_1) and transverse (T_2) relaxation times of the NRs can be affected by collisions of the nitroxide with dissolved oxygen. The first observation of oxygen-induced NR broadening in various solvents was reported by Povich (Povich, 1975). Two years later the T_2 oximetry method was proposed and applied to follow mitochondrial respiration in samples containing about 100 liver cells (Backer et al., 1977). T_1-sensitive EPR oximetry was developed by Hyde et al (Froncisz et al., 1985; Hyde et al., 1990). T_2 oximetric applications using NRs are preferred for biological applications in low viscosity solutions because T_2 is close to the collision rate resulting in optimal EPR sensitivity. On the other hand, T_1 oximetry may have an advantage for applications in highly viscous environments and for nitroxide-labeled macromolecules due to the fact that the nitroxide $T_1 >> T_2$. T_1 oximetric methods include pulsed saturation-recovery, continuous wave saturation and rapid passage displays (Hyde & Subszynski, 1989).

The typical effect of oxygen on the low-field component of the EPR spectrum of NRs is shown in Figure 1 for the five-membered ring pyrroline CTPO nitroxide. In the absence of oxygen each spectral component of the triplet spectrum shows additional superhyperfine splittings with 12 protons of four methyl groups and proton at carbon C-4 of the heterocycle. An increase in oxygen concentration results first in the broadening of the lines of the superhyperfine structure followed by the increase of the enveloped EPR linewidth, ΔH_{pp}.

Note that most six- and five-membered ring NRs with larger numbers of protons in the radical heterocycle do not reveal superhyperfine structure even in the absence of oxygen,

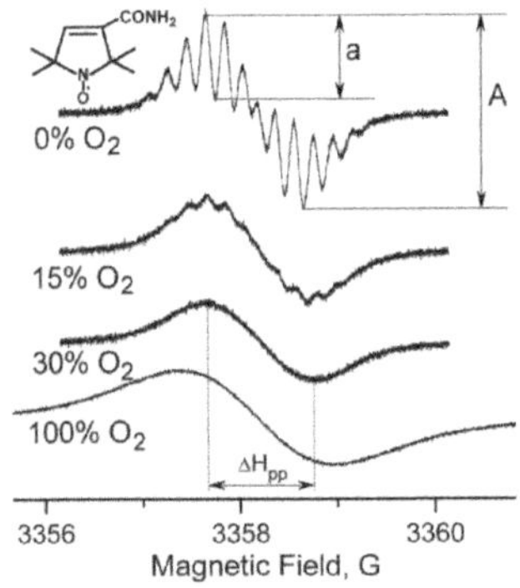

Figure 1. Chemical structure of the CTPO nitroxide and the EPR spectra of its low-field spectral component measured in aqueous solutions at various oxygen concentrations at 37 ºC.

e.g. TEMPO radical of the piperidine type (see Fig. 2). The oxygen-induced line-broadening effects vary slightly with the NR structure and are about $\Delta H_{PP}/\Delta[O_2] \approx 500$ mG/100% oxygen or 450 mG/mM of oxygen (see Fig. 1 and 2).

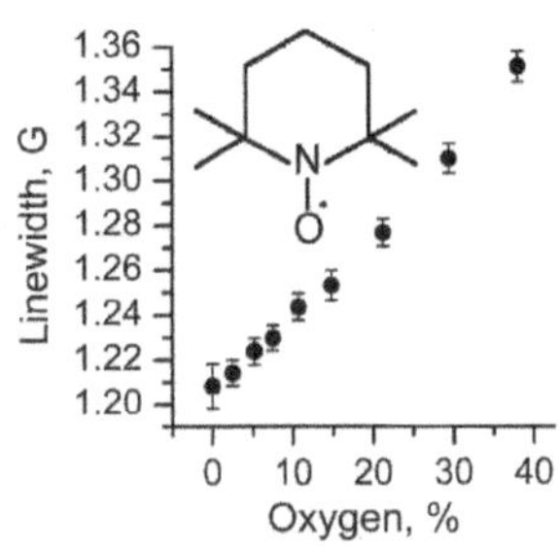

Figure 2. Chemical structure of TEMPO NR and oxygen-induced broadening of its low-field spectral line measured at 24 ºC. Linewidth at 100% oxygen is equal to 1.70 G (not shown).

The well-developed chemistry of the NRs (Volodarsky et al., 1994; Hideg et al., 2005; Karoui et al., 2010) allows manipulation of their structure and properties, including charge, presence of hydrophilic or hydrophobic groups, and ability to be targeted. Small neutral NRs normally easily penetrate cellular membranes and are equally distributed throughout the intracellular and extracellular environments. Conversely, charged NRs will not cross the plasma membrane and thus can be used to measure oxygen concentrations in the extracellular compartment (Glockner et al., 1993; Baker et al., 1997). NRs encapsulated in liposomes or linked to carrier molecules can be used to achieve organ or tissue selectivity (Gallez et al., 1993). In general, NRs have low toxicity and can be administered to an animal by infusion or by intraperitoneal, intravenous or intratissue injection. To protect NRs against reduction and enhance their oxygen sensitivity Liu et al. (Liu et al., 1994) encapsulated NRs in proteinaceous microspheres filled with an organic liquid. The authors used encapsulated NR to measure the changes in oxygen concentration *in vivo* for 70 min after intravenous injection of the microspheres into a mouse. Potential concern of this approach is the leakage, as well as release of the organic liquid from the microspheres when they are eventually metabolized, resulting in possible toxicity to the tissue.

Figure 3 demonstrates an example of the L-band EPR oximetry application using TEMPO nitroxide (see Fig.2 for the structure) for monitoring ischemia-induced oxygen depletion in isolated rat hearts (Zweier & Kuppusamy, 1988; Kuppusamy et al., 1994). The spectra showed a gradual decrease in the linewidth over the duration of ischemia approaching the linewidth observed in the absence of oxygen. Oxygen consumption observed in cardiopleged hearts subjected to global ischemia (Kuppusamy et al., 1994) was significantly slower than in noncardiopleged hearts (Zweier & Kuppusamy, 1988). This would be expected since contractile function of cardiopleged hearts is arrested. The oxygen consumption data for cardiopleged heart were obtained from spectral-spatial EPRI which allowed for spatially resolved oxygen mapping but required long acquisition times (16 min in Fig. 3a). PEDRI provides a faster alternative for oxygen mapping. Application of PEDRI for myocardium oxygen mapping was first demonstrated in perfused sheep heart using a high concentration, 4 mM, of Fremy's salt nitroxide (Grucker & Chambron, 1993).

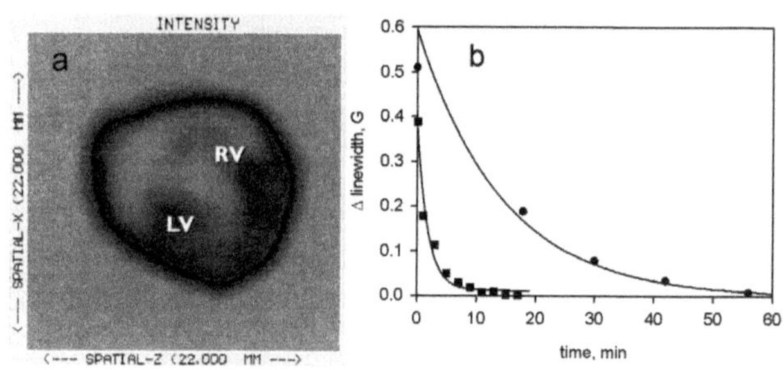

Figure 3. (a). Cross-sectional transverse 2D-spatial L-band EPR image of the isolated rat heart preloaded with 1 mM TEMPO (field of view, 22×22 mm²; acquisition time, 16 min). The structure of the left ventricle (LV) and right ventricle (RV) are observed. (b). Changes in L-band EPR linewidth of TEMPO spin label induced by oxygen depletion during ischemia in cardiopleged (•) and noncardiopleged (■) hearts. Adapted from (Zweier & Kuppusamy, 1988; Kuppusamy et al., 1994), copyright National Academy of Sciences, U.S.A.

Low sensitivity of the NRs to oxygen tensions below 40 mmHg and concentration-induced line broadening complicate quantitative oximetric applications of the NRs. Typically, the self-broadening effect is about 100-200 mG/mM of NRs, and may interfere with accurate oxygen measurements at NR concentrations above 100 μM. Application of deuterated NRs provides significant enhancement in sensitivity to low oxygen concentrations due to the narrowing linewidth, ΔH_{pp} (Kuppusamy et al., 1994; Gallez et al., 1996a; Velan et al., 2000). Alternatively, measurement of "the depth of resolution" of the NRs superhyperfine structure (e.g. parameter a/A for the CTPO shown in the Fig.1) provides similar or even better sensitivity to oxygen than perdeuterated NRs, and has been used for oxygen measurements in cellular and enzymatic systems (Backer et al., 1977; Sarna et al., 1980; Lai et al., 1982). However, both approaches still suffer from concentration-induced line broadening which is difficult to disentangle from oxygen-induced broadening. To overcome this problem, Halpern et al. proposed selectively deuterated nitroxide with only one hydrogen hyperfine splitting, mHCTPO (see Scheme 1) (Halpern et al., 1994). Increasing NR concentration but not oxygen results in narrowing of hydrogen hyperfine splitting. This allows for discrimination between oxygen- and concentration-induced line broadening, and therefore, for quantitative oxygen detection in living tissues with sensitivity about 10 mmHg. The authors demonstrated an efficiency of low-field 250 MHz EPR oximetry in combination with a mHCTPO probe to report oxygen concentration in murine FSa and NFSa fibrosarcomas 7 cm deep in tissues of a living animal. An oximetric 2D (1spectral/1 spatial) spectral image of the tumor was also obtained allowing, in principle, direct assessment of tumor hypoxia to determine the usefulness of radiation and chemotherapy adjuvants directed to hypoxic cell compartments. In another application, cell-permeable perdeuterated NR, PDT (see Scheme 1), was applied to measure oxygenation level in radiation-induced fibrosarcoma (RIF-1) in mice (Kuppusamy et al., 1998). The data showed a 3-fold lower level of oxygenation of the tumor tissue compared with that of the normal muscle.

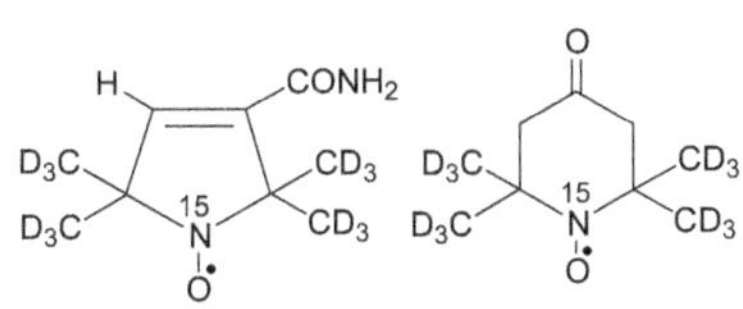

Scheme 1. Chemical structures of the NRs referred to in the text: 4-protio-3-carbamoyl-2,2,5,5-tetraperdeuteromethyl-3-pyrrolinyl-1-^{15}N-oxy (mHCTPO) and 4-oxo-2,2,6,6-tetramethylpiperidine-d_{16}-1-^{15}N-oxyl (perdeuterated tempone, PDT).

Distribution of the NRs in living tissue can be measured using EPRI techniques as demonstrated in Fig. 3a. Moreover, spectral–spatial imaging can be performed which contains a complete spectral profile, as a function of field, at each spatial voxel element. Because the spatial and spectral dimensions are fully separable, information about local linewidth, and hence local oxygen concentration, can be derived independently from local spin density (Velan et al., 2000; Kuppusamy & Zweier, 2004). To date most NRs imaging applications have been performed using CW EPRI. Nitroxide imaging by time-domain pulsed EPR had not been attempted until recently because of the short spin-spin relaxation times, typically under 1 microsecond. Nevertheles, recent advances in RF electronics have enabled the pulses on the order of 10-50 ns (Murugesan et al., 1997) and improved spectrometer recovery times, therefore providing an opportunity for *in vivo* pulsed EPR imaging of the nitroxides. Figure 4 shows oxygen mapping of a tumor-bearing anesthetized mouse obtained by pulsed EPRI using PDT nitroxide (Scheme 1) (Hyodo et al., 2009). Strong heterogeneity of tumor oxygenation and significant hypoxic regions (<10 mmHg) characteristic of tumor pathogenesis were observed demonstrating the capability of *in vivo* functional nitroxide imaging using time-domain pulsed EPR.

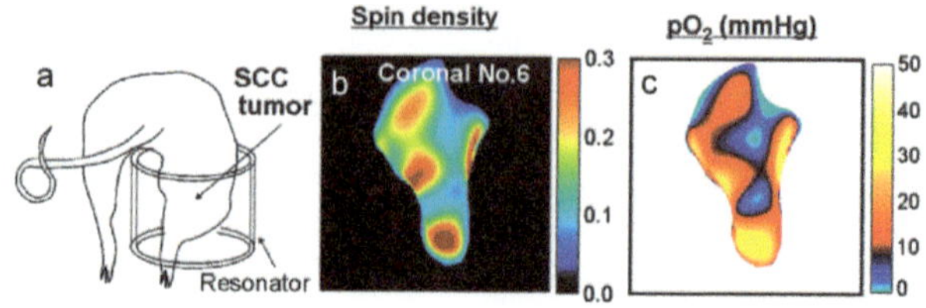

Figure 4. 3D-pulsed EPR imaging of tumor oxygenation using PDT nitroxide. (a) Schematic illustration of tumor-implanted mouse setting in the smaller parallel coil resonator. (b) Coronal slice views selected from 3D-spin density map of the nitroxide in tumor bearing leg, and (c) corresponding oxygen image calculated from line width distribution of PDT. Adapted from (Hyodo et al., 2009), copyright 2009, Elsevier Inc.

EPR applications of other oxygen-sensitive paramagnetic materials include soluble trityl radicals (Krishna et al., 2002; Bobko et al., 2009) and particulate probes such as lithium phthalocyanine particles (Liu et al., 1993; Presley et al., 2006) and carbonaceous materials (chars, coals, carbon blacks) (Clarkson et al., 1998). It should be noted that particulate probes such as lithium phthalocyanine and synthetic char are suitable for measurements of oxygen partial pressure in place of implantation whereas soluble probes such as nitroxides and trityl compounds more suitable for imaging experiments.

2.2. *In vivo* evaluation of redox state using nitroxide probes

Regulation of tissue redox status is important for maintenance of normal physiological conditions in the living body. Disruption of redox homoeostasis may lead to oxidative stress and can induce many pathological conditions such as cancer, neurological disorders and ageing. The intracellular thiols, and particularly the redox couple of glutathione, GSH, and its disulfide form, GSSG, are considered the major regulators of the intracellular redox state (Schafer & Buettner, 2001). Therefore, noninvasive spectroscopic evaluation and imaging of tissue redox status and, in particularly, GSH redox status, could have clinical applications (Kuppusamy & Krishna, 2002; He et al., 2004; Swartz et al., 2007; Hyodo et al., 2008; Ojha et al., 2008; Roshchupkina et al., 2008; Bobko et al., 2012).

2.2.1. Nitroxides as redox-sensitive paramagnetic probes

NRs introduced into biologically relevant systems are predominantly observed in the radical and hydroxylamine forms and exist in redox equilibrium as shown in Scheme 2. The reduction of NRs to EPR-silent hydroxylamines in many cases is an unfavorable factor that limits their applications. On the other hand, the EPR-measured rates of NR reduction depend on overall tissue redox status allowing for the differentiation of normal and pathological states (Kuppusamy et al., 2002; Ojha et al., 2008; Bobko et al., 2012). The reduction of cell-permeable

NRs to hydroxylamines is primarily intracellular and, therefore to a great extent is determined by intracellular redox status. On the other hand, cell-impermeable nitroxides allow for detection of reducing capacity of extracellular microenvironment.

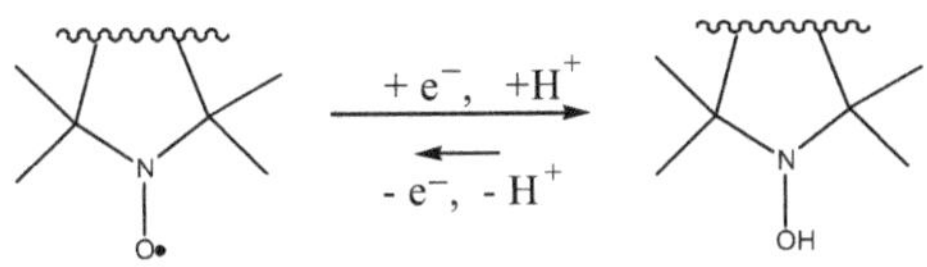

Scheme 2. Illustration of the nitroxide/hydroxylamine redox couple. In general, for most biologically relevant samples one-electron reduction of the nitroxides prevails and the equilibrium is strongly shifted towards the hydroxylamine (Kocherginsky & Swartz, 1995).

Cancer and ischemic heart disease, two leading causes of mortality in the United States, represent pathologic conditions with compromised redox state. Figure 5 demonstrates the application of cell-permeable CP nitroxide (see Scheme 3) for EPR *in vivo* assessment of myocardial tissue redox status in a mouse model of regional ischemia (Zhu et al., 2007). Authors observed about two-fold ischemia-induced increase of the overall myocardial reducing capacity in the area at risk. Conversely, the CP reduction rates in reperfused hearts were about 30% lower than in pre-ischemic hearts. The latter correlated with oxygen overshoot in post-ischemic hearts. Both lower redox potential and higher oxygen may result in elevated production of reactive oxygen species and myocardial damage. Authors observed that ischemic preconditioning normalizes post-ischemic myocardial oxygenation and tissue redox status which may contribute to the mechanisms of myocardial protection.

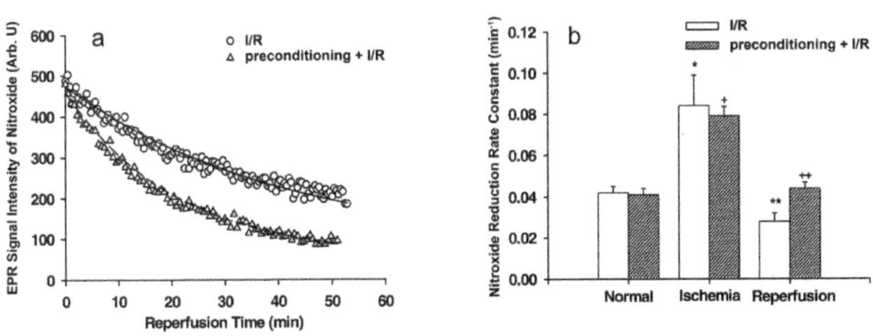

Figure 5. (a). *In vivo* EPR measurements of myocardial tissue redox status in a mouse model of 30 min regional ischemia and 60 min reperfusion (I/R). 5 μl of a 10 mM solution of CP probe was injected in the area at risk. EPR signal intensity decays are shown for the control (○) and preconditioned (Δ) groups measured during reperfusion. (b) Reduction rate constants of the CP nitroxide before, during, and after ischemia in control I/R and preconditioned I/R mice. Adapted from (Zhu et al., 2007), copyright 2007, the American Physiological Society.

Tumor cells are known to generate significant alterations in the redox status. This status is an important determinant in the response of the tumor to certain chemotherapeutic agents, radiation, and bioreductive hypoxic cell cytotoxins (Cook et al., 2004). Figure 6 illustrates the application of cell-impermeable RSG nitroxide (see Scheme 3) designed to access both

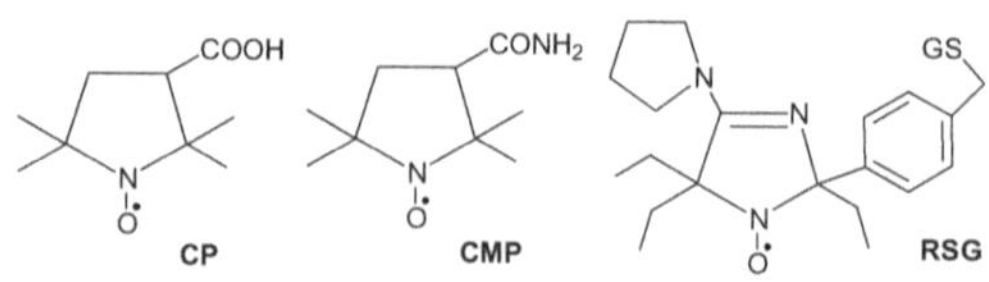

Scheme 3. Representative structures of the NRs used for *in vivo* evaluation of tissue redox status, cell-permeable pyrrolidine NRs, CP and CMP, and cell-impermeable imidazoline nitroxide, RSG, bound to hydrophilic tripeptide, glutathione (GSH).

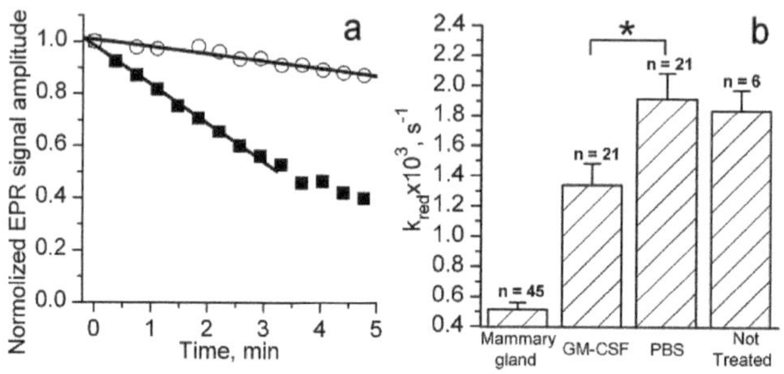

Figure 6. (a) EPR signal decay of the NR probe RSG measured *in vivo* after probe injection (10 μL, 10 mM) in mammary gland (o) and mammary tumor (■) of PyMT mouse. (b) The reduction rates of RSG nitroxide, k_{red}, in extracellular media of normal mammary glands and variously-treated tumors (GM-CSF: Granulocyte Macrophage Colony-Stimulating Factor; *$p<0.02$). From ref. (Bobko et al., 2012) with permission from John Wiley & Sons, Ltd.

extracellular tumor redox and pH (Bobko et al., 2012), latter measurements being discussed in the Section 2.3. The extracellular reducing capacity of tumor tissue was found to be about four-fold higher compared with that of the normal mammary gland (Fig. 6). Inhibition of tumor angiogenesis with granulocyte-macrophage colony-stimulating factor (GM-CSF) resulted in significant "normalization" of the tumor redox status (Fig. 6b) as well as correlating with a decrease in tumor growth and metastases (Eubank et al., 2009).

EPR studies using cell-permeable nitroxides support significantly higher reducing capacity of the tumor tissues. Figure 7 demonstrates L-band EPRI application for redox mapping of the tumor in living mice (Kuppusamy & Krishna, 2002; Kuppusamy et al., 2002) using CMP nitroxide. A significant decrease in CMP reduction rates after treatment with a GSH depleting agent clearly demonstrates a central role of GSH in tissue redox homeostasis. Note that in general, appreciable chemical reduction of the NRs by GSH is not observed (Finkelstein et al., 1984; Glebska et al., 2003; Bobko et al., 2007b). However GSH significantly contributes to the reduction of the NRs indirectly by acting as a secondary source of reducing equivalents (Takeshita et al., 1999; Kuppusamy et al., 2002; Bobko et al., 2007b).

Application of the NRs as redox-sensitive contrast agent for EPRI, MRI and PEDRI (also termed as Overhauser magnetic resonance imaging, OMRI) was recently discussed (Hyodo et al., 2008). In summary, NRs provide a useful tool for quantitative assessment and mapping of the redox environment in living tissues.

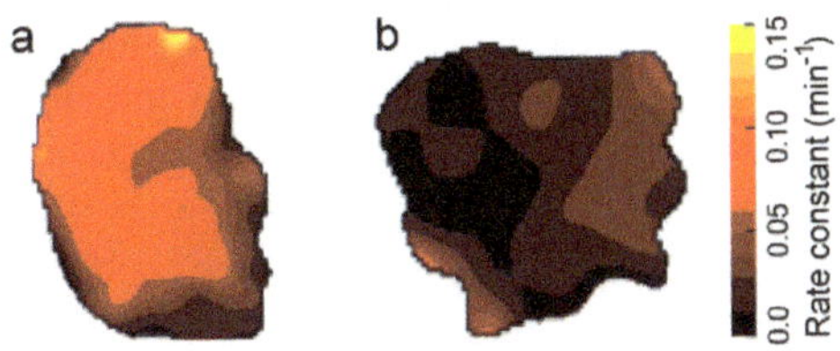

Figure 7. Redox mapping of untreated **(a)** and BSO-treated **(b)** mice tumors. 2-D spatial map of pseudo-first order rate constants of the CMP reduction in RIF-1 tumor implanted in the upper leg of a mouse was obtained using *in vivo* L-band EPRI. The rate of reduction was significantly slower in tumor of mouse treated for 6 hours with 2.25 mmol/kg of L-buthionine-S,R-sulfoximine (BSO), a GSH depleting agent. Adapted from (Kuppusamy et al., 2002), copyright 2002, American Association for Cancer Research.

2.2.2. *Nitroxides as glutathione-sensitive paramagnetic probes*

The redox couple of glutathione, GSH, and its disulfide form, GSSG, is considered the major regulator of the intracellular redox state (Schafer & Buettner, 2001). Therefore, GSH redox status *in vivo* might be a useful indicator of disease risk in humans. The use of fluorometric, photometric, and chromatographic assays for GSH measurements is mostly limited to *in vitro* or *ex vivo* systems due to the invasiveness involved and the limited depth of penetration of light (Ellman, 1959; Tietze, 1969; Yamashita & Rabenstein, 1989; Patsoukis & Georgiou, 2005). NMR based approaches normally detect endogenous GSH but have somewhat low sensitivity and the spectral assignment is complicated due to the overlapping of numerous resonances (Livesey et al., 1992; Willis & Schleich, 1995; Terpstra et al., 2003; Potapenko et al., 2005).

EPR spectroscopy in combination with thiol-specific nitroxides allows for determination of the accessible thiol groups in various biological macromolecules, such as human plasma low-density lipoproteins (Kveder et al., 2003) and erythrocyte membranes (Soszynski & Bartosz, 1997). This approach normally requires purification of the sample from the unbound label and can not be used *in vivo*. Moreover, application of thiol-specific mononitroxides for EPR measurement of glutathione or cysteine is hardly possible due to insignificant EPR spectral changes of the label upon binding to low-molecular-weight compounds. The latter limitation was overcome by the development of paramagnetic analogs of Ellman's reagent, namely disulfide nitroxyl biradicals, RSSR, shown in Scheme 4 (Khramtsov et al., 1989; 1997; Roshchupkina et al., 2008).

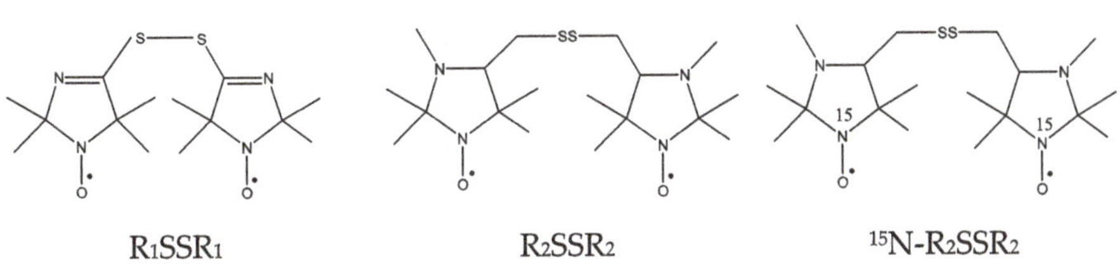

Scheme 4. Structures of the disulfide nitroxyl biradicals, R_1SSR_1, R_2SSR_2 and ^{15}N-R_2SSR_2.

EPR spectra of the RSSR labels are significantly affected by intra-molecular spin exchange between two radical fragments resulting in appearance of "biradical" spectral components in addition to the conventional triplet spectral pattern of the mononitroxide. Figure 8a demonstrates the typical changes of the EPR spectra of R_2SSR_2 upon GSH addition consistent with splitting of the biradical disulfide bond and formation of two monoradicals due to the reaction of thiol-disulfide exchange, RSSR + GSH → RSSG + RSH. Figure 8b shows the corresponding kinetics of the monoradical component increase allowing for the calculation of the rate constant. The RSSR labels being lipophilic compounds diffuse easily across cellular membranes where they reacts with intracellular GSH providing a reliable EPR approach for determination of GSH *in vitro* (Weiner, 1995) and *in vivo* (Roshchupkina et al., 2008; Bobko et al., 2012). The approach is based on a dominant contribution of the GSH in the intracellular pool of fast-reacting thiols.

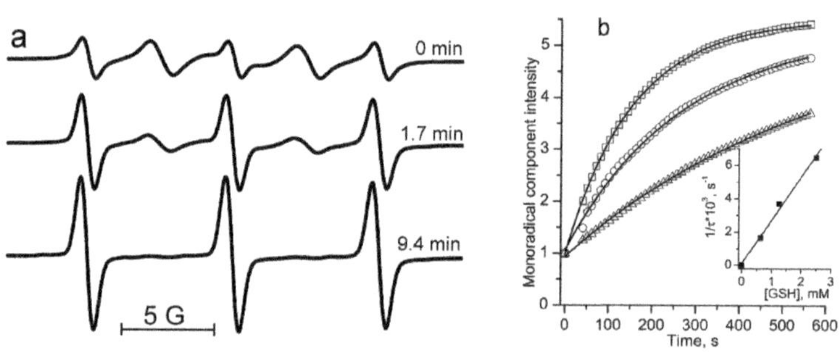

Figure 8. (a). The EPR spectra of 100 μM R_2SSR_2 measured at various time points after incubation with 2.5 mM GSH in 0.1 M Na-phosphate buffer, pH 7.2, at 34°C. (b). The kinetics of the increase of amplitude of low-field monoradical component of the EPR spectrum after addition of GSH: 0.625 mM (Δ), 1.25 mM (○) and 2.5 mM (□). Lines represent the best fit of the experimental kinetics to the monoexponents. Insert: the dependence of inverse time constant of the exponential kinetics, $1/\tau$, on [GSH]. The linear regression ($1/\tau = k_{obs} \times$ [GSH]) provides the observed rate constant value of the reaction between GSH and R_2SSR_2, k_{obs}(pH 7.2, 34°C) = 2.8 ± 0.2 M^{-1} s^{-1}. From the ref. (Bobko et al., 2012) with permission from John Wiley & Sons, Ltd.

The convenient time window of the reaction of the imidazolidine RSSR labels, R_2SSR_2 (Khramtsov et al., 1997) and ^{15}N-R_2SSR_2 (Roshchupkina et al., 2008) allows for quantitative measurement of GSH content by the analysis of their EPR spectral change kinetics. The kinetics approach looses the attractive simplicity of the static EPR measurements performed *in vitro* using fast-reacting R_1SSR_1 label but gains a decisive advantage by using low concentrations of the label compared to the GSH content. This important advantage makes the approach less invasive and, therefore, applicable *in vivo*. Recently we demonstrated their applicability for *in vivo* GSH detection in ovarian tumor allograft (Roshchupkina et al., 2008) and breast cancer tumors (Bobko et al., 2012) in mice using L-band EPR spectroscopy.

Figure 9a shows typical *in vivo* L-band EPR spectra measured immediately after R_2SSR_2 probe intratumoral injection and 150 s after injection at the time point of maximal monoradical signal amplitude. Comparison of the integral intensities of the spectra shows less than 10% loss of

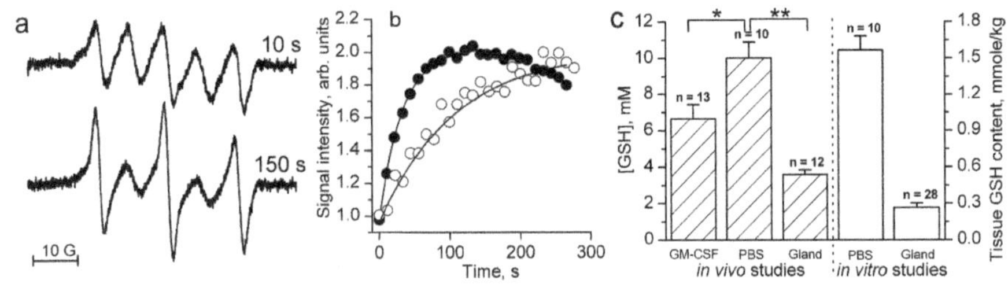

Figure 9. (a) L-band EPR spectra of the R_2SSR_2 probe measured *in vivo* in mammary tumor of female FVB/N mice 10 s and 150 s after probe injection. (b) Kinetics of the monoradical peak intensity change measured in tumor (•) and normal mammary gland (o). The solid lines are the fits of the initial part of the kinetics by the monoexponent $I_m(t) = I_m^{max} - (I_m^{max} - I_m^{t=0})\exp(-k_{obs}[GSH]\cdot t)$ supposing $k_{obs}=2.8\ M^{-1}\ s^{-1}$ (see Fig. 8b) and yielding [GSH] =10.7 mM and 3.3 mM for the tumor and mammary gland, correspondingly. (c) Intracellular [GSH] measured using R_2SSR_2 probe *in vivo*, and tissue GSH contents measured *in vitro* in normal mammary glands and mammary tumors of female FVB/N mice (*$p<0.01$, **$p<0.002$). From (Bobko et al., 2012) with permission from John Wiley & Sons, Ltd.

the total EPR signal, supporting an insignificant contribution of the reduction and justifying approximation of the initial exponential signal increase by the reaction of R_2SSR_2 with GSH. Figure 9b shows the typical kinetics of the monoradical spectral peak intensity change measured in mammary tumor and normal mammary gland of female FVB/N mice. The analysis of the kinetics calculates the GSH concentration to be three-fold higher in tumor compared with that in mammary gland. Figure 9c compares GSH contents measured *in vivo* with those measured in tissue homogenates from the sacrificed animals. While quantitative comparison of *in vitro* and *in vivo* data is difficult and requires knowledge of the relative intracellular aqueous volume of the total tissue volume (Vaupel, 2009), qualitatively *in vitro* measurements confirm significantly higher GSH levels in tumors. The observed decrease of intracellular GSH measured *in vivo* upon treatment with GM-CSF correlates with decrease of extracellular reducing capacity (cf. figs. 6b and 9c) therefore supporting interrelationship between extracellular and intracellular tumor redox states.

Intracellular GSH has been shown to be one of the major factors modulating tumor response to a variety of commonly used anti-neoplastic agents, such as cisplatin (Rabik & Dolan, 2007). The ^{15}N-R_2SSR_2 probe has been used to compare *in vivo* concentrations of GSH in cisplatin-resistant and cisplatin-sensitive tumors in ovarian tumor-bearing mice (Roshchupkina et al., 2008). The use of ^{15}N-substituted probe resulted in decreasing the number of EPR spectral lines and about a two-fold increase in the signal-to-noise ratio compared with nonsubstituted R_2SSR_2 label which is important for *in vivo* applications. The obtained two-fold higher levels of GSH in cysplatin-resistant tumors are in agreement with previously reported data (Lee et al., 1989). GSH depletion was supposed to re-sensitize drug-resistant ovarian cancer cells to cisplatin. Figure 10 shows EPR imaging of the R_2SSR_2 probe in cisplatin-resistant ovarian xenograft tumors confirming a significantly lower reducing capacity of the tumors in NCX-4040-treated mice compared with untreated animals, apparently due to GSH depletion by NCX-4040 drug (Bratasz et al., 2008).

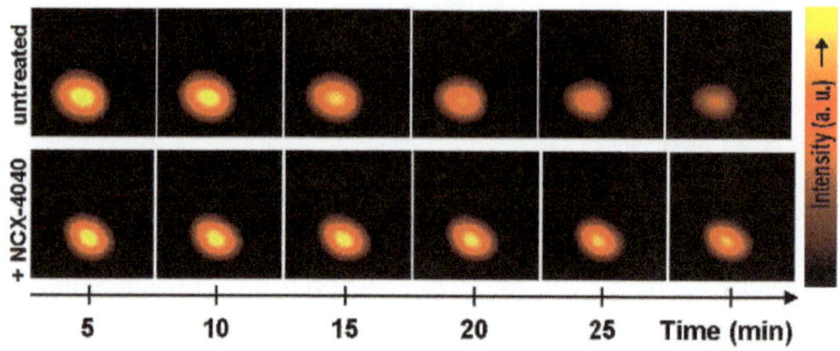

Figure 10. *In vivo* L-band EPR redox images of cisplatin-resistant ovarian xenograft tumors in mice obtained after intrarumoral injection of R_2SSR_2 probe. The images show the probe distribution in the tumor, while the tumor itself is not visible. Tumor treated with a NCX-4040 drug shows a significantly lower capacity to reduce the R_2SSR_2 probe when compared with untreated tumors. Adapted from (Bratasz et al., 2008) with permission.

As can be concluded from figures 9b and 10, the EPR signal changes of the R_2SSR_2 probe show biphasic character: (i) comparatively fast monoradical signal increase due to the reaction with GSH, and (ii) slow signal decay due to bioreduction. A comparatively long EPRI acquisition time of about 5 min did not allow us to resolve the first phase of the kinetics from images given in Figure 10. With the development of faster imaging techniques, e.g. functional PEDRI discussed in the section 2.3.3, paramagnetic disulfide RSSR nitroxides can be used as dual function intracellular GSH and redox imaging probes.

In summary, it seems likely that EPR and EPR-based imaging approaches to visualize redox processes in living tissues, and in particular to assess intracellular GSH, will become increasingly utilized and valuable tools.

2.3. *In vivo* spectroscopy and imaging of pH using nitroxide probes

Spatially and temporarily addressed pH measurements *in vivo* are of considerable clinical relevance due to the critical role pH status plays in physiology and pathophysiology of living organisms. *In vivo* pH assessment using ^{31}P-NMR and inorganic phosphate, P_i, suffers from low intrinsic NMR sensitivity and lack of functional resolution (about 0.2-0.3 pH units) (Gillies et al., 1982; Pietri et al., 2001). It is assumed that the chemical shift of endogenous P_i reflects intracellular pH_i because of the generally higher fraction of intracellular volume, and 2-3-fold higher intracellular Pi concentrations over extracellular ones (2-3 mM and ca. 1 mM, respectively). Because of these problems, exogenous pH probes are being designed for NMR spectroscopy to improve detection of myocardial acidosis (Pietri et al., 2001) and extracellular pH in tumors (Gillies et al., 1994; 2004). Upon application of exogenous probes, low-field EPR spectroscopy has higher sensitivity compared with NMR for the same probe concentration, and reasonable depth of penetration in living tissues.

2.3.1. *pH effect on the EPR spectra of stable nitroxides*

The first pH effect on EPR spectra of stable NRs was observed in strong acids and was related to protonation of the nitroxyl fragment itself (Hoffman & Eames, 1969; Malatesta & Ingold, 1973). Imino nitroxyl radicals (Ullman & Osiecki, 1970) were apparently the first reported stable NRs with spectral sensitivity in physiological range but did not find applications as pH probes due to the complexity of their EPR spectra and rapid reduction to EPR silent products in biological fluids (Woldman et al., 1994; Haseloff et al., 1997; Bobko et al., 2004). The observed pH effects on the NRs of piperidine and pyrrolidine types with ionizable functional groups were impractically small (Hsia & Boggs, 1972; Quintanilha & Mehlhorn, 1978; Nakaie et al., 1981; Mathew & Dodd, 1985; Khramtsov & Weiner, 1988; Saracino et al., 2002) due to the long distance between radical center and ionizable group.

Stable imidazoline and imidazolidine NRs have been proven to be the most useful spin probes for EPR spectroscopy and imaging of pH (Khramtsov et al., 2000, Khramtsov, 2005) due to the large effect of pH on their EPR spectra and large number of structures synthesized. Scheme 5 illustrates the chemical origin of the pH effect on EPR spectra of these types of NR. Protonation results in an EPR-detected difference in hyperfine splitting, a_N, and g-factor ($\Delta a_N \approx$ 1 G and $\Delta g \approx 0.0002$) between R and RH^+ forms (Khramtsov et al., 1982). For NR with an equilibrium constant K_a, the ratio of concentrations of these two forms is described by the Henderson-Hasselbalch equation, $[H^+]=K_a[RH^+]/[R]$, providing the basis for EPR measurements of pH. One of the great strengths of the technique is that it is ratiometric, the pH measurement being independent of the NR concentration but dependent on the ratio, $[RH^+]/[R]$. In general, spectral simulation is required for accurate $[RH^+]/[R]$ determination. In practice, two convenient spectral parameters can be used as pH markers: (i) the ratio of peak intensities of RH^+ and R spectral components resolved upon detection by high-frequency EPR and partly resolved in X-band EPR (9.5 GHz) spectra, and (ii) nitrogen hyperfine splitting, a_N, measured as a distance between unresolved spectral components, and being used almost exclusively as a highly sensitive pH marker in numerous applications. Note that sensitivity of the hyperfine splitting to pH values depends both on EPR frequency and spectrometer settings (e.g., modulation amplitude) and can be optimized (Khramtsov et al., 2004).

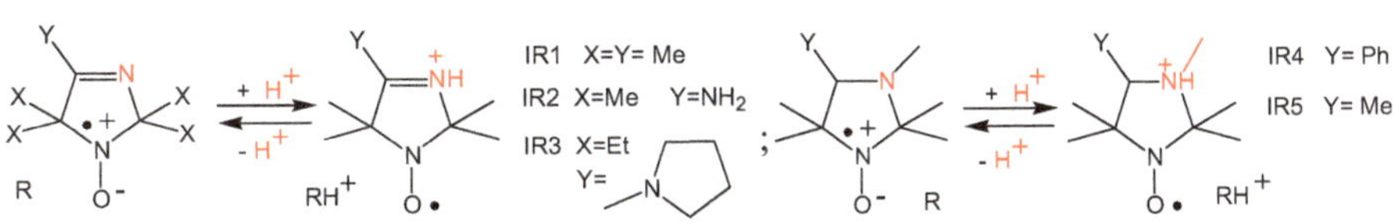

Scheme 5. Reversible protonation of the nitrogen atom N-3 of the imidazoline and imidazolidine NRs, and chemical structures of the IR1-IR5 referred in the Figure 11. Two main resonance structures are shown illustrating higher unpaired electron density on nitrogen atom N-1 in the unprotonated form.

Up to the present a wide variety of pH-sensitive NRs have been developed with different ranges of pH sensitivity, labeling groups, lipophilicity and stability towards bioreduction (Khramtsov & Weiner, 1988; Khramtsov & Volodarsky, 1998; Kirilyuk et al., 2004; 2005; Voinov et al., 2005; Bobko et al., 2012). These spin pH probes, together with low-field EPR-based techniques, offer unique opportunities for non-invasive pH assessments in living animals in compartments with widely varying pH ranges. Figure 11 exemplifies a set of nitroxide pH probes that cover all pH ranges from acidic values observed in stomach and ischemic tissues to alkaline values characteristic of mitochondria. The potential applications are enormous, as tumors and ischemic areas may have acidic pH values compared to surrounding tissues, local areas of infection or inflammation can exhibit specific localized pH reductions allowing infection to be imaged and localized.

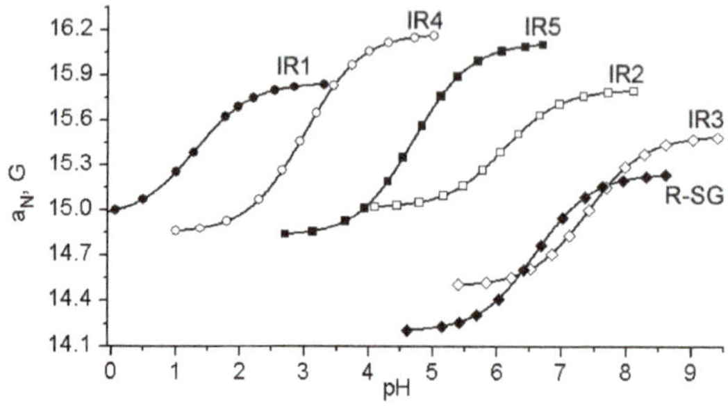

Figure 11. Nitroxides as molecular pH sensors. pH dependencies of nitrogen hyperfine splitting, a_N, obtained using X-band EPR for the imidazoline and imidazolidine nitroxides IR1-IR5 (Scheme 5) and RSG (Scheme 3).

2.3.2. *In vivo pH monitoring using pH-sensitive nitroxides*

Low-field EPR spectroscopy using spin pH probes has been shown to be a valuable tool for *in vivo* pH monitoring in organs and small animals such as rodents (Gallez et al., 1996b; Mader et al., 1996; Khramtsov et al., 2000; Foster et al., 2003; Potapenko et al., 2006; Bobko et al., 2012; Komarov et al., 2012). Figure 12a shows the structure of the nitroxide IR6 with

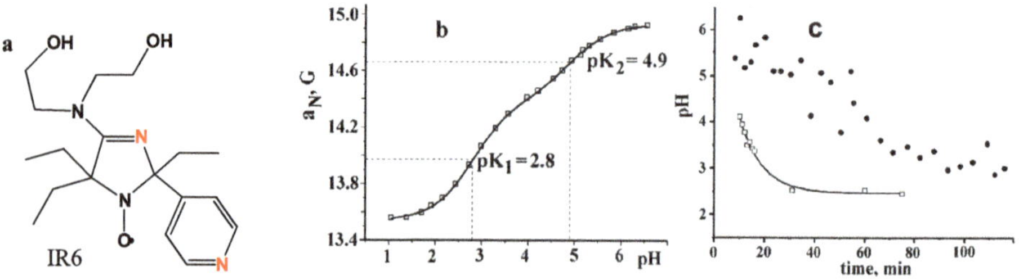

Figure 12. *In vivo* EPR monitoring of stomach acidity. (a) Chemical structure of imidazoline nitroxide IR6. Two protonatable nitrogen atoms are shown in red. (b) The pH dependence of hyperfine splitting, a_N, measured from 300 MHz EPR spectra of IR6. The solid line was calculated according to the standard titration equation for the compound with two ionizable groups. The extended pH sensitivity range from pH 1.8 to pH 6 is ideally optimized for monitoring stomach acidity. (c) Changes in the stomach acidity of living rat measured by longitudinally detected EPR spectrometer operated at 304 MHz excitation frequency, after giving 3 ml of gavage containing 5 mM IR6 probe alone (□) or with 50 mM bicarbonate (•). From (Potapenko et al., 2006) with permission, copyright 2006, Elsevier Inc.

properties optimized for monitoring stomach acidity (Potapenko et al., 2006). The presence of bulky groups in the vicinity of the NO fragment enhanced IR6 stability towards reduction, resulting in about four fold increase of its lifetime *in vivo* compared with that for previously used IR3 (Gallez et al., 1996b; Foster et al., 2003). The presence of two ionizable groups in the IR6 structure, imino nitrogen N-3 and pyridine, extended its range of pH sensitivity (see Fig. 12b). The hydrophilic character of pyridine, hydroxy, and amino groups of the IR6 probe prevents its penetration through biomembranes and probe redistribution from the stomach. Fig. 12c demonstrates real-time measurements of the stomach acidity in living rats using 300 MHz EPR and IR6 probe (Potapenko et al., 2006). Long lifetime of the IR6 probe *in vivo* allows for monitoring drug-induced perturbation of stomach acidity and its normalization afterwards for periods of 1 h or longer, therefore supporting applicability of pH sensitive NRs to the studies of drug pharmacology and disease in living animals.

The acidic extracellular pH (pH_e) in tumors has a number of important consequences, playing a role in tumor initiation, progression, and therapy (Gillies et al., 2004). To be used for monitoring of tumor extracellular tissue acidosis, the nitroxide pH probe has to (i) possess pH-sensitive spectral properties in the pH range from about 6.5 to 7.2, (ii) have enhanced stability to survive a reducing tumor microenvironment, and (iii) not penetrate cellular membranes to ensure probe targeting to extracellular space. These properties were achieved by synthesizing RSG nitroxide (Scheme 3) with $pK_a \approx 6.6$ (Fig. 11) being ideally fitted for detection of extracellular tumor tissue acidity. The bulky ethyl substitutes at positions 2 and 5 around the NO fragment of the RSG are introduced to enhance its stability towards bioreduction (Kirilyuk et al., 2004). The binding of the radical to hydrophilic tripeptide, GSH, prevents probe diffusion across the plasma membrane, and therefore, enforces probe localization to extracellular aqueous volumes (Woldman et al., 2009). Recently the RSG probe has been used to monitor tissue pH_e in mice bearing breast cancer tumors for assessment of therapeutic effectiveness of various treatments (Bobko et al., 2012). It was observed that tumor pH_e is about 0.4 pH units lower than in normal mammary gland tissue in agreement with microelectrode data (see Table 1). Note that RSG nitroxide can be used as a dual function pH and redox probe allowing for concurrent *in vivo* monitoring of both parameters (see section 2.2.1, Fig. 6).

	pH values measured in tumors and mammary glands		
	L-band EPR	pH microelectrode	PEDRI
Tumor	6.60±0.07 (n=3)	6.70±0.05 (n=3)	6.7±0.1
Gland	6.98±0.05 (n=3)	7.01±0.05 (n=3)	7.1±0.1

Table 1. Average pH values measured in tumor and mammary gland tissues in wild type C57Bl/6 mice using different techniques (Bobko et al., 2012).

The enhanced stability of the RSG probe allowed for development of an EPR approach for monitoring ischemia-induced acidosis in isolated perfused rat hearts (Komarov et al., 2012). RSG probe demonstrated excellent EPR signal stability in the heart while 90% of the previously used IR2 probe (Khramtsov, 2005) was reduced in myocardial tissue within 5 minutes. As seen in Figure 13 ischemic preconditioning improved pH homeostasis during the global no-flow ischemia. Similar kinetics of myocardial acidification observed by the cell-impermeable RSG and cell-permeable IR2 probes support fast pH equilibration between intracellular and extracellular spaces in agreement with data obtained previously using glass microelectrodes and ^{31}P-NMR (Asimakis et al., 1992; Lundmark et al., 1999).

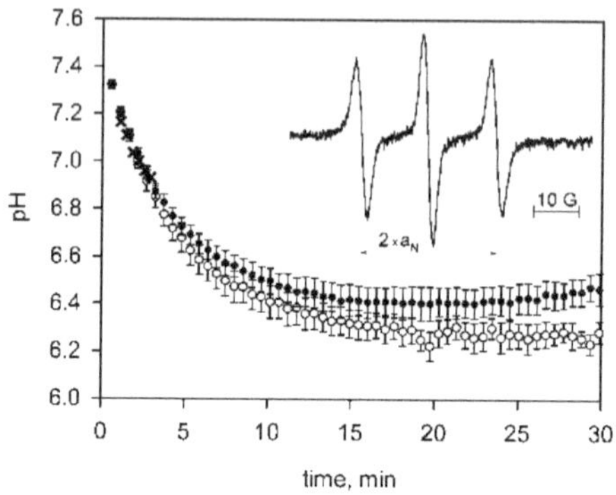

Figure 13. Ischemia-induced myocardial acidosis measured during global no-flow ischemia in control (○) and ischemic preconditioned perfused rat hearts (•) using L-band EPR and nitroxide pH probes. The data obtained during first five minutes of ischemia using IR2 probe are denoted by (×) symbols. Preconditioned hearts were subjected to two episodes of five minutes ischemia followed by five minutes of reperfusion. The probes (1.5 mM) were infused into the hearts through side arm of the perfusate line upon onset of global ischemia. Data are mean ± S.E.; n = 6. From (Komarov et al., 2012) with permission from John Wiley & Sons, Ltd.

2.3.3. pH mapping of living tissues

Spatially-resolved *in vivo* pH measurement is critically important to allow for identification of the area with compromised pH homeostasis. Low-field CW EPRI allows for pH mapping of living tissues in small animals such as rodents. However loss in EPR sensitivity at low frequency aggravated by necessity of acquisition of numerous spectral projections with a variable strength of field gradients significantly limits spatial and temporal resolution of the approach. Application of low-field EPR spectra detection at high modulation amplitude provides significant improvement in spectral intensity and functional sensitivity allowing easy conversion of the 'position' image of the low- or high-field spectral component to 'pH map' with a good spatial (0.2 mm) and functional (0.2 pH units) resolution as was demonstrated for phantom samples (Khramtsov et al., 2004, 2005).

Figure 14 demonstrates *in vivo* application of CW EPRI for tumor pH mapping. 3D EPR pH map was superimposed with co-registered MRI measured in right hind leg of C3H mouse bearing squamous cell carcinoma after i.v. injection of RSG probe (Goodwin et al., 2012). EPR-measured pH_e values show higher heterogeneity in tumor compared with normal tissue with mean pH_e, 6.67, significantly lower than mean tissue pH_e in healthy leg, 7.17.

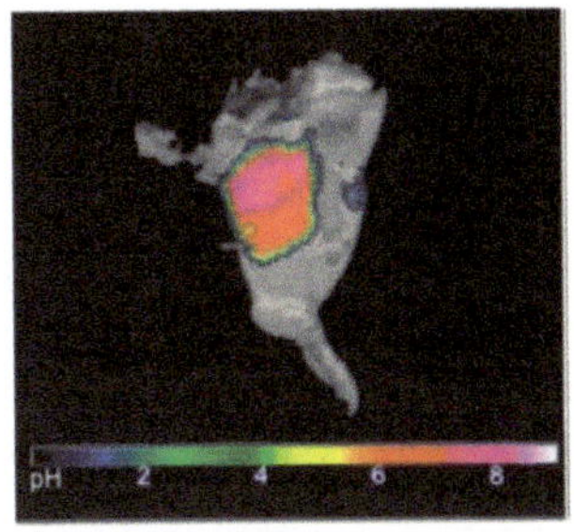

Figure 14. Co-registered EPR pH map and MRI of right hind leg of C3H mouse bearing squamous cell carcinoma. A 4D (1×spectral, 3×spatial) CW EPRI (750 MHz) of low field and high field EPR peaks were performed using the following parameters: scan time, 0.3 s; modulation amplitude, 1.5 G; G_{max}, 16 G/cm; FOV, 5.09 cm; projections, 576, total acquisition time, ≈6 min. EPR pH map was superimposed with Varian 7T MRI image. Adapted from (Goodwin et al., 2012).

Recently we made significant progress towards pH mapping of living tissue using functional PEDRI approaches (Khramtsov et al., 2010; Efimova et al., 2011). Fig. 15 demonstrates PEDRI pH mapping of a tumor-bearing mouse after injection of deuterated analog of the RSG radical in tumor and mammary gland. The deuterated probe (see Fig. 15) has a narrower linewidth (ΔH_{pp}=1.2 G) than RSG (ΔH_{pp}=2.1 G), and therefore higher spectral intensity. The EPR signal of the deuterated RSG probe is easily saturated by RF irradiation which is of critical importance for PEDRI experiments. In total, more than one order decrease in acquisition time was achieved for PEDRI pH mapping (24.8 s in Fig. 15)

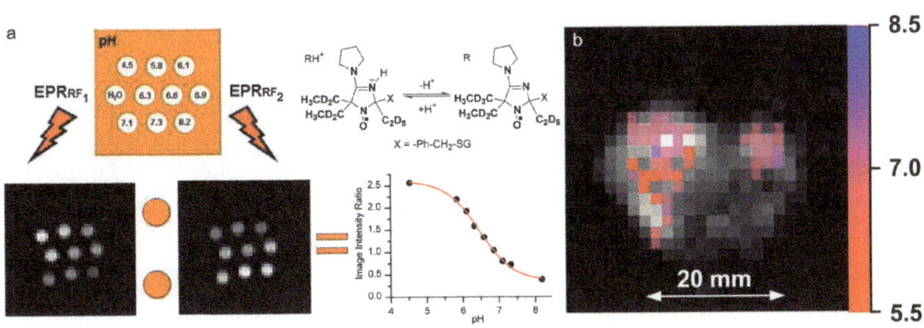

Figure 15. (a) Illustration of variable radio frequency PEDRI approach for the case of pH-mapping. Two PEDRI images of the phantom were acquired during EPR irradiation (8.4 s) at two pre-selected EPR frequencies, RF_1=559.3 MHz and RF_2=562.1 MHz which correspond to EPR resonances of RH^+ and R forms of deuterated RSG probe. pH values of 1 mM solutions of RSG are indicated on the phantom picture. Calibration curve was calculated as pH dependence of the ratio of signal amplitudes, $I(RF_1)/I(RF_2)$. (b) pH_e mapping of living mouse by PEDRI. pH map (in color) was superimposed with low-field MRI (gray scale) showing the transvers view of the mouse. The pH probe was injected into the number 4 mammary gland containing tumor (left) and in the number 9 normal mammary gland (right) before PEDRI/MRI acquisitions. The pH map was calculated from two PEDRI images, acquired during EPR irradiation at two EPR frequencies, RF_1 and RF_2. Total acquisition, 24.8 s. From (Bobko et al., 2012) with permission from John Wiley & Sons, Ltd.

compared with CW EPRI. Two areas of probe localization correspond to tumor (Fig. 15b left) and mammary gland (Fig. 15b right) with higher pH heterogeneity in tumor. Mean pH_e values in tumor and mammary gland were found to be in agreement with EPR spectroscopy and microelectrode data measured in the same mice (see Table 1). The PEDRI functional approach may find applications for pH mapping of other living tissues and has potential for applications to humans.

2.4. *In vivo* assessment of nitric oxide using nitronyl nitroxides

The measurement of nitric oxide *in vivo* is of paramount importance due to its key role as a signaling molecule in numerous physiological and pathophysiological processes. Methods for NO detection in biological systems are limited because of its short half-life and low concentration in living systems. Fe(II)-dithiocarbamate complexes (Mordvintcev et al., 1991), widely used for NO detection by EPR spectroscopy, have an important advantage of relative stability of the paramagnetic adduct with NO. However, disadvantages of the approach include the preparation of metal-chelator complexes from two components, which are accompanied by both iron and dithiocarbamate moiety toxicities. Application of simple low-molecular weight nitronyl nitroxides (NNRs) as NO-sensitive probes seems to be very attractive. The scavenging of nitric oxide by NNRs has been used for NO detection by EPR spectroscopy and to antagonize biological actions of nitric oxide (Akaike et al., 1993; Joseph et al., 1993; Woldman et al., 1994; Blasig et al., 2002; Cao & Reith, 2002). The detection of NO using NNR is based on the specific radical-radical reaction (see Fig.16a) followed by EPR-measured spectra changes.

The physiological effects of NNR cannot be entirely explained by their specific radical-radical reactions with nitric oxide due to the rapid reduction of NNR to its corresponding hydroxylamines. The performed mechanistic studies of the NNR reaction with NO in a reducing environment (Bobko et al., 2004) demonstrate an ability of the NNR to react with NO in the presence of the reducing agent, ascorbate, as shown in Figure 16b. It has been shown that equilibrium between radical, NNR, and its hydroxylamine is normally strongly shifted towards diamagnetic hydroxylamine, therefore no initial EPR spectra were observed in the presence of ascorbate (Fig. 16b). However NO generation and its consequent reaction with NNR results in accumulation of paramagnetic NNR and appearance of the corresponding EPR spectra (Fig. 16b, 3min). During the reaction, NNR is transformed to INR followed by characteristic changes of EPR spectral pattern (Fig. 16b, 6 and 9 min). After the reaction was complete, INR was reduced in its diamagnetic form by ascorbate (Fig. 16b, 20 min). The presence of fluorine atoms in the structure of fNNR allowed us to monitor accumulation of diamagnetic hydroxylamine product by ^{19}F NMR spectroscopy providing additional proof of the redox-sensitive mechanism of the NNR reaction with NO. The corresponding scheme of the reactions provides a plausible mechanism explaining the antagonistic action of NNR against NO in a reducing environment, a phenomenon well documented *in vivo* (for details see ref. (Bobko et al., 2004)).

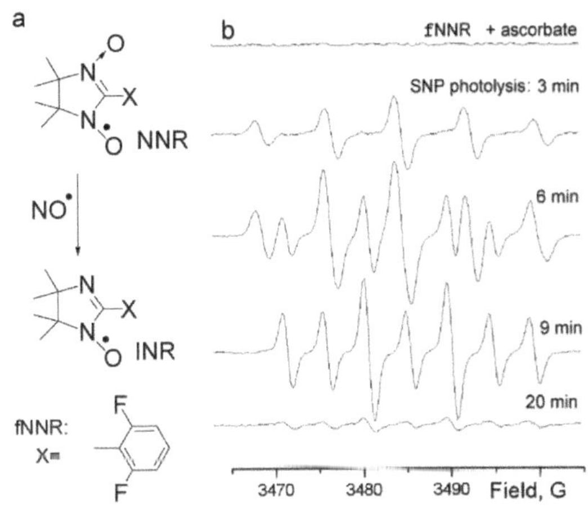

Figure 16. (a) The scheme of the reaction of nitronyl nitroxides, NNR, with NO forming iminonitroxides, INR. (b) NO-induced EPR spectral changes measured for fluorinated NNR derivative, fNNR, in the presence of reducing agent, ascorbate. Nitric oxide was generated upon photolysis of sodium nitroprusside (SNP). The spectra were measured in the mixture of 3.33 mM fNNR, 3.33 mM ascorbate and 33.3 mM SNP before and after SNP photolysis. From (Bobko et al., 2004) with permission, copyright 2003, Elsevier Inc.

The application of NNR for NO detection by EPR in biological systems with physiologically low rates of NO generation is limited due to the very rapid reduction of NNR, and particularly of INR, into diamagnetic EPR-silent product (Woldman et al., 1994; Haseloff et al., 1997; Bobko et al., 2004), so equilibrium radical concentration is lower than threshold of EPR detection. An alternative approach for NO detection using NNR is based on application of fluorinated NNRs, such as fNNR, in combination with ^{19}F NMR spectroscopy. An *in vivo* accumulation of the product of the reaction of fNNR with NO, hydroxylamine of fINR can be measured by ^{19}F NMR. Low NO concentrations in living tissues and low intrinsic NMR sensitivity make this approach hardly applicable *in vivo*. Nevertheless, stability of the diamagnetic fINR product allows for its *in vivo* accumulation followed by *ex vivo* ^{19}F NMR measurement. The *ex vivo* ^{19}F NMR measurement of fluorinated hydroxylamine products in blood samples taken from normotensive and hypertensive rats 1 hr after intraperitoneal injection of fNNR positively correlated with the levels of nitrite/nitrate evaluated by Griess method and negatively correlated with the blood pressure (Bobko et al., 2005). While ^{19}F NMR spectroscopy allows *ex vivo* evaluation of NO production, the development of NNR structures with higher stability for EPR detection *in vivo* is desirable. Several attempts by our group and others to protect NNR against reduction via encapsulation in liposomal (Woldman et al., 1994; Bobko et al., 2010) or dendrimeric (Rosen et al., 2003) structures resulted in significant lifetime increase but further work has to be done to obtain NNR probes stable enough for *in vivo* applications.

3. Conclusion

In recent decades functional EPR spectroscopy and imaging applications have moved closer to biomedical applicability. The bottleneck of *in vivo* EPR-based applications is the requirement of appropriate paramagnetic probes. NRs and triarylmethyl radicals, TAMs, represent two main classes of soluble paramagnetic materials used for EPR spectroscopy and imaging applications (Khramtsov & Zweier, 2010). TAMs have advantages over nitroxides in extraordinary stability toward tissue redox processes, longer relaxation time and narrower line width making them particularly attractive for imaging applications. Applications of TAM radicals as functional probes include EPR oximetry (Ardenkjaer-Larsen et al., 1998; Golman et al., 2000; Krishna et al., 2002) and recently reported sensitivity to the superoxide anion (Kutala et al., 2004; Kutala et al., 2008) and pH (Bobko et al., 2007a; Dhimitruka et al., 2008; Driesschaert et al., 2012). On the other hand NRs have tremendous advantage over TAMs in well-developed chemistry. As a consequence, the nitroxides provide a wide range of available structures that vary in solubility and tissue redistribution, spectral and functional sensitivity, ability to be targeted and lifetimes in living tissues. Specific nitroxides provide EPR-based spectroscopy and imaging with the capacity for functional temporally and spatially resolved mapping of physiologically relevant parameters of living tissues, such as oxygen, redox status, glutathione content and pH. The nitroxides were the first compounds applied to EPR oximetry and are still a useful tool for *in vivo* tissue oxygen mapping. The most well known reaction of the nitroxides, their reduction to EPR-silent hydroxylamines, provides information on redox state in living tissues. The disulfide biradical nitroxides were developed for *in vivo* assessment of intracellular glutathione concentration. The imidazoline nitroxide spin pH probes allowed for EPR spectroscopy and imaging of tissue pH. NanoSized *P*articles with *I*ncorporated *N*itroxides, or nano*SPIN*s, may serve as a future platform for stabilized nitroxide-based biosensors (Woldman et al., 2009). In particular, encapsulation of NO-sensitive nitronyl nitroxides may protect the sensing probe from biological reductants while retaining sensitivity to a small analyte, molecule of nitric oxide. *In summary*, the area of nitroxide biomedical application as functional paramagnetic probes is truly emerging.

Author details

Valery V. Khramtsov
Division of Pulmonary, Allergy, Critical Care & Sleep Medicine, Department of Internal Medicine, The Ohio State University, Columbus, OH, USA

Acknowledgement

This work was partly supported by NIH Grant EB014542-01A1. VVK thanks Dr. Andrey Bobko for technical assistance and helpful discussion.

4. References

Akaike, T., Yoshida, M., Miyamoto, Y., Sato, K., Kohno, M., Sasamoto, K., Miyazaki, K., Ueda, S. and Maeda, H. (1993). Antagonistic action of imidazolineoxyl N-oxides against endothelium-derived relaxing factor/.NO through a radical reaction. *Biochemistry,* Vol. 32, No 3, pp. 827-32.

Ardenkjaer-Larsen, J. H., Laursen, I., Leunbach, I., Ehnholm, G., Wistrand, L. G., Petersson, J. S. and Golman, K. (1998). EPR and DNP properties of certain novel single electron contrast agents intended for oximetric imaging. *J Magn Reson,* Vol. 133, No 1, pp. 1-12.

Asimakis, G. K., Inners-McBride, K., Medellin, G. and Conti, V. R. (1992). Ischemic preconditioning attenuates acidosis and postischemic dysfunction in isolated rat heart. *Am J Physiol,* Vol. 263, No 3 Pt 2, pp. H887-94.

Bacic, G., Nilges, M. J., Magin, R. L., Walczak, T. and Swartz, H. M. (1989). *In vivo* localized ESR spectroscopy reflecting metabolism. *Magn Reson Med,* Vol. 10, No 2, pp. 266-72.

Backer, J. M., Budker, V. G., Eremenko, S. I. and Molin, Y. N. (1977). Detection of the kinetics of biochemical reactions with oxygen using exchange broadening in the ESR spectra of nitroxide radicals. *Biochim Biophys Acta,* Vol. 460, No 1, pp. 152-6.

Baker, J. E., Froncisz, W., Joseph, J. and Kalyanaraman, B. (1997). Spin label oximetry to assess extracellular oxygen during myocardial ischemia. *Free Radic Biol Med,* Vol. 22, No 1-2, pp. 109-15.

Berliner, L. J., Ed. (1998). Spin Labeling. The Next Millennium. New York, London, Plenum Press, 423 p.

Blasig, I. E., Mertsch, K. and Haseloff, R. F. (2002). Nitronyl nitroxides, a novel group of protective agents against oxidative stress in endothelial cells forming the blood-brain barrier. *Neuropharmacology,* Vol. 43, No 6, pp. 1006-14.

Bobko, A. A., Bagryanskaya, E. G., Reznikov, V. A., Kolosova, N. G., Clanton, T. L. and Khramtsov, V. V. (2004). Redox-sensitive mechanism of no scavenging by nitronyl nitroxides. *Free Radic Biol Med,* Vol. 36, No 2, pp. 248-58.

Bobko, A. A., Sergeeva, S. V., Bagryanskaya, E. G., Markel, A. L., Khramtsov, V. V., Reznikov, V. A. and Kolosova, N. G. (2005). 19F NMR measurements of NO production in hypertensive ISIAH and OXYS rats. *Biochem Biophys Res Commun,* Vol. 330, No 2, pp. 367-70.

Bobko, A. A., Dhimitruka, I., Zweier, J. L. and Khramtsov, V. V. (2007a). Trityl radicals as persistent dual function pH and oxygen probes for *in vivo* electron paramagnetic resonance spectroscopy and imaging: concept and experiment. *J Am Chem Soc,* Vol. 129, No 23, pp. 7240-1.

Bobko, A. A., Kirilyuk, I. A., Grigor'ev, I. A., Zweier, J. L. and Khramtsov, V. V. (2007b). Reversible Reduction of Nitroxides to Hydroxylamines: the Roles for Ascorbate and Glutathione. *Free Rad. Biol. Med.,* Vol. 42 No. 3, pp. 404-12.

Bobko, A. A., Dhimitruka, I., Eubank, T. D., Marsh, C. B., Zweier, J. L. and Khramtsov, V. V. (2009). Trityl-based EPR probe with enhanced sensitivity to oxygen. *Free Radic Biol Med,* Vol. 47, No 5, pp. 654-8.

Bobko, A. A., Ivanov, A. and Khramstov, V. V. (2010). Discriminating detection of NO and HNO using encapsulated nitronyl nitroxides. *Free Radic Biol Med,* Vol. 49 (Suppl.1), p. S109.

Bobko, A. A., Eubank, T. D., Voorhees, J. L., Efimova, O. V., Kirilyuk, I. A., Petryakov, S., Trofimiov, D. G., Marsh, C. B., Zweier, J. L., Grigor'ev, I. A., Samouilov, A. and Khramtsov, V. V. (2012). *In vivo* monitoring of pH, redox status, and glutathione using L-band EPR for assessment of therapeutic effectiveness in solid tumors. *Magn Reson Med,* doi: 10.1002/mrm.23196.

Bratasz, A., Selvendiran, K., Wasowicz, T., Bobko, A., Khramtsov, V. V., Ignarro, L. J. and Kuppusamy, P. (2008). NCX-4040, a nitric oxide-releasing aspirin, sensitizes drug-resistant human ovarian xenograft tumors to cisplatin by depletion of cellular thiols. *Journal of Translational Medicine,* Vol. 6, No 9. doi: 10.1186/1479-5876-6-9.

Cao, B. J. and Reith, M. E. (2002). Nitric oxide scavenger carboxy-PTIO potentiates the inhibition of dopamine uptake by nitric oxide donors. *Eur J Pharmacol,* Vol. 448, No 1, pp. 27-30.

Chan, H. C., Glockner, J. F. and Swartz, H. M. (1989). Oximetry in cells and tissues using a nitroxide-liposome system. *Biochim Biophys Acta,* Vol. 1014, No 2, pp. 141-4.

Clarkson, R. B., Odintsov, B. M., Ceroke, P. J., Ardenkjaer-Larsen, J. H., Fruianu, M. and Belford, R. L. (1998). Electron paramagnetic resonance and dynamic nuclear polarization of char suspensions: surface science and oximetry. *Phys Med Biol,* Vol. 43, No 7, pp. 1907-20.

Cook, J. A., Gius, D., Wink, D. A., Krishna, M. C., Russo, A. and Mitchell, J. B. (2004). Oxidative stress, redox, and the tumor microenvironment. *Semin Radiat Oncol,* Vol. 14, No 3, pp. 259-66.

Dhimitruka, I., Bobko, A. A., Hadad, C. M., Zweier, J. L. and Khramtsov, V. V. (2008). Synthesis and characterization of amino derivatives of persistent trityl radicals as dual function pH and oxygen paramagnetic probes. *J Am Chem Soc,* Vol. 130, No 32, pp. 10780-7.

Driesschaert, B., Marchand, V., Levêque, P., Gallez, B. and Marchand-Brynaert, J. (2012). Phosphonated Triarylmethyl Radical as Probe for Measurement of pH by EPR. *Chem. Commun.,* Vol. 48, pp. 4049-51. DOI: 10.1039/C2CC00025C.

Efimova, O. V., Sun, Z., Petryakov, S., Kesselring, E., Caia, G. L., Johnson, D., Zweier, J. L., Khramtsov, V. V. and Samouilov, A. (2011). Variable Radio Frequency Proton-Electron Double-Resonance Imaging: application to pH mapping of aqueous samples. *J Magn Reson,* Vol. 209, No 2, pp. 227-232.

Ellman, G. L. (1959). Tissue sulfhydryl groups. *Arch Biochem Biophys,* Vol. 82, No 1, pp. 70-7.

Eubank, T. D., Roberts, R. D., Khan, M., Curry, J. M., Nuovo, G. J., Kuppusamy, P. and Marsh, C. B. (2009). Granulocyte macrophage colony-stimulating factor inhibits breast cancer growth and metastasis by invoking an anti-angiogenic program in tumor-educated macrophages. *Cancer Res,* Vol. 69, No 5, pp. 2133-40.

Finkelstein, E., Rosen, G. M. and Rauckman, E. J. (1984). Superoxide-dependent reduction of nitroxides by thiols. *Biochim. Biophys. Acta,* Vol. 802, No, pp. 90-98.

Foster, M. A., Grigor'ev, I. A., Lurie, D. J., Khramtsov, V. V., McCallum, S., Panagiotelis, I., Hutchison, J. M., Koptioug, A. and Nicholson, I. (2003). *In vivo* detection of a pH-sensitive nitroxide in the rat stomach by low-field ESR-based techniques. *Magn Reson Med,* Vol. 49, No 3, pp. 558-67.

Froncisz, W., Lai, C. S. and Hyde, J. S. (1985). Spin-label oximetry: kinetic study of cell respiration using a rapid-passage T1-sensitive electron spin resonance display. *Proc Natl Acad Sci U S A,* Vol. 82, No 2, pp. 411-5.

Gallez, B., Debuyst, R., Demeure, R., Dejehet, F., Grandin, C., Van Beers, B., Taper, H., Pringot, J. and Dumont, P. (1993). Evaluation of a nitroxyl fatty acid as liver contrast agent for magnetic resonance imaging. *Magn Reson Med,* Vol. 30, No 5, pp. 592-9.

Gallez, B., Bacic, G., Goda, F., Jiang, J., O'Hara, J. A., Dunn, J. F. and Swartz, H. M. (1996a). Use of nitroxides for assessing perfusion, oxygenation, and viability of tissues: *in vivo* EPR and MRI studies. *Magn Reson Med,* Vol. 35, No 1, pp. 97-106.

Gallez, B., Mader, K. and Swartz, H. M. (1996b). Noninvasive measurement of the pH inside the gut by using pH-sensitive nitroxides. An *in vivo* EPR study. *Magn Reson Med,* Vol. 36, No 5, pp. 694-7.

Gillies, R. J., Alger, J. R., den Hollander, J. A. and Shulman, R. G. (1982). Intracellular pH measured by NMR: methods and results, In: *Intracellular pH: Its measurement, regulation and utilization in cellular functions*. R. Nuccitelli, Deamer, D. W. (Ed.): 79-104, New York, Alan R.Liss.

Gillies, R. J., Liu, Z. and Bhujwalla, Z. (1994). 31P-MRS measurements of extracellular pH of tumors using 3-aminopropylphosphonate. *Am J Physiol,* Vol. 267, No 1 Pt 1, pp. C195-203.

Gillies, R. J., Raghunand, N., Garcia-Martin, M. L. and Gatenby, R. A. (2004). pH imaging. A review of pH measurement methods and applications in cancers. *IEEE Eng Med Biol Mag,* Vol. 23, No 5, pp. 57-64.

Glebska, J., Skolimowski, J., Kudzin, Z., Gwozdzinski, K., Grzelak, A. and Bartosz, G. (2003). Pro-oxidative activity of nitroxides in their reactions with glutathione. *Free Radic Biol Med,* Vol. 35, No 3, pp. 310-6.

Glockner, J. F., Norby, S. W. and Swartz, H. M. (1993). Simultaneous measurement of intracellular and extracellular oxygen concentrations using a nitroxide-liposome system. *Magn Reson Med,* Vol. 29, No 1, pp. 12-8.

Golman, K., Petersson, J. S., Ardenkjaer-Larsen, J. H., Leunbach, I., Wistrand, L. G., Ehnholm, G. and Liu, K. J. (2000). Dynamic *in vivo* oxymetry using overhauser enhanced MR imaging. *J Magn Reson Imaging,* Vol. 12, No, pp. 929-938.

Goodwin, J., Koda, S., Ohfuchi, M., Pawlak, A., Yasui, H., Khramstov, V. and Hirata, H. (2012). Four-dimensional spectral spatial pH mapping of mouse tumour using Continuous Wave-Electron Paramagnetic Resonance imaging (CW-EPRI) and pH sensitive imidazoline nitroxide, In: *Proceedings of 20th Annual Meeting of the International Society for Magnetic Resonance in Medicine,* # 1708, Melbourne, Australia.

Grucker, D. and Chambron, J. (1993). Oxygen imaging in perfused hearts by dynamic nuclear polarization. *Magn Reson Imaging,* Vol. 11, No 5, pp. 691-6.

Halpern, H. J., Yu, C., Peric, M., Barth, E., Grdina, D. J. and Teicher, B. A. (1994). Oxymetry deep in tissues with low-frequency electron paramagnetic resonance. *Proc Natl Acad Sci U S A*, Vol. 91, No 26, pp. 13047-51.

Haseloff, R. F., Zollner, S., Kirilyuk, I. A., Grigor'ev, I. A., Reszka, R., Bernhardt, R., Mertsch, K., Roloff, B. and Blasig, I. E. (1997). Superoxide-mediated reduction of the nitroxide group can prevent detection of nitric oxide by nitronyl nitroxides. *Free Radic Res*, Vol. 26, No 1, pp. 7-17.

He, G., Samouilov, A., Kuppusamy, P. and Zweier, J. L. (2002). *In vivo* imaging of free radicals: applications from mouse to man. *Mol Cell Biochem*, Vol. 234-235, No 1-2, pp. 359-67.

He, G., Kutala, V. K., Kuppusamy, P. and Zweier, J. L. (2004). *In vivo* measurement and mapping of skin redox stress induced by ultraviolet light exposure. *Free Radic Biol Med*, Vol. 36, No 5, pp. 665-72.

Hideg, K., Kalai, T. and Sar, C. P. (2005). Recent results in chemistry and biology of nitroxides. *J. Heterocycl. Chem.*, Vol. 42, No, pp. 437-450.

Hoffman, B. M. and Eames, T. B. (1969). Protonated nitroxide free radical. *J Am Chem Soc*, Vol. 91, No, pp. 2169-2170.

Hsia, J. C. and Boggs, J. M. (1972). Influence of pH and cholesterol on the structure of phosphatidylethanolamine multibilayers. *Biochim Biophys Acta*, Vol. 266, No 1, pp. 18-25.

Hubbell, W. L., Cafiso, D. S. and Altenbach, C. (2000). Identifying conformational changes with site-directed spin labeling. *Nat Struct Biol*, Vol. 7, No 9, pp. 735-9.

Hyde, J. S. and Subszynski, W. K. (1989). Spin label oximetry., In: *Spin Labeling: Theory and Application.* L. J. Berliner and J. Reubens (Ed.), 8: 399-425, New York, Plenum Press.

Hyde, J. S., Jin, J.-J., Felix, J. B. and Hubbell, W. L. (1990). Advances in spin label oximetry. *Pure & Applied Chem.*, Vol. 62, No, pp. 255-260.

Hyodo, F., Murugesan, R., Matsumoto, K., Hyodo, E., Subramanian, S., Mitchell, J. B. and Krishna, M. C. (2008). Monitoring redox-sensitive paramagnetic contrast agent by EPRI, OMRI and MRI. *J Magn Reson*, Vol. 190, No 1, pp. 105-12.

Hyodo, F., Matsumoto, S., Devasahayam, N., Dharmaraj, C., Subramanian, S., Mitchell, J. B. and Krishna, M. C. (2009). Pulsed EPR imaging of nitroxides in mice. *J Magn Reson*, Vol. 197, No 2, pp. 181-5.

Joseph, J., Kalyanaraman, B. and Hyde, J. S. (1993). Trapping of nitric oxide by nitronyl nitroxides: an electron spin resonance investigation. *Biochem Biophys Res Commun*, Vol. 192, No 2, pp. 926-34.

Karoui, H., Le Moigne, F., Ouari, O. and Tordo, P. (2010). Nitroxide Radicals: Properties, Synthesis and Applications, In: *Stable Radicals. Fundamental and Applied Aspects of Odd-Electron Compounds.* R. G. Hicks (Ed.): 173-229, Chichester, UK, John Wiley & Sons, Ltd.

Khramtsov, V. V., Weiner, L. M., Grigor'ev, I. A. and Volodarsky, L. B. (1982). Proton exchange in stable nitroxyl radicals. EPR studies of the pH of aqueous solutions. *Chem Phys Lett*, Vol. 91, No, pp. 69-72.

Khramtsov, V. V. and Weiner, L. M. (1988). Proton exchange in stable nitroxyl radicals: pH sensitive spin probes, In: *Imidazoline nitroxides*. L. B. Volodarsky (Ed.), 2: 37-80, Boca Raton, FL., CRC Press.

Khramtsov, V. V., Yelinova, V. I., Weiner, L. M., Berezina, T. A., Martin, V. V. and Volodarsky, L. B. (1989). Quantitative determination of SH groups in low- and high-molecular-weight compounds by an electron spin resonance method. *Anal Biochem,* Vol. 182, No 1, pp. 58-63.

Khramtsov, V. V., Yelinova, V. I., Glazachev Yu, I., Reznikov, V. A. and Zimmer, G. (1997). Quantitative determination and reversible modification of thiols using imidazolidine biradical disulfide label. *J Biochem Biophys Methods,* Vol. 35, No 2, pp. 115-28.

Khramtsov, V. V. and Volodarsky, L. B. (1998). Use of imidazoline nitroxides in studies of chemical reactions. ESR measurements of the concentration and reactivity of protons, thiols and nitric oxide, In: *Spin labeling. The next Millennium*. L. J. Berliner (Ed.), 14: 109-180, New York, Plenum Press.

Khramtsov, V. V., Grigor'ev, I. A., Foster, M. A., Lurie, D. J. and Nicholson, I. (2000). Biological applications of spin pH probes. *Cell Mol Biol,* Vol. 46, No 8, pp. 1361-74.

Khramtsov, V. V., Grigor'ev, I. A., Foster, M. A., Lurie, D. J., Zweier, J. L. and Kuppusamy, P. (2004). Spin pH and SH probes: enhancing functionality of EPR-based techniques. *Spectroscopy,* Vol. 18, No, pp. 213-225.

Khramtsov, V. V. (2005). Biological imaging and spectroscopy of pH. *Curr. Org. Chem.,* Vol. 9, No, pp. 909-923.

Khramtsov, V. V., Caia, G. L., Shet, K., Kesselring, E., Petryakov, S., Zweier, J. L. and Samouilov, A. (2010). Variable Field Proton-Electron Double-Resonance Imaging: Application to pH mapping of aqueous samples. *J Magn Reson,* Vol. 202, No 2, pp. 267-73.

Khramtsov, V. V. and Zweier, J. L. (2010). Functional *in vivo* EPR Spectroscopy and Imaging Using Nitroxide and Trityl Radicals, In: *Stable Radicals: Fundamentals and Applied Aspects of Odd-Electron Compounds*. R. Hicks (Ed.): 537-566, Chichester, UK., John Wiley & Sons, Ltd.

Khramtsov, V. V. and Komarov, D. A. (2011). Nanospin Probes and Applications to Cardiology, In: *Nanomedicine and the Cardiovascular System*. R. J. Hunter and V. R. Preedy (Ed.): 200-220, Enfield, NH, Science Publishers and CRC Press.

Kirilyuk, I. A., Bobko, A. A., Grigor'ev, I. A. and Khramtsov, V. V. (2004). Synthesis of the tetraethyl substituted pH-sensitive nitroxides of imidazole series with enhanced stability towards reduction. *Org Biomol Chem,* Vol. 2, No 7, pp. 1025-30.

Kirilyuk, I. A., Bobko, A. A., Khramtsov, V. V. and Grigor'ev, I. A. (2005). Nitroxides with two pK values--useful spin probes for pH monitoring within a broad range. *Org Biomol Chem,* Vol. 3, No 7, pp. 1269-74.

Kocherginsky, N. and Swartz, H. M. (1995). Nitroxide spin labels. Reactions in biology and chemistry, CRC Press, Boca Raton, New York, London, Tokyo.

Komarov, D. A., Dhimitruka, I., Kirilyuk, I. A., Trofimiov, D. G., Grigor'ev, I. A., Zweier, J. L. and Khramtsov, V. V. (2012). Electron paramagnetic resonance monitoring of ischemia-induced myocardial oxygen depletion and acidosis in isolated rat hearts using soluble paramagnetic probes. *Magn Reson Med,* doi: 10.1002/mrm.23251.

Krishna, M. C., English, S., Yamada, K., Yoo, J., Murugesan, R., Devasahayam, N., Cook, J. A., Golman, K., Ardenkjaer-Larsen, J. H., Subramanian, S. and Mitchell, J. B. (2002). Overhauser enhanced magnetic resonance imaging for tumor oximetry: coregistration of tumor anatomy and tissue oxygen concentration. *Proc Natl Acad Sci U S A,* Vol. 99, No 4, pp. 2216-21.

Kuppusamy, P., Chzhan, M., Vij, K., Shteynbuk, M., Lefer, D. J., Giannella, E. and Zweier, J. L. (1994). Three-dimensional spectral-spatial EPR imaging of free radicals in the heart: a technique for imaging tissue metabolism and oxygenation. *Proc Natl Acad Sci U S A,* Vol. 91, No 8, pp. 3388-92.

Kuppusamy, P., Afeworki, M., Shankar, R. A., Coffin, D., Krishna, M. C., Hahn, S. M., Mitchell, J. B. and Zweier, J. L. (1998). *In vivo* electron paramagnetic resonance imaging of tumor heterogeneity and oxygenation in a murine model. *Cancer Res,* Vol.58, No 7, pp.1562-8.

Kuppusamy, P. and Krishna, M. C. (2002). EPR Imaging of tissue redox status. *Current topics in Biophysics,* Vol. 26, No, pp. 29-34.

Kuppusamy, P., Li, H., Ilangovan, G., Cardounel, A. J., Zweier, J. L., Yamada, K., Krishna, M. C. and Mitchell, J. B. (2002). Noninvasive imaging of tumor redox status and its modification by tissue glutathione levels. *Cancer Res,* Vol. 62, No 1, pp. 307-12.

Kuppusamy, P. and Zweier, J. L. (2004). Cardiac applications of EPR imaging. *NMR Biomed,* Vol. 17, No 5, pp. 226-39.

Kutala, V. K., Parinandi, N. L., Zweier, J. L. and Kuppusamy, P. (2004). Reaction of superoxide with trityl radical: implications for the determination of superoxide by spectrophotometry. *Arch Biochem Biophys,* Vol. 424, No 1, pp. 81-8.

Kutala, V. K., Villamena, F. A., Ilangovan, G., Maspoch, D., Roques, N., Veciana, J., Rovira, C. and Kuppusamy, P. (2008). Reactivity of superoxide anion radical with a perchlorotriphenylmethyl (trityl) radical. *J Phys Chem B,* Vol. 112, No 1, pp. 158-67.

Kveder, M., Krisko, A., Pifat, G. and Steinhoff, H. J. (2003). The study of structural accessibility of free thiol groups in human low-density lipoproteins. *Biochim Biophys Acta,* Vol. 1631, No 3, pp. 239-45.

Lai, C. S., Hopwood, L. E., Hyde, J. S. and Lukiewicz, S. (1982). ESR studies of O_2 uptake by Chinese hamster ovary cells during the cell cycle. *Proc Natl Acad Sci U S A,* Vol. 79, No 4, pp. 1166-70.

Lebedev, O. A. and Kayanovskii, S. N. (1959). Catalytic oxidation of aliphatic amines with hydrogen peroxide. *Trudy po Khimii i Khim Technologii (Gorkii),* Vol. 8, No, pp. 649-652.

Lee, F. Y., Vessey, A., Rofstad, E., Siemann, D. W. and Sutherland, R. M. (1989). Heterogeneity of glutathione content in human ovarian cancer. *Cancer Res,* Vol. 49, No 19, pp. 5244-8.

Liu, K. J., Gast, P., Moussavi, M., Norby, S. W., Vahidi, N., Walczak, T., Wu, M. and Swartz, H. M. (1993). Lithium phthalocyanine: a probe for electron paramagnetic resonance oximetry in viable biological systems. *Proc Natl Acad Sci U S A,* Vol. 90, No 12, pp. 5438-42.

Liu, K. J., Grinstaff, M. W., Jiang, J., Suslick, K. S., Swartz, H. M. and Wang, W. (1994). *In vivo* measurement of oxygen concentration using sonochemically synthesized microspheres. *Biophys J*, Vol. 67, No 2, pp. 896-901.

Livesey, J. C., Golden, R. N., Shankland, E. G., Grunbaum, Z., Wyman, M. and Krohn, K. A. (1992). Magnetic resonance spectroscopic measurement of cellular thiol reduction-oxidation state. *Int J Radiat Oncol Biol Phys*, Vol. 22, No 4, pp. 755-7.

Lundmark, J. A., Trueblood, N., Wang, L. F., Ramasamy, R. and Schaefer, S. (1999). Repetitive acidosis protects the ischemic heart: implications for mechanisms in preconditioned hearts. *J Mol Cell Cardiol*, Vol. 31, No 4, pp. 907-17.

Mader, K., Gallez, B., Liu, K. J. and Swartz, H. M. (1996). Non-invasive *in vivo* characterization of release processes in biodegradable polymers by low-frequency electron paramagnetic resonance spectroscopy. *Biomaterials*, Vol. 17, No 4, pp. 457-61.

Malatesta, V. and Ingold, K. U. (1973). Protonated nitroxide radicals. *J Am Chem Soc*, Vol. 95, No, pp. 6404-6407.

Mathew, A. E. and Dodd, J. R. (1985). Synthesis of substituted, 2,2,5,5-tetramethylpyrrolidin-1-oxyl spin labels. pH sensitivity studies. *J. Heterocycl. Chem.*, Vol. 22, No, pp. 225-228.

Mordvintcev, P., Mulsch, A., Busse, R. and Vanin, A. (1991). On-line detection of nitric oxide formation in liquid aqueous phase by electron paramagnetic resonance spectroscopy. *Anal Biochem*, Vol. 199, No 1, pp. 142-6.

Murugesan, R., Cook, J. A., Devasahayam, N., Afeworki, M., Subramanian, S., Tschudin, R., Larsen, J. A., Mitchell, J. B., Russo, A. and Krishna, M. C. (1997). *In vivo* imaging of a stable paramagnetic probe by pulsed-radiofrequency electron paramagnetic resonance spectroscopy. *Magn Reson Med*, Vol. 38, No 3, pp. 409-14.

Nakaie, C. R., Goissis, G., Schreier, S. and Paiva, A. C. (1981). pH dependence of EPR spectra of nitroxides containing ionizable groups. *Braz J Med Biol Res*, Vol. 14, No 2-3, pp. 173-80.

Neiman, M. B., Rozatzev, E. G. and Mamedova, Y. G. (1962). Free radical reactions involving no unpaired electrons. *Nature*, Vol. 196, No 3, pp. 472-474.

Ojha, N., Roy, S., He, G., Biswas, S., Velayutham, M., Khanna, S., Kuppusamy, P., Zweier, J. L. and Sen, C. K. (2008). Assessment of wound-site redox environment and the significance of Rac2 in cutaneous healing. *Free Radic Biol Med*, Vol. 44, No 4, pp. 682-91.

Patsoukis, N. and Georgiou, C. D. (2005). Fluorometric determination of thiol redox state. *Anal Bioanal Chem*, Vol. 383, No 6, pp. 923-9.

Pietri, S., Martel, S., Culcasi, M., Delmas-Beauvieux, M. C., Canioni, P. and Gallis, J. L. (2001). Use of diethyl(2-methylpyrrolidin-2-yl)phosphonate as a highly sensitive extra- and intracellular 31P NMR pH indicator in isolated organs. Direct NMR evidence of acidic compartments in the ischemic and reperfused rat liver. *J Biol Chem*, Vol. 276, No 3, pp. 1750-8.

Potapenko, D. I., Bagryanskaya, E. G., Grigoriev, I. A., Maksimov, A. M., Reznikov, V. A., Platonov, V. E., Clanton, T. L. and Khramtsov, V. V. (2005). Quantitative determination of SH groups using 19F NMR spectroscopy and disulfide of 2,3,5,6-tetrafluoro-4-mercaptobenzoic acid. *Magn Reson Chem*, Vol. 43, No 11, pp. 902-9.

Potapenko, D. I., Foster, M. A., Lurie, D. J., Kirilyuk, I. A., Hutchison, J. M., Grigor'ev, I. A., Bagryanskaya, E. G. and Khramtsov, V. V. (2006). Real-time monitoring of drug-induced changes in the stomach acidity of living rats using improved pH-sensitive nitroxides and low-field EPR techniques. *J Magn Reson,* Vol. 182, No 1, pp. 1-11.

Povich, M. J. (1975). Electron Spin Resonance Oxygen Broadening. *J. Phys. Chem.,* Vol. 79, No, pp. 1106-1109.

Presley, T., Kuppusamy, P., Zweier, J. L. and Ilangovan, G. (2006). Electron paramagnetic resonance oximetry as a quantitative method to measure cellular respiration: a consideration of oxygen diffusion interference. *Biophys J,* Vol. 91, No 12, pp. 4623-31.

Quintanilha, A. T. and Mehlhorn, R. J. (1978). pH gradients across thylakoid membranes measured wit a spin-labeled amine. *FEBS Lett.,* Vol. 91, No, pp. 104.

Rabik, C. A. and Dolan, M. E. (2007). Molecular mechanisms of resistance and toxicity associated with platinating agents. *Cancer Treat Rev,* Vol. 33, No 1, pp. 9-23.

Rosen, G. M., Porasuphatana, S., Tsai, P., Ambulos, N. P., Galtsev, V. E., Ichikawa, K. and Halpern, H. J. (2003). Dendrimeric-Containing Nitronyl Nitroxides as Spin Traps for Nitric Oxide: Synthesis, Kinetic, and Stability Studies. *Macromolecules,* Vol. 36, No, pp. 1021-1027.

Roshchupkina, G. I., Bobko, A. A., Bratasz, A., Reznikov, V. A., Kuppusamy, P. and Khramtsov, V. V. (2008). *In vivo* EPR measurement of glutathione in tumor-bearing mice using improved disulfide biradical probe. *Free Rad. Biol. Med.,* Vol. 45, No, pp. 312-320.

Saracino, G. A. A., Tedeschi, A., D'Errico, G., Improta, R., Franko, L., Ruzzi, M., Corvaia, C. and Barone, V. (2002). Solvent polarity and pH effects on the magnetic properties of ionizable nitroxide radicals: a combined computational and experimental study of 2,2,5,5-tetramethyl-3-carboxypyrrolidine and 2,2,6,6-tetramethyl-4-carboxypiperidine nitroxides. *J. Phys. Chem. A.,* Vol. 106, No, pp. 10700-10706.

Sarna, T., Duleba, A., Korytowski, W. and Swartz, H. (1980). Interaction of melanin with oxygen. *Arch Biochem Biophys,* Vol. 200, No 1, pp. 140-8.

Schafer, F. Q. and Buettner, G. R. (2001). Redox environment of the cell as viewed through the redox state of the glutathione disulfide/glutathione couple. *Free Radic Biol Med,* Vol. 30, No 11, pp. 1191-212.

Soszynski, M. and Bartosz, G. (1997). Decrease in accessible thiols as an index of oxidative damage to membrane proteins. *Free Radic Biol Med,* Vol. 23, No 3, pp. 463-9.

Swartz, H. M. (2004). Using EPR to measure a critical but often unmeasured component of oxidative damage: oxygen. *Antioxid Redox Signal,* Vol. 6, No 3, pp. 677-86.

Swartz, H. M., Khan, N. and Khramtsov, V. V. (2007). Use of electron paramagnetic resonance spectroscopy to evaluate the redox state *in vivo*. *Antioxid Redox Signal,* Vol. 9, No 10, pp. 1757-71.

Takeshita, K., Hamada, A. and Utsumi, H. (1999). Mechanisms related to reduction of radical in mouse lung using an L-band ESR spectrometer. *Free Radic Biol Med,* Vol. 26, No 7-8, pp. 951-60.

Terpstra, M., Henry, P. G. and Gruetter, R. (2003). Measurement of reduced glutathione (GSH) in human brain using LCModel analysis of difference-edited spectra. *Magn Reson Med,* Vol. 50, No 1, pp. 19-23.

Tietze, F. (1969). Enzymic method for quantitative determination of nanogram amounts of total and oxidized glutathione: applications to mammalian blood and other tissues. *Anal Biochem,* Vol. 27, No 3, pp. 502-22.

Ullman, E. F. and Osiecki, J. H. (1970). Stable free radicals. VIII. New imino, amidino and carbomoyl nitroxides. *J Org Chem,* Vol. 35, No, pp. 3623-3631.

Vaupel, P. (2009). Pathophysiology of Solid Tumors, In: *The Impact of Tumor Biology on Cancer Treatment and Multidisciplinary Strategies.* M. Molls, P. Vaupel, C. Nieder and M. S. Anscher (Ed.): 51-92, Berlin, Heidelberg, Springer.

Velan, S. S., Spencer, R. G., Zweier, J. L. and Kuppusamy, P. (2000). Electron paramagnetic resonance oxygen mapping (EPROM): direct visualization of oxygen concentration in tissue. *Magn Reson Med,* Vol. 43, No 6, pp. 804-9.

Voinov, M. A., Polienko, J. F., Schanding, T., Bobko, A. A., Khramtsov, V. V., Gatilov, Y. V., Rybalova, T. V., Smirnov, A. I. and Grigor'ev, I. A. (2005). Synthesis, structure and X-band (9.5 GHz) characterization of the new series of pH-sensitive probes: N-N-disubstituted 4-amino-2,2,5,5-tetramethyl-3-imidazoline 1 Oxyls. *J.Org.Chem.,* Vol. 70, No, pp. 9702-9711.

Volodarsky, L. B., Reznikov, V. A. and Ovcharenko, V. I. (1994). Synthetic chemistry of stable nitroxides, CRC Press, Boca Raton, Fla.

Weiner, L. M. (1995). Quantitative determination of thiol groups in low and high molecular weight compounds by electron paramagnetic resonance. *Methods Enzymol,* Vol. 251, No, pp. 87-105.

Willis, J. A. and Schleich, T. (1995). 13C-NMR spectroscopic studies of 2-mercaptoethanol-stimulated glutathione synthesis in the intact ocular lens. *Biochim Biophys Acta,* Vol. 1265, No 1, pp. 1-7.

Woldman, Y., Khramtsov, V. V., Grigor'ev, I. A., Kiriljuk, I. A. and Utepbergenov, D. I. (1994). Spin trapping of nitric oxide by nitronylnitroxides: measurement of the activity of no synthase from rat cerebellum. *Biochem Biophys Res Commun,* Vol. 202, No 1, pp. 195-203.

Woldman, Y. Y., Semenov, S. V., Bobko, A. A., Kirilyuk, I. A., Polienko, J. F., Voinov, M. A., Bagryanskaya, E. G. and Khramtsov, V. V. (2009). Design of liposome-based pH sensitive nanoSPIN probes: nano-sized particles with incorporated nitroxides. *Analyst,* Vol. 134, No 5, pp. 904-10.

Yamashita, G. T. and Rabenstein, D. L. (1989). Determination of penicillamine, penicillamine disulfide and penicillamine-glutathione mixed disulfide by high-performance liquid chromatography with electrochemical detection. *J Chromatogr,* Vol. 491, No 2, pp. 341-54.

Yasui, H., Matsumoto, S., Devasahayam, N., Munasinghe, J. P., Choudhuri, R., Saito, K., Subramanian, S., Mitchell, J. B. and Krishna, M. C. (2010). Low-field magnetic resonance imaging to visualize chronic and cycling hypoxia in tumor-bearing mice. *Cancer Res,* Vol. 70, No 16, pp. 6427-36.

Zhu, X., Liu, B., Zhou, S., Chen, Y. R., Deng, Y., Zweier, J. L. and He, G. (2007). Ischemic preconditioning prevents *in vivo* hyperoxygenation in postischemic myocardium with preservation of mitochondrial oxygen consumption. *Am J Physiol Heart Circ Physiol,* Vol. 293, No 3, pp. H1442-50.

Zweier, J. L. and Kuppusamy, P. (1988). Electron paramagnetic resonance measurements of free radicals in the intact beating heart: a technique for detection and characterization of free radicals in whole biological tissues. *Proc Natl Acad Sci U S A,* Vol. 85, No 15, pp. 5703-7.

Chapter 12

Quantitative Determination of Thiol Status of Proteins and Cells by Nitroxyl Biradical ˙RS-SR˙

Lev Weiner

Additional information is available at the end of the chapter

Dedicated to the memory of Leonid Volodarsky

1. Introduction

Sulfhydryl (thiol, SH) groups of proteins and of low-molecular weight compounds, such as glutathione (GSH) and cysteine (Cys) play important roles in numerous biological processes. In the last decades, interest in the redox state of SH groups in proteins has grown because thiol-disulfide exchange has been found to play an important role in protein folding and to influence protein stability (1-3).

In cells, the ratio/equilibrium between oxidized and reduced forms of glutathione and between cysteine and cystine (main cells antioxidants) affect thiol balance and redox status of cells and proteins (4). In general, spectroscopic and chromatographic methods are used for quantitative determination of low-molecular thiols and sulfhydryl groups in proteins. Optical methods are employed to detect absorption or fluorescence, which appears on results of interaction between reagents and free SH groups. However, samples must be optically transparent, so preliminary homogenation and centrifugation of biological samples and other procedures are necessary (4). Chromatographic methods, especially HPLC (4), cannot be used for express analysis of thiol status of biological samples.

Among the optical methods for determining the free thiol groups the method proposed by Ellman (5) is definitely in the first place. This approach is based on the thiol-disulfide exchange reaction between the disulfide containing reagent (5,5'-dithiobis-(2-nitrobenzoic acid, DTNB), **ES-SE**, and the free thiol, **SH-T**:

$$\text{ES-SE} + \text{SH-T} \rightarrow \text{ES-ST} + \text{ES}^- \quad (1)$$

The resulting product, mono-thiol, **ES**$^-$, has a characteristic optical spectrum (λ_{max}= 412 nm) with a known extinction coefficient, ε=14 150 $M^{-1}cm^{-1}$. It is the simplicity of this method that has determined its widespread use (more than 13,500 citations for 50 years!). However, this method suffers from all the drawbacks typical for other optical methods: the impossibility of measuring in a colored, scattering, and turbid media, i.e. in real biological systems. In addition, the sensitivity of this method is often insufficient.

In 1987 we had a project including reversible modification of SH-group in NADPH-cytochrome P-450 reductase. We decided to get a paramagnetic analogue of the Ellman reagent, stable nitroxyl biradical, containing disulfide bond (6). We hoped that, if successful, the biradical would enter the free thiol/biradical thiol-disulfide exchange reaction, which could be followed by ESR. Our colleagues, Vladimir Martin and Tatyana Berezina from the team of Prof. Leonid Volodarsky (Institute of Organic Chemistry, Novosibirsk), synthesized biradical for the task (6,7)

In contrast to the known at that time disulfide containing spin label, [(1-Oxyl-2,2,5,5-tetramethylpyrroline-3-methyl) methanethiosulfonate],MTSSL, (8), our biradical allowed us to kill two birds with one stone: (a) to measure the kinetics of chemical modification of available SH groups in the protein (by appearance of free radical in solutions) and (b) using a traditional technique, i.e. gel- filtration or dialysis, to get the spin-labeled protein after incubation with our probe.(See eq.2 and Fig. 1))

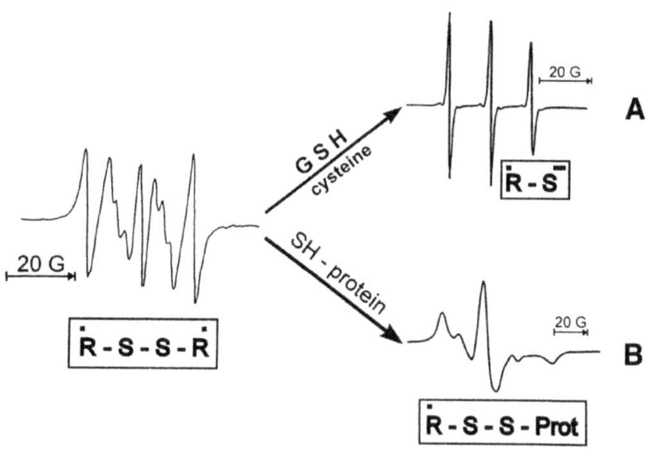

Figure 1. ESR spectra of biradical and its reaction products. **(A)** – the monoradical formed on a reaction with low-molecular weight thiol or free SH group in the protein and **(B)** - the immobilized radical formed on a reaction with protein-linked SH group.

This approach combines advantages of the methodology developed by Ellman (5) that makes use of thiol – disulfide exchange reaction (see eq. 1) and of ESR, that provides high sensitivity and possibility of carrying out work in colored and/or turbid and scattering media, such as cells, tissue culture, blood, etc.

2. Use of SNRs for determination of Thiol status in cells

For this purpose the symmetrical biradical containing disulfide bond, bis(2,2,5,5-tetramethyl-3-imidazoline-1-oxyl-4-il)-disulfide, ·**RS-SR**· was synthesized:

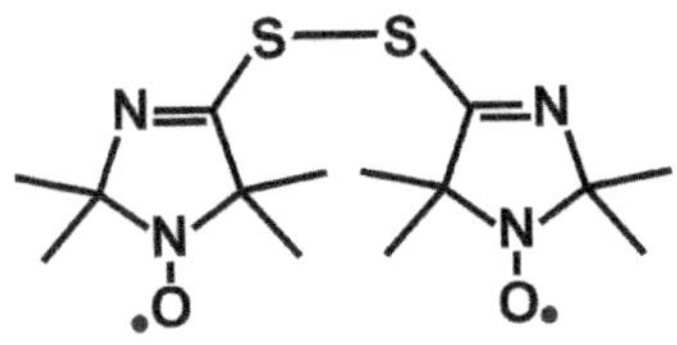

•RS—SR•

The observed ESR spectrum of ·**RS-SR**· is typical for symmetrical biradical with intermediate character of exchange between two unpaired electrons (9) (see fig.2A).

In the presence of a free thiol group the reaction of thiol-disulfide exchange takes place:

$$\bullet RS-SR\bullet + HS-A \rightleftarrows \bullet RS-SA + \bullet R-SH \tag{2}$$

$$\bullet RS-SA + HS-A \rightleftarrows AS-SA + \bullet R-SH \tag{3}$$

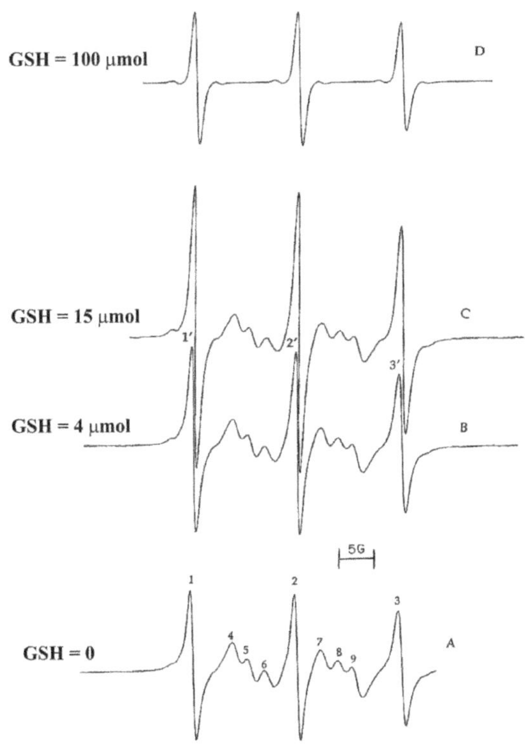

Figure 2. The effect of GSH on ESR spectra of ·**RS-SR**· (100 μM) in PBS, pH=7.5. Spectra **A, B, C** were carried out at: *gain* 5^x10^4, modulation amplitude 1 G, microwave power 10 mW; Spectrum **D**- *gain* 3.2^x10^3, modulation amplitude 1 G, microwave power 10 mW.

The exchange integral, J, was estimated: J = 3.6 a_N (9,10). The absence of any change in the ESR spectrum up to 80°C can be interpreted in terms of existence of a single average conformation of ·**RS-SR**· in solution.

Figure 2 shows the effect of reduced glutathione, GSH, on the ESR spectrum of ·**RS-SR**· : with increasing GSH concentration, the biradical spectral components (1,2,3,4,5,6,7,8,9) decrease with simultaneous increase of monoradical components (1′,2′3′). Thus nine broadened components of biradical decrease with concomitant appearance of three narrow components of two monoradicals as a result of the sequential reactions:

Note that the integral intensity of the ESR spectrum of ·**RS-SR**· remains unchanged. The peak intensities of the monoradical components **(·R-SH)** resulting from reactions (2,3) are about 17-fold higher than those of the corresponding biradical components (whose position in the field coincides with that of monoradicals). These phenomena provide the physical basis for the proposed method.

Four reviews give a detailed description of the physical and chemical background for the practical use of the biradical ·**RS-SR**· for thiols meaurements (10-13).

The proposed methodology allowed quantitative assessment of glutathione concentrations in mouse erythrocytes (7), in hamster ovary cells (10,14) and various types of malignant cells (15,16). ·**RS-SR**· is a hydrophobic molecule and , therefore, can easily cross biological membranes and penetrate into cells.

In contrast to conventional methods, our approach is non-invasive and suitable for work with *intact* cells and tissues. It is also extremely sensitive, permitting determination of GSH concentrations in as few as 100 cells (11,14). The method was also used successfully to measure GSH in an isolated reperfused heart (11). Using the biradical ·**RS-SR**· in conjunction with the spin trap DMPO, we were able to demonstrate that the efficacy of oxygen radical generation, stimulated by redox active quinones, correlated with GSH levels and the induction of expression of GSH transferase in cancer cells (15,16). The biradical method was also successfully applied to the monitoring of GSH levels in cancer cells treated with allicine, an active component of garlic, which can arrest the proliferation of cancer cells (17).

The biradical method was also used for direct determination of the catalytic activity of acetylcholinesterase in homogenates of the heads of individual larvae of the bollworm *Heliothis armigera,* following the rate of hydrolysis of acetylthiocholine by monitoring reduction of biradical by the thiocholine produced , according to eq.2(10,18).

Note that synthesis of new disulfide containing SNR is still in progress (19,20). Using the new disulfide containing biradical, the glutathione level (by L-band ESR spectrometer) in tumors in nude mice was measured. This "improved" biradical contains N-15 where deuterium substitutes for hydrogen atoms. This approach enhances the method sensitivity (20).

3. Determination of availability of Thiol groups in proteins

Traditionally, both alkylation and acylation spin labels have been used for chemical modification of proteins using SNRs. After incubation of nitroxyl radical with protein, the modified protein is separated from the free SNR by gel filtration, dialysis or precipitation. The use of biradical ·**RS-SR**· permits direct measurement of the rate of protein modification by the monitoring the appearance of the free monoradical, ·**R-SH,** in solution, (Fig. 1,eq. 2) thereby providing "visible" information about the rate of thiol-disulfide reaction (eq. 2) and, consequently, about the availability of the thiol group in the protein.

3.1. ESR study of the alcohol dehydrogenase free SH groups

Figure 3 illustrates the kinetics of chemical modification of the thermophilic alcohol dehydrogenase *from Thermoanaerobacter brockii* (TBADH) by biradical ·**RS-SR**· (10, 21). The high reaction rate suggests that when modified, the free thiol group is highly accessible. Modification of TBAD by [2-^{14}C] iodoacetic acid and identification of the labeled peptide indicated that the thiol group labeled was that of Cys 203. In the presence of coenzyme, NADP$^+$, the rate of modification falls (Fig.3, line C) providing evidence that NADP$^+$ interacts with Cys 203. The kinetics of chemical modification of TBADII after removal of Zn^{2+} from its active site by treatment with phenanthroline were almost two-fold higher than for the native enzyme (Fig.3, curve B). As follows from HPLC analysis of the radio-labeled peptide in apo-TBADH, Cys 37, which serves as a ligand for Zn^{2+} in the active site, is available for modification. Using apo-TBADH double-labeled at both Cys 203 and Cys 37 with biradical,

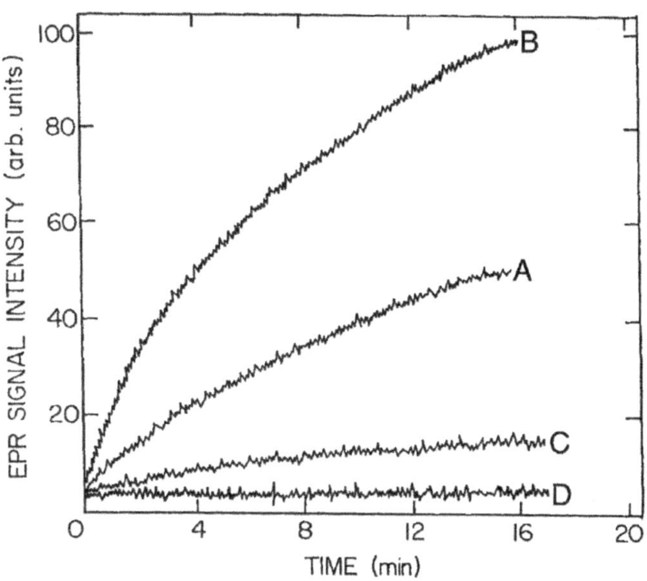

Figure 3. Kinetics of modification of TBADH by biradical ·**RS-SR**·. The peak intensity of the monoradical component, ·**R-SH** (released to solution as a result of reaction between biradical and TBADH (see eq.2)) was monitored. **A**- native TBAD; **B**-apo-TBAD; **C**-TBADH in the presence of NADP$^+$; **D**-TBADH pretreated with *p*-chloromercury benzoate. (from ref. 21)

we were able to estimate the distance between radicals covalently bound at these two cysteines (r~ 10 Å). Later, our colleagues in Weizmann Institute obtained an X-ray structure of TBADH at 2.5 Å resolution (22). The 3D structure has revealed that Cys 203 is indeed a surface residue which is occluded by the coenzyme $NADP^+$ (Fig.4)

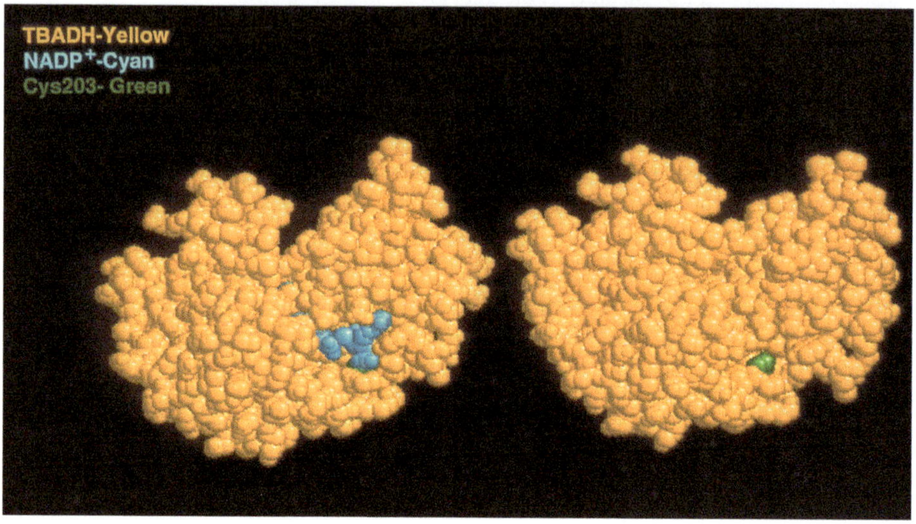

Figure 4. Representation of the monomer of TBADH (PDB entry 1ykf). Individual residues are represented as spheres colored in yellow, the $NADP^+$ cofactor and Cys 203 are colored in cyan (left) and green (right), respectively.

3.2. ESR study of the alliinase's SH groups

Alliinase (Cys sulfoxide lyase, alliin lyase, C-S lyase; EC 4.4.1.4) from garlic (Allium sativum) is an enzyme that uses pyridoxal-5′-phosphate (PLP) as a cofactor to catalyze the conversion of a nonprotein amino acid alliin , Sally cysteine Sulfoxide, to allicin (diallyl thiosulfinate), pyruvate, and ammonia, as shown in the following scheme:

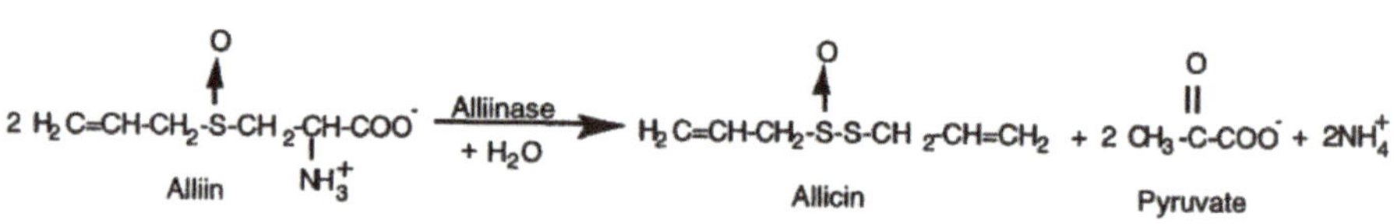

Scheme 1.

Allicin, a product of the enzymatic reaction of alliinase with alliin, is a well-characterized, biologically active compound of garlic. It is responsible for the pungent odor and for a variety of biological effects attributed to garlic preparations, including antimicrobial, anticancer, antiatherogenic, and other activities (23, 24)

Incubation of native alliinase either with 4,4′-dithiodipyridine (DTP) or with 5,5′-dithio-bis-(2-nitrobenzoic acid) (Ellman reagent) in the presence of 6M guanidine- HCl provided evidence for the existence of two free cysteine residues in the alliinase molecule.

To identify the free cysteine residues, alliinase was modified by treatment with N-(4-dimethylamino-3, 5-dinitrophenyl) maleimide (DDPM) and digested with trypsin, chymotrypsin or pepsin (27). Peptides in digests containing the nitrophenyl chromophore were separated and detected on a 360- nm absorbance profile using reversed-phase HPLC. By analyzing the trypsin and chymotrypsin digests, we were able to identify a single (but different) Cys-containing peptide in each case. In the case of trypsin, it was a peptide containing a sequence with Cys220, and in the case of chymotrypsin the peptide contained the sequence with Cys350 (27). Treatment with pepsin made it possible to identify both of these free cysteine residues simultaneously in one digest. These experimental findings (predating the alliinase structure determination) provided direct confirmation that the two free thiols in the alliinase molecule (27).

Using ESR spectroscopy, we examined the availability of the free —SH groups of alliinase for chemical modification with the disulfide containing biradical ·**RS-SR**· (27). The rate of the thiol–disulfide exchange reaction was monitored by ESR assay of the monoradical **·R—SH** released in this reaction, according to Eq. 2.

Figure 5 shows the increase in peak intensity of the ESR signal for the reaction between the biradical and the native alliinase. These data demonstrate that the kinetics of modification occur at two different rates. Pretreatment of alliinase with *p*-chloro mercury benzoate dramatically inhibited the modification kinetics (data not shown). Figure 6 shows the ESR spectrum of alliinase modified by the biradical (after 5 h of incubation followed by removal of the excess reagent by gel filtration) at 120 K. This spectrum is typical of a nitroxyl stable radical in a frozen solution. The degree of modification obtained by double integration of the ESR spectrum was 1.61 ± 0.15 per subunit of alliinase. To estimate the distance between two labeled cysteine residues, we used the empirical parameter d_1/d (see Fig.6), which characterizes the dipole–dipole interaction between unpaired electrons of two nitroxyl groups, as proposed by Kokorin *et al* (28) and commonly used to estimate the distance between two radicals covalently bound to proteins (29, 30). In the absence of dipole–dipole interactions between the radicals, a value $d_1/d < 0.4$ is expected. The value obtained for d_1/d obtained in our experiment was 0.38, and the distance between the labeled cysteines Cys 220 and Cys 350 was estimated to be larger than 22 Å.

Alliinase has been crystallized and its three-dimensional structure solved (25-27). The enzyme is a homodimeric glycoprotein belonging to the fold-type I family of PLP-dependent enzymes.

As shown earlier with biochemical methods, the enzyme subunit contains two free thiols: Cys220 in the PLP-binding domain 2 and Cys350 in the C-terminal part of domain 1 (C7 and

C8, respectively) (Figs7 and 8) located relatively far from the active site and from the substrate-binding area. As shown in Figure 8(A), Cys220 is located on the surface of the alliinase molecule, while Cys350 is in a more buried location but is still water-accessible. The free thiol groups of Cys220 and Cys350 have different relative orientations with respect to each other (Fig. 8B), which might affect their chemical modification rates by ·**RS-SR**· (see Fig. 5). Distances between all the cysteines involved in disulfide bonds and the free thiols in the alliinase dimer range between 15 and 68 Å, and do not allow rearrangement of disulfide bonds in the native state.

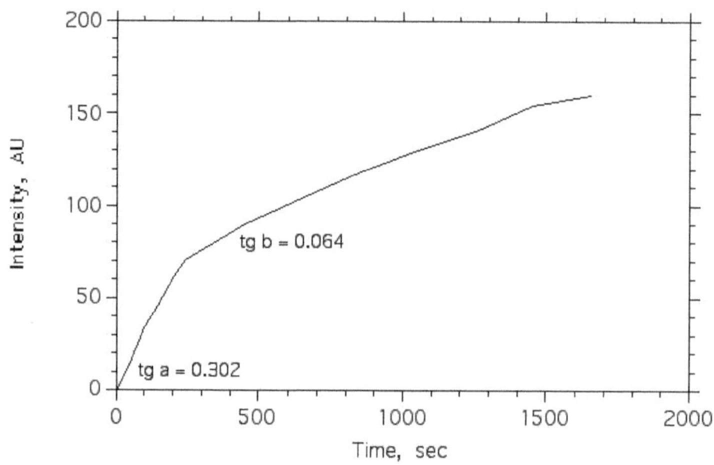

Figure 5. Kinetics of the nitroxyl biradical modification of alliinase. Peak intensity of the ESR spectrum of the monoradical (·**R-SH)** component that was released into solution as a result of the thiol–disulfide exchange between SH groups of alliinase and the biradical ·**RS-SR**· (see eq. 2) (from ref. 27)

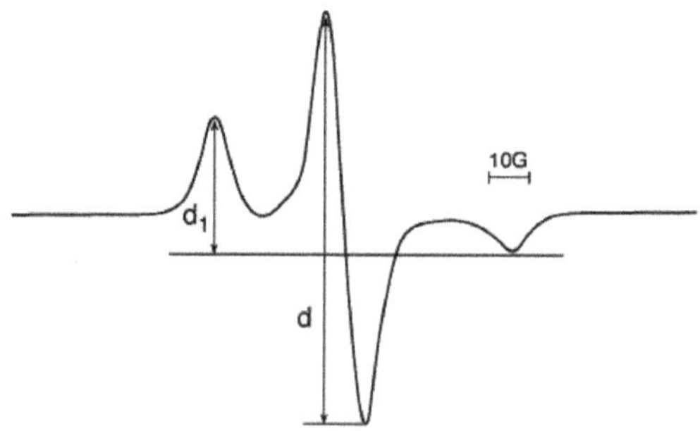

Figure 6. ESR spectrum of the alliinase-biradical conjugate. The conjugate (12 μM) was measured in a PBS/glycerol (70/30) mixture at 120 K. ESR conditions: microwave power, 10 mW; modulation amplitude, 1.25 G;

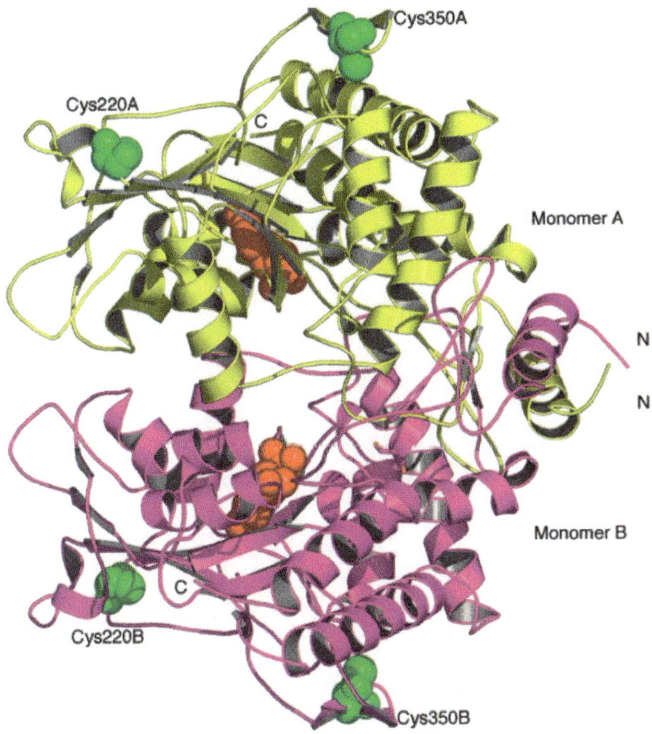

Dimeric structure of alliinase is depicted in the illustration. Yellow and magenta indicate monomers A and B, respectively. Pyridoxal-5'-phosphate groups are shown as red spheres. The dimer is rotated 180^0 around a vertical axis with respect to A. Free cysteines are shown as green spheres

Figure 7. Distribution of cysteines in a monomer of alliinase from garlic (*Allium sativum*). (from ref. 27)

4. Reversible modification of Thiol groups in proteins

Free thiol groups, whether intrinsic or introduced by site-directed mutagenesis are convenient targets for introduction of stable nitroxyl radicals, SNRs, into proteins. Now, this approach, named site directed spin labeling is very popular, because it can give information about structure (mobility) of different parts of the protein globule (31-33). "Classical" SNRs used for modification of thiol groups, such as NR-labeled derivatives of iodoacetamide and maleimide, yield strong covalent S-C bonds which do not permit release of the spin label from the protein. Chemical modification using a disulfide-containing SNRs permits subsequent demodification by a low-molecular weight thiol such as mercaptoethanol, reduced glutathione, cysteine or dithiothreitol. Such demodification, performed in conjunction with simultaneous measurements of activity and of structural characteristics, allows evaluation of the contribution of the group modified to the stability and 3D structure of the protein studied. Berliner et al. (8) used the spin label MTSSL, for reversible chemical modification of Cys 25 in papain. We made use of biradical ·**RS-SR**· for the reversible chemical modification of NADPH-cytochrome P-450 reductase from rat liver (6,34), human hemoglobin (7), *Torpedo californica* acetylcholinesterase (*Tc*AChE) (35-37), TBADH (10,21) and allinase from garlic (27).

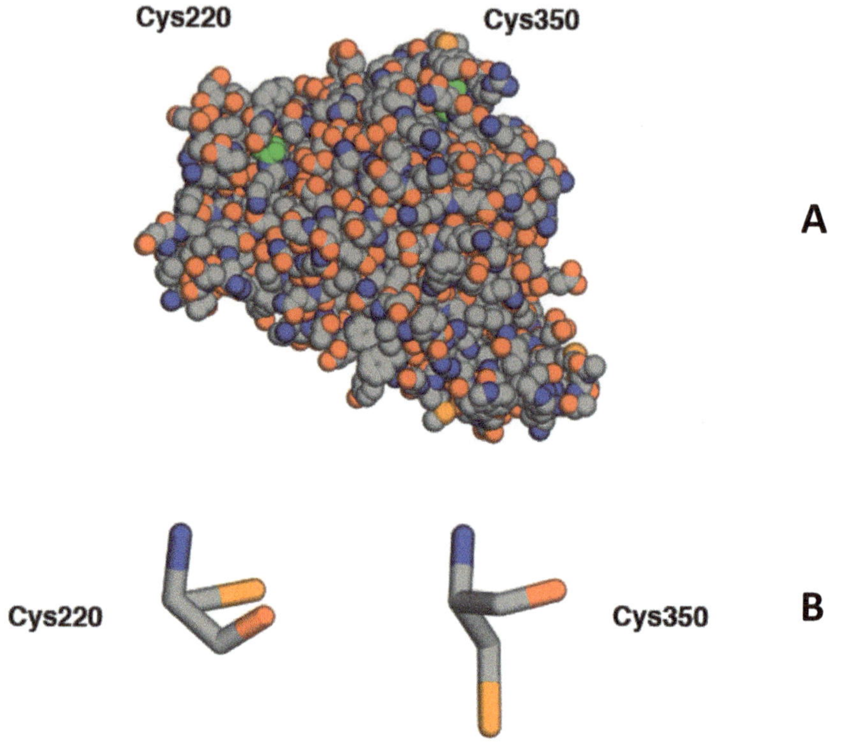

Figure 8. Free thiols of alliinase. (**A**) Relative locations of Cys220 and Cys350 (green) on the surface of the alliinase monomer. (**B**) Respective orientations of CysS220 and Cys350 relative to A. (from ref. 27)

4.1. Acetylcholinesterase from *Torpedo Californica* (*Tc*AChE)

Cys 231, a deeply buried residue in *Tc*AChE, was modified by biradical (eq.4) to yield a catalytically inactive species, even though it is not involved in the active site of the enzyme, which is a serine hydrolase (35-37).

$$\cdot\text{RS-SR}\cdot + \text{HS-}\text{(PR)} \rightleftharpoons \cdot\text{RS-S-}\text{(PR)} + \cdot\text{R-SH} \qquad (4)$$

$$\cdot\text{RS-S-}\text{(PR)} + \text{HS-G} \rightleftharpoons \cdot\text{RS-SG} + \text{HS-}\text{(PR)} \qquad (5)$$

Demodification of spin labeled protein by GSH (see eq.5), with concomitant release of the free monoradical spin label, done by ESR control, did not result in recovery of enzymatic activity. The use of a wide repertoire of physicochemical and biochemical techniques subsequently established that both the modified and demodified enzymes had assumed a partially unfolded, molten globule, **MG**, conformation (38,39). In such cases, where the chemical modification induced unfolding of the protein, there was a concomitant complete disappearance of ellipticity in the near-UV of the CD spectrum (λ_{min} = 280 nm), red shift of the maximum of intrinsic fluorescence spectrum (**333 nm → 341 nm**) and enzyme inactivation. However, changes of the secondary structure were very modest (35). Chemical modification Cys 231 by organo-mercurials nitroxyl radical: (2,2,5,5-Tetramethyl-4-[2-(chloromercuri)phenyl]-3-imidazoline-1-oxy1 (**HgR**·) and the natural thiosulfinate, allicin (see sheme 1), transforms *Tc*AChE to a quasi-native (**N***) state (36,40). Note, that modification by **HgR**· and allicin was also reversible and modified AChE was demodified by treatment with reduced glutathione (eq 5). Demodification of the organomercurial (or allicin) modified enzyme with GSH shortly after modification leads to regeneration of the physicochemical characteristics of the native enzyme as well as to *Tc*AChE reactivation. The modified enzyme in **N*** state is, however, metastable, and is converted spontaneously and irreversibly, at room temperature, with $t_{1/2} \simeq$ 100 min, to an **MG** state. Using the developed approach we were able to describe the conformational states of *Tc*AChE, transitions between these states (**N**- native, **U**-unfolded state), stimulation transition to MG state by biological membranes (41,42) as well as stabilization of **N** and **N*** states by chemical and pharmacological chaperons (40,43)

$$\mathbf{N} \Leftrightarrow \mathbf{N^*} \Rightarrow \mathbf{MG} \Leftrightarrow \mathbf{U}$$

Scheme 2.

4.2. TBADH and NADPH-cytochrome P-450 reductase

Chemical modification of both TBADH and NADPH-cytochrome P-450 reductase by biradical ·**RS-SR**· (6,10,21,34)) also led to their inactivation (see Fig. 9)

However, in both cases removal of the bound spin label by treatment with the free thiol according to eq,5, resulted in immediate reactivation (Fig. 9). Spectroscopic measurements showed that modification had changed neither the tertiary nor secondary structure of the proteins and could be protected by affine inhibitor $NADP^+$.

4.3. Alliinase from garlic (*Allium sativum*)

We have shown recently that modification of Cys 220 and Cys 350 of alliinase with ·**RS-SR**· does not change its enzymatic activity (27). In this case chemical modification of both free cysteine residues was found to leave both the secondary and the tertiary structure of the enzyme unchanged. This might be attributable to the marked thermodynamic and structural

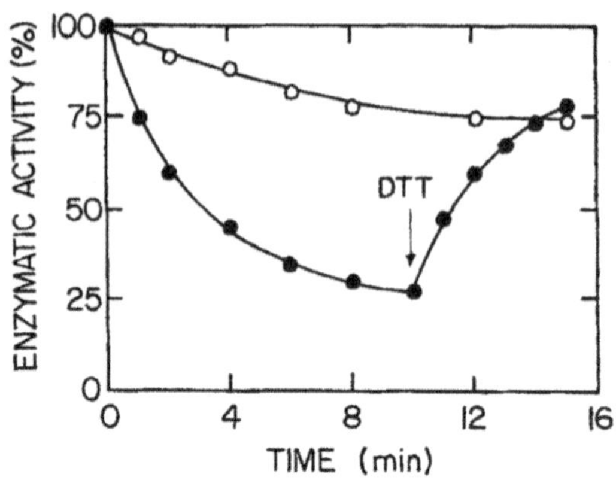

Figure 9. Influence of $NADP^+$ and DTT on effect of biradical ·**RS-SR**· on TBADH activity. Enzyme was incubated with the biradical without (closed circles) and with (open circles) $NADP^+$. DTT was introduced at 10 min (shown by arrow). (from ref. 21)

stability of alliinase, as well as the relatively long distances from modified free cysteines to the active center of the enzyme (see fig 7). This experimental finding permits one to use cysteines of alliinase for covalent binding with antibodies for targeted delivery of enzyme and for site-specific allicin generation to inhibit cancer cells proliferation (44,45).

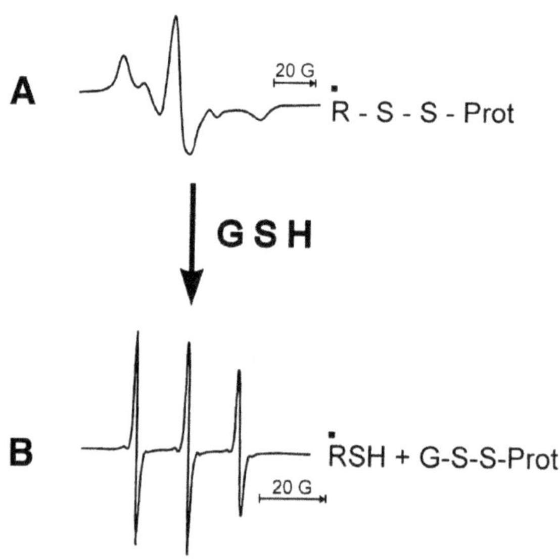

Figure 10. ESR spectrum of Alliinase-radical conjugate. Protein concentration was 13 μM in 10 mM PBS buffer, pH7.6. **(A)** - ESR spectrum of modified protein, ESR conditions: microwave power 10 mW; modulation amplitude 1 G; gain $2x10^5$. **(B)**- ESR spectrum of sample **(A)** 4 min after addition of 0.2 mM of glutathione. ESR conditions were the same as in **(A)**, but the gain was $3.2x10^4$.

5. Conclusions

For quantitate determination of sulfhydryl groups in low molecular weight compounds and proteins the symmetrical biradical containing disulfide bond, bis(2,2,5,5-tetramethyl-3-imidazoline-1-oxyl-4-il)-disulfide, or **·RS-SR·** , was synthesized. The biradical, **·RS-SR·** permits quantitative determination of glutathione and cysteine in biological systems ('thiol status') by use of ESR. Note that any *routine* CW ESR spectrometer can be used for the measurements! This non-invasive express method has very high sensitivity: possibility of measuring < 10^{-12} mol of SH groups in a sample. Thus, concentrations of glutathione and cysteine in living cells were determined under physiological conditions, as well as in cells, including cancer cells, treated by pro-and anti-oxidants.

Different from the traditional spin label method, the **·RS-SR·** usage makes it possible to assess the rate of modification of SH groups in proteins (availability) and the influence of substrates, inhibitors, coenzymes, other proteins, artificial and biological membranes, pH, etc. in direct experiments. Modification of SH groups in proteins by **·RS-SR·** is *reversible*, which permits application of this approach, coupled with site directed spin labeling, for evaluation of stability and unfolding of proteins and different parts of the protein globule.

Author details

Lev Weiner
Weizmann Institute of Science, Rehovot, Israel

6. References

[1] Kim, P.S. and Baldwin, R.L.(1990) Intermediates in the folding reactions of small proteins. Ann.Rev.Biochem. 59, 631-660

[2] Freedman, R.B. (1995) The formation of protein disulphide bonds. Curr. Opin. Struct. Biol. 5, 85-91

[3] Eichler, J., Kreimer, D.I., Varon, L., Silman, I., and Weiner, L. (1994) A 'molten globule' of Torpedo acetylcholinesterase undergoes thiol-disulfide exchange. J. Biol. Chem., 269, 30093-30096

[4] Methods in Enzymology: Biothiols,v.251 A,B (Packer, L.,Ed.) 1995.

[5] Ellman G.L. (1959) Tissue sulfhydryl groups. Arch. Biochem. Biophys. 82 , 70–77

[6] Elinova, V.I., Slepneva, I.A., Weiner, L.M. Affinity modification of NADPH-Cytochrome P450 Reductase by reversible inhibitors.In: Cytochrome P-450, Biochemistry and Biophysics (I. Shuster, ed.) pp.411-414, 6th International Conference Vienna, Austria 1988., London, Taylor and Francis,1989.

[7] Khramtsov, V.V., Elinova, V.I., Weiner, L.M., Berezina, T.A. Martin, V.V., and Volodarsky, L.B. (1989) Quantitative determination of SH groups in low- and high-molecular-weight compounds by an electron spin resonance method. Anal. Biochem., 182, 58-63

[8] Berliner,LJ., Grunwald,J., Hankovszky, O. and Hideg, K., (1982) A novel reversible thiol-specific spin label: papain active site labeling and inhibition. Anal. Biochem. 119, 450-455

[9] Luckhurst, G.R. Biradicals as Spin Probes. In: Spin Labeling, Theory and Applications, (L.J. Berlinrer, ed) p.113, Academic Press, New York, (1976)

[10] Weiner, L.M. (1995) Quantitative determination of thiol groups in low and high molecular weight compounds by electron paramagnetic resonance. Methods in Enzymology, 251, 87-105

[11] Nohl, H., Stolze, K., and Weiner, L.M. (1995) Noninvasive measurement of thiol levels in cells and isolated organs. Methods in Enzymology 251, 191-203

[12] Khramtsov VV, and Volodarsky LB. Use of imidazoline nitroxides in studies of chemical reactions: ESR measurements of concentration and reactivity of protons, thiols and nitric oxide. In: Biological Magnetic Resonance (Berliner LJ, editor). Vol. 14. Spin labelling: the next millennium New York: Plenum Press; 1998, pp. 109–179

[13] Weiner, L.M. (2007 Stable nitroxyl radicals as pH, thiol and electron transfer probes Appl. Magn. Res. 31, 357-373

[14] Weiner, L.M., Hu, H., and Swartz, H. (1991) EPR method for the measurement of cellular sulfhydryl groups. FEBS Letters, 290, 243-246

[15] Pinkus, R., Weiner, L., and Daniel, V. (1995) Role of quinone-mediated generation of hydroxyl radicals in the induction of glutathione S-transferase gene expression. Biochemistry 34, 81-88.

[16] Pinkus, R., Weiner, L., and Daniel, V. (1996) Role of oxidants and antioxidants in the induction of AP-1, NF-kappa B, and glutathione S-transferase gene expression J. Biol. Chem. 271, 13422-13429

[17] Hirsch K, Danilenko, M, Giat J, Miron, T, Rabinkov, A, Wilchek, M, Mirelman, D, Levy, J, and Sharoni, Y. (2000) Effect of purified allicin, the major ingredient of freshly crushed garlic, on cancer cell proliferation. Nutrition and Cancer, 38, 245-254

[18] Khramtsov, V.V., Gorunova, T.E., and Weiner, L.M.: (1991) Monitoring of enzymatic activity *in situ* by EPR Biochem. Biophys. Res. Commun., 179, 520-527

[19] Ionita,P., Caragheorgheopol,A., Gilbert,B.C., and Chechik.V. (2004) Mechanistic Study of a Place-Exchange Reaction of Au Nanoparticles with Spin-Labeled Disulfides. Langmuir 20, 11536-11544

[20] Roshchupkina GI, Bobko AA, Bratasz A, Reznikov VA, Kuppusamy P, and Khramtsov VV. (2008) *In vivo* EPR measurement of glutathione in tumor-bearing mice using improved disulfide biradical Free Radic Biol Med. 45,312-330

[21] Peretz, M., Weiner, L., and Burstein, Y. Protein Science (1997) Cysteine reactivity in *Thermoanaerobacter brockii* alcohol dehydrogenase 6, 1074-1083

[22] Korkhin,Y, Kalb(Gilboa), A. J., Peretz, M., Bogin, O., Burstein, Y., and Frolow,F. (1998) NADP-dependent bacterial alcohol dehydrogenases: crystal structure, cofactor binding and cofactor specificity of the ADHs of *Clostridium beijerinckii* and *Thermoanaerobacter brockii*. J. Mol. Biol. 278, 965-979

[23] Rahman, M.S. (2007) Allicin and other functional active components in garlic: Health benefits and bioavailability. Int. J. Food Properties 10,245–268

[24] Rabinkov A, Miron T, Konstantinovski L, Wilchek M, Mirelman D, and Weiner, L.(1998) The mode of action of Allicin: Trapping of radicals and interaction with thiol containing proteins. Biochim Biophys Acta 1379, 233–244

[25] Kuettner EB, Hilgenfeld R, Weiss MS.(2002) The active principle of garlic at atomic resolution. J Biol Chem 277, 46402–46407

[26] Shimon LJ, Rabinkov A, Shin I, Miron T, Mirelman D, Wilchek M, and Frolow F.(2007) Two structures of alliinase from Alliium sativum L.: apo form and ternary complex with aminoacrylate reaction intermediate covalently bound to the PLP cofactor. J. Mol. Biol. 366,611–625

[27] Weiner, L., Shin, I; Shimon, LJW., Miron, T., Wilchek, M., Mirelman, D., Frolow, F. and Rabinkov. A. (2009) Thiol-disulfide organization in alliin lyase (alliinase) from garlic (*Allium sativum*). Protein Science 18, 196-205

[28] Kokorin AI, Zamaraev KI, Grigorian GL, Ivanov VP, and Rozantsev EG., (1972) Measurement of the distance between paramagnetic centers in solid solutions of nitrosyl radicals, biradicals and spin-labelled proteins. Biofozika, 17,31-39

[29] Mevorat-Kaplan K., Weiner, L. and Sheves M., J. Phys. Chem. B. (2006) Spin labeling of *Natronomonas pharaonis halorhodopsin*: probing the cysteine residues environment. J. Phys. Chem B, 110, 8825-8831

[30] Schiemann, O. and Prisner,T. (2007) Applications of electron paramagnetic resonance to distance measurements in biomolecules. Quarterly Reviews of Biophysics; 40, 1-53

[31] Altenbach,C., Marti, T., Khorana,H.G., and Hubbell, W.L.(1990) Transmembrane protein structure: spin labeling of bacteriorhodopsin mutants. Science, 248, 1088-1092

[32] Hubbell, W.L. and Altenbach,C. (1994) Investigation of structure and dynamics in membrane proteins using site-directed spin labeling. Curr.Opin. Struct.Biol., 4, 566- 573

[33] Eliash, T., Weiner,L., Ottolenghi, M., and Sheves,M.(2001) Specific binding sites for cations in bacteriorhodopsin. Biophys. J. 81, 1155-1162

[34] Elinova, V.I., Weiner, L.M., Slepneva, I.A., and Levina, A.S. (1993) Reversible Modification of Cysteine Residues of NADPH-Cytochrome P450 Reductase. Biochem. Biophys. Res. Commun. 193, 1044-1048

[35] Dolginova, E., Roth, E., Silman, I., Weiner, L.(1992) Chemical modification of *Torpedo* acetylcholinesterase by disulfides: Appearance of a "molten globule" state . Biochemistry, 31, 12248-12254

[36] Kreimer, D., Dolginova, E., Raves, M., Sussman, J., Silman, I., and Weiner, L.M. (1994) A metastable state of *Torpedo californica* acetylcholinesterase generated by modification with organomercurials.Biochemistry, 33, 14407-14418

[37] Kreimer, D.I., Shin, I., Shnyrov, V.L., Villar, E., Silman, I., and Weiner, L.(1996) Two partially unfolded states of *Torpedo californica* acetylcholinesterase. Protein Sci. 5,1852-1864

[38] Arai, M. and Kuwajima, K.:(2000) Role of the molten globule state in protein folding . Adv. Protein Chem. 53, 209-282

[39] Englander, S.W. (2000) Protein folding intermediates and pathways studied by hydrogen exchange. Annu Rev. Biophys. Biomol. Struct. 29, 213-238

[40] Millard, CB., Shnyrov, VL.,Newstead, S., Shin, I., Roth, E., Silman, I and Weiner, L.(2003) Stabilization of a metastable state of *Torpedo californica* acetylcholinesterase by chemical chaperones. Protein Sci. 12, 2337-2347,

[41] Shin, I., Kreimer, D.I., Silman, I. & Weiner, L. (1997). Membrane-promoted unfolding of acetylcholinesterase: A possible mechanism for insertion into the lipid bilayer. Proc. Natl. Acad. Sci. USA, 94, 2848-2852

[42] Shin, I., Silman,I., Bon, C., and Weiner ,L. (1998) Liposome-catalyzed unfolding of acetylcholinesterase from Bungarus fasciatus. Biochemistry, 37, 4710-4716

[43] Weiner, L., Shnyrov, VL., Konstantinovskii, L., Roth, E., Ashani, Y.,and Silman,I. (2009) Stabilization of *Torpedo Californica* Acetylcholinesterase by Reversible Inhibitors. Biochemistry 48, 563-574

[44] Appel, E ., Vallon-Eberhard, A., ; Rabinkov, A .,Brenner, O., Shin, I., Sasson, K., Osherov, N., Jung, S., and Mirelman, D. Antimicrob. Agents Chemotherap. 54, 898-906 (2010)

[45] Appel, E ., Rabinkov, A., Neeman, M., Kohen, F., and. Mirelman, D.(2011) Conjugates of daidzein-alliinase as a targeted pro-drug enzyme system against ovarian carcinoma. J. Drug. Target. 19, 326-335

Chapter 13

Use of Spin Trap Technique for Kinetic Investigation of Elementary Steps of RAFT-Polymerization

Anatoly Filippov, Elena Chernikova, Vladimir Golubev, Ganna Gryn'ova, Ching Yeh Lin and Michelle L. Coote

Additional information is available at the end of the chapter

1. Introduction

The development of controlled free-radical polymerization has made possible the synthesis of polymer structures with exquisitely controlled molecular weight distributions and architectures, unattainable by conventional free-radical polymerization. The reversible addition-fragmentation chain-transfer (RAFT) process (Barner-Kowollik, 2008) is one of the most promising techniques. Control is achieved by reversibly storing the majority of the propagating species as dormant dithioester compounds. The principal reaction steps are shown in Scheme 1, where radical P_n is a propagating polymer radical of chain length *n* and M is the monomer.

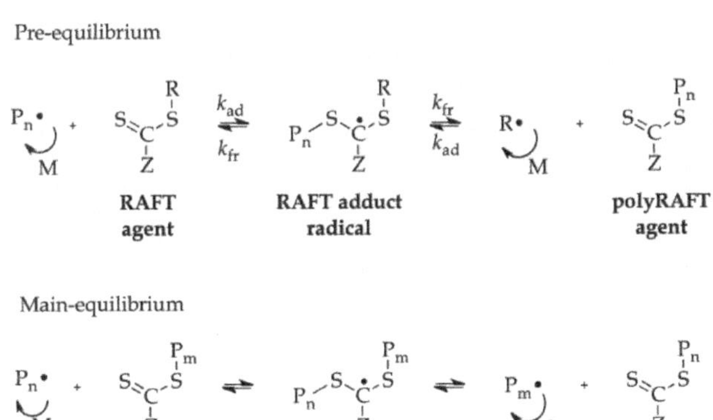

Scheme 1. RAFT Reaction Scheme

In the early stages of the RAFT process these radicals undergo chain transfer with the initial RAFT agent, releasing leaving groups R• that are chosen to be capable of reinitiating polymerization. The main equilibrium is a symmetrical process whereby a polymeric propagating radical reacts with a corresponding polymeric RAFT agent, forming an equivalent polymeric RAFT agent and an equivalent propagating radical. In this way a relatively small population of propagating radicals is constantly interchanging with a much larger population of the dormant species. For a successful process it is necessary to choose a RAFT agent such that the rate of addition of the propagating radical to the RAFT agent is considerably faster than the propagation rate, but not so favored that the (reverse) fragmentation reaction is inhibited. It is also necessary to choose a leaving group R such that it fragments preferentially from the intermediate radical in the pre-equilibrium, but at the same time is not overly stabilized (compared with the propagating radical) and thus remains capable of reinitiating polymerization. Accurate and precise kinetic data can be extremely helpful when choosing appropriate RAFT agents that meet these criteria.

Despite extensive studies of the RAFT mechanism over the last decade or so, some questions remain unresolved. Since the first publications of the kinetic features of the RAFT process, there has been an ongoing debate about the origin of retardation and inhibition effects in polymerizations mediated by cumyl dithiobenzoate and related dithiobenzoate RAFT agents (Barner-Kowollik et al, 2006, Klumperman et al, 2010). On the one hand, a kinetic model that assumes that cross- and self-termination reactions of intermediate radicals occur with diffusion-controlled rate coefficients, similar to those for the bimolecular termination of propagating polymer radicals, can be successfully fitted to experimentally-determined overall reaction rates and ESR-derived radical concentrations (Kwak et al, 2002). However, this model predicts that significant concentrations of the termination products should be produced even under standard RAFT conditions and these are not detected in significant quantities in the resulting polymer except under forcing conditions (Ah Toy et al, 2004). Furthermore, fragmentation rate coefficient ($k_{fr} = 10^4\ s^{-1}$) and equilibrium constants ($K=k_{ad}/k_{fr}=$ 55 mol L^{-1}, where k_{ad} is the addition rate coefficient) obtained under this model are incompatible with those predicted from quantum chemistry and radical storage experiments (Barner-Kowollik et al., 2003). Many of these problems can be addressed if instead one assumes a model, in which termination of the intermediate radical is not kinetically significant under normal polymerization conditions (i.e., $\langle k_t \rangle <$ ca. 10^4 L $mol^{-1}\ s^{-1}$). In this case, model fitting to experimental kinetic data indicates that the intermediate radicals are much more stable ($K = 1.06 \times 10^7$ L mol^{-1}) and that there is no significant amount of termination products (Feldermann et al., 2004). However, this so-called slow fragmentation model also predicts intermediate radical concentrations that are incompatible with the available ESR data, measured for polymerizing systems. Resolving these inconsistencies and finding the correct kinetic model for RAFT polymerization would improve our understanding of the process and assist in its optimization and control.

To eliminate the existing contradictions, several new models have been proposed recently. For example, Buback has suggested that the products of the termination reactions involving radical intermediates could interact with propagating radicals and regenerate the radical intermediates (Buback et al., 2007); however, no direct evidence for this reaction has been provided. Moreover, the above-described reaction may occur only in the case of dithiobenzoates; in the case of trithiocarbonate-mediated polymerization, which is also characterized by retardation and in which inhibition phenomena are typical, this reaction is not possible. Another proposal is that there is a sharp dependence of the rate constant of termination of radical intermediates on the chain length (Konkolewicz et al., 2008). In this way, the majority of termination products would have low chain lengths and would not be detectable in the termination products of the polymer. They would also not contribute significantly enough to influence the fitted equilibrium constant, which could remain compatible with the quantum chemical calculations. At the same time, the termination could still reduce the intermediate radical concentration, particularly given that the ESR studies were carried out in the presence of large quantities of initiator. However, this hypothesis has been criticized on the basis that there are no grounds to believe that the chain length dependence of the termination rate coefficient would differ so appreciably from the known dependence of the bimolecular termination rate coefficient for propagating radicals (Klumperman et al, 2010). Nonetheless, recent tantalizing experiments with macroazoinitiators and macroRAFT agents do indicate that rate retardation phenomena can be avoided when low molecular weight radicals are eliminated (Ting et al, 2011).

Thus, the analysis of the published data shows that the general scheme of RAFT polymerization presented above (Scheme 1) cannot fully describe this complex process. To resolve these issues, we have been using ESR spectroscopy to measure rate coefficients. An advantage of ESR is that this procedure allows direct observation of the formation of radical intermediates and confirmation of their structure (Chernikova et al., 2004, Hawthorne et al., 1999, Golubev et al., 2005). This is possible because radical intermediates are less reactive than other radicals involved in polymerization and their steady-state concentrations are sufficient for their direct detection with modern radiospectrometers. However, the direct use of ESR spectroscopy for experimental measurements of the individual rate coefficients of elementary steps of addition-fragmentation reactions is a very complicated task due to the simultaneous participation of many radical species, of various chemical structures, in the polymerization process. The extraction of the individual rate coefficients in this case has necessarily involved the fitting of some type of assumed kinetic model to the experimental data. This approach is sensitive to the choice of the kinetic model and requires searching for dozens of kinetic parameters. Experimentally, one can reduce the dependence on model-based assumptions by carrying out experiments in situations where the kinetic effects of these assumptions are minimal, or by studying the reactions in isolation, usually on much simpler model compounds. A promising example of the former approach is a laser flash photolysis technique recently introduced by Buback et al. (Buback et al, 2006), which has been used to measure the fragmentation rate and equilibrium constants for S-S-bis(methyl-2-propionate)-trithiocarbonate mediated polymerization of butyl acrylate in toluene at 30°C. A drawback of this technique is that the measurements take place in a polymerizing system where radicals and RAFT agents of various chain lengths may contribute to the detected intermediate radical concentrations; in their previous work, the authors did not confirm that the contribution of the intermediates formed in pre-equilibrium (reaction 1, Scheme 1) to the hyperfine structure of the observed ESR spectra was negligibly small. Nonetheless, subsequent quantum-chemical calculations suggest that convergence with respect to chain length is rapid in this particular system and the experimental value does correspond well to the converged theoretical value (Lin et al., 2009). More generally though, there are serious problems when applying this technique to dithiobenzoates, which result from their absorbance and decomposition at the wavelength of the laser. Alternative and complementary techniques are therefore desirable.

In this chapter we describe how the mechanism and kinetics of the elementary events of RAFT polymerization can be studied using an experimental approach based on the combination of the ESR spectroscopy and the spin-trapping technique. The collected experimental data are compared with the values computed with the aid of quantum-chemical methods. The experimental data reported in Section 3 is taken partly from our recent publications as noted (Chernikova et al., 2010, Golubev et al., 2011); the data in Section 4 is new. The theoretical calculations in Section 5 are likewise taken from these publications or our other relevant theoretical studies (Coote et al., 2006, Izgorodina et al., 2006, Lin et al., 2009, 2011); where necessary additional calculations have been performed to convert the theoretical data to the same conditions (temperature, solvent) as the experiments for consistent comparison.

2. Use of the spin-trapping technique to investigate the RAFT mechanism

2.1. Choice of Spin Trap

Spin traps are inhibitors that can rapidly capture active radicals (with spin-trapping constants $k_{RT} = 10^5$–10^8 L mol^{-1}s^{-1} (Denisov, 1971)) and convert them into new kinetically or thermodynamically stable radicals (spin adducts), preferably whilst retaining information about their nature. As opposed to active radicals, whose steady-state concentration in conventional liquid-phase reactions is low and, therefore, cannot be detected by ESR-spectroscopy, spin adducts can be accumulated in amounts sufficient for direct quantitative analysis. The choice of the spin trap and its application technique depend on the particular reaction being studied. Since the 1970s, the spin trap technique has been applied for investigation of a wide range of liquid-phase radical reactions, including radical polymerization reactions (Golubev et al., 2001, Golubev, 1994). However, until very recently, the application of this method for investigation of RAFT polymerization had not been reported. Below, we will consider the use of spin traps for investigation of RAFT polymerization.

Nitrosocompounds and nitrones are spin traps that convert carbon-centered radicals to (relatively) stable nitroxide radicals. An example of such a reaction is shown below, where the active radical $R^{\bullet}$ reacts with the spin trap 2-methyl 2-nitrosopropane (MNP) forming a nitroxide radical as the so-called spin adduct (reaction 1).

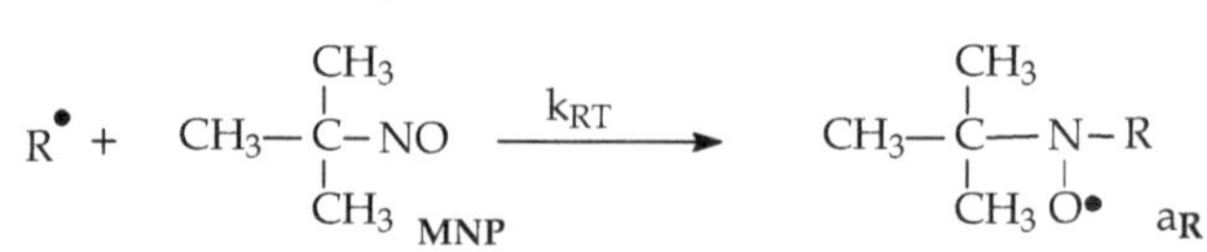

(1)

The ESR spectrum of this nitroxide is strongly dependent on the chemical nature of the trapped radical R•, and hence it may serve for qualitative identification of the radicals formed in the reaction media. When MNP is used as a spin trap, its adducts with carbon-centered tertiary or secondary radicals (i.e., the propagating radicals of numerous vinyl monomers) remain stable even at elevated temperatures (50–70°C), while its adducts with primary radicals (e.g., benzyl radical) are less stable and can be observed only at room temperature. Radicals with an unpaired electron on electronegative atoms (–O·, –S·, >N·) or hydrogen atom form non-stable adducts; in order to detect them it is necessary to decrease the temperature to between –10 and –30°C. MNP has an additional advantage in that it can also serve as an initiator. When irradiated by visible light, this compound can decompose with an appreciable rate, forming the active tert-butyl radical and nitric oxide (reaction 2) (Golubev, 1994).

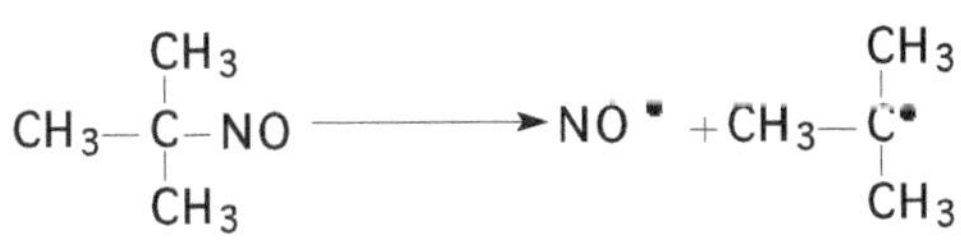

(2)

Thus, MNP can be used as a spin trap and a photoinitiator simultaneously. Moreover, because it functions in the visible region, it is possible to avoid the problems caused by the UV-absorbance and the associated decomposition of dithiobenzoate compounds.

When C-phenyl-N-tert-butyl nitrone (PBN) is used as a spin trap, its adducts are usually more stable but less information is provided by their spectra. Indeed, all radical adducts of PBN give essentially the same spectrum, independent of R — a triplet of doublets resulting from the hyperfine splitting on the N nucleus and β-H (reaction 3).

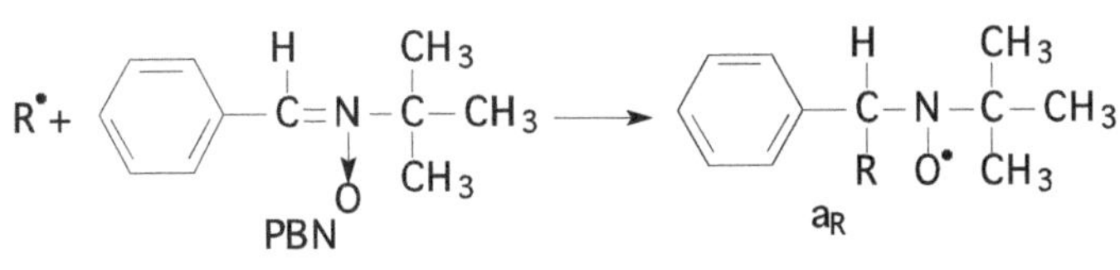

(3)

The radical adducts of PBN therefore cannot be distinguished by means of ESR-spectroscopy. Nonetheless, PBN has proven useful when it comes to estimating the general number of radicals in the system due to the superior stability of its spin adducts. Indeed, even the radicals bearing unpaired electrons on electronegative atoms (such as N, O and S) can be captured effectively by PBN, are stable at room temperature and can be detected by modern ESR-spectrometers.

2.2. Kinetic measurements using spin traps

Radical reactions generally proceed as rapid chain processes and the direct determination of the individual reaction rate coefficients is usually difficult. By reacting competitively with the active species, spin traps break these chains and generate populations of stable detectable radicals. The stage at which the chain reaction is halted is determined by the rate constants and concentrations of reagents (primarily of the spin trap). At high spin trap concentrations (0.5 mol L^{-1} or above), the trap captures initiating radicals, but, as the concentrations decrease, the products of the deeper stages of the process are detected. This phenomenon underlies the use of spin traps for the study of the mechanism and kinetics of elementary stages of chain radical reactions. As mentioned above, the introduction of RAFT agents into polymerization mixture results in appearance of numerous new reactions involving different active radicals. When a spin trap is used, these radicals are captured rapidly and adducts of many kinds emerge in such system. They affect the resulting ESR-spectrum of the sample to the extent that it is not usually comprehensible. Thus, to acquire the values of kinetic constants of elementary stages of RAFT polymerization, one should further simplify the whole system so that the elementary reactions can be studied separately. To achieve this one can separate these reactions in time (e.g. using pulsed irradiation that separates initiation from the subsequent downstream reactions), and use simpler compounds, which would eliminate some of the competing reactions.

In the general scheme of the RAFT polymerization, the addition reaction of P_n with the RAFT agent leads to formation of the intermediate adduct radical (see Scheme 1). The reverse reaction (regardless of which C–S bond is broken) is the decomposition (fragmentation) of the intermediate. If the intermediate is sufficiently stable, we can detect it using ESR and study its formation. However, in many cases, the stability of the intermediate is low, and its decomposition product radical is observed instead. In this case, only the overall substitution reaction is observed. Naturally, such a classification relying on the possibility or impossibility of the direct observation of intermediates is conditional. We have studied the addition of a variety of model radicals to both low molecular weight and polymeric RAFT agents, leading to the formation of intermediates of various stabilities ranging from very stable species (lifetimes of the order of 10 min) to species that cannot be observed via ESR-spectroscopy. Our results are summarized in the following section.

3. Experimental case studies

To model RAFT polymerization processes we have initially examined the interaction of the tert-butyl radical with various low molecular weight RAFT agents in a non-reactive solvent, such as benzene (Chernikova et al., 2010, Golubev et al., 2011). This system was chosen because, in the presence of visible light, MNP undergoes photolysis releasing tert-butyl radical (reaction 2), which is reactive enough to initiate the chain transfer process. Processes that occur in the system during irradiation are depicted in Scheme 2. In the absence of monomer, the tert-butyl radical is capable of interacting with the spin trap to form di-tert-butyl nitroxide (DTBN), or with the RAFT agent. The latter reaction leads to formation of an unstable radical intermediate, which decomposes, forming either the initial reactants, or a new RAFT agent and a new radical R, which is also captured by the spin trap. The reactivity of the radical intermediate is too low to allow its interaction with MNP. By dosing light irradiation one can "switch" the initiation on and off at any point. Thus, it becomes possible to study a chain transfer reaction that is essentially identical to the pre-equilibrium of RAFT polymerization (Scheme 1). The rate of the chain transfer process (i.e. substitution) can be compared with that of the concurrent spin capture reaction, for which the rate constant is well known.

The success of the experiments relies on two important conditions. First, the spin trap and its adducts must be stable in the presence of the RAFT agent. This is not a priori evident, since traps and related nitroxides are very reactive compounds. They are readily involved not only in radical reactions but also in redox reactions, and their stability depends on the acidity of the medium and the temperature. We have shown that MNP and its spin adducts do meet this condition for all of our experiments. Second, one must ensure that the accumulation of spin adducts obeys a linear rate law in order to make a correct comparison of the rates of described competitive reactions. Only if this condition is met, it is acceptable to neglect spin adduct termination reactions. If nitric oxide is present in the system in a significant concentration, spin adducts start to undergo undefined side termination reactions (see below). As MNP is photolyzed, the concentration of DTBN generally reaches its steady-state limit, when the rates of spin adduct formation and termination equalize.

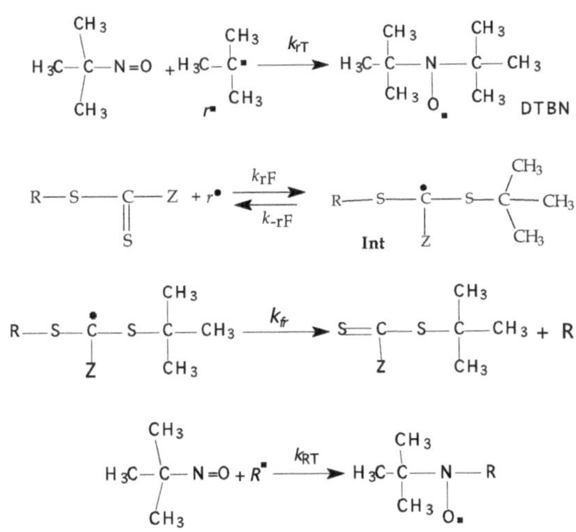

Scheme 2. Reactions occurring during irradiation of 2-methyl 2-nitrosopropane (MNP) in the presence of di-tert-butyl nitroxide (DTBN) and a RAFT agent (S=C(Z)SR)

3.1. Model reaction of tert-butyl radical with tert-butyl dithiobenzoate

The simplest model system is the one where tert-butyl radical is both the attacking radical and leaving group of the RAFT agent (Scheme 2). Since the participating radicals are identical to one another, the reaction scheme is simplified significantly. One such example is the decomposition of MNP in the presence of tert-butyl dithiobenzoate (TB) (i.e. the RAFT agent with Z=phenyl and R=tert-butyl in Scheme 2 above). When the photolysis time is short, the following reactions occur in the system:

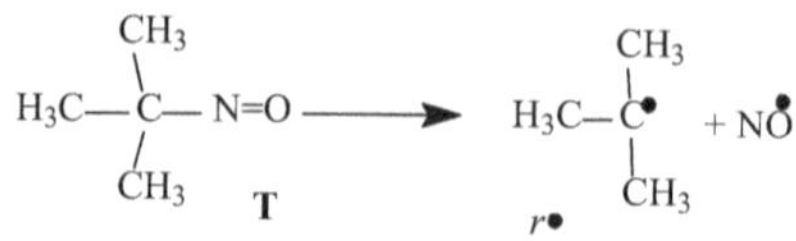
(4)

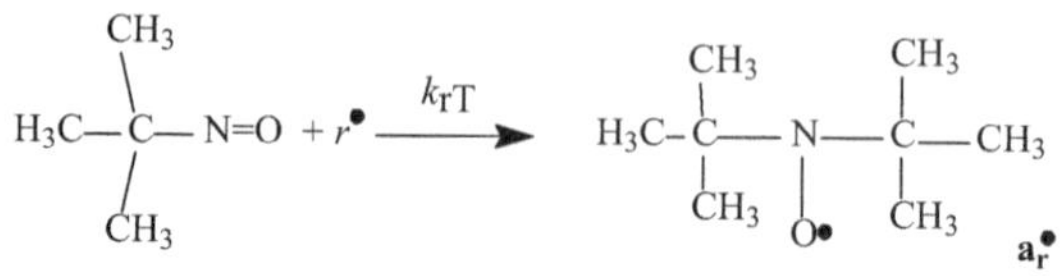
(5)

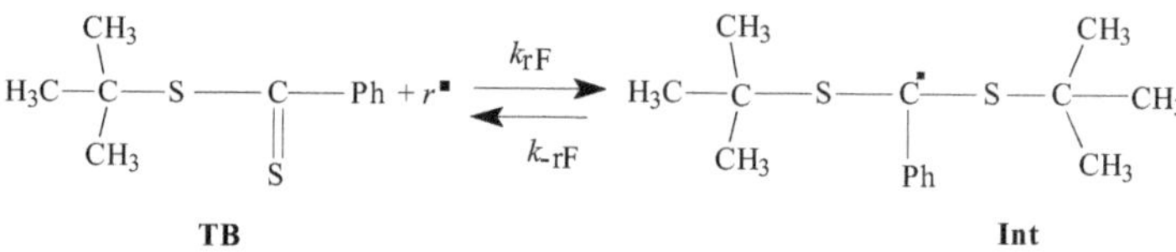
(6)

In principle, the radical species formed in this system can also participate in various termination processes:

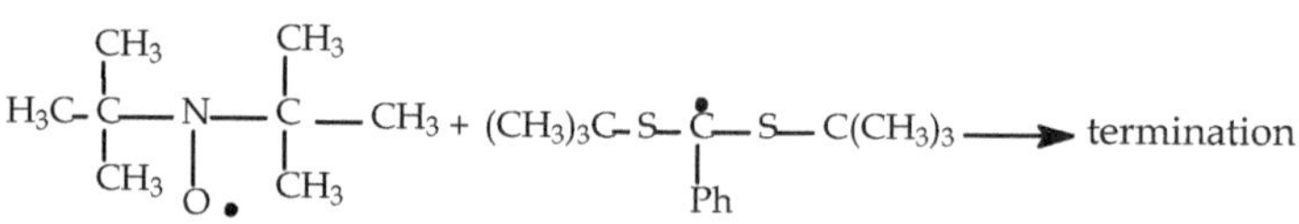
(7)

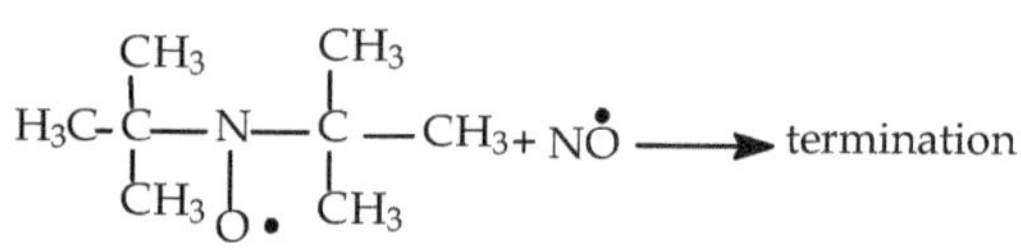
(8)

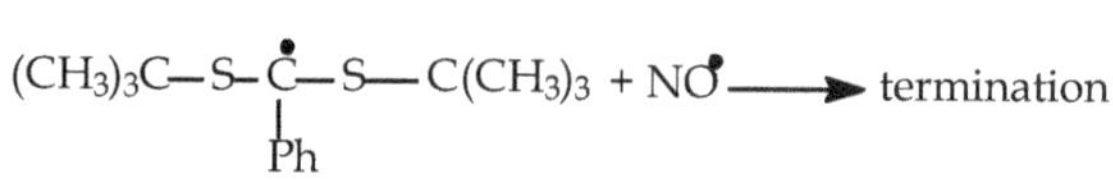
(9)

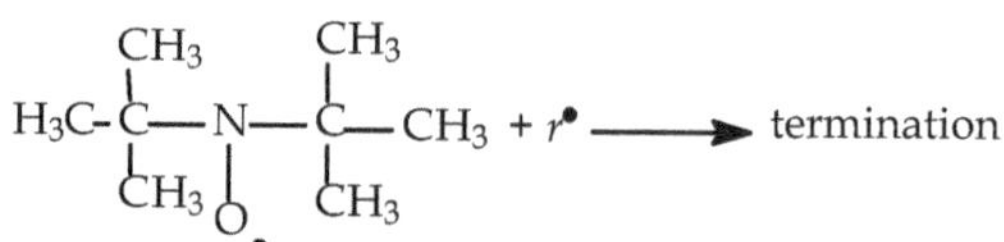
(10)

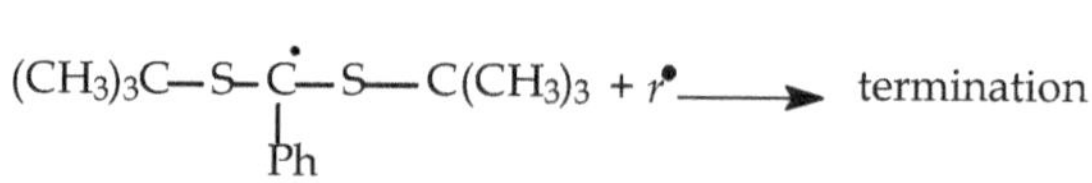
(11)

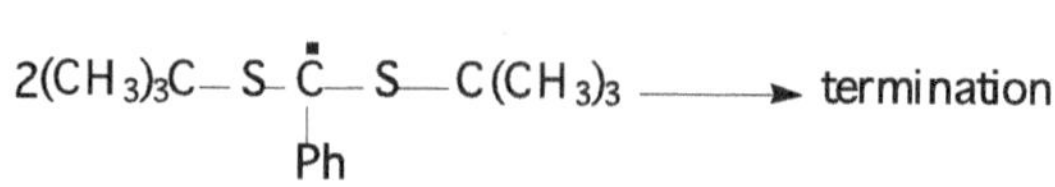
(12)

However, due to the relatively high stability of the nitroxide radicals, many of these reactions can be assumed to be negligible. In particular, termination reactions involving the spin adducts and RAFT intermediate radicals are unlikely. Termination of radical species with NO are possible, but only when the irradiation lasts for a long time, causing an appreciable portion of MNP to decay. Such conditions are not fulfilled in our experiments; hence this process can also be neglected. Also, reaction of DTBN with the tert-butyl radical affects the kinetics only when MNP has decayed significantly. Cross- and self-termination of the radical intermediate is possible in our system, in which case the rate of the intermediate consumption would not obey the first-order law.

Now let us consider the experimental data obtained for the system containing TB (Chernikova et al., 2010). A typical ESR-spectrum, observed after the irradiation of TB and MNP benzene solution, is shown in Fig. 1. There appears to be a superposition of two different signals: DTBN triplet with A_N = 15.1 G and a multiplet (g =2.0041 G, $A_{\delta H}$ = 0.42 G, A_{oH} = 3.65 G, A_{mH} = 1.34 G, A_{pH} = 3.99 G), attributed to the radical intermediate Int.

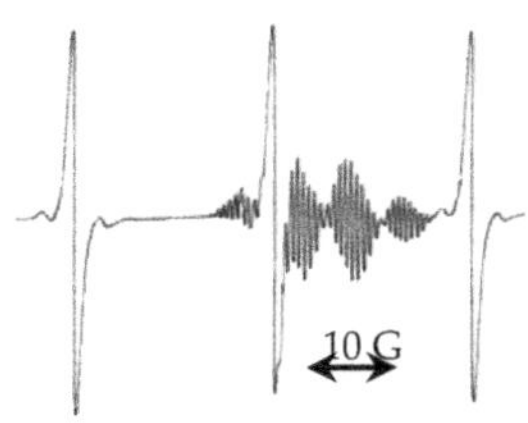

Figure 1. ESR-spectrum observed on irradiation by visible light of the system TB–MNP–benzene at 20°C.

During irradiation, both the DTBN and the Int concentrations increase as a result of the trapping of tert-butyl radicals by MNP and TB, respectively (reactions 5, 6). Upon the completion of photolysis, the concentration of a_r increases (Fig. 2a, curves 1 and 1′), while the concentration of Int decreases (Fig. 2a, curves 2 and 2′). This is because the intermediate radical has a finite lifetime and decomposes releasing tert-butyl radicals (reverse of reaction 6), which are subsequently trapped by MNP forming DTBN (reaction 5). Since the irradiation time is noticeably shorter than the intermediate lifetime (e.g., in the present work the irradiation time is 5 – 10 s whilst the intermediate decays over several hundreds of seconds), we can consider the processes of intermediate accumulation and consumption separately. All side reactions of DTBN (Scheme 2) can also be excluded due to short period of irradiation. Given this assumption, one can express the ratio between the rates of formation of DTBN (a_r) and Int during irradiation with a simple formula:

$$\left(d[a_r]/dt\right)/\left(d[Int]/dt\right)=\left(k_{rT}[T]/k_{rF}[F]\right), \tag{13}$$

where T and F are the spin trap and RAFT agent (TB) respectively. Since the photolysis time is very short, it is legitimate to neglect any termination processes. This means that the concentrations of both species formed during irradiation are proportional to their accumulation rates. Thus, the ratio of DTBN to Int at the instant the light is switched off provides a measure of the relative rates of tert-butyl radical addition to MNP versus to TB. That is,

$$\frac{[a_r]_0}{[\text{Int}]_0} = \frac{(d[a_r]/dt)}{(d[\text{Int}]/dt)} = \frac{k_{rT}[T]_0}{k_{rF}[F]_0} \tag{14}$$

In this equation, $[a_r]_0$ and $[\text{Int}]_0$ are obtained by the extrapolation of the kinetic curves up to zero time, $[T]_0$ and $[F]_0$ are preset, and $k_{rT} = 3.3\times10^6$ L mol^{-1} s^{-1} (Doba et al., 1977), and hence k_{rF} can be obtained.

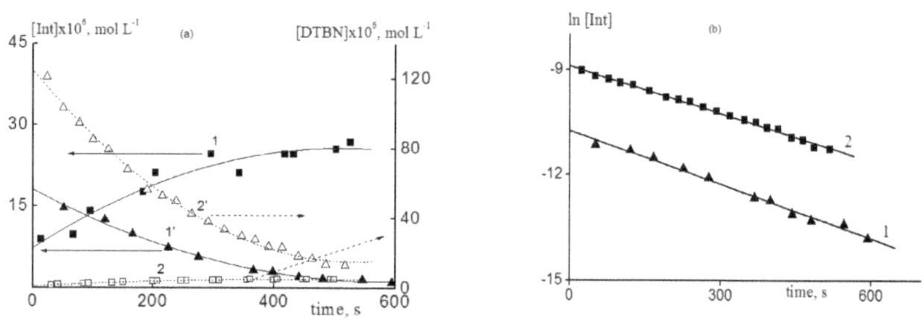

Figure 2. (a) Kinetic curves of accumulation of DTBN and consumption of intermediate radical Int in the system TB–MNP–benzene after switch-off of MNP photolysis: [MNP] = 10^{-2} mol L^{-1}and [TB] × 10^2= 1 (1 and 1′ respectively) and [TB] × 10^2=9 (2 and 2′ respectively) at 20°C; (b) Semi-logarithmic plot of the kinetic curves 1′ and 2′ presented in Fig. 2, [TB] × 10^2= 1 mol L^{-1} (1) and 9 mol L^{-1}(2).

To calculate the starting concentrations, the kinetic curves are extrapolated to the zero time of photolysis. These experimental data can thus be used to estimate the rate constant of addition of tert-butyl radical to TB. Over the range of TB concentrations 10^{-2} – 10^{-1} mol L^{-1} and at [MNP] = 10^{-2} mol L^{-1} the average value of the rate coefficient of tert-butyl radical addition to TB was found to be equal to $(5\pm1)\times10^6$ L mol^{-1}s^{-1}. This value is in a good agreement with the data for the addition of various oligomeric and polymeric radicals to various RAFT agents (Barner-Kowollik, 2008, Kwak et al, 2002). Under the assumption that a decrease in the concentration of the intermediate after stoppage of illumination is associated solely with its fragmentation (reverse reaction 9), it is possible to estimate the rate constant of the fragmentation of the intermediate from its post-illumination decay curve, plotted in semi-logarithmic coordinates (Fig 2). From these curves a value of k_{dec}=k_{-rF}= $(5 \pm 1) \times 10^{-3}$ s^{-1} is obtained. If the released tert-butyl radical reacts iteratively with TB (addition reaction 6), then the concentration of the radical intermediate would increase, prolonging its decomposition. As a result we would underestimate the rate constant of intermediate decomposition. This is where the presence of the spin trap offers an enormous advantage. By reacting selectively with the tert-butyl radicals, it largely prevents their re-addition to the RAFT agent. However, it is necessary to introduce correction coefficient p:

$$p = \frac{R_{rF}}{R_{rT} + R_{rF}} = \frac{k_{rF}[F]}{k_{rT}[T] + k_{rF}[F]} \quad (15)$$

which, for [TB]=[MNP]=10^{-2} mol L^{-1} corresponds to a value of 0.6, giving a corrected value of $k_{dec}=k_{-rF}=(8\pm 2)\times 10^{-3}\ s^{-1}$ (Golubev et al., 2011).

If termination reactions (9, 11, 12) occur, they should lead to a more rapid disappearance of the radical intermediate; i.e., our estimate of k_{dec} will be the upper bound to this value. However, the observed linear dependence of the logarithm of the concentration of Int versus time (Fig. 2b) unambiguously indicates that side reactions involving the intermediate can be ignored under the chosen experimental conditions. Nevertheless, we have shown that the self-termination of radical intermediates (reaction 12) could occur in principle in the TB–MNP system, though the rate constant of this reaction was found to be very low ($(6.5 \pm 3.0) \times 10^2$ L mol^{-1} s^{-1}) (Chernikova et al., 2010).

3.2. Model reaction of tert-butyl radical with di-tert-butyl trithiocarbonate

Another model system in which the attacking and leaving radicals are the same is when MNP interacts with di-tert-butyltrithiocarbonate (TC). In this case, the reaction of tert-butyl radical with TC (reaction 16) also gives rise to an intermediate radical that can be detected by means of ESR-spectroscopy (Golubev et al., 2005).

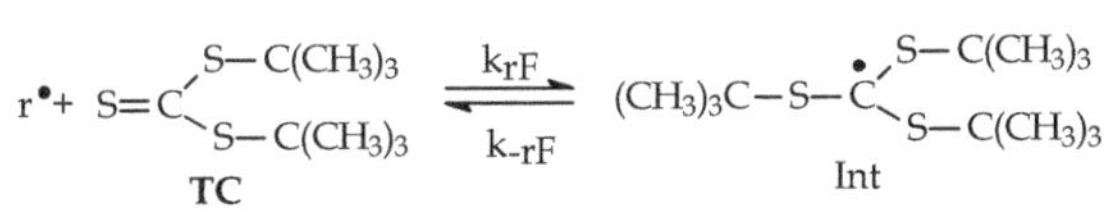

(16)

Once again, we neglect all radical termination processes while considering the kinetics of chain transfer reactions. Fig. 3 illustrates a typical ESR-spectrum, observed after the irradiation of TC and MNP benzene solution, which is the superposition of a triplet corresponding to DTBN and a complex multiplet, attributed to radical intermediate Int (reaction 16) (Golubev et al., 2011).

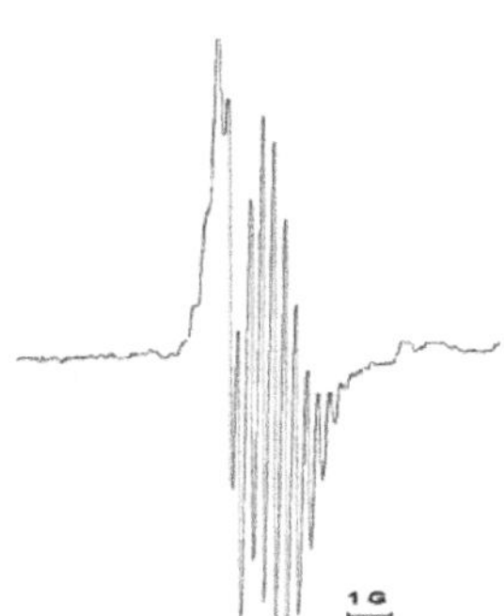

Figure 3. ESR-spectrum observed on irradiation by visible light of the system TC–MNP–benzene at 20°C ([TC]=1 mol L^{-1}, [MNP]=5×10^{-3} mol L^{-1}).

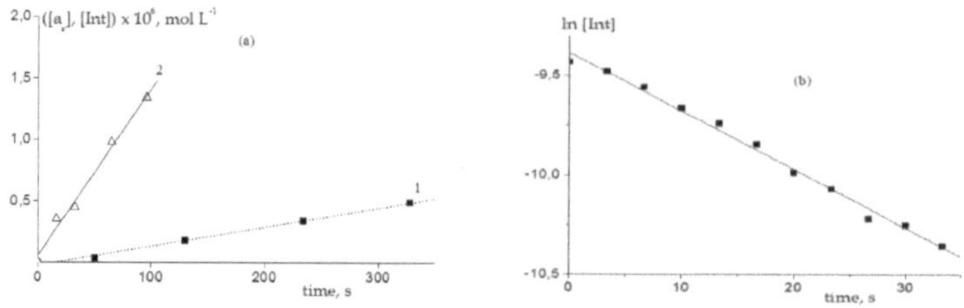

Figure 4. Kinetic curves of accumulation of DTBN (1) and intermediate radical (2) during the photolysis of MNP and TC in benzene solution (a); semi-logarithmic plot of kinetic curve of radical intermediate consumptionafter switch-off of MNP photolysis (b). ([TC] = 1 mol L^{-1}, [MNP]=0.05 mol L^{-1}), T = 20°C.

As follows from Fig.4, the replacement of the phenyl substituent on the RAFT agent with an additional thiyl group results in a decreased rate constant for addition of the tert-butyl radical to TC: $k_{rF} = 2.2 \times 10^5$ L $mol^{-1}s^{-1}$ (Fig. 4a), and an increase in its fragmentation rate ($k_{dec} \sim 2 \times 10^{-2}$ s^{-1}) (Fig. 4b, accordingly). These changes result from both a decreased stability of the intermediate radical and an increased resonance stabilization of the C=S π bond. The phenyl group is much more effective than the lone pair donor sulfur at stabilizing the unpaired electron on the intermediate radical (which already has two other lone pair donors), but much less effective at stabilizing the C=S π bond of the RAFT agent than the sulfur, which can stabilize through resonance structures such as S=C–S ↔$^-$S–C$^+$=S).

3.3. Model reaction of tert-butyl radical with benzyl dithiobenzoate

When the tert-butyl leaving group is replaced in the RAFT agent structure by benzyl, one should expect the formation of a stable intermediate radical after addition of the tert-butyl radical to the RAFT agent, e.g. benzyl dithiobenzoate (BB):

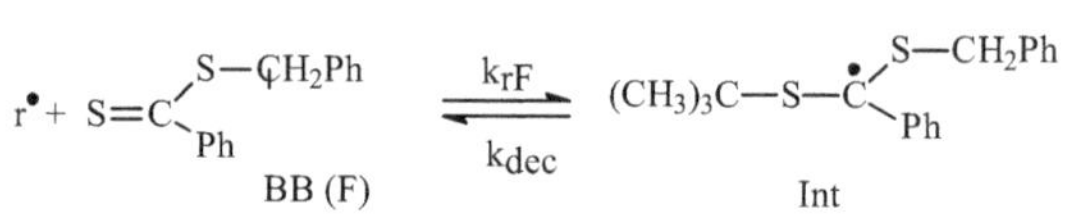

(17)

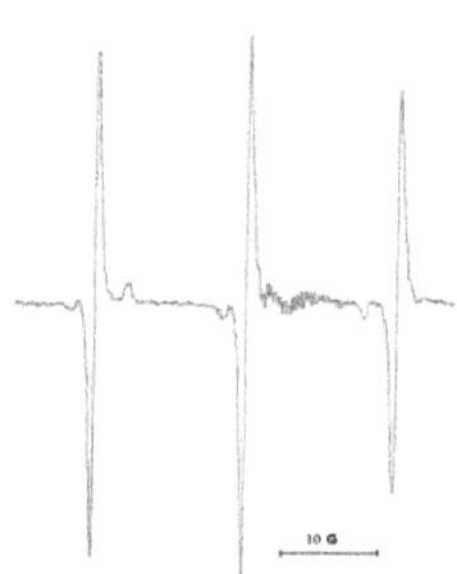

Figure 5. ESR-spectrum observed on irradiation by visible light of the system BB–MNP–benzene at 20°C ([BB]=1 mol L^{-1}, [MNP]=5×10^{-3} mol L^{-1}).

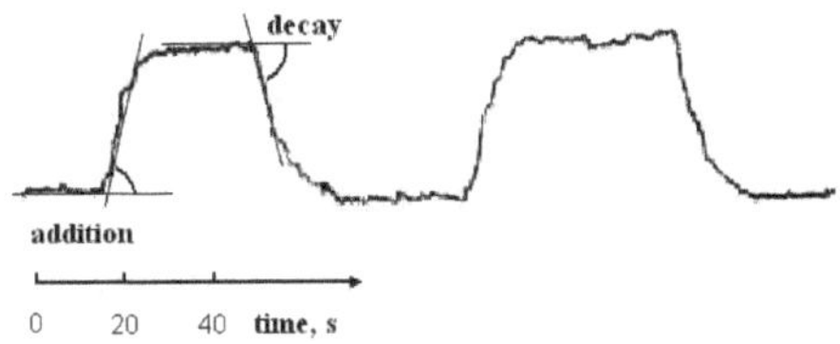

Figure 6. Continuous kinetic curve of accumulation and decomposition of the intermediate radical measured during periodic illumination of MNP – BB – benzene system. [MNP] = 0.22 mol L^{-1}, [BB] = 0.61 mol L^{-1}, T = 25°C.

Instead, the stability of the radical intermediate decreases dramatically. In this system the ESR-spectrum of the radical intermediate can only be observed under conditions of continuous photolysis. The spectrum obtained is shown in Fig. 5, and corresponds to the typical superposition of the DTBN triplet and the radical intermediate multiplet (Golubev et al., 2011). For the quantitative determination of the rate coefficients of addition and fragmentation reactions in the investigated system a special technique has been employed (Fig. 6). The magnetic intensity was adjusted to the point at which the most intense component of the intermediate spectrum appeared. When the light was switched on, the signal of a value proportional to the intensity of the intermediate spectrum was obtained; after the light was switched off, rapid loss of the intermediate occurred (Fig. 6).

The absence of signals corresponding to the spin adduct of MNP with the benzyl radical in the spectra indicates that the initial tert-butyl radical, rather than the benzyl radical, fragments preferentially during decomposition of the intermediate and reaction 18 does not occur. This is very surprising, given the greater radical stability of the benzyl radical and the results for the corresponding trithiocarbonate (see section 3.4 below), and the theoretical predictions (see section 5).

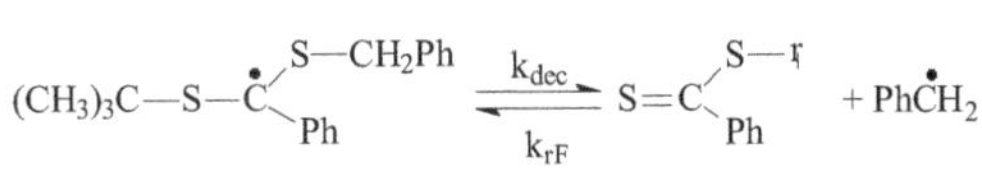

(18)

To evaluate the addition rate coefficient in reaction 17, the obtained kinetic curves were calibrated relative to the rate of formation of DTBN, which has been determined in an independent experiment. Estimation of k_{rF} value for addition of the tert-butyl radical to BB yielded k_{rF}= 2.2 × 10^5 L mol^{-1} s^{-1}. The rate coefficient of the fragmentation reaction of the intermediate k_{dec} was estimated to be approximately equal to 9 × 10^{-1} s^{-1} (Golubev et al., 2011).

3.4. Model reaction of tert-butyl radical with dibenzyltrithiocarbonate

Another pattern is observed for the system containing dibenzyl trithiocarbonate (BC). The addition of the tert-butyl radical to BC followed by the capture of the released benzyl radical by MNP may be schematically outlined as follows. The tert-butyl radical and DTBN are formed in a manner similar to that described above.

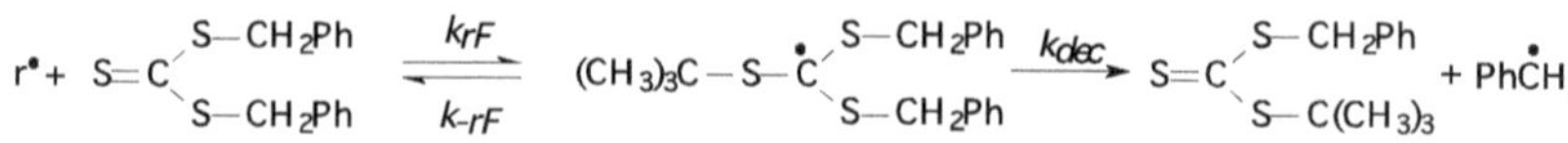
(19)

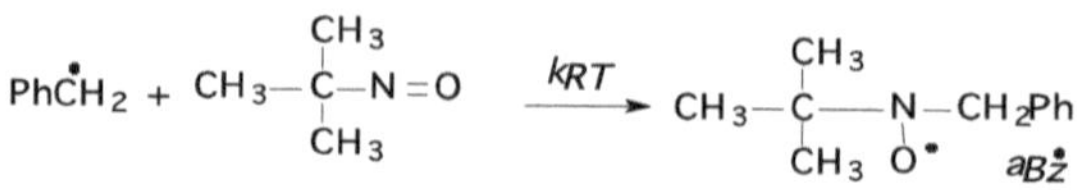
(20)

The spectrum of the sample containing MNP and BC in benzene that was irradiated with the visible light is the superposition of the above triplet corresponding to DTBN and the adduct a_{Bz} formed by interaction of MNP and benzyl radical (Fig. 7, seven lines with a ratio of intensities of 1 : 2 : 1 : 2 : 1 : 2 : 1) (Golubev et al., 2011). The second, fourth, and sixth lines of the spectrum are combined lines, while the other lines arise from adduct a_{Bz}. The lifetime of the intermediate in this system is very small, and it cannot be detected via ESR spectroscopy under any conditions. An examination of the spectrum makes it possible to determine the concentrations of adducts a_r and a_{Bz} separately and, thus, to calculate k_{rF}, which, under these conditions, is no longer the constant of addition but the constant of substitution of the tert-butyl radical with the benzyl fragment of the RAFT agent:

$$[a_r]/[a_{Bz}] = k_{rT}[T]/k_{rF}[F] \tag{21}$$

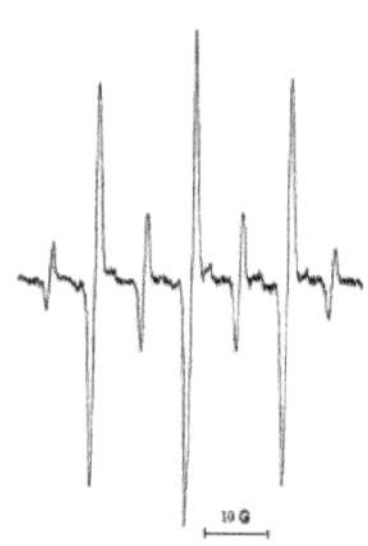

Figure 7. ESR spectra of DTBN and MNP adduct with benzyl radical observed for MNP – BC – benzene system.

The kinetic curves are shown in Fig. 8; calculations yield $k_{rF} = (2.8 \pm 0.3) \times 10^6$ L mol^{-1} s^{-1}.

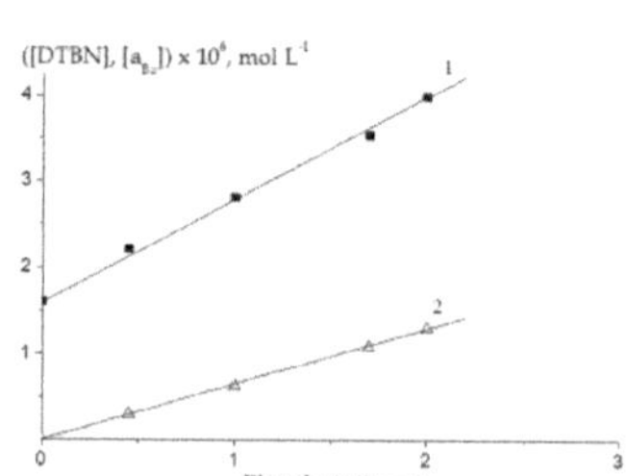

Figure 8. Kinetic curves of accumulation of tert-butyl (1) and benzyl (2) spin adducts in the system BC-MNP-benzene ([BC]=0.01 mol L^{-1}, [MNP]=0.02 mol L^{-1}) during the irradiation by the visible light.

3.5. Model reactions of tert-butyl radical with polymeric RAFT agents

During RAFT polymerization the initial low molecular mass RAFT agent converts into a polymeric one, which ensures the control of the polymerization process up to high conversions. It is known that the efficiency of polymeric RAFT agents is typically one or two orders of magnitude higher than that of low molecular mass agents. To explore whether our spin trapping experiments on model compounds could detect this increase, we compared the reactivity of polymeric RAFT agents (prepared via a polymerization of a number of monomers mediated by dithiobenzoates and trithiocarbonates) in model reaction with the tert-butyl radical (Golubev et al., 2011).

During the photolysis of the benzene solutions of MNP containing polystyrene trithiocarbonate (PSC), tert-butyl radical r• is generated (reaction 4) and captured by MNP with the accompanying formation of DTBN (reaction 5). We also observe the addition of r• to PSC leading to the formation of intermediate Int and its decomposition with the subsequent release of polystyrene radical P_m• (22), and the capture of this radical by MNP (23):

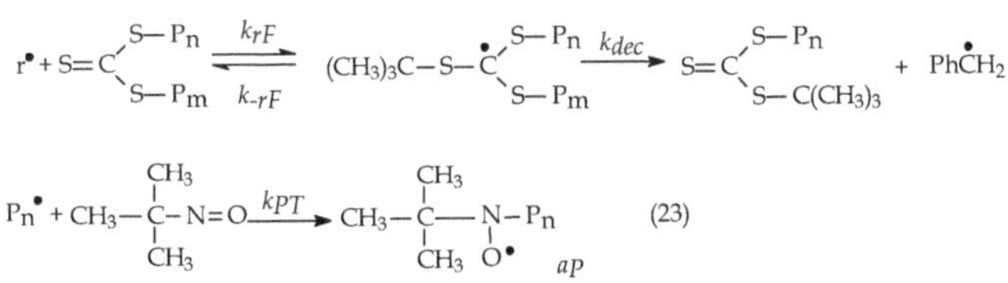

(22)

(23)

The ESR spectra of the photolyzed system (Fig. 9), along with the spectrum of DTBN (a_r), shows a triplet of doublets (A_N= 14.9 G and A_H= 3.5 G) corresponding to adduct a_P with the polystyrene radical (Golubev et al., 2011). The spectrum of the intermediate is not observed under any conditions, thereby implying that the intermediate is very unstable and decomposes rapidly. As seen from Fig. 10, the kinetics of accumulation of adducts a_r and a_P during photolysis is linear. Thus, from the known equation:

$$[a_r]/[a_P] = k_{rT}[T]/k_{rF}[F] \qquad (24)$$

one can obtain k_{rF} = (2.0 ±0.4)×10^7 L mol^{-1} s^{-1}.

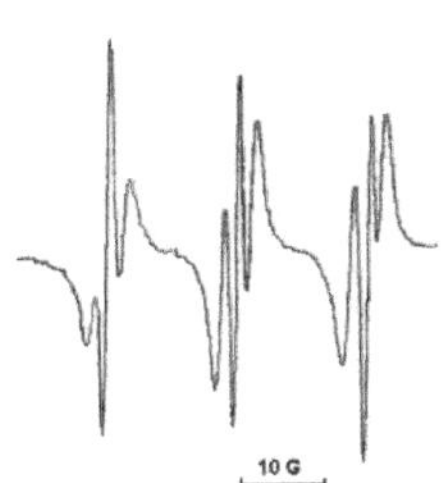

Figure 9. ESR spectra of DTBN and MNP adduct with polystyrene radical observed for MNP – PSC – benzene system

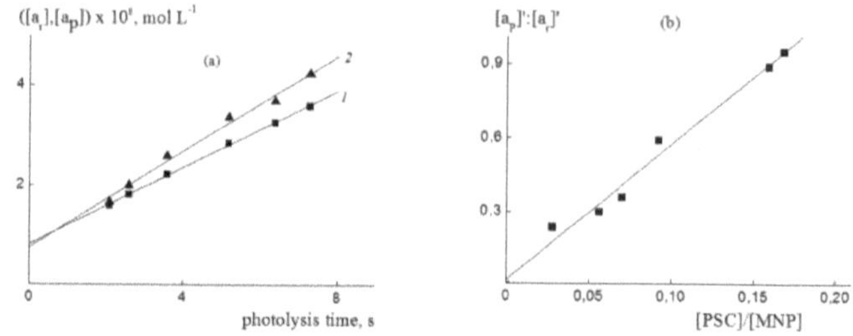

Figure 10. Kinetic curves of accumulation of tert-butyl a_r (1) and polystyrene a_p (2) spin adducts in the system PSC-MNP-benzene during the irradiation by visible light (a) and dependence of molar ratio of adducts a_p/a_r from molar ratio of PSC/MNP (b).

In a similar manner, the kinetics of addition of r• to the polymeric RAFT agent prepared through the polymerization of styrene mediated by BB (PSB) was studied. The ESR spectrum of the photolyzed system is identical to those given in Fig.10 and no intermediate radical could be observed. In this case, $k_{rF} = (4.1 \pm 0.2) \times 10^7$ L mol^{-1} s^{-1}. With consideration of the above evidence, the measured constants should be attributed to the substitution reaction rather than the addition, and this is what we find (see below). It is natural to anticipate that the intermediate arising from addition of tert-butyl radical to polymeric RAFT agents with poor leaving groups (e.g., polyacrylate or poly(vinyl acetate)), would tend to release the initial tert-butyl radical rather the more active polymer radicals. In experiments, this effect should manifest itself as a reduction in the substitution constant. Indeed, for poly(butyl acrylate) dithiobenzoate (PBAB), the value of k_{rF} was an order of magnitude lower than that for PSB: ($k_{rF} = (4.5 \pm 2.0) \times 10^6$ L mol^{-1} s^{-1}. Parameters of the ESR spectrum for the MNP adduct with the poly(butyl acrylate) radical (A_N= 14.2 G and A_H= 2.7 G) differ appreciably from the corresponding parameter for the styrene adduct.

For the azeotropic copolymerization of styrene with *n*-butyl acrylate (87 : 13, mol %) in the presence of dithiobenzoate, the propagating radical will contain the styrene terminal unit with a much higher probability. Therefore, one can expect that a polymeric substituent in polymeric RAFT agent will predominantly contain the styrene terminal unit. Naturally, the values of the substitution constant, like the ESR parameters of the adduct, coincide with the corresponding values of PSB: $k_{rF} = 5.7 \times 10^7$ L mol^{-1} s^{-1}, A_N= 14.9 G, and A_H= 3.5 G. This demonstrates another application of the spin-trapping technique for structural studies of the polymers synthesized by RAFT polymerization. The nature of the terminal unit added to the RAFT agent during the synthesis may be directly assessed from the ESR parameters of the MNP adduct with the polymeric radical.

The values of substitution constants for polymeric trithiocarbonates were estimated in a similar manner. Table 1 summarizes the values of substitution (addition) constants obtained for all polymeric RAFT agents under study and parameters characterizing the activity of the monomers. There is a clear correlation between these characteristics. A sharp increase in the efficiency of polymeric RAFT agents relative to that of low molecular mass RAFT agents was observed in dozens of systems for both homo- and copolymerization of various monomers mediated by RAFT agents (Barner-Kowollik, 2008). The model studies performed in the present work have confirmed the general character of this phenomenon.

Polymeric RAFT-agent	k_{rF}×10^{-5}, L mol^{-1} s^{-1}	k_{rF}×10^{-5}, L mol^{-1} s^{-1} (Low molecular weight analog)
PSB	420	2.2
PBAB	45	
copoly(butyl acrylate/styrene) dithiobenzoate	570	
PSC	250	28
poly(butyl acrylate) trithiocarbonate	120	
poly(vinyl acetate) trithiocarbonate	50	

Table 1. Substitution rate coefficients for the reactions of tert-butyl radical with polyRAFT agents.

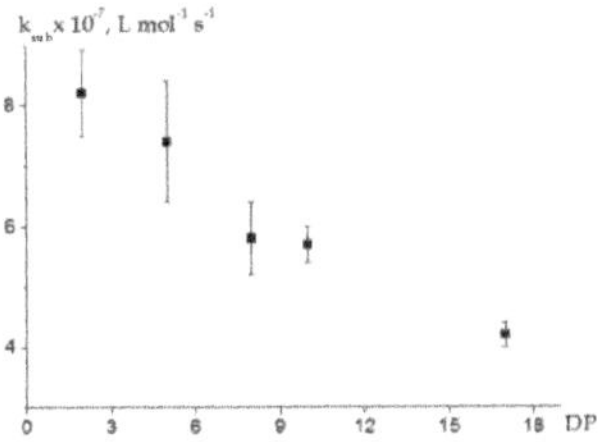

Figure 11. Dependence of rate coefficient of substitution (k_{sub}) reaction between PSB and tert-butyl radical on the degree of polymerization (DP).

To further probe the nature of the chain length effects, we have applied the spin trap technique to polystyrene dithiobenzoate of different polymerization degrees (2–17 monomer units). As seen in Fig. 11, the results obtained indicate the slight chain length dependence of rate coefficient of substitution reaction between PSB and tert-butyl radical. Similar regularities have been observed in the model reaction of tert-butyl radical with polystyrene trithiocarbonate of various chain lengths (the results are not given here).

This means that for small conversions the equal radical reactivity principle is not applicable and the radicals of different length can participate in the elementary steps of RAFT process with different rate coefficients. Strong chain length effects have also been observed previously in our quantum-chemical calculations, though these focused on the equilibrium constants of the individual addition-fragmentation reactions, rather than the overall substitution reactions (Coote et al., 2006, Izgorodina et al., 2006, Lin et al., 2011). It is impossible to observe radical intermediate in our experiments due to its instability, which makes it difficult to estimate rate coefficients for addition and fragmentation reactions and, as a consequence, to show whether both of them depend on the chain length in the same way. In the section 5 we use theory to calculate equilibrium constants for the substitution process and determine if these follow similar trends to those of the substitution rate coefficients.

4. Extending the scope

To date, the spin trapping method has been used to study the reaction of low molecular weight radicals such as tert-butyl with low molecular weight and polymeric RAFT agents. However, to prove that this technique is fully applicable to polymerization studies, one would also like to model the reactions of polymeric radicals with polymeric RAFT agents. Unfortunately, the inclusion of the monomer in the MNP model systems described above overcomplicates the kinetics because the tert-butyl radicals formed during the irradiation of MNP do not react with the monomer solely. MNP and RAFT agents would readily interact with tert-butyl radicals too. Thus, the resulting product mixture would consist of adducts of the tert-butyl radical, propagating species and the leaving radical of RAFT agent. The interpretation of the ESR-spectrum of such a "cocktail" of species is very difficult. Instead, we have considered a new experimental system based on the thermal decay of cyclohexylperoxodicarbonate (CHPC):

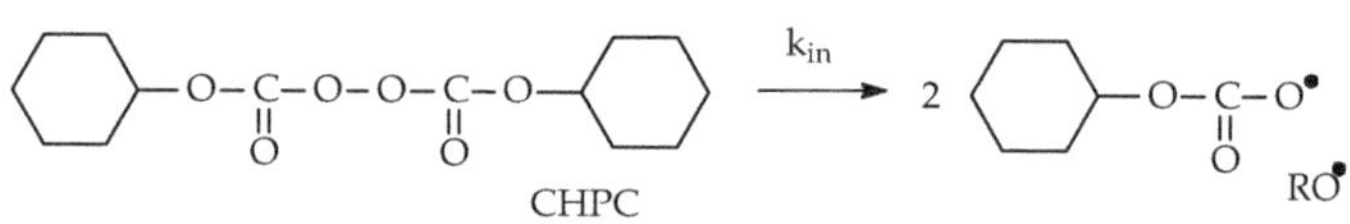
(25)

This initiator has a number of advantages. First, its decay rate constant is appreciable (k_{in}~10^{-7} s^{-1}) even at ambient temperature. Second, the resulting radical bears an unpaired electron on the electronegative oxygen atom. Whilst this radical can react with both monomer and MNP, its adduct with MNP is unstable and decays rapidly to initial reagents. The scheme of reactions occurring in the system MNP-styrene-CHPC-PSB is given below.

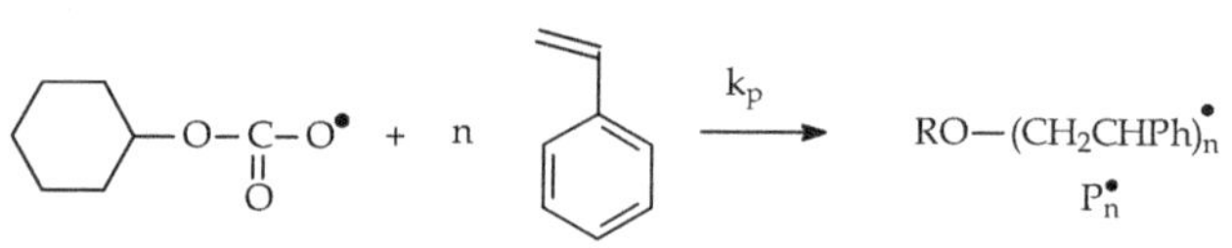
(26)

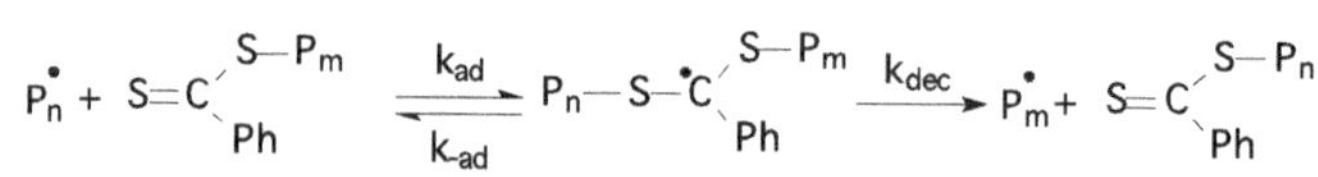
(27)

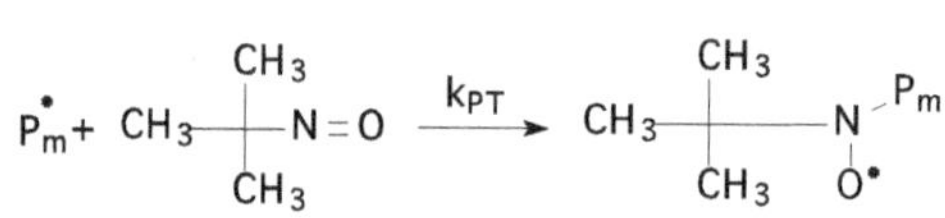
(28)

The initiating radical reacts with the monomer, generating propagating species, which can further react with MNP and RAFT agent. The former reaction leads to the MNP-based adduct of polystyrene, while the latter results in formation of unstable radical intermediate that decays to form either the initial propagating species or the new ones, able to reinitiate the process. The new radical is also captured by the spin trap and the new stable adduct is formed.

At this point, one may object to the model we propose on the basis that both the leaving and the attacking radicals are styrene propagating radicals, and the ESR-spectra of their adducts with MNP should be indistinguishable. However, as shown earlier (Golubev, 1994), the ESR-spectrum of styrene solution of MNP and CHPC changes slightly when the concentration of MNP is augmented. The system containing a small concentration of MNP ($<10^{-3}$ mol L^{-1}) gives rise to a triplet of doublets with A_N=14.6 G and A_H=3.4 G. When the quantity of spin trap is increased greatly ($\geq$0.3 mol L^{-1}), the signal changes its hyperfine interaction parameters: A_N=14.3 G and A_H=2.8 G. The further increase in spin trap concentration up to 1 mol L^{-1} does not result in any further changes to the ESR-spectrum. It was shown by means of kinetic calculations that the first spectrum should be attributed to oligomeric styrene spin adduct (n≥2) and the second – to the adduct of so-called "unimer" propagating radical P_1, containing only one monomer unit. These ESR-spectra have similar spectral parameters, and when they occur together they cannot be resolved. However, the characteristics of the resulting spectrum can be used to evaluate the molar fractions of these adducts in their mixture. To accomplish this, one should obtain spectra of each adduct separately. It is then possible to perform simple algebraic operations and construct the set of spectra with pre-determined fractions of these adducts.

Fig. 12 depicts the lower-field doublets, typical for unimer (Fig. 12, curve *1*) and oligomer (Fig. 12, curve *2*). Based on these data we can construct a set of model spectra, each with a known fraction of the unimer adduct. For each simulation, two parameters were measured: the line magnitude ratio (L_1/L_2) and the resulting constant of hyperfine splitting on styrene α-proton. The data obtained are shown in Fig. 13. The suggested approach allows evaluation of the molar fraction of unimer adduct from the real spectrum to within 10% accuracy.

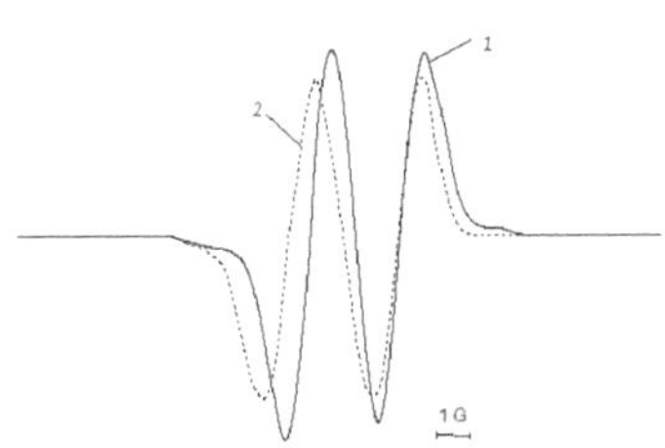

Figure 12. The ESR-spectra of unimer (1) and oligomer (2) styrene adducts with MNP.

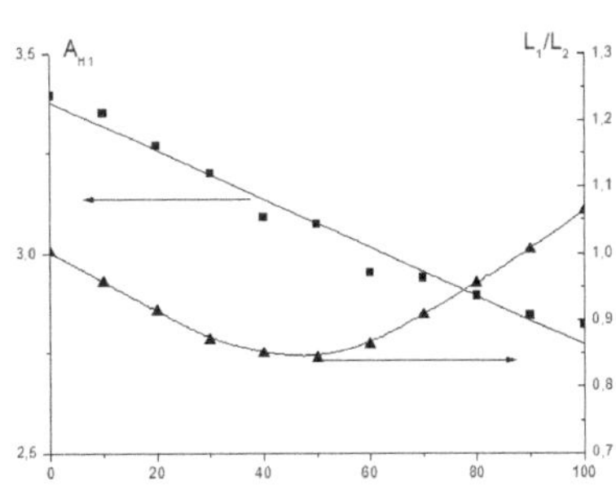

Figure 13. The parameters of model ESR-spectra as a function of unimer adduct part.

Let us return to the MNP–CHPC–PSB–styrene system. When the concentration of MNP is extremely high (over 0.3 mol L^{-1}), the propagating styryl radical P_n is one unit long, so we can designate it as P_1. The simplified reaction scheme may be written as below (Scheme 3). Now, a question arises as to whether this scheme is complete. In particular, if any $RO^{\bullet}$ generated after the CHPC decay does not react with MNP, is it safe to say that this radical does not interact with polymeric RAFT agent? Indeed, when the reaction is carried out in an inert solvent like benzene, the system CHPC–MNP–PSB generates polystyrene adducts with MNP. This means that $RO^{\bullet}$ is able to react with polymeric RAFT agent. When the experiment is repeated but with PBN in place of MNP, the radical accumulation is doubled, since PBN can capture the radicals bearing an unpaired electron on oxygen. So, in the absence of monomer, an appreciable portion of the formed $RO^{\bullet}$ can attack the polymeric RAFT agent. However, when this process is run in bulk styrene, we can neglect this reaction, because the reaction with monomer is dominant.

For the sake of simplicity we assume the rate coefficients of spin trapping of unimer and oligomer propagating radicals to be equal. Then the rate constant of substitution of oligomer radical by a unimer one can be expressed as

$$k_{sub} = \frac{k_{PT}[T](1-\chi(a_1))}{\chi(a_1)[P_n Z]} \tag{29}$$

where $\chi(a_1)$ is the molar fraction of unimer radical adduct and k_{PT}=4.6×10^4 L mol^{-1} s^{-1}.We have carried out experiments using PSB of M_n=2900 D; the conditions and the results are shown in Table 2.

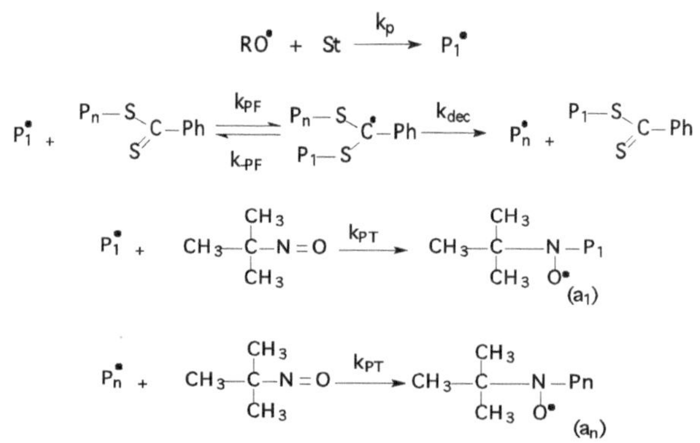

Scheme 3. Simplified reaction scheme for MNP–CHPC–PSB–styrene system.

[MNP], mol L^{-1}	[PSB], mol L^{-1}	[CHPC], mol L^{-1}	$\chi(a_1)$	k_{sub}, Lmol^{-1} s^{-1}
0.370	0.010	0.01	0.94	1×10^5
0.316	0.029	0.005	0.71	2×10^5

Table 2. The experimental data obtained for systems containing MNP, PSB and CHPC in styrene

The values obtained for substitution rate constants correlate with the data from the kinetic analysis of RAFT polymerization. The chain transfer constant, known as $C_{tr}=k_{sub}/k_p$, lies between 2000 and 6000 for styrene polymerization in the presence of PSB at 80°C, while the obtained data yield C_{tr} in the range of 1300–2500 at 24°C. Such a good agreement between values obtained by very different methods cross-validates these techniques.

5. Quantum-chemical calculations

To help validate the kinetic results from the spin-trapping technique, we now compare the experimental results with the corresponding predictions from high-level ab initio molecular orbital theory calculations. Unlike experimental approaches, quantum chemistry offers the opportunity to calculate the rate and equilibrium coefficients for any (chemically-controlled) individual reaction within a complicated multistep processes directly, assuming only the Schrödinger equation and the values of a few fundamental constants (such as the masses and charges of the electron, proton and neutron). In this way, quantum chemical predictions are independent of any kinetic model-based assumptions and may thus be used to test their validity. Admittedly, because the many-electron Schrödinger equation does not have a simple analytical solution, numerical approximations must be made, introducing a potential source of error instead. Whilst very accurate methods exist, these usually require large amounts of computer power, with the computational cost scaling with the size of the system. Nonetheless, in recent years, rapid and continuing advances in computational resources and the development of cost-effective accurate quantum-chemical methods have made possible the chemically accurate prediction of rate coefficients for oligomeric reactions relevant to conventional and controlled radical polymerization processes (Coote et al., 2009, Lin et al., 2010).

The calculations described in this section all use the same high-level ab initio molecular orbital theory methods, which are described in full in the original studies (Coote et al., 2006, Izgorodina et al., 2006, Lin et al., 2011). Owing to their computational cost, we have focused only on the calculation of the equilibrium constant for the addition-fragmentation reaction $K=k_{ad}/k_{fr}$. All geometries and frequencies are calculated using relatively low-cost DFT procedures which have been shown to be suitable for this purpose due to systematic error cancellation (Coote et al., 2002). We always conduct a thorough search of conformational space and have developed an efficient algorithm for this purpose called Energy-Directed Tree Search (Izgorodina et al., 2007). Having located our geometries, we then calculate very accurate energies using an ONIOM approximation (Izgorodina et al., 2006, 2007, Lin et al., 2009) to the W1 method, which in turn approximates CCSD(T) calculations at the infinite basis set limit via extrapolation procedures and has kJ accuracy provided single reference methods are appropriate (Martin et al., 1999). To this end, we divide the chemical reaction into a series of layers: an inner core that captures the reaction center; a core that includes the reaction center and all alpha substituent effects; the full system. In forming the core and inner core deleted substituents are replaced with hydrogens to maintain chemical valency and all geometries are relaxed. The inner core is then studied with W1 and a slightly lower-cost composite method, G3(MP2)-RAD (Henry et al., 2003), the core is studied with G3(MP2)-RAD and RMP2/6-311+G(3df,2p) and the full system is studied with RMP2/6-311+G(3df,2p). The W1 energy on the full system is then approximated as the sum of the RMP2/6-311+G(3df,2p) energy, plus a correction for RMP2 versus G3(MP2)-RAD error (as measured for the core) and a further correction for the G3(MP2)-RAD versus W1 error (as measured for the inner core). In this way RMP2 is used only to measure the remote substituent effects (primarily dispersion effects) for which it is well suited, and the more difficult task of modeling the reaction itself is performed using higher-level methods. Having obtained geometries, frequencies and energies for reactants and products, equilibrium constants are then calculated using the standard textbook formulae for the statistical thermodynamics of an ideal gas under the harmonic oscillator and rigid rotor approximation; further corrections are then made by treating low frequency torsional modes as hindered internal rotations. Finally, gas-phase results are corrected to the solution phase by means of a thermodynamic cycle in which the free energies of solvation are calculated using continuum models such as PCM (Miertus et al., 1981).

Table 3 summarizes the calculated and experimental values of the equilibrium constant $K(=k_{ad}/k_{fr})$ for the addition of the tert-butyl (*t*Bu) radical to the various low molecular weight RAFT agents. The calculations in this table were performed for the same conditions (20°C, benzene solution) as the experiments; the energies in solution in this case were calculated using our W1-ONIOM procedure described above with PCM-UAHF. Further computational details are available in the original reference (Chernikova et al., 2011); data for BB and BC differ slightly from this previous work as the conversion from the gas-phase to solution-phase standard state was accidentally applied twice in this earlier study. In cases where the leaving group(s) on the RAFT agent is not *t*Bu, there are multiple addition/fragmentation reactions possible and all pathways were considered computationally. As noted above, the fragmentation of the *t*Bu radical is observed exclusively in the experiments when the RAFT agent is BB and fragmentation of the Bz radical is observed exclusively when the RAFT agent is BC. In this latter case, the experiments were unable to measure the equilibrium constant for the addition/fragmentation reactions because the intermediate radical was too short-lived and only the substitution rate was reported.

Examination of Table 3 shows that, in general, there is good accord between the computed and measured equilibrium constants, though with one important exception. For TB the agreement is excellent (within a factor of 1.5), while for TC the theoretical calculations underestimate experiment by approximately 2 orders of magnitude, an error still within the reasonable limits of uncertainty of both techniques. However, for BB the calculated and experimental equilibrium constants show very good agreement (within a factor of 2), but theory predicts that fragmentation of the benzyl radical is preferred whilst experiment detects *t*Bu radical exclusively. Although the experimental observation is difficult to argue with, the conclusion that *t*Bu undergoes fragmentation more readily than benzyl radical is

RAFT Agent	Intermediate Radical	Leaving Group	Experimental			Theory
			k_{ad}	k_{fr}	K	K
TB	*t*Bu–SC•(Ph)S–*t*Bu	*t*Bu	5×10^{6}	8×10^{-3}	6.2×10^{8}	8.9×10^{8}
TC	*t*Bu–SC•(S*t*Bu)S–*t*Bu	*t*Bu	2.2×10^{5}	2×10^{-2}	1.1×10^{7}	5.3×10^{4}
BB	*t*Bu–SC•(Ph)S–CH_2Ph	*t*Bu	2.2×10^{5}	9×10^{-1}	2.4×10^{5}	3.5×10^{8}
		CH_2Ph	—	—	—	3.6×10^{3}
BC	*t*Bu–SC•(S*t*Bu)S–CH_2Ph	*t*Bu	2.8×10^{5}	—	—	1.8×10^{2}
		CH_2Ph	—	[Obs.]	—	4.6×10^{-2}

Table 3. Comparison of theoretical and experimental equilibrium constants (K, L mol^{-1} 20°C, benzene solution)

not easy to accept at face value. In particular, as is well known, the benzyl radical (whose radical stabilization energy is 61.0 kJ mol^{-1}) is significantly more stable than *t*Bu (23.0 kJ mol^{-1}) and should thus fragment preferentially (Coote et al., 2010). Indeed for BC, experiment does detect the preferential fragmentation of benzyl (instead of *t*Bu), a result in accord with theory. At present we are unable to explain the exclusive detection of *t*Bu radical in the experimental study of BB, but one tantalizing possibility is that there may be another (as yet undetected) reaction channel available to the benzyl radical (but not the *t*Bu radical) in the presence of the dithiobenzoate RAFT agent BB that is not significant in the presence of the trithiocarbonate RAFT agent BC. Such a process, if it exists, may help to resolve the apparent inconsistencies in the experimental and theoretical data for dithiobenzoate mediated polymerizations of styrene, and further work on this problem is currently underway.

To study the effects of chain length on the RAFT equilibrium, we considered the reactions of *t*Bu radical with dithiobenzoates bearing various oligomers of polystyrene and poly(methyl methacrylate) as leaving groups. We also considered the addition of these same oligomers to the TB RAFT agent. The substitution (or chain transfer) reactions studied experimentally are the overall reaction that results when pairs of corresponding polyRAFT and TB addition reactions are subtracted. The calculated equilibrium constants (calculated new as part of the present work but with some species taken from our earlier studies) are shown in Figure 14. As in the data above, calculations were performed using our W1-ONIOM method for energies, the PCM-UAHF solvation model but, due to the larger size of the species, the harmonic oscillator approximation was used for the gas-phase partition functions. This latter approximation adds an additional uncertainty of approximately one order of magnitude to our results but should not affect the qualitative trends.

From Fig. 14 we see that chain length effects on the equilibrium constant for the RAFT agent, in reactions with a constant attacking radical, are relatively small. However, not surprisingly, the equilibrium constant is much more sensitive to the nature of the attacking radical, especially with respect to its terminal and penultimate units where K values can stretch over as much as 10 orders of magnitude with trends that are very specific to the reaction under study. However, beyond the penultimate position convergence is quite rapid. With respect to reactions involving the *t*Bu radical with polymeric RAFT agents, it is clear that the former is a much poorer leaving group than either the polystyryl or polymethyl methacrylate attacking radicals and this is consistent with the experimental observation that they undergo rapid substitution.

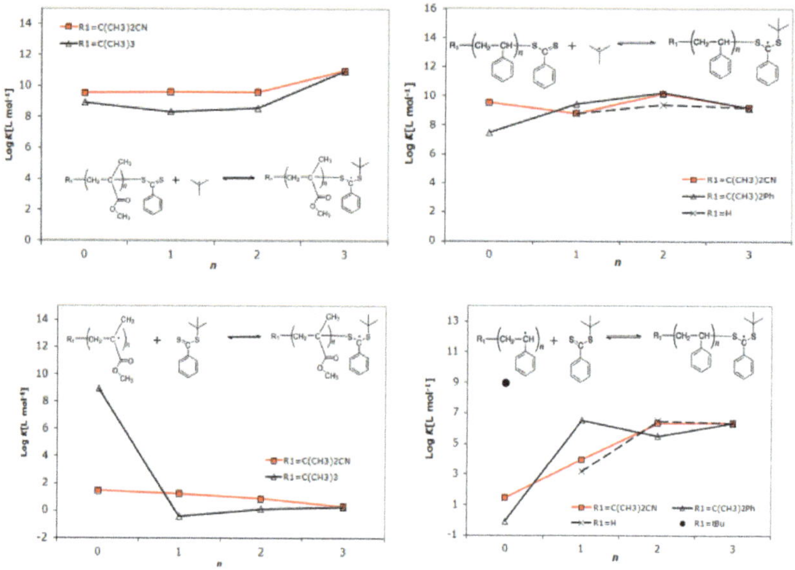

Figure 14. Chain length dependence of the equilibrium constant (K, L mol^{-1}, 20°C, benzene solution) in RAFT polymerization of methyl methacrylate and styrene

6. Conclusion

The most convincing argument for the currently accepted mechanism of RAFT polymerization is the direct monitoring of the radical intermediates formation by means of ESR (Chernikova et al., 2004, Hawthorne et al., 1999, Golubev et al., 2005, Kwak et al, 2002). The concentrations of these intermediates and their changes during the process provide valuable data for the kinetic modeling of RAFT polymerization. If the radical intermediates are relatively stable, they are accumulated in concentrations sufficient to be observed by ESR. However, if the intermediates are active, one cannot detect them in a reaction mixture. The use of spin traps can help to overcome this disadvantage and thereby extend the possibilities for the kinetic investigations of this process. We have found that the MNP and C-phenyl-N-butyl nitrone spin traps and their adducts are stable in the presence of RAFT agents, and this allows us to determine the nature of the trapped radicals from their ESR spectra. Through careful design of the reaction conditions (especially the use of low concentrations of active radicals) spin traps allow us to simplify the kinetic analysis by outcompeting certain side reactions, in particular termination and re-addition processes.

Reactions of addition and fragmentation are the principal stages in RAFT process. Two extreme cases can occur in different systems: radical addition to the RAFT agent resulting in formation of a stable intermediate, which one can observe via ESR (the model for this case is the TB–MNP system); or formation of an unstable intermediate that does not accumulate in the system and immediately decomposes (BC–MNP system). In the first case one can say that two different reactions of addition and fragmentation occur sequentially; in the second case – only the overall radical substitution is observed. In terms of the application of spin traps it is important to note that there are at least two different radicals in these reactions: one that is reacting, and another that is forming. In some cases, the forming radical may be an intermediate that is not caught by the spin trap.

For the successful application of spin-trap technique, particularly, for the estimation of kinetic constants of these reactions, the two following requirements must be met. The first requirement is that the ESR-spectra of spin trap adducts with reacting and leaving radicals should be different enough to determine quantitatively the contribution of each adduct. This requirement is easily met in the model studies presented here, and more generally in the first stage of RAFT polymerization where the propagating radical interacts with typical initial RAFT agents. We also showed that to some extent short oligomers of different chain lengths could also be discriminated. However, for study of the main equilibrium in RAFT, where the attacking and leaving radicals are essentially identical, this is more problematic and other strategies are required. The second requirement for successful application of the spin-trap technique is knowledge of the rate coefficients for the various spin trapping reactions. Unfortunately, there are not always reliable data available for these constants. At present we use the average value presented by Golubev et al. (2001). The development of accurate methods for the determination of trapping constants remains an ongoing challenge.

We have shown that the spin trapping technique can readily be used to measure substitution (and hence chain transfer) constants with a high level of accuracy. We have also used it to measure rate coefficients for the individual addition and fragmentation rate reactions that are in very good agreement with theoretical calculations for most systems studied to date. However, determination of these rate coefficients is still problematic for some systems, particularly if the intermediate radical is short lived. To measure these rate coefficients we use the method of competitive reactions (MCR), where the comparative reaction is the spin trapping. At the moment there is a lack of reliable information about the kinetics of these reactions. As a result, the determination of kinetic parameters is limited by the accuracy of the reference reaction, which is often unknown. In the literature the reproducibility of experimental data is usually mentioned, but the absolute accuracy is not. In the present work we have assumed that the absolute error accounts for half an order of magnitude at least; in particular cases the error could be appreciably higher. This is one of the reasons why we use extensively the photolysis of MNP: for this reaction there is a reliable value of rate coefficient of tert-butyl radical spin trapping. The other challenge is that, in estimating kinetic constants, one also has to assume that other reaction channels are not significant. The generally good agreement with theoretical calculations suggests that these assumptions are likely to be reasonable most of the time, though the unusual results for BB are also a warning that these assumptions need to be continually re-evaluated for each new system studied. Nonetheless, it has to be stressed that the presence of the spin trap, by outcompeting most side reactions, does greatly minimize the impact of kinetic assumptions on the accuracy of the kinetic results, particularly when compared with standard polymerization conditions.

The spin trap technique therefore has much to offer the radical polymer field. For instance, the application of the spin traps has helped to solve a number of problems in complex-radical polymerization and copolymerization (Golubev et al., 1978). They have also helped to address controversies related to spontaneous initiation and unusual inhibition of some polymerization systems (Zaremski et al., 1999). We hope that the results presented in this work will encourage the further development and application of these techniques for RAFT polymerization and other controlled radical polymerization processes.

Author details

Anatoly Filippov, Elena Chernikova, Vladimir Golubev, Ganna Gryn'ova,
Ching Yeh Lin and Michelle L. Coote
Moscow State University named by M.V. Lomonosov, Russian Federation
ARC Centre of Excellence for Free-Radical Chemistry and Biotechnology,
Research School of Chemistry, Australian National University, Australia

Acknowledgement

MLC gratefully acknowledges generous allocations of time on the National Facility of the National Computational Infrastructure, support from the Australian Research Council (ARC) under its Centres of Excellence program, and an ARC Future Fellowship. EC gratefully acknowledges financial support from the Russian Foundation for Basic Research (project 11-03-00640).

7. References

Ah Toy, A., Vana, P., Davis, T. P. & Barner-Kowollik, C. (2004). Reversible Addition Fragmentation Chain Transfer (RAFT) Polymerization of Methyl Acrylate: Detailed Structural Investigation via Coupled Size Exclusion Chromatography–Electrospray

Ionization Mass Spectrometry (SEC–ESI-MS), *Macromolecules, vol.* 37 (No. 3), pp. 744–751

Barner-Kowollik, C., Coote, M. L., Davis, T. P. et al. (2003). The reversible addition-fragmentation chain transfer process and the strength and limitations of modeling: comment on the "Magnitude of the fragmentation rate coefficient. *Journal of Polymer Science, Part A: Polymer Chemistry, vol.* 41 (No. 18), pp. 2828–2832

Barner-Kowollik, C., Buback, M., Charleux, B. et al. (2006). Mechanism and kinetics of dithiobenzoate-mediated RAFT polymerization. I. The current situation. *Journal of Polymer Science, Part A: Polymer Chemistry, vol. 44 (No. 20),* pp. 5809–5831

Barner-Kowollik, C. (Ed.), (2008). *Handbook of RAFT Polymerization.* Wiley-VCH, ISBN 978-3-527-31924-4, Weinheim.

Buback, M., Hesse, P., Junkers, T., Vana, P. (2006). Determination of Addition and Fragmentation Rate Coefficients in RAFT Polymerization via Time-Resolved ESR Spectroscopy after Laser Pulse Initiation. *Macromolecular Rapid Communications, vol. 27* (No. 3), pp. 182–187

Buback, M., Janssen, O., Oswald, R. et al. (2007). A missing reaction step in dithiobenzoate-mediated RAFT polymerization. In: *Radical polymerization: kinetics and mechanism*. Ed. Buback M., v. Herk A.M., pp. 158-167, Wiley-VCH, ISBN 10 3-527-32056-3, 13 978-3-527-32056-1, Weinheim.

Chernikova, E., Morozov, A., Leonova, E. et al. (2004). Controlled Free-Radical Polymerization of n-Butyl Acrylate by Reversible Addition-Fragmentation Chain Transfer (RAFT) in the Presence of tert-Butyl dithiobenzoate. A Kinetic Study. *Macromolecules, vol.* 37 (No. 17), 6329–6339

Chernikova, E., Golubev, V., Filippov, A. et al. (2010). Use of spin traps to measure the addition and fragmentation rate coefficients of small molecule RAFT-adduct radicals. *Polymer Chemistry,* vol. 1 (No. 9), 1437–1440

Coote, M.L., Wood, P.F., Radom, L. (2002). Methyl radical addition to C=S double bonds: kinetic versus thermodynamic preferences. *Journal of Physical Chemistry, A,* vol. 106 (No. 50), pp. 12124–12138

Coote, M L, Izgorodina, E I, Krenske, E H, Busch, M. & Barner-Kowollik, C. (2006). Quantum chemical mapping of initialization processes in RAFT polymerization. *Macromolecular Rapid Communications,* vol. 27 (No. 13), pp. 1015–1022

Coote, M.L. (2009). Quantum-chemical modeling of free-radical polymerization. *Macromolecular Theory and Simulations,* vol. 18 (No. 7-8), pp. 388-400

Coote, M.L., Lin, C.Y., Beckwith, A.L.J., Zavitsas, A.A. (2010). A comparison of methods for measuring relative radical stabilities of carbon-centred radicals. *Physical Chemistry Chemical Physics,* vol. 12 (No. 33), pp. 9597–9610

Cossi, M.; Rega, N.; Scalmani, G.; Barone, V. (2003). Energies, structures, and electronic properties of molecules in solution with the C-PCM solvation model. *Journal of Computational Chemistry,* vol. 24 (No. 6), pp. 669 -681

Denisov, E. T. (1971), *Rate Constants of Homolytic Liquid-Phase Reactions.* Nauka, Moscow [in Russian].

Doba, T., Ichikawa, T., & Yoshida, H. (1977). Kinetic Studies of Spin-trapping Reactions. I. The Trapping of the t-Butyl Radical Generated by the Photodissociation of 2-Methyl-2-nitrosopropane by Several Spin-trapping Agents. *Bulletin of the Chemical Society of Japan.* 50, p. 3158

Feldermann, A., Coote, M. L., Stenzel, M. H. et al. (2004). Consistent experimental and theoretical evidence for long-lived intermediate radicals in living free radical polymerization. *Journal of American Chemical Society*, vol. 126 (No. 48), pp. 15915-15923

Golubev, V. B.; Zubov, V. P.; Georgiev, G. S.; Stoyachenko, I. L.; Kabanov, V. A. (1978). Mechanism of alternating copolymerization of polar and hydrocarbon polymers. *Journal of Polymer Science: Polymer Chemistry,* vol. 11 (No. 10), pp. 2463–2470

Golubev, V. B. (1994). The mechanism of alternating polymerization. *Polymer Science*, Series A, vol. 36 (No. 2), pp. 244–250

Golubev, V. B., Maksimenko, O. O. & Zubov, V. P. (2001). A Kinetic Study of Styrene Polymerization by the Spin-Trap Method . *Polymer Science*, Series A, vol. 43 (No. 10), pp. 1242–1249

Golubev, V. B., Chernikova, E. V., Leonova, E. A. & Morozov, A. V. (2005). Controlled Reversible Addition-Fragmentation Chain-Transfer Polymerization: ESR Spectra and Structure of Radical Intermediates. *Polymer Science*, Series A, vol. 47 (No. 5), pp. 678–688

Golubev, V. B., Filippov, A. N., Chernikova, E. V., Coote, M. L., Lin, C. Y., & Gryn'ova, G. (2011). The Use of Spin Traps for the Kinetic Investigation of Elementary Events of Pseudoliving Radical Reversible Addition Fragmentation Chain Transfer Polymerization. *Polymer Science Ser. C*, vol. 53 (No. 1), pp. 14–26

Hawthorne, D. G., Moad, G., Rizzardo E., & Thang S. H. (1999). Living Radical Polymerization with Reversible Addition–Fragmentation Chain Transfer (RAFT): Direct ESR Observation of Intermediate Radicals. *Macromolecules,* vol. 32 (No. 16), pp. 5457–5459

Henry, D. J. Sullivan, S. B. Radom. (2003).L. G3-RAD and G3X-RAD: Modified Gaussian-3 (G3) and Gaussian-3X (G3X) procedures for radical thermochemistry. *Journal of Chemical Physics*, vol. 118 (No. 11), pp. 4849–4860

Izgorodina, E I, Coote, M. L. (2006). Is the addition-fragmentation step of the RAFT polymerisation process chain length dependent? *Macromolecular Theory and Simulations,* vol. 15 (No. 5), pp. 394-403.

Izgorodina, E. I. Coote, M. L. (2006). Reliable low-cost theoretical procedures for studying addition-fragmentation in RAFT polymerization. *Journal of Physical Chemistry,* A, vol. 110 (No. 7), pp. 2486–2492

Izgorodina, E.I., Lin, C.Y., Coote, M.L. (2007). Energy-directed tree search: an efficient systematic algorithm for finding the lowest energy conformation of molecules. *Physical Chemistry Chemical Physics*, vol. 9 (No. 20), pp. 2507–2516

Izgorodina, E. I. Brittain, D. R. B. Hodgson, J. L. Krenske, E. H. Lin, C. Y. Namazian, M. Coote, M. L. (2007). Should contemporary density functional theory methods be used to study the thermodynamics of radical reactions? *Journal of Physical Chemistry* A, vol. 111 (No. 42), pp. 10754–10768

Klamt, A.; Schueuermann, G. (1993). COSMO: a new approach to dielectric screening in solvents with explicit expressions for the screening energy and its gradient. *Journal of Chemical Society*, Perkin Transactions 2, (No. 5), pp. 799-805.

Klamt, A. (2005) *COSMO-RS: from Quantum Chemistry to Fluid- Phase Thermodynamics and Drug Design*. Elsevier Science, ISBN 0-444-51994-7, Amsterdam.

Klumperman, E. T. A, Van Dungen, J. P., Heuts A. & Monteiro M. J. (2010). RAFT-Mediated Polymerization—A Story of Incompatible Data? *Macromolecular Rapid Communications*, vol. 31 (No. 21), pp.1846–1862

Konkolewicz, D., Hawkett, B. S., Gray-Weale, A. & Perrier, S. (2008). RAFT Polymerization Kinetics: Combination of Apparently Conflicting Models. *Macromolecules*, vol. 41 (No. 17), pp. 6400–6412

Kwak, Y., Goto, A., Tsujii, Y. et al. (2002). A Kinetic Study on the Rate Retardation in Radical Polymerization of Styrene with Addition–Fragmentation Chain Transfer. *Macromolecules*, vol. 35 (No. 8), pp. 3026–3029

Lin, C.Y. & Coote, M. L. (2009). How well can theory predict addition-fragmentation equilibrium constants in RAFT polymerization? *Australian Journal of Chemistry*, vol. 62 (No. 11), pp. 1479-1483

Lin, C. Y., Hodgson, J. L., Namazian, M., Coote, M. L. (2009). Comparison of G3 and G4 theories for radical addition and abstraction reactions. *Journal of Physical Chemistry*, A, vol. 113 (No. 15), pp. 3690–3697

Lin, C.Y., Izgorodina, E.I., Coote, M.L. (2010). First principles prediction of the propagation rate coefficients of acrylic and vinyl esters: are we there yet? *Macromolecules*, vol. 43 (No. 1), pp. 553–560

Lin, C.Y., Coote, M. L. (2011). An Ab Initio Investigation of the Chain Length Dependence of the Addition-Fragmentation Equilibria in RAFT Polymerization. *Australian Journal of Chemistry*, vol. 64 (No. 6), pp. 747-756

Martin, J. M. L. De Oliveira, G. J. (1999). Towards standard methods for benchmark quality ab initio thermochemistry—W1 and W2 theory. *Journal of Chemical Physics*, vol. 111 (No. 4), pp. 1843–1857

Miertus S., Scrocco E., Tomasi J. (1981). Electrostatic interaction of a solute with a continuum. A direct utilizaion of AB initio molecular potentials for the prevision of solvent effects. *Chemical Physics*, vol. 55 (1), pp. 117-129

Ting, S. R. S., Davis, T.P., & Zetterlund P.B. (2011). Retardation in RAFT Polymerization: Does Cross-Termination Occur with Short Radicals Only? *Macromolecules*, vol. 44 (No. 11), pp. 4187–4193

Zaremski, M.Yu.; Plutalova, A.V.; Garina, E.S.; Lachinov, M.B.; Golubev V.B. (1999). On the Mechanism and Kinetics of TEMPO-Mediated Radical Copolymerization. *Macromolecules*, vol. 32 (No. 19), pp. 6359–6362

Chapter 14

Platinum Complexes with Bioactive Nitroxyl Radicals: Synthesis and Antitumor Properties

Vasily D. Sen', Alexei A. Terentiev and Nina P. Konovalova

Additional information is available at the end of the chapter

1. Introduction

Nitroxyl radicals (NRs), which are sometimes called "organic nitrogen oxides" exhibit a wide range of biological activities, e.g., hemodynamic effect, protection against ionizing radiation, suppression of oxidative stress in different types of pathology (Soule et al., 2007; Wilcox, 2010). Already in early studies of the simplest NRs, their antitumor activity was demonstrated on a model tumor, leukemia La (Konovalova et al., 1964) and their cytotoxicity was shown for HeLa cells (Klimek, 1966). Subsequent studies included: 1) in-depth studies of antitumor activity of simple NRs (TEMPOL, TEMPO), 2) trials of therapeutical efficiency of NRs used in combination with the clinically approved anticancer drugs, and 3) synthesis and studies of hybrid compounds with NRs covalently bound to anticancer pharmacophores. Recent studies have shown that simple nitroxyls affect the cell viability through a redox-mediated signaling and induce a multifactor cell death response, including oxidative damage, cell cycle arrest and apoptosis (Gariboldi et al., 1998; 2000; 2003; Suy et al., 2005).

Nitroxide TEMPOL potentiates the cytotoxicity of doxorubicin in the culture of tumor cells with multidrug resistance (Gariboldi et al., 2006), and reduces its cardiotoxicity in rats (Monti et al, 1996). Combinations of low doses of nitroxyl TEMPO and doxorubicin or mitoxantrone exhibit additive or synergistic cytotoxicity, depending on the type of the tumor cells (Suy et al., 2005). In experiments on mice, nitroxyls at low doses (0.25–10 mg/kg) were shown to decrease toxicity of anticancer drugs substantially (Konovalova et al., 1991).

A considerable amount of research has been carried out on hybrid compounds containing NRs linked covalently to an anticancer pharmacophore. Nitroxyl derivatives of (thio)phosphamides (Shapiro et al., 1971; Emanuel et al., 1976; Sosnovsky & Paul, 1984;

Sosnovsky & Li, 1985a), cyclophosphamide (Tsui et al., 1982), actinomycin D (Sinha et al., 1979), ethylenimino triazines (Emanuel & Konovalova, 1992), nitrosoureas (Raikov et al., 1985; Sosnovsky & Li, 1985b; Emanuel et al., 1986; Sen', 1993), 5-fluorouracil (Emanuel et al., 1985; Sen' et al., 1989), daunorubicin (Emanuel et al., 1982) were synthesized and studied. In comparison with the parent compounds, nitroxyl derivatives of the cytostatic drugs possess lower overall toxicity in animal studies and higher values of the *half inhibitory* concentrations IC_{50} in cell cultures. At the same time, they exhibit higher chemotherapeutic indexes (are effective in a wider range of doses) and, in the cases studied, are characterized by fewer side effects. For example, ruboxyl, a nitroxyl derivative of daunorubicin, is 8-fold less toxic to mice than the parent compound. At optimal doses, ruboxyl is more effective in experimental animal tumors and has no cardiotoxicity (Emanuel et al., 1982, 1992). After a successful phase II clinical trials (1991), its further study was interrupted due to financial problems.

Over the past 30 years, platinum complexes occupy leading positions among drugs for cancer chemotherapy. The antitumor activity of cisplatin (CP) was discovered in 1960s, and in 1978 it was approved for clinical use (Kelland, 2007). The subsequent search for improved cisplatin analogues resulted in introduction of carboplatin (1989) and oxaliplatin (2002) into clinical practice. About 15 other complexes, for various reasons, have been rejected in clinical trials. Currently, JM216 (satraplatin), picoplatin, and nanopolymer ProLindac, bearing the oxaliplatin moiety, are subject to clinical trials (Wheate et al., 2010) (Fig. 1).

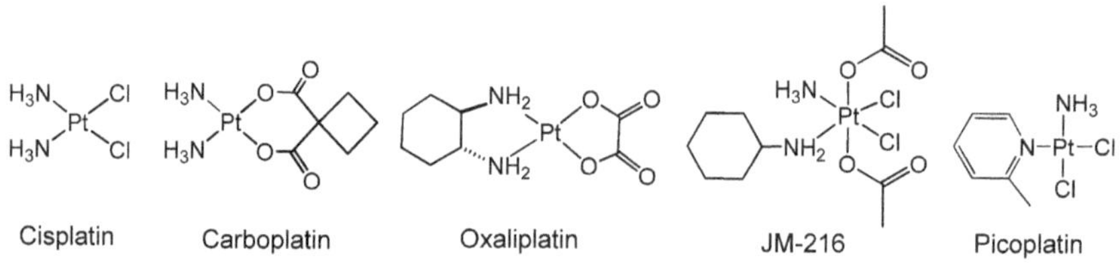

Figure 1. Platinum anticancer drugs which are in clinical use and undergoing clinical trials.

Cisplatin and other bivalent platinum complexes are effective against a number of human tumors. They are used in almost half of the treatment regimes in combinations with other anticancer drugs (Wheate et al., 2010). Complexes of bivalent platinum are highly reactive and, therefore, they are highly toxic drugs. To avoid acute toxicity, cisplatin is administered by continuous infusion of a very dilute solution (Blokhin & Perevodchikova, 1984). Another disadvantage of cisplatin is a rapid development of tumor resistance to this drug (Koeberle et al., 2010).

Complexes of Pt(IV), being chemically more inert than Pt(II) complexes, are characterized by moderate toxicity, and are suitable for oral administration. Complexes like satraplatin can pass through the digestive tract where they are absorbed into the bloodstream. With the bloodstream they reach organs and tissues, interact with cellular targets, and thus provide an antitumor effect (Kelland, 1999). Complexes of Pt(IV) are prodrugs (drug precursors) that, after entering into the cell or on the way to it, are reduced to corresponding active Pt(II) derivatives causing cytotoxic effect. At the same time, Pt(IV) complexes are potent inhibitors of proliferation of tumor cells including those resistant to cisplatin. Recent advances in the study of anticancer platinum amino complexes are summarized in a number of reviews (Kelland, 2007; Wheate et al., 2010; Klein & Hambley, 2009; Koeberle et al., 2010; Bonetti et al., 2009).

This review focuses mainly on the authors' data on synthesis and studies of new highly active platinum compounds with low toxicity, *viz,* Pt(II) and Pt(IV) complexes with biologically active aminonitroxyl radicals. In addition to biological activity, the advantage of such compounds is their paramagnetism which gives an opportunity to use them as spin labels in the study of the mechanism of antitumor action. The work involved the synthesis of platinum-nitroxyl complexes (PNCs), studies of their structure, physico-chemical properties and interactions with the main target, DNA. Studies of cytotoxic properties of PNCs, their impact on cell cycle and cell death were carried out using *in vitro* cultured tumor cells. Studies of antitumor activity, development of the tumor resistance, and synergistic antitumor effects of combinations of new complexes with cisplatin were performed in animal model tumor leukemia P388.

2. Synthesis of PNCs

2.1. Pt(II) complexes

Most of platinum complexes with high antitumor activity are non-ionic compounds with the *cis*-arrangement of the amino ligands (Hambley, 1997) (Fig. 1). Synthesis of complexes with two identical amino ligands or one diamino ligand is similar to the synthesis of cisplatin and is generally outlined in Fig. 2.

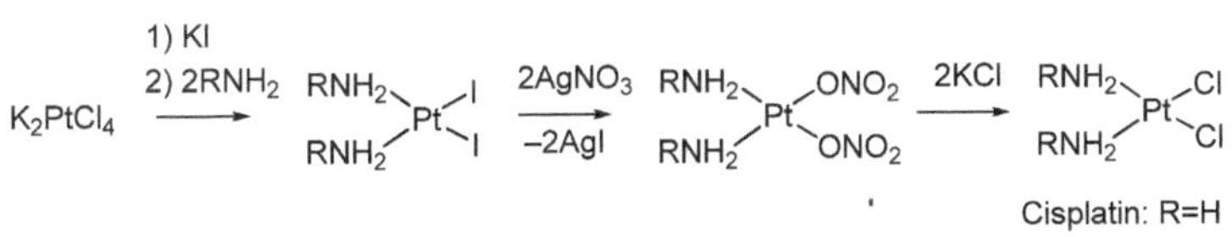

Figure 2. Synthesis of Pt(II) *cis*-diamino complexes.

To achieve good yields, diiodo complexes cis-[Pt(RNH_2)$_2$$I_2$] were prepared in the first step of the synthesis. The diiodo complexes were then converted to the target complexes by exchange reaction via water soluble dinitrato complexes (Fig. 2) (Dhara, 1970).

Complexes with two bulky amino ligands, such as compounds **1** (see Fig. 4), poorly bind to the DNA target (Section 3). Presumably, this is the main cause of their weak antitumor activity. The search for the active complexes led to the development of synthesis of mixed-ligand *cis*-diamino complexes containing only one bulky aminoligand (Giandomenico et al., 1995). The key intermediate in this synthesis is the salt K[Pt(NH_3)Cl_3], which can be obtained in several ways (Oksanen & Leskela, 1994; Oksanen, 1997; Potucka et al., 2011). The described method (Giandomenico et al., 1995) was modified by us for the preparation of PNCs with general formulae **2** and **3** (Fig. 3, 4).

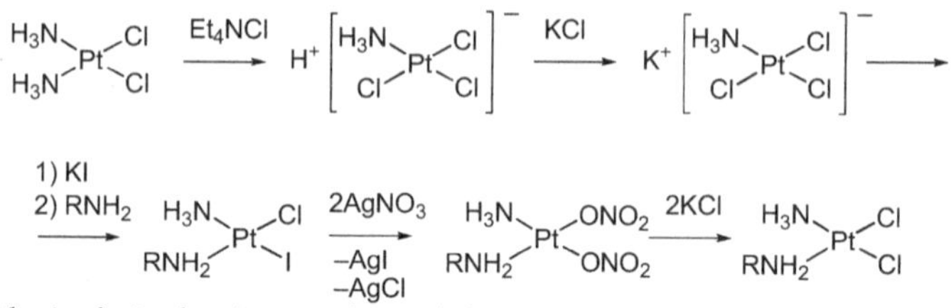

Figure 3. Synthesis of mixed amine complexes of platinum (II) (R – nitroxyl radical).

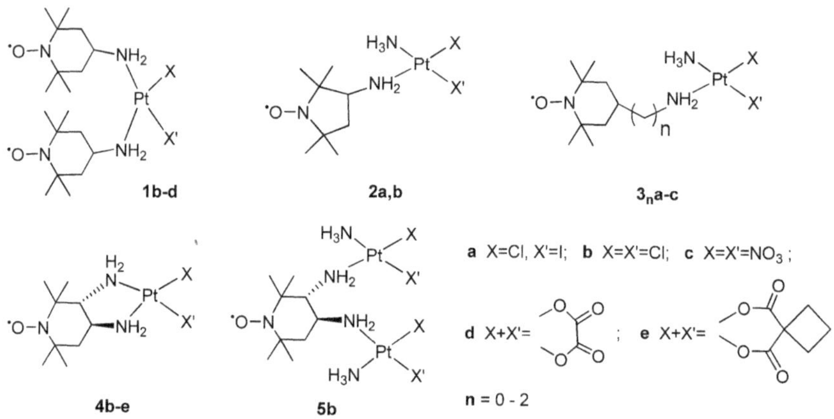

Figure 4. Platinum(II) amino nitroxyl complexes.

The formulae of complexes obtained are shown in Fig. 4 (Sen' et al., 1996, 1998, 2000). Two biradical complexes of type **1** were synthesized previously (Mathew et al., 1979; Claycamp et al., 1986). For the preparation of compounds of general formula **4**, which are structural analogues of oxaliplatin, we synthesized the first nitroxyl with two vicinal amino groups, *trans*-3,4-diamino-2,2,6,6-tetramethylpiperidine-1-oxyl (Sen', 1989). Binuclear complex **5b** was also obtained based on this radical.

2.2. Pt(IV) complexes

Complexes of Pt(IV) with mixed amino ligands can be obtained only by oxidation of the Pt(II) precursors. According to the published method (Giandomenico et al., 1995), the starting Pt(II) complexes **6** are oxidized with an excess of H_2O_2 under relatively harsh conditions (70 °C, ≥ 2 h). Under these conditions, the oxidation of Pt(II)-nitroxyl complexes leads to the formation of significant amount of by-products, probably as a result of oxidation of nitroxyl radical by Pt(IV) at elevated temperature. We found that catalytic amounts of salts of tungstic acid strongly accelerate the reaction so that preparative oxidation under mild conditions (0 – 20 °C) is limited only by the rate of dissolution of the starting complex and takes from 0.5 to 2.5 hours.

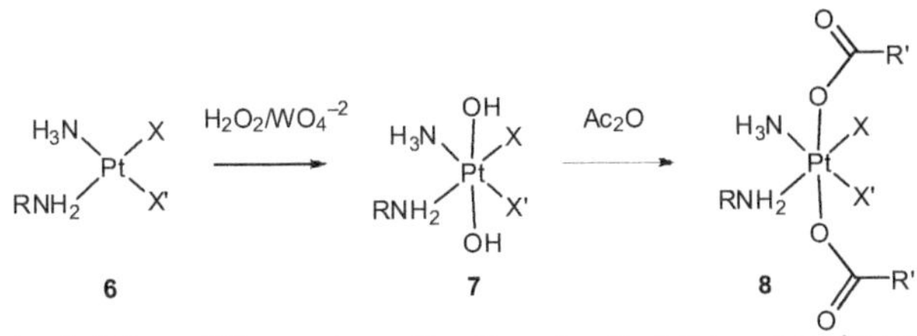

Figure 5. Synthesis of platinum(IV) complexes, R - nitroxyl radical, R '- carboxylic acid residue.

This significantly increases the reaction selectivity and the yield of the target products. *trans*-Dihydroxo complexes **7** resulted from the oxidation are of interest themselves. Their acylation with organic acid anhydrides leads to the *trans*-dicarboxylate derivatives **8** (Fig. 5) (Sen' et al., 2003, 2006).

The method described allows one to introduce different amines and exchange the so-called leaving X-ligands at the step of preparation of Pt(II) complexes (Fig. 3) and incorporate various carboxylate ligands with the alkyl residue R' of different length at the final step (Fig. 5). Thus, we can obtain the amino complexes of Pt(IV), which differ in chemical activity, solubility in water and aqueous-lipid distribution. The formulae of Pt(IV) complexes are shown in Fig. 6.

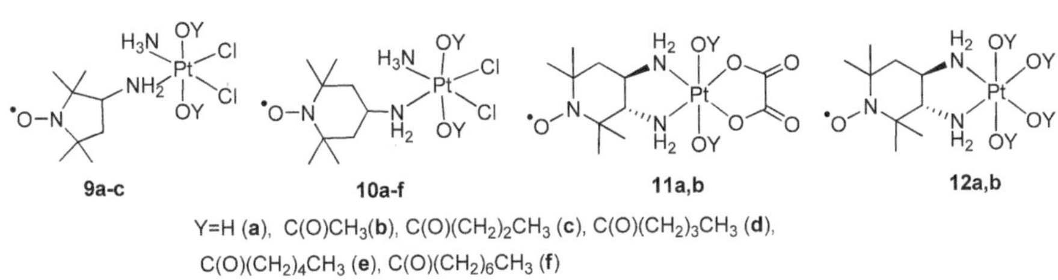

Figure 6. Platinum(IV) amino nitroxyl complexes.

The structure of PNCs was proved by elemental analysis and spectroscopic data (Sen' et al., 1996, 1998, 2000, 2003, 2006). For complexes **2b**, **4d**, and **10a**, the structures are determined by X-ray crystallography (Sen' et al., 2000, 2003; Chekhlov, 2005).

3. Interaction of PNCs with DNA

Reactivity of Pt(II) diamine complexes **4** depends strongly on the nature of leaving X-ligands. Pseudo-monomolecular rate constants for X-ligands hydrolysis in complexes **4c**, **4d**, cisplatin, and **4e** at 25 °C in 0.08 *M* NaOH are > 10^{-2}, $1.2 \cdot 10^{-4}$, $1.9 \cdot 10^{-5}$, and $2.9 \cdot 10^{-7}$ s^{-1}, respectively, *i.e.*, differ by five orders of magnitude (Shugalii et al., 1998). Therefore, the reaction of complexes with S- or N-donor groups can proceed either through the step of preliminary hydrolysis with the formation of an active intermediate aqua complex (Fig. 7, path *A*), or by the direct substitution of the X-ligands (path *B*).

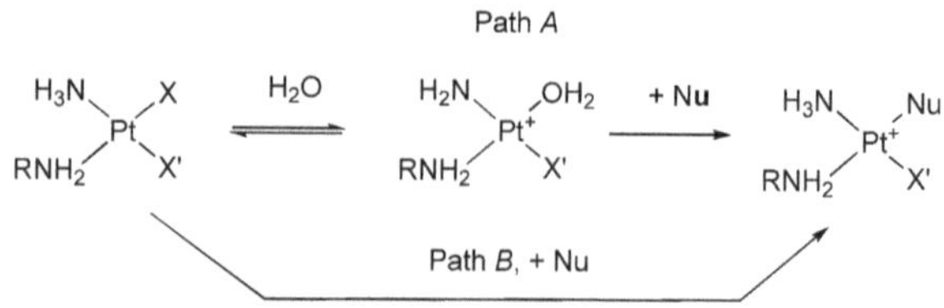

Figure 7. Reaction of Pt(II) complexes with nucleophilic atom Nu of a target molecule.

For relatively easily hydrolyzable complexes, including cisplatin, the reaction proceeds through the path *A* (Alderden et al., 2006), but for complexes of type **4e** (X + X' = cyclobutane dicarboxylate) there is an evidence for direct substitution of the X-ligands by N-donor groups (Frey, 1993).

It is known that cisplatin and its analogues bind mainly to the guanine and adenine bases of DNA with the formation of cross-links, thus perturbing the structure of DNA (Kelland, 2007; Wheate et al., 2010). Analysis of the EPR spectra of DNA modified with PNCs, together with hydrolytic determination of platinated DNA bases, showed (Shugalii et al., 1998) that complexes **3ob** and **4d** form predominantly (≥ 95%) bidentate intrastrand adducts with DNA. In adducts formed by both complexes, the rotation of the nitroxyl radicals is equally slow (Fig. 8) (correlation time $\tau \sim 10^{-8}$ s^{-1}).

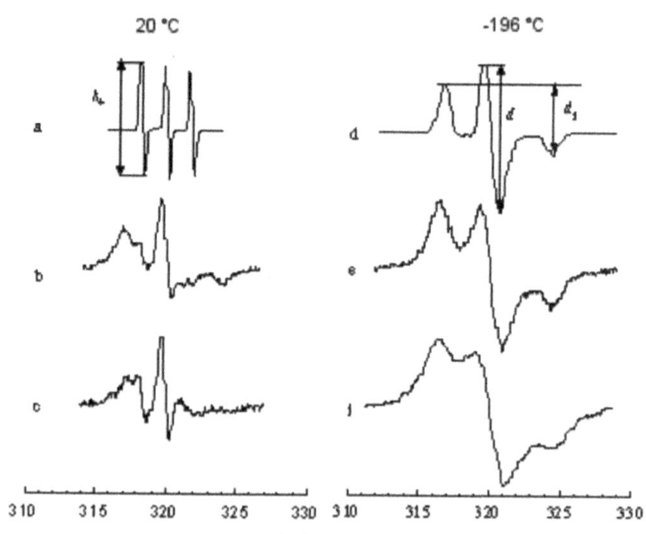

Figure 8. EPR spectra of free **4d** in water (a) and ethanol (d), and DNA modified by **4d** (b, e), and **3ob** (c, f) in aqueous 0.01 M $NaHCO_3$ at Pt-to-nucleotide ratio *r* =0.16, 37°C, 24 h. Spectra were recorded at magnetic modulation 0.32 mT, microwave power 5 (20°C) and 0.5 mW (–196°C), field scan 20 mT, scan time 8 min, time constant 1 s.

This result can be explained by immobilization of the radical moiety in the major DNA groove for the complex 3_0**b** adducts, and by the immobilization and/or by rigid structure of the double bound radical moiety in adducts formed by complex **4d**. Adducts formed by complexes 3_1**b** and 3_2**c** whose radical moiety is separated from the Pt atom by the methylene or ethylene bridge, are characterized by an order of magnitude lower values of the parameter τ. Presumably, this phenomenon is related to partial release of nitroxyl moiety from relatively shallow major groove of DNA, which increases its rotational mobility (Sen', 2002).

Exciting opportunities for the instrumental use of PNCs were shown by Dunham et al., 1998. An adduct of 3_0**a** complex with synthetic DNA fragment containing 11 base pairs was synthesized. Its structure in solution was determined by NMR based on the dependence between paramagnetic broadenings of protons of DNA bases and distances between bases and nitroxyl radical. Formation of the adduct was proved to result in the bending of DNA molecule that forms the angle ~80° with respect to the major groove, whereas the minor groove becomes strongly broaden.

The ability of complexes **1** – **5** to bind to the isolated DNA *in vitro* was determined under standard conditions and characterized by a parameter *r*, which is equal to the number of linked labels per one nucleotide. In the series of complexes with the same amino ligand **4c**–**e**, parameter *r* grows with the increase in the rate of hydrolysis of X-ligands. Platinating activity of compounds with different amino ligands depends on the total volume of these ligands and/or their linear sizes. Bulky biradical complex **1c** and complexes 3_1**b** and 3_2**c**, whose sizes are enlarged due to a methylene or ethylene bridge, bind to DNA 5 to 10 times less efficiently than cisplatin or complexes **4c**, **d** (Fig. 9).

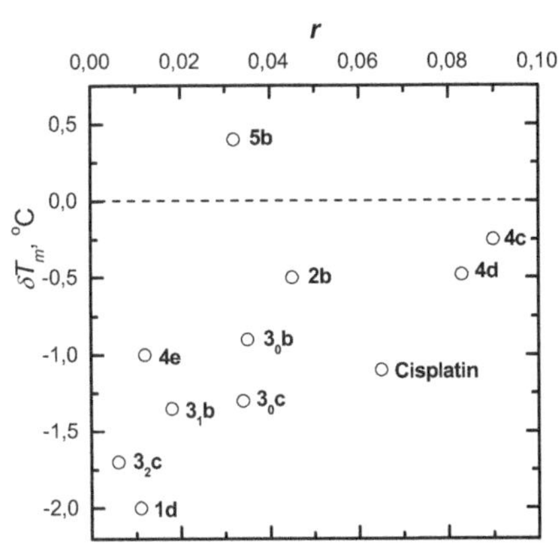

Figure 9. Relationship between the platinating activity (*r*) and specific destabilization of the DNA duplex (δT_m) for PNCs. The platination of DNA was carried out in 0.01 *M* $NaHCO_3$ over a period of 24 h at 37 °C at the initial molar ratio r_{in} = 0.1.

Ordinates in Fig. 9 are the values of specific destabilization of the DNA duplex, δT_m, that corresponds to the decrease in the DNA melting temperature due to formation of one adduct per 100 nucleotides. These values were calculated according to the formula

$$\delta T_m=(T_m' - T_m)/100r,$$

where T_m and T_m' are the melting points for unmodified and platinated DNA, respectively. The data in Fig. 9 show that the low-activity complexes (with low *r*–values) cause the greatest disorder in the DNA duplex. Presumably, adducts of these complexes are readily recognized by the repair machinery. This is in agreement with the data on low antitumor activity of such complexes (see below). The binuclear complex **5b** stabilizes DNA due to a predominant (~70%) formation of the interstrand crosslinks interfering with the thermal dissociation of the DNA duplex. It is interesting that bi- and trinuclear platinum amino complexes exhibit cytotoxic properties different from that of mononuclear ones, in particular, the former are active against cisplatin-resistant cells (Farrell et al., 1999).

4. Cytotoxicity of PNCs in tumor cell cultures

A simplified mechanism of cytotoxic effect of cisplatin and its analogs includes the transport of the complexes into the cell, their activation by the hydrolysis of leaving ligands (Cl^-, carboxylates), penetration into the nucleus, and formation of adducts with DNA (Kelland, 2007; Wheate et al., 2010). The DNA lesions are either repaired, or initiate a complex process of programmed cell death, *i.e.*, apoptosis. In addition, it is known that cisplatin, directly or indirectly, causes the generation of reactive oxygen species. This process is important for the initiation of apoptosis (Miyajima et al., 1997; Bragado et al., 2007) and may also be responsible for side effects, *e.g.* nephrotoxicity (Tsutsumishita et al., 1998).

Nitroxyl radicals are antioxidants, which can react with active radicals not only stoichiometrically, but also act as catalysts of redox reactions and mimetics of enzymatic systems. For example, in aqueous medium they perform superoxide dismutation through the reduction of radical $HO_2^\bullet$ by nitroxyl radical and the oxidation of radical $O_2^{\bullet -}$ by oxoammonium cation (Sen' et al., 1976, 2009; Goldstein et al., 2003) (Fig. 10).

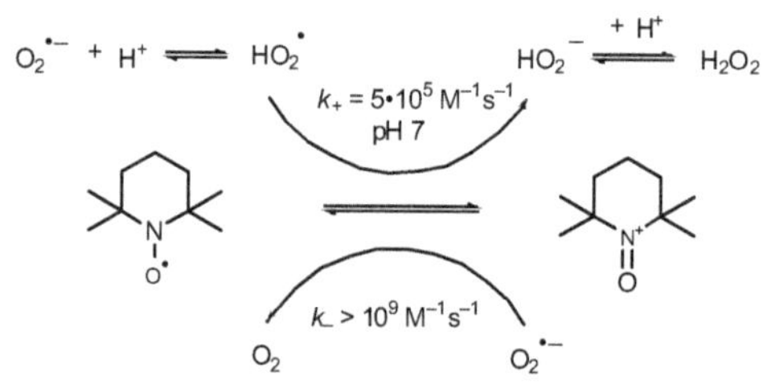

Figure 10. Superoxide dismutase-like activity of nitroxyl radicals

Interestingly, the nitroxyl based catalysis of dismutation of $HO_2^\bullet$ radical, generated in organic compounds undergoing oxidation, is carried out by the pair of nitroxyl radical/hydroxylamine (Denisov, 1996) (Fig. 11) Thermodynamic data are presented in support of the latter mechanism in organic medium. The measured constants of forward and reverse reactions at 50 °C are equal to $10^4 - 10^5\ M^{-1}\cdot s^{-1}$ (Denisov, 1996). Existence of two mechanisms for different media is not excluded. It looks reasonable that in an aqueous

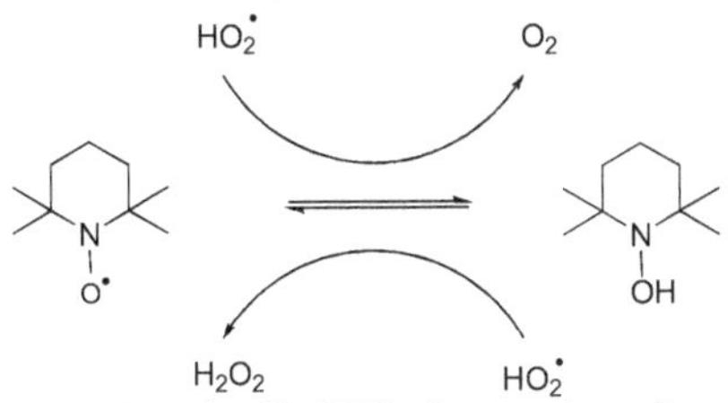

Figure 11. Catalytic disproportionation of radical $HO_2^•$ in organic medium

medium the preferred process is an electron transfer followed by a thermodynamically favorable hydration of oxoammonium cation (Fig. 10), while in an organic medium more typical reactions are the redox processes involving a hydrogen atom transfer (Fig. 11). Therefore, in biphasic aqueous-organic systems present in biological objects, both mechanisms are possible.

Like other antioxidants, under certain conditions, nitroxyls may exhibit pro-oxidant activity. The structure and concentration of nitroxyls, the medium properties, and other hard-to-identify factors can determine their anti– or pro-oxidant effect. At submillimolar concentrations, nitroxyls, as a rule, exhibit antioxidant properties and protect cells from apoptosis (Wilcox, 2010). At millimolar concentrations, nitroxyls are cytotoxic toward cultured tumor cells (Gariboldi et al., 1998, 2000, 2003, 2006; Suy et al., 2005) and are active against model animal tumors (Konovalova et al., 1964; Suy et al., 2005). Nitroxyl radicals cause cell death both in the wild type and p53 mutant cells (Suy et al., 2005). The study of interplay of platinum and nitroxyl pharmacophores combined in one molecule is of interest also in connection with the recent discussions on application of antioxidants and redox-active agents in tumor chemotherapy (Seifried, 2003; Wondrak, 2009).

To elucidate the interaction between platinum and nitroxyl pharmacophores, we studied the effect of 4-amino-2,2,6,6-tetramethylpiperidine-1-oxyl (4-amino-TEMPO) on the toxicity of cisplatin to HeLa cells. According to the published approach (Reynolds & Maurer, 2005; Chou, 2006), the dose-response relationships were determined for individual agents and their mixtures at a fixed concentrations ratio. These data were transformed into the combination index (CI) – dead cells fraction (*fa*) dependence (Fig. 12). In a wide range of fraction affected *fa*, values of log (CI) are positive, which indicates antagonism of the agents. In the range of relatively high fraction affected, corresponding to millimolar concentrations of 4-amino-TEMPO, a transition to negative values of log (CI) takes place which means the mutual reinforcement of cytotoxicity, i.e., synergy. These results are consistent with the antioxidant properties of nitroxyls at low and their pro-oxidant properties at high concentrations.

Data on the cytotoxicity of PNCs also reflects the antagonism of platinum and nitroxyl pharmacophores. Complexes of platinum (II) **2b** and **3$_0$b**, containing nitroxyl radicals of different structure, are significantly less toxic to tumor cells compared to cisplatin (Table 1).

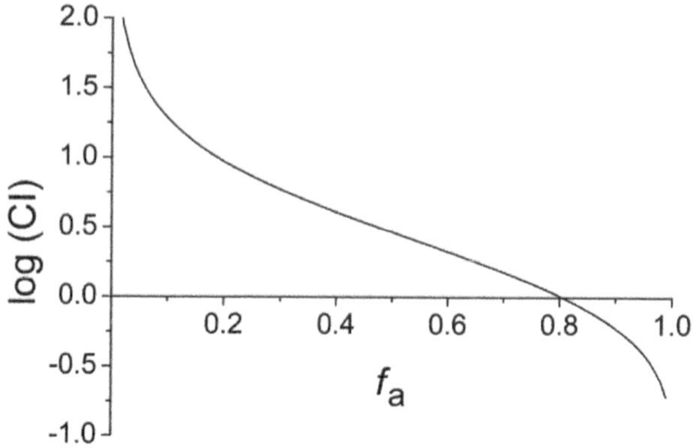

Figure 12. Combination index [log (CI)] versus fraction affected (f_a) of HeLa cells for combination of 4-amino-2,2,6,6-tetramethylpiperidine-1-oxyl and cisplatin.

Cell line	IC$_{50}$, µM										
	Cis-platin	**2b**	**3ob**	JM216	**9b**	**9c**	**10b**	**10c**	**10d**	**10e**	**10f**
HeLa	14.8	125	112	14.4	>200*	13.4	200	4.18	2.45	0.23	0.09
H1299	66.7	>150*	>150*	38.8	>200*	25.4	220	24.6	16.6	1.36	0.69

Table 1. 50% Inhibitory concentrations for PNCs in HeLa and H1299 cells; MTT assay, 24 h of exposure. *Evaluation of IC$_{50}$ is limited by low solubility of complexes.

Platinum(IV)-nitroxyl complexes **9b** and **10b**, being structurally close to JM-216, are also considerably less toxic to cells compared to the latter. Elongation of axial ligands Y increases both lipophilicity and cytotoxicity (complexes **9c** and **10c-f**), which is obviously due to the increased accumulation of complexes in cells (Fig. 13).

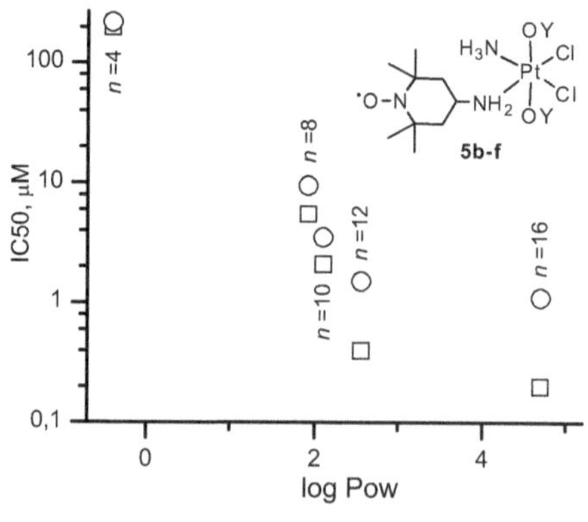

Figure 13. Relationship between lipophilicity (log Pow) and cytotoxicity (IC$_{50}$) of complexes **10b-f** in HeLa (□) and H1299 (O) cell lines; ***n*** means the total number of carbon atoms in carboxylic residues Y.

The H1299 cells are less sensitive to platinum complexes. Unlike the HeLa cells, H1299 cells do not contain p53 protein because of mutations of p53 gene in both alleles (Mitsudomi et al., 1992). Since p53 protein plays a key role in the process of apoptosis in response to DNA damages (Vousden & Prives, 2009), the observed lower sensitivity of H1299 cells to platinum complexes compared to HeLa cells can be related to lack of p53 function.

Our further study was focused on complex **10d** since it combined high cytotoxicity with sufficient solubility in water. The effect of complex **10d** and cisplatin on the cell cycle of HeLa and H1299 cells was studied (Fig. 14). According to flow cytofluorimetry data, both complexes cause approximately fivefold increase in the number of HeLa cells in the subG1 fraction, thus indicating induction of cell death. Accumulation of HeLa cells in early S phase was also observed, which suggests that cell death is induced after cell cycle arrest during DNA synthesis. For H1299 cells, some increase in the S phase population and two-fold decrease in G2/M phases population was demonstrated, which shows cytostatic activity of the complexes without significant cell death.

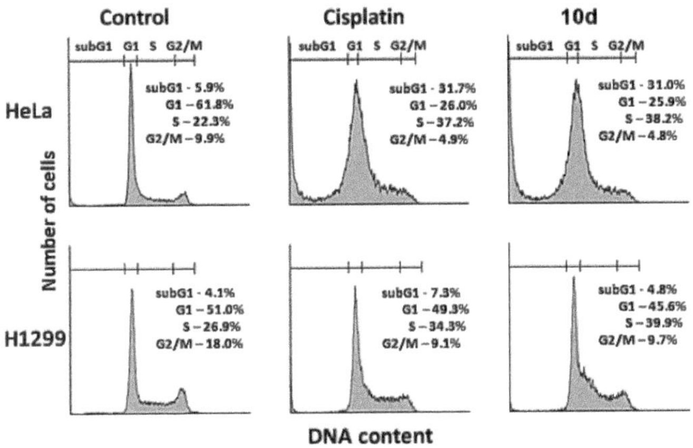

Figure 14. Cell cycle analysis of HeLa and H1299 cells treated with cisplatin and complex **10d** (concentration IC_{50}, 24 h).

Cell death found in flow cytofluorimetry experiments was shown to be apoptotic. Both cisplatin and complex **10d** cause in HeLa cells characteristic for apoptosis morphological changes of cell nuclei and internucleosomal cleavage of DNA leading to electrophoretic DNA laddering (Fig. 15a-d).

As it was discussed above, cisplatin and its analogues form adducts with DNA that, when are not repaired, trigger the tumor suppressor protein p53 (Alderden et al., 2006; Kelland, 2007; Wheate et al., 2010). Unlike cisplatin, the **10d** complex does not cause increase of the p53 protein expression in MCF7 cells containing wild-type p53 gene (Fig. 15e). This finding indicates differences in the mechanism of cytotoxic action of these two complexes. Interestingly, on the rat glioma C6 cells, the simple nitroxyl Tempol was shown to cause apoptosis without elevation of p53 protein levels (Gariboldi et al., 2003). Many compounds

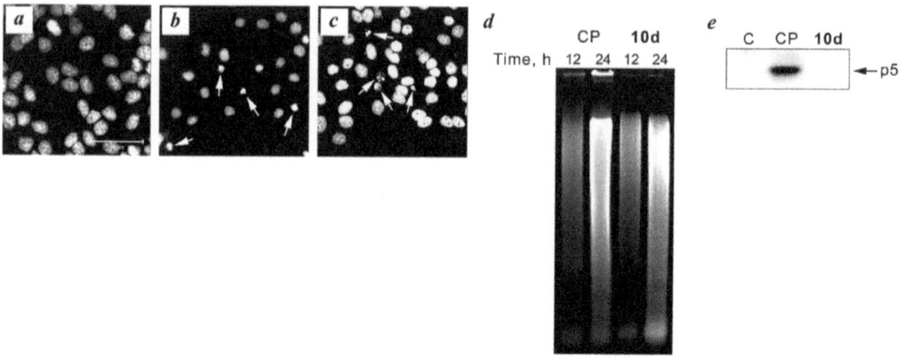

Figure 15. The mechanism of cytotoxicity of platinum complexes. *a—c* DAPI staining of DNA in HeLa cells in the control (a) and after 24 h exposure to cisplatin (CP) (b) or complex **10d** (c); arrows indicate the fragmented nuclei of the apoptotic cells. *d* Agarose gel electrophoresis of HeLa cells DNA after 12 h and 24 h exposure to CP and complex **10d**. *e* Immunoblotting of MCF7 cell lysates with antibody to p53 in the control (C) and after 6 h exposure to CP and complex **10d**.

including nitroxyls (Sui at al., 2005) and platinum complexes (Gorczyca et al., 1993; Kalimutho et al., 2011; Roubalová et al., 2010) induce apoptosis both in cells with wild-type p53 gene and in p53-deficient cells. However, p53-independent apoptosis of tumor cells harboring wild-type p53 gene, to our knowledge, was observed for the first time.

5. Antitumor activity of PNCs in animal experimental tumors

Data on cytotoxicity revealed from cell culture studies and antitumor activity observed in animal tumor models do not correlate for platinum complexes. A striking example of this phenomenon is carboplatin, the known antitumor chemotherapy drug possessing negligible cytotoxicity *in vitro*. Carboplatin has been approved for clinical use because of favorable efficiency to toxicity ratio found in animal tumor models (Jakupec et al., 2008).

5.1. Pt(II) complexes

Toxicity and biological activity of Pt(II) diamino complexes depend on the structure of both the carrier amino ligands and the leaving groups, the latter being replaced during metabolism and target binding (Ho et al., 2003). The biradical complexes **1** containing two bulky amino ligands poorly bind to DNA *in vitro* (see above) and exhibit low toxicity and antitumor activity *in vivo* (Table 2).

Complexes **2** and **3** bearing one bulky substituent are close in their properties to cisplatin. Their LD_{50} values are only 1.5 to 2.5-fold higher than that of cisplatin. They efficiently platinate DNA and exhibit antitumour activity comparable with cisplatin. Influence of the nitroxyl structure on the complex activity can be seen when comparing complexes **2b** and **3**$_{0}$**b** that differ only in the size of nitroxyl cycle. Compound **3**$_{0}$**b** is both more toxic and more active against leukemia P388. For complexes **4**, the correlation between the rate of leaving ligands

Complex	LD_{50}, mg·kg^{-1} (mM·kg^{-1})	Single dose, mg·kg^{-1}	*ILS*, %
1b	570 (0.94)	190	106 (0)
1c·H_2O	500 (0.74)	166	79(1)
1d	380 (0.61)	127	76 (0)
2b	27 (0.061)	6.8	237 (1)
3$_0$**b**	15 (0.033)	3.8	292 (2)
4b	80 (0.18)	16	189 (0)
4c	11 (0.022)	–	–
4d·$2H_2O$	50 (0.10)	11	132 (0)
4e	500 (0.95)	133	202 (2)
9a	45 (0.095)	15	133 (0)
9b	100 (0.180)	34	247 (0)
10a·$2H_2O$	27 (0.052)	9	270 (4)
10b	46 (0.080)	7.5	220 (4)
10d	4.5 (0.007)	1.5	120 (1)
11b	260 (0.38)	87	290 (1)
Cisplatin	12 (0.040)	3.0	245(1)

Table 2. Toxicity (LD_{50}) and antileukemic (P388) activity (*ILS*) of PNCs. LD_{50} is a dose which is lethal to 50% of healthy mice. Increase in the life span $ILS = [100(T/C - 1)]$, where T and C the average life-time (days) of treated and control animals, respectively. The numbers of animals survived for more than 60 days in the group of six animals are given in brackets.

hydrolysis and the toxicity was established (Sen' et al., 1996; Shugalii et al., 1998). Readily hydrolyzable **4c** has the highest toxicity whereas **4e**, the most slowly hydrolyzable compound among complexes **4**, exhibits the lowest toxicity, but, like carboplatin, it possesses good antitumor activity only at high doses. Compound **4d**, the structural analogue of oxaliplatin, is approximately 2-fold less toxic compared to the latter. The observed decrease in toxicity might be due to the influence of the nitroxyl group. The data presented demonstrate that, among Pt(II) complexes with amino nitroxyl radicals, high antitumor activity *in vivo* is characteristic for those that contain no more than one bulky amino ligand, platinate DNA with high efficiency, and cause moderate destabilization of DNA duplex (Fig. 9, Table 2).

5.2. Pt(IV) complexes

Toxicity of Pt(IV) complexes **9** and **10** varies in the wide range depending on the structure of axial ligands. Compounds **9a,b** and **10a,b** are 1.6 to 3-fold less toxic compared to the corresponding Pt (II) analogues, **2** and **3** (Table 2). As in the case of divalent complexes, the piperidine oxyl derivatives **10** are both more toxic and more active against leukemia P388 than pyrrolidine oxyl derivatives **9**.

An important feature of PNCs is found in comparative study of development of tumor resistance in leukemia P388 to complex **10a** and cisplatin (Sen' et al., 2003; Goncharova et al., 2011). The development of resistance was induced by sequential inoculation of tumor cells from animals treated with equitoxic doses of drugs. The tumor acquired resistance (≤ 20% of the sensitivity of the parent tumor) to cisplatin at the 4th and to complex **10a** at the 10th generation of tumor (Fig. 16). This data demonstrate that the resistance to complex **10a** develops 2.5-fold slower than that to cisplatin.

Interesting results were observed when PNCs and cisplatin were used in combination at low doses (1/10 to 1/20 of LD_{50}) for leukemia P388 treatment. Individual compounds in the same doses caused low *ILS*–indices with no cured animals, but their combinations cured up to 100% of mice (Fig. 17).

Complexes **10a** and **10b** containing piperidinoxyl moiety exhibit higher antitumor activity compared to that of complex **9b**, both as single agents and in combination with cisplatin. Thus, small difference in the structure of nitroxyl radicals in these PNCs has a significant influence on their antitumor activity *in vivo*.

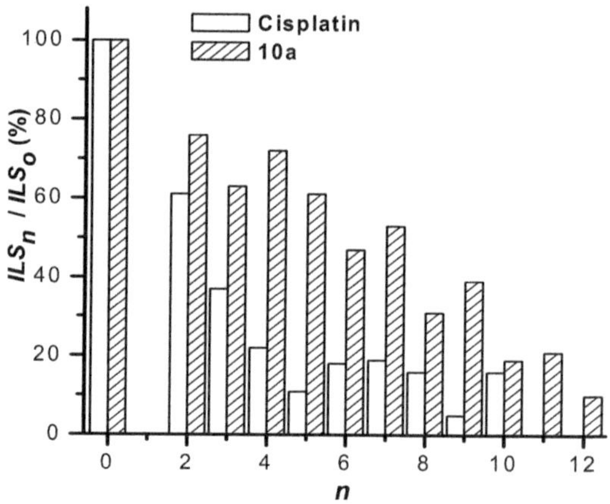

Figure 16. Development of resistance to cisplatin and complex **10a** in a series of successive transplant generations (***n***). ILS_0 is the increase in the life span of treated animals bearing the sensitive (parent) generation of leukemia P388.

It is known that reduction potentials $E_{1/2}$(>N$^+$=O/>N–O$^•$) of nitroxyls of piperidine series are, on the average, approximately 0.1 V lower than that of radicals of pyrrolidine series (Goldstein et al., 2006; Manda et al., 2007; Sen' & Golubev, 2009). Therefore, piperidinoxyls are oxidized by $HO_2^•$ radical more readily (Fig. 10), and they are more efficient superoxide dismutase mimetics compared to pyrrolidinoxyls (Goldstein et al., 2006). At the same time, *in vivo*, pyrrolidinoxyls undergo reduction to corresponding hydroxylamines about tenfold slower than piperidinoxyls (Komarov et al., 1994). Along with possible differences in pharmacokinetics of the complexes, these features of the redox properties of nitroxyls, presumably, affect the biological activity of PNCs.

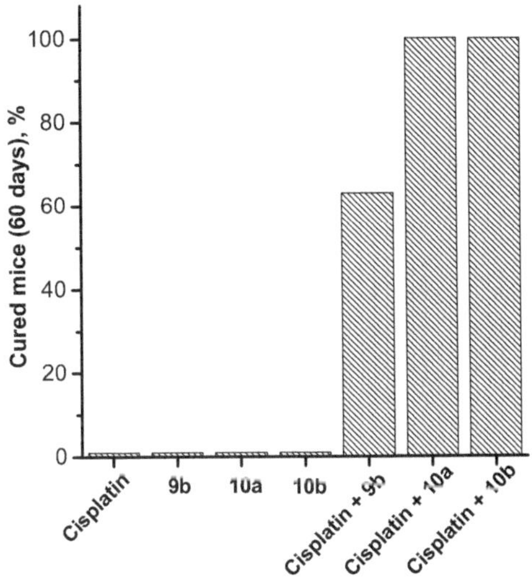

Figure 17. Synergy for the antitumor effect of cisplatin (0.6 mg/kg) combined with complexes **9b** (2,3 mg/kg), **10a** (1,4 mg/kg) or **10b** (5,0 mg/kg) against leukemia P388 (days of treatment 1, 3, 5, 7).

6. Conclusion

Recent studies (Wondrak, 2009; DeNicola et al., 2011) show that modulation of the redox state of cancerous cells could provide a new approach to suppression of tumor growth. Effects of nitroxyl radicals on the redox processes in normal cells and their cytotoxicity in tumor cells are documented in many examples (Gariboldi et al., 2000; Suy et al., 2005; Wilcox, 2010). Presumably, nitroxyls may affect the tumor cell viability through a redox-mediated signaling, which ultimately activate apoptosis.

On the other hand, the influence of nitroxyl radical on activity of anticancer agent, when they are used in combination or are covalently linked in one molecule, appears to depend on local concentration of radicals. At low concentrations of nitroxyls, which corresponds to rather low therapeutic doses of hybrid compounds, radicals are likely to impair the oxidative stress caused by tumor process and an anticancer agent itself. Published data show that *in vivo* the overall and/or specific toxicity may be reduced for hybrid compounds without significant loss of antitumor activity. On the other hand, high concentrations of nitroxyls, which can be implemented in combinations with anticancer agents, increase the intrinsic oxidative stress in tumor cells and exert a selective cytotoxic effect.

The known active anticancer complexes like cisplatin, oxaliplatin, and satraplatin bear in their structure redox-inert amino ligands. We synthesized structurally close analogs, i.e., platinum-nitroxyl complexes, amino ligands of which hold a wide spectrum of redox activity and are able to modulate biological properties of the new compounds. Their physicochemical properties, interaction with DNA, cytotoxicity *in vitro* and antitumor activity *in vivo* were studied. A limited correlation was found between the ability of new complexes to form adducts with DNA, thermal stability of these adducts and antitumor activity of the complexes. PNCs are characterized by features that distinguish them from platinum complexes with usual alkylamines. An antagonism of platinum and nitroxyl pharmacophore was observed in cell culture. As compared to cisplatin or satraplatin, structurally close nitroxyl derivatives are approximately 10 times less cytotoxic. This may be explained by a moderate inhibition of p53-dependent apoptosis due to the antioxidant properties of nitroxyl radicals. These *in vitro* findings do not correlate with *in vivo* data, and antitumor activities of some PNCs and cisplatin are approximately the same. The relatively slow development of resistance to PNCs and the presence of synergy for the combinations of PNCs and cisplatin imply the differences in the mechanism of antitumor action of cisplatin and PNCs. Due to antioxidant properties (Sen' et al., 2000), nitroxyls in PNCs may ameliorate side effects typical for cisplatin, such as nephrotoxicity and neurotoxicity. For example, nitroxyl derivative of daunorubicin, ruboxyl, has virtually no cardiotoxicity that limits therapeutical doses of the parental compound (Emanuel & Konovalova, 1992). Reduced side effects, in turn, contribute to the better survival and, as a result, the higher efficacy of tumor chemotherapy *in vivo*. Our and literature data show that nitroxyls are promising modulators of the activity of anticancer agents and, as such, could be approved for clinical use.

Author details

Vasily D. Sen', Alexei A. Terentiev and Nina P. Konovalova
Institute of Problems of Chemical Physics, Russian Academy of Sciences, Chernogolovka, Russian Federation

Acknowledgement

This work was partly supported by the Russian Foundation for Basic Research (Project No. 09-03-01187).

7. References

Alderden, R.A., Hall, M.D., & Hambley, T.W. (2006). The Discovery and Development of Cisplatin. *J. Chem. Education.*, Vol.83, No.5, pp. 728–734.

Blokhin, N.N., & Perevodchikova, N.I. (1984). *Khimioterapiya opukholevykh zabolevanii* (*Chemotherapy of Neoplastic Diseases*), Meditsina, Moscow, (in Russian).

Bonetti, A., Leone, R., Muggia, F.M., & Howell, B. (Eds.). (2009). *Platinum and other Heavy Metal Compounds in CancerChemotherapy,* Humana Press, ISBN 978-1-60327-458-6, New York.

Bravado, P., Armesilla, A., Silva, A., & Porras, A. (2007). Apoptosis by Cisplatin Requires p53 Mediated p38*a* MAPK Activation through ROS Generation. *Apoptosis*, Vol.12, No.9, pp. 1733–1742.

Chekhlov, A.N. (2005). Crystal and Molecular Structures of (*trans*-3,4-Diamino-2,2,6,6-tetramethylpiperidine-1-oxyl-N^3,N^4)(oxalato-O,O')platinum(II). *Russ. J. Coord. Chem.* Vol.31, No.3, pp. 179–182.

Chou, T.C. (2006). Theoretical Basis, Experimental Design, and Computerized Simulation of Synergism and Antagonism in Drug Combination Studies. *Pharmacol. Rev.*, Vol.58, No.3, pp. 621-681.

Claycamp, H.G., Shaw, E.I., & Zimbrick, J.D. (1986) Spin-Label Substitutions in Cisplatin Reduce Toxicity and Interaction with Radiation. *Radiation Res.*, Vol.106, No.2, pp. 141–146.

DeNicola, G.M., Karreth, F.A., Humpton, T.J., Gopinathan, A., Wei C., Frese, K., Manga, D., Yu K.H, Yeo, C.J., Calhoun, E.S., Scrimieri, F., Winter, J.M., Hruban, R.H., Iacobuzio-Donahue, C., Kern, S.E., Blair, I.A., & Tuveson, D.A. (2011). Oncogene-Induced Nrf2 Transcription Promotes ROS Detoxification and Tumorigenesis. *Nature*, Vol.475, 7 July, pp. 106–109.

Denisov, E.T. (1996) Cyclic Mechanisms of Chain Termination in the Oxidation of Organic Compounds. *Russ. Chem. Rev.* Vol.65, No.6, pp. 505–520.

Dhara, S.C. (1970). A Rapid Method for the Synthesis of cis-[$Pt(NH_3)_2Cl_2$]. *Indian J. Chem.*, Vol.8, No.1, p. 193.

Dunham, Sh.U., Dunham, S.U., Turner, C.J., & Lippard, S.J. (1998). Solution Structure of a DNA Duplex Containing a Nitroxide Spin-Labeled Platinum d(GpG) Intrastrand Cross-Link Refined with NMR-Derived Long-Range Electron-Proton Distance Restraints. *J. Am. Chem. Soc.*, Vol.120, No.22, pp. 5395-5406.

Emanuel, N.M., Konovalova, N.P., & Djachkovskaya, R.F. (1976). Toxicity, Antitumor Activity, and Pharmacokinetics of Spin-labeled ThioTEPA Analogs. *Cancer Treat. Rep.* Vol. 60, No.11, pp. 1605–1609.

Emanuel, N.M., Konovalova, N.P., Povarov, L.S., Shapiro, A.B., Dyachkovskaya, R.F., Suskina, V.I., & Denisova, L.K. (1982). 13-(1-Oxyl-2,2,6,6-tetramethylpiperidin-4-diyl)hydrazone Rubomicyn Hydrochloride with a Paramagnetic Center and a Method of Producing Same. *US Pat.* 4332934, http://worldwide.espacenet.com/publicationDetails/biblio?DB=EPODOC&II=3&ND=3&adjacent=true&locale=en_EP&FT=D&date=19820601&CC=US&NR=4332934A&KC=A.

Emanuel, N.M., Rosenberg, A.N., Golubev, V.A., Bogdanov, G.N., Vasil'eva, L.V., & Konovalova, N.P. (1985). 5-Fluorouracyl Nitroxyl Derivatives Possessing Antitumor Activity. *US Pat.* 4558127, http://worldwide.espacenet.com/publicationDetails/biblio?DB=EPODOC&II=1&ND=3&adjacent=true&locale=en_EP&FT=D&date=19851210&CC=US&NR=4558127A&KC=A.

Emanuel, N.M., Sen', V.D., Golubev, V.A., Bogdanov, G.N., Vasil'eva, L.V., & Konovalova, N.P. (1986). Nitroxyl Derivatives of Nitrosourea Exhibiting Antitumor and Mutagenic Activity. *SSSR Pat.* 1259650, http://worldwide.espacenet.com/publicationDetails/biblio?DB=EPODOC&II=8&ND=3&adjacent=true&locale=en_EP&FT=D&date=19900623&CC=SU&NR=1259650A1&KC=A1.

Emanuel, N.M., & Konovalova, N.P. (1992). Nitroxyl Radicals for Cancer Chemotherapy, In: *Bioactive Spin Labels*, Zhdanov, R.I. (Ed.). pp. 439–460, Springer, ISBN 3-54-53811-9, Berlin.

Farrell, N., Qu, Y., Bierbach, U., Valsecchi, M., & Menta, E. (1999). Structure-Activity Relationships Within Di- and Trinuclear Platinum Phase-I Clinical Anticancer Agents, In: *Cisplatin: Chemistry and Biochemistry of a Leading AnticancerDrugs*, B. Lippert, (Ed.), pp. 479–496, Verlag Helvetica Chimica Acta, ISBN 3-906390-20-9, Zurich.

Frey, U., Ranford, J.D., & Sadler, P.J. (1993). Ring-opening Reactions of the Anticancer Drug Carboplatin: NMR Characterization of cis-[Pt(NH_3)$_2$(CBDCA-O)(5'-GMP-N^7)] in Solution. *Inorg. Chem.*, Vol.32, No.8, pp. 1333–1340.

Gariboldi, M.B., Lucchi, S., Caserini, C., Supino, R., Oliva, C., & Monti, E. (1998). Antiproliferative Effect of the Piperidine Nitroxide Tempol on Neoplastic and Nonneoplastic Mammalian Cell Lines. *Free Rad. Biol. Med.*, Vol.24, No.6, pp. 913–923.

Gariboldi, M.B., Rimoldi, V., Supino, R., Favini, E., & Monti, E. (2000). The Nitroxide Tempol Induces Oxidative Stress, p21WAF1/CIP1, and Cell Death in HL60 Cells. *Free Rad. Biol. Med.*, Vol.29, No.7, pp. 633–641.

Gariboldi, M.B., Ravizza, R., Petterino, C., Castagnaro, M., Finocchiaro, G., & Monti, E. (2003). Study of *in vitro* and *in vivo* Effects of the Piperidine Nitroxide Tempol—a Potential New Therapeutic Agent for Gliomas. *Eur. J. Cancer*, Vol.39, No.6, pp. 829–837.

Gariboldi, M.B., Terni, F., Ravizza, R., Meschini, S., Marra, M., Condello, M., Arancia, G., & Monti, E. (2006). The Nitroxide Tempol Modulates Anthracycline Resistance in Breast Cancer Cells. *Free Rad. Biol. Med.*, Vol.40, No.8, pp. 1409–1418.

Giandomenico, C.M., Abrams, M.J., Murrer, B.A., Vollano, J.F., Rheinheimer, M.I., Wyer, S.B., Bossard, G.E., Higgins, J. D. (1995). Carboxylation of Kinetically Inert Platinum(IV) Hydroxy Complexes. An Entrée into Orally Active Platinum(IV) Antitumor Agents. *Inorg. Chem.*, Vol.34, No.5, pp. 1015–1021.

Goldstein, S., Merenyi, G., Russo, A., & Samuni, A. (2003). The Role of Oxoammonium Cation in the SOD-Mimic Activity of Cyclic Nitroxides. *J. Am. Chem. Soc.* Vol.125, No.3, pp. 789-795.

Goldstein, S., Samuni, A., Hideg, K., & Merenyi, G. (2006). Structure-Activity Relationship of Cyclic Nitroxides as SOD Mimics and Scavengers of Nitrogen Dioxide and Carbonate Radicals. *J. Phys. Chem. A.* Vol.110, No.10, pp. 3679-3685.

Gorczyca, W., Gong, J., Ardelt, B., Traganos, F., & Darzynkiewicz, Z. (1993). The Cell Cycle Related Differences in Susceptibility of HL-60 Cells to Apoptosis Induced by Various Antitumor Agents. *Cancer Res.*, Vol.53, No.13, pp. 3186–3192.

Hambley T.W. (1997). The Influence of Structure on the Activity and Toxicity of Pt Anti-cancer Drugs. *Coord. Chem. Rev.*, Vol.166, November 1997, pp. 181–223.

Ho, Y.P., Au-Yeung, S.C., To, K.K. (2003). Platinum-based Anticancer Agents: Innovative Design Strategies and Biological Perspectives. *Med. Res. Rev.*, Vol.23, No.5, pp. 633-655.

Huang, J.C., Zamble, D.B., Reardon, J.T., Lippard, S.J., & Sancar, A. (1994). HMG-Domain Proteins Specifically Inhibit the Repair of the Major DNA Adduct of the Anticancer

Drug Cisplatin by Human Excision Nuclease. *Proc. Natl. Acad. Sci. USA*. Vol.91, No.22, pp. 10394-10398.

Jakupec, M.A., Galanski, M., Arion, V.B., Hartinger C.G., & Keppler, B.K. (2008). Antitumour Metal Compounds: More than Theme and Variations. *Dalton Trans.*, No.2, pp. 183–194.

Kalimutho, M., Minutolo, A., Grelli, S,. Federici, G., & Bernardini, S. (2011). Platinum-(IV)-Derivative Satraplatin Induced G2/M Cell Cycle Perturbation via p53-p21(waf1/cip1)-independent Pathway in Human Colorectal Cancer Cells. *Acta Pharmacol. Sin.*, Vol. 32, No.11, pp. 1387–1396.

Kelland, L. R. (1999). Cisplatin-based Anticancer Agents, In: *Uses of Inorganic Chemistry in Medicine*, Farrell, N.P. (Ed.). pp. 109–121, Springer, ISBN 0-85404-444-2, Berlin.

Kelland, L.R., (2007). The Resurgence of Platinum-Based Cancer Chemotherapy. *Nature Rev.*, Vol.7, No.8, pp. 573–584.

Klein, A.V., & Hambley, T.W. (2009). Platinum Drug Distribution in Cancer Cells and Tumors, *Chem. Rev.*, Vol.109, No.10, pp. 4911–4920.

Klimek, M., (1966). Toxicity of N-Ethylmaleimide and Di-t-Butyl Nitroxide Radiosensitizers towards HeLa S3 Cells in a Single Cell Cultures. *Nature*. Vol.209, No.5029, pp. 1256–1257.

Koeberle, B., Tomicic, M.T., Usanova, S., & Kaina, B. (2010). Cisplatin Resistance: Preclinical Findings and Clinical Implications, *Biochim. Biophys. Acta*, Vol.1806, No.2, pp. 172–182.

Komarov, A.M., Joseph, J., & Lai, C.S. (1994). *In Vivo* Pharmacokinetics of Nitroxides in Mice. *Biochem. Biophys. Res. Comm.* Vol.201, No.2, pp. 1035–1042.

Konovalova, N.P., Bogdanov, G.N., Miller, V.B., Rozantsev, E.G., Neiman, M.B., & Emanuel, N.M. (1964). Antitumor Activity of Stabile Free Radicals. *Dokl. Akad. Nauk SSSR*, Vol.157, No.3, pp. 707–709 [*Dokl. Chem.* (Engl. Transl.), 1964].

Konovalova, N.P., Diatchkovskaya, R.F., Volkova, L.M., & Varfolomeev, V.N. (1991). Nitroxyl radicals decrease toxicity of cytostatic agents. *Anticancer Drugs*. Vol.2, No.6, pp. 591–595.

Manda, S., Nakanishi, I., Ohkubo, K., Yakumaru, H., Matsumoto, K., Ozawa, T., Ikota, N., Fukuzumi, S., & Anzai, K. (2007). Nitroxyl Radicals: Electrochemical Redox Behaviour and Structure–Activity Relationships. *Org. Biomol. Chem.*, Vol.5, No.24, pp. 3951–3955.

Mathew, A., Bergquist, B., & Zimbrick, J. (1979). Synthesis of Spin-labelled Platinum Drugs and Interaction with Deoxyguanidine. *Chem. Comm.*, No.2, pp. 222–224.

Mitsudomi, T., Steinberg, S.M., Nau, M.M., Carbone, D., D'Amico, D., Bodner, S., Oie, H.K., Linnoila, R.I., Mulshine, J.L., Minna, J.D., et al. (1992). p53 Gene Mutations in Non-small-cell Lung Cancer Cell Lines and their Correlation with the Presence of ras Mutations and Clinical Features. *Oncogene*, Vol.7, No.1, pp. 171-180.

Miyajima, A., Nakashima, J., Yoshioka, K., Tashibana, M., Tazaki, H., & Murai, M. (1997). Role of Reactive Oxygen Species in *cis*-Dichlorodiammineplatinum-induced Cytotoxicity on Bladder Cancer Cells. *British J. Cancer*, Vol.76, No.2, pp. 206–210.

Monti, E., Cova, D., Guido, E., Morelli, R., & Oliva, C. (1996). Protective Effect of the Nitroxide Tempol against the Cardiotoxicity of Adriamycin. *Free Rad. Biol. Med.* Vol.21, No.4, pp. 463–470.

Oksanen, M., & Leskela, M. (1994). Synthesis of Ammonium Trichloromonoammineplatinate(II) Improved through Control of Temperature. *Acta Chim. Scand.*, Vol.48, No.2, pp. 485-489.

Oksanen, M. (1997). A Novel Characterization of the Structure of "Cossa's Salt" $K_x(NH_4)_{1-x}[PtCl_3(NH_3)]\cdot H_2O$ by Crystallographic Comparison with the Stoichiometric Compounds $K[PtCl_3(NH_3)]\cdot H_2O$ and $NH_4[PtCl_3(NH_3)]\cdot H_2O$. *Inorg. Chim. Acta.*, Vol.269, No.1, pp. 53-60.

Potucka, L., Houskova, J., Syslova, K., Hamtil, R., Kacer, P., & Cerveny, L. (2011). HPLC Method for the Determination of the Purity of $K[Pt(NH_3)Cl_3]$, a Precursor of the Platinum Complexes with Cytostatic Activity. *Anal. Lett.* , Vol.44, No.10–12, pp. 2182-2193.

Raikov, Z., Todorov, D., Ilarionova, M., Demirov, G., Tsanova, T., & Kafalieva, D. (1985). Synthesis and Study of Spin-Labeled Nitrosoureas. *Cancer Biochem. Biophys.*, Vol.7, No.2, pp. 343–348.

Reynolds, C.P. & Maurer, B.J. (2005). Evaluating Response to Antineoplastic Drug Combinations in Tissue Culture Models. *Methods Mol. Med.*,Vol.110, pp. 173–183.

Roubalová, E., Kvardová, V., Hrstka, R., Borilová, S., Michalová, E., Dubská, L., Müller, P., Sova, P., & Vojtesek, B. (2010). The Effect of Cellular Environment and p53 Status on the Mode of Action of the Platinum Derivative LA-12. *Invest. New Drugs*, Vol.28, No.4, pp. 445-453.

Satoh, M., Kashihara, N., Fujimoto, S., Horike, H., Tokura, T., Namikoshi, T., Sasaki, T., & Makino, H. (2003). A Novel Free Radical Scavenger, Edarabone, Protects Against Cisplatin-induced Acute Renal Damage in vitro and in vivo. *J. Pharmacol. Exp. Ther.*, Vol.305, No.3, pp. 1183-1190.

Sen', V.D., Golubev, V.A., Kulyk, I.V., & Rozantsev, E.G. (1976). Mechanism of Reaction of Hydrogen Peroxide with Oxopiperidine Salts and Piperidinoxyl Radicals. *Russ. Chem. Bull.*, Vol.25, No.8, pp. 1647–1654.

Sen', V.D., Golubev, V.A., & Bogdanov, G.N. (1989a). Nitroxylcarbamoyl Derivatives of 5-Fluorouracyl, *Russ. Chem. Bull.*, Vol.38, No.10, pp. 2132–2135.

Sen', V.D., (1989b). Synthesis of 3,4-Diamino-2,2,6,6-tetramethylpiperidine-1-oxyl, *Russ. Chem. Bull.*, Vol.38, No.9, pp. 1928–1931.

Sen', V.D., (1993). Synthesis of Nitroxide Derivatives of Alkylnitrosourea, *Russ. Chem. Bull.*, Vol.42, No.3, pp. 548–551.

Sen', V.D., Golubev, V.A., Volkova, L.M., & Konovalova, N. P. (1996). Synthesis and Antitumor Activity of Platinum(II) Complexes with *trans*-3,4-Diamino-2,2,6,6-tetramethylpiperidine-1-oxyl. *J. Inorg. Biochem.*, Vol.64, No.1, pp. 69–77.

Sen', V.D., Kulikov, A.V., Shugalii, A.V., & Konovalova, N.P. (1998). Synthesis of Dinitroxyl cis-Diaminoplatinum(II) Complexes and their Interaction with DNA. *Russ. Chem. Bull.*, Vol.47, No.8, pp. 1598–1602.

Sen', V.D., Rukina, N.A., Tkachev, V.V., Pis'menskii, A.V., Volkova, L.M., Goncharova, S.A., Raevskaya, T.A., Tikhomirov, A.G., Gorbacheva, L.B., & Konovalova, N.P. (2000). Synthesis, Structure, and Biological Activity of Mixed-ligand Platinum(II) Complexes with Aminonitroxides. *Russ. Chem. Bull.*, Vol.49, No.9, pp. 1613–1619.

Sen', V.D., Shugalii, A.V., & Kulikov, A.V. (2002). DNA Modification by *cis*-Diaminoplatinum(II) Complexes with Aminonitroxide Ligands. *Russ. Chem. Bull.*, Vol.51, No.6, pp. 11058–1064.

Sen', V.D., Tkachev, V.V., Volkova, L.M., Goncharova, S.A., Raevskaya, T.A., & Konovalova, N.P. (2003). Synthesis, Structure, and Antitumor Properties of Platinum(IV) Complexes with Aminonitroxyl Radicals. *Russ. Chem. Bull.*, Vol.52, No.2, pp. 421–426.

Sen', V.D., Golubev, V.A., Lugovskaya, N.Yu., Sashenkova, T.E., & Konovalova N.P. (2006). Synthesis and Antitumor Properties of New Platinum(IV) Complexes with Aminonitroxyl Radicals. *Russ. Chem. Bull.*, Vol.55, No.1, pp. 62–65.

Sen', V.D., & Golubev, V.A., (2009). Kinetics and Mechanism for Acid-catalyzed Disproportionation of 2,2,6,6-Tetramethylpiperidine-1-oxyl. *J. Phys. Org. Chem.*, Vol.22, No.2, pp. 138–143.

Shapiro, A.B., Kropacheva, A.A., Suskina, V.I., Rozynov, B.V., & Rozantsev, E.G. (1971). Mass-spectrometric Study of Paramagnetic Derivatives of Ethylenephosphoramides. *Russ. Chem. Bull.*, Vol.20, No.4, pp. 780–783.

Shugalii, A.V., Kulikov, A.V., Lichina, M.V., Golubev, V.A., & Sen', V.D. (1998). Binding of Platinum-Diaminonitroxyl Complexes to Animal DNA. *J. Inorg. Biochem.*, Vol.69, No.1-2, pp. 67–77.

Seifried, H.E., McDonald, S.S., Anderson, D.E., Greenwald, P., & Milner, J. A. (2003). The Antioxidant Conundrum in Cancer. *Cancer Res.*, Vol.63, August 1, pp. 4295–4298.

Sinha, B.K., Cox, M.G., Colin F. Chignell, C.F., & Cysyk, R.L. (1979). Synthesis and Biological Properties of N2-Substituted Spin-Labeled Analogs of Actinomycin D. *J. Med. Chem.*, Vol.22, No.9, pp. 1051–1055.

Siomek, A., Tujakowski, J., Gackowski, D., Rozalski, R., Foksinski, M., Dziaman, T., Roszkowski, K., & Olinski, R. (2006). Severe Oxidatively Damaged DNA after Cisplatin Treatment of Cancer Patients. *Int. J. Cancer*, Vol.119, No.9, pp. 2228-2230.

Sosnovsky, G., Paul, B.D. (1984). In the Search for New Anticancer Drugs. 9. Synthesis and Anticancer Activity of Spin-Labeled Analogues of N,N:N',N':N",N"-Tri-1,2-Ethanediylphosphoric Triamide and N,N:N',N':N",N"-Tri-1,2-Ethanediylphosphorothioic Triamide. *J. Med. Chem.*Vol.27, No.6, pp. 782–788.

Sosnovsky, G., Li, S.W. (1985a). In the Search for New Anticancer Drugs X. N,N:N',N' -Bis-(1,2-Ethanediyl) -N"- (1-oxyl-2,2,6,6-tetramethyl-4-piperidinylaminocarbonyl) Phosphoric Triamide - a New Potential anticancer Drug of High Activity and Low Toxicity. *Life Sci.*, Vol.36, No.15, pp. 1473-1477.

Sosnovsky, G., Li, S.W. (1985b). In the Search for New Anticancer Drugs XII. Synthesis and Biological Evaluation of Spin Labeled Nitrosoureas. *Life Sci.*, Vol.36, No.15, pp. 1479-1483.

Soule, B.P., Hyodo, F., Matsumoto, K., Simone, N. L., Cook, J.A., Krishna, M.C., & Mitchell, J.B. (2007). The Chemistry and Biology of Nitroxide Radicals. *Free Rad. Biol. Med.* Vol.42, No.11, pp. 1632–1650.

Suy, S., Mitchell, J. B., Samuni, A., Mueller, S., & Kasid, U. (2005). Nitroxide Tempo, a Small Molecule, Induces Apoptosis in Prostate Carcinoma Cells and Suppresses Tumor Growth in Athymic Mice. *Cancer*, Vol.103, No.6, pp. 1302–1313.

Reynolds, C.P. & Maurer, B.J. (2005). Evaluating Response to Antineoplastic Drug Combinations in Tissue Culture Models. *Methods Mol. Med.*,Vol.110, pp. 173–183.

Tsui, F.P., Robey, F., Engle, T.W., Ludeman, S.M., & Zon, G. (1982). Synthesis and study of a spin-labeled cyclophosphamide analogue, 3-(1-oxyl-2,2,6,6-tetramethyl-4-piperidinyl)cyclophosphamide. *J. Med. Chem.*, Vol.25, No.9, pp. 1106–1110.

Tsutsumishita, Y., Onda, T., Okada, K., Takeda, M., Endou, H., Futaki, S., & Niwa, M. (1998). Involvement of H_2O_2 in Cisplatin-Induced Nephrotoxicity. *Biochem. Biophys. Res. Comm.*,Vol.242, No.2, pp. 310–312.

Vousden, K.H. & Prives, C. (2009). Blinded by the Light: The Growing Complexity of p53. *Cell*, Vol.137, No.3, pp. 413-431.

Wheate, N. J., Walker, S., Craig, G. E., & Oun, R. (2010). The Status of Platinum Anticancer Drugs in the Clinic and in Clinical Trials. *Dalton Trans.*, Vol.39, No.35, 8113–8127.

Wilcox, C.S. (2010). Effects of Tempol and Redox-cycling Nitroxides in Models of Oxidative Stress. *Pharmacol. Ther.*, Vol.126, No.2, pp. 119–145.

Wondrak, G.T. (2009). Redox-Directed Cancer Therapeutics: Molecular Mechanisms and Opportunities. *Antioxid. Redox Signal.*,Vol.11, No.12, pp. 3013–3069.

Chapter 15

Fluorescent Nitrones for the Study of ROS Formation with Subcellular Resolution

Stefan Hauck, Yvonne Lorat, Fabian Leinisch, Christian Kopp, Jessica Abrossinow and Wolfgang E. Trommer

Additional information is available at the end of the chapter

1. Introduction

Reactive oxygen species (ROS) are the prize we are paying for an energy-efficient life under oxygen. They play a role in many diseases such as atherosclerosis, hypertension, ischemia-reperfusion injury, inflammation, type-2 diabetes, certain neurodegenerative diseases, even cancer and, certainly, aging (Kohen & Nyska, 2002). Among these ROS are several radicals such as the hydroxyl, peroxy, alkoxy as well as the superoxide anion radical (Boveris, 1977). Direct measurement of the radicals is hampered by their short half lifes in the range of nano- to microseconds. They can, however, be trapped by addition to nitrones leading to relatively stable nitroxides with $t_{½}$ of minutes (Janzen, 1971). Respiring mitochondria are a major source of ROS, particularly of the superoxide anion radical, which is formed in complexes I and III (Cadenas & Davies, 2000; Dröge, 2002; Inoue et al., 2003; Turrens, 2003).

A couple of years ago we have described the synthesis and first application of a fluorescent nitrone composed of *tert*-butyl-nitrone and a *p*-nitro-stilbene moiety, which can be used for the detection of ROS with subcellular resolution. Short-lived radicals add to the nitrone under formation of a nitroxide radical which then quenches the fluorescence of the *p*-nitro-stilbene (Hauck et al., 2009).

Similar double labels have previously been synthesized or at least suggested, partially in collaboration with us. Likhtenshtein and Hideg suggested in the eighties to couple nitroxides to a fluorophore which will become fluorescent only after reduction of the nitroxide moiety in viable biological systems (Bystryak et al., 1986). Hideg used fluorescent pyrrolines which can be oxidized to nitroxides mainly by singlet oxygen (Kalai et al., 1998). Rosen and collaborators investigated a fluorescent nitrone much like ours composed of nitrobenzene instead of nitrostilbene. The authors did not observe a decrease in the fluorescence upon reaction of the nitrone with α-hydroxyethyl radicals which was probably due to an only very small concentration of the nitroxide formed in this reaction (Pou et al.,

1995). To the best of our knowledge, this work has not been continued. Somewhat different approaches for the detection and trapping of ROS were taken by Bottle et al., 2003, Heyne et al., 2007 and Blough et al., 1988.

Here we describe some extended studies with the *p*-nitrostilbene-*tert*-butyl-nitrone (1) under a variety of physiological conditions *i.e.*, employing inhibitors of components of the respiratory chain, the F_1-F_0-ATP synthase and the membrane potential as well as similar studies with a corresponding coumaryl-styryl-*tert*-butyl-nitrone derivative (2) and finally, a third compound based on 4-pyrrolidine-1,8-naphthylimido-methylbenzene as fluorophore (3).

2. Results and discussion

2.1. *p*-Nitrostilbene-*tert*-butyl-nitrone, 1

Fig. 1 shows the reaction of the previously employed fluorescent nitrone 1 with the hydroxyl radical under formation of the corresponding nitroxide. The hydroxyl radical was generated by the Fenton reaction and the product extracted from the aqueous phase with degassed ethyl acetate yielding an EPR spectrum composed of 6 lines typical for hydroxyl radical adducts (Hauck et al., 2009).

When cultures of various cell lines, *i.e.*, HeLa, COS 7 and CHO, were incubated with 1 and subsequently washed, it had almost exclusively accumulated in mitochondria as shown by co-staining with tetramethylrhodamine ethylester (TMRE, Farkas et al., 1989) and confocal laser spectroscopy (Fig. 2). Without any further addition the fluorescence slowly decreased to almost none within about 20 min. However, generation of hydroxyl radicals by the Fenton reaction (Walling, 1975) reduced this time to 20 sec (Fig. 3, Hauck et al., 2009). Similar results were obtained in presence of the complex I and III inhibitors, rotenone and antimycin A, but within a timeframe of 40 to 60 sec. Under these conditions both complexes are known to produce substantial amounts of the superoxide anion radical (Dlaskova et al., 2008; Han et al., 2001). In these early studies a quasi-confocal microscope was used (Axiovert 440, equipped with an ApoTome, Carl Zeiss, Jena) which has broad bandwidth filters, only. Thus, not allowing for monitoring small shifts in the emission spectra of 1 upon addition of the radical as had been observed in cell-free controls.

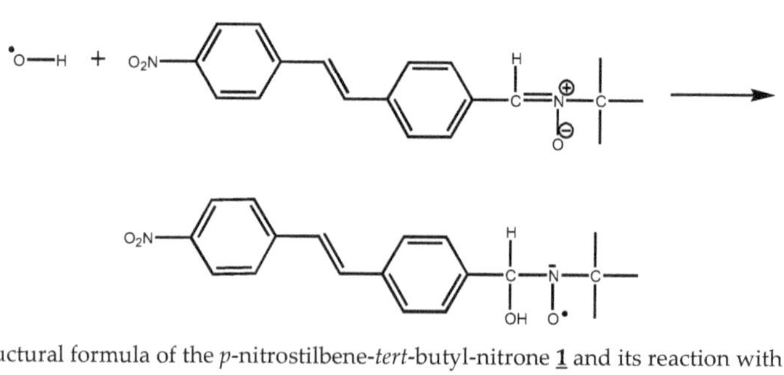

Figure 1. Structural formula of the *p*-nitrostilbene-*tert*-butyl-nitrone 1 and its reaction with the hydroxyl radical

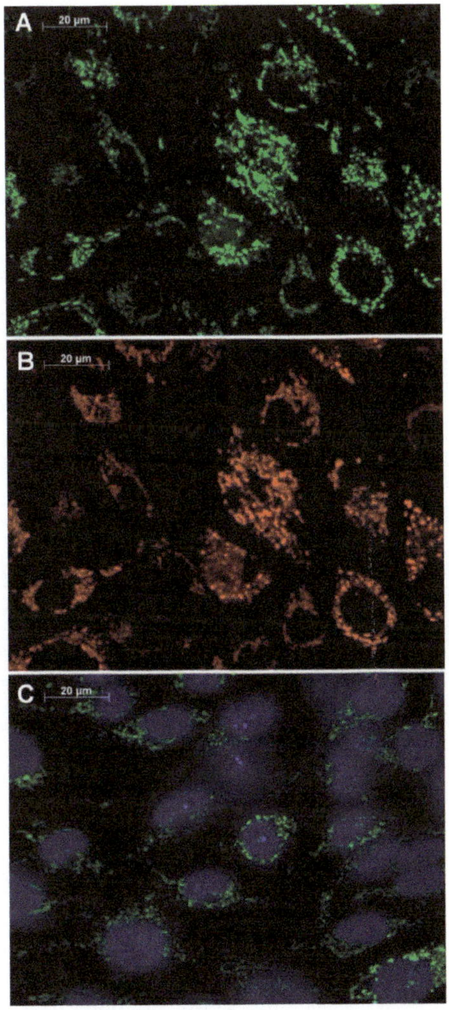

Figure 2. COS cells incubated with 1 (**A**) or with TMRE (**B**). The same cells were stained for DNA in the nuclei with DAPI (**C**, λ_{abs} = 340 nm, λ_{em} = 504 nm), and also with 1 (λ_{abs} = 388 nm, λ_{em} = 504 nm)

The accumulation of 1 in mitochondria came unexpected but is possibly due to an effect previously observed with so-called Skulachev ions, derivatives of ubiquinone or plastoquinone with long hydrophobic side chains composed of up to 10 isoprene units and a positively charged triphenylphosphonium group at the end, *e.g.*, SkQ1 as introduced by Vladimir Skulachev. The positive charge is not localized but spread over the aromatic rings, thus allowing for membrane permeation due to the already negative potential of cells with respect to their outside and even more so of mitochondria, the only human organelle with a negative potential with respect to the cytoplasm. Once inside, they are trapped by the opposite potential outside (Severin et al., 2010). Similar effects have been observed with certain amphiphilic dipolar compounds to which 1 could belong (M.V. Skulachev, personal communication).

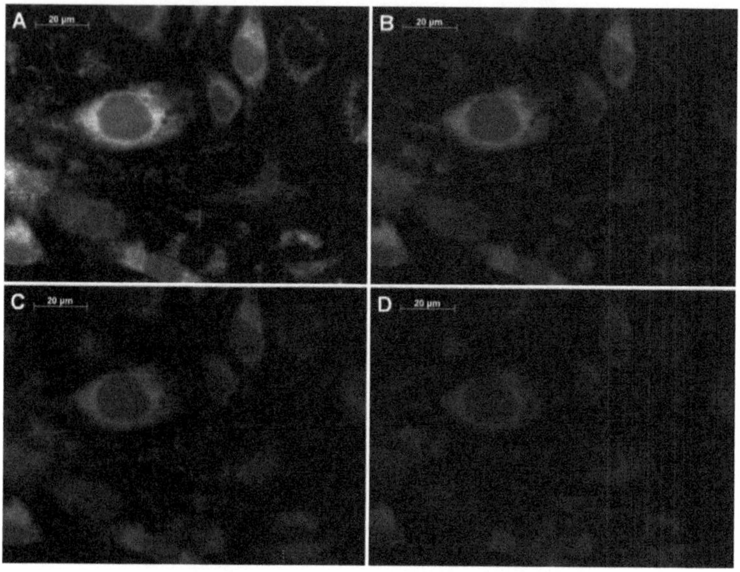

With kind permission from Springer Science + Business Media: Appl. Magn. Reson., p-Nitrostilbene-*tert*-butyl-nitrone: A novel fluorescent spin trap for the detection of ROS with subcellular resolution, vol. 36 (2009), p. 145, Stefan Hauck, Yvonne Lorat, Fabian Leinisch, Wolfgang E. Trommer, Fig. 10

Figure 3. Fluorescence decay of *p*-nitrostilbene-*tert*-butyl-nitrone 1 in CHO cells: **A**: 0 sec, **B**: 5 sec, **C**: 10 sec, **D**: 20 sec after addition of 20 µl each of 10 mM Fe^{2+} and 10 mM H_2O_2 (Fenton reaction)

The fluorescence half-life of 1 in mitochondria was studied under a variety of conditions. Data are summarized in Table I together with those from the coumaryl derivative 2. But why look for another double label? 1 has an absorption maximum at 383 nm which is well separated from the emission at 568 nm. However, only recently have confocal laser microscopes been equipped with lasers of 405 nm, most commercial instruments operate at 480 nm and above, rather outside the absorption range of 1. Moreover, stilbenes as fluorophores pose yet another problem, *cis-trans* isomerization upon irradiation leading to a substantial shift in emission wavelength which is accompanied by photobleaching and also recovery rendering interpretation of data more complex (Fig. 6).

2.2. Coumaryl-styryl-*tert*-butyl-nitrone, 2

2-(4-Trifluoromethyl-2-oxo-2H-chromen-7-yl)-*E*-vinyl-1-(N-(1,1-dimethyl))-phenyl-4-nitrone 2 was synthesized according to the scheme shown in Fig. 4.

Unexpectedly, the absorbance and fluorescence properties of 1 and 2 do not differ very much as shown in Fig. 5, A & B.

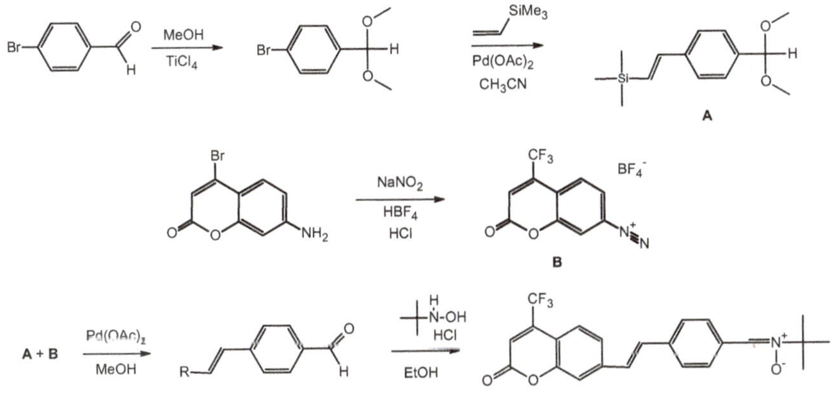

Figure 4. Synthetic scheme for the synthesis of 2

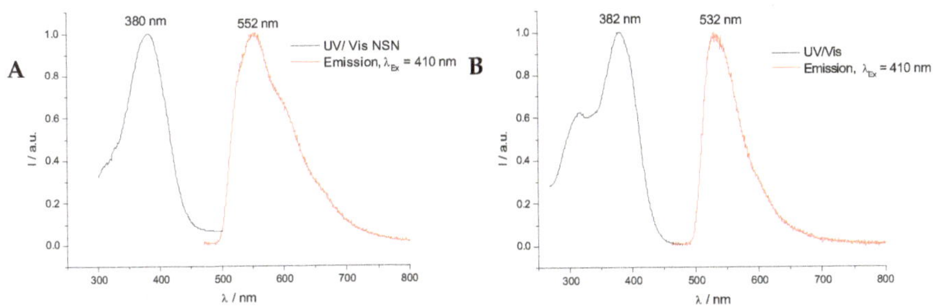

Figure 5. Absorption and emission spectra of **A**, *p*-nitrostilbene-*tert*-butyl-nitrone 1 and **B**, cumaryl-styryl-*tert*-butyl-nitrone 2, excitation at 410 nm

Whereas 1 undergoes substantial photobleaching upon irradiation at 310 nm for 1 min (Fig. 6 B), the fluorescence of 2 is almost completely stable under these conditions (Fig. 6 A). However, 1 recovers rather quickly reaching about 40 % of the initial fluorescence within 4 min (Fig. 6 C).

Unfortunately, 2 turned out to be far more toxic to cells in comparison with 1. Growing adherent MCF-7 cells, a human breast adenocarcinoma cell line, were exposed to these compounds for three days. After fixation of viable cells with trichloroacetic acid, the damaged cells were removed by washing and cell proteins of the remaining cells were stained with sulforhodamine and measured photometrically at 570 nm (Skehan et al., 1990). In a rough estimate assuming sigmoidal behavior in a semi-logarithmic plot the half maximal dosis, IC_{50}, of 1 was about 300 μM as compared to 40 μM for 2. Simultaneous irradiation at 366 nm for 3 min further reduced this value to 30 μM. The known toxicity of coumarin may play a role here (Fig. 7, A & B; Oodyke, 1974). Therefore spin-trapping experiments in the presence of inhibitors were primarily carried out with 1. However, 2 may be employed as well by using lower concentrations of 10 to 20 μM.

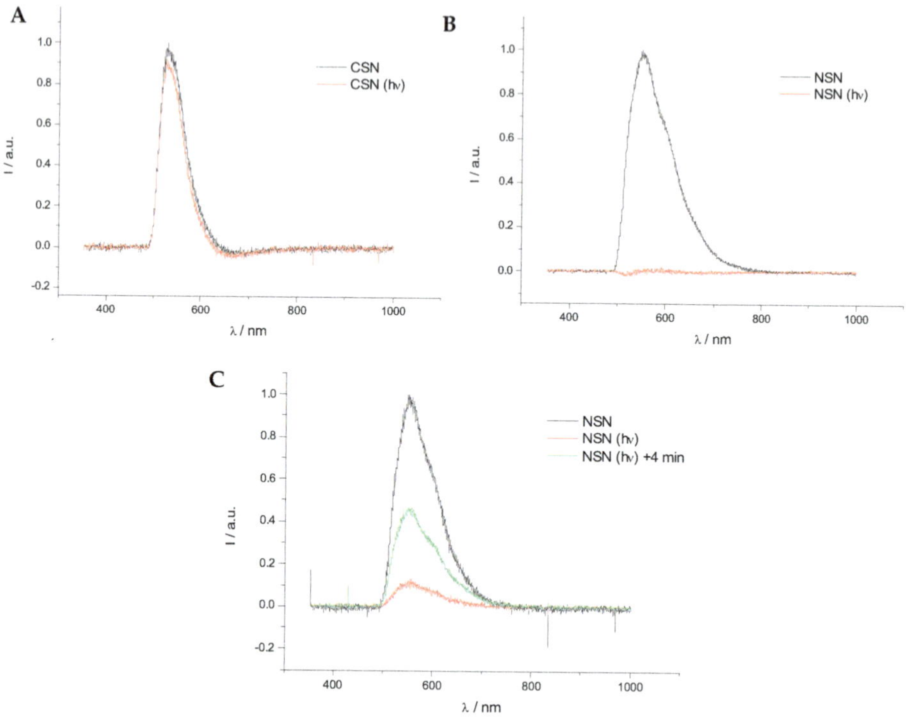

Figure 6. Photobleaching of 2 (CSN, **A**) und 1 (NSN, **B**) upon irradiation at 310 nm for 1 min. The recovery of 1 with time is shown in panel **C**.

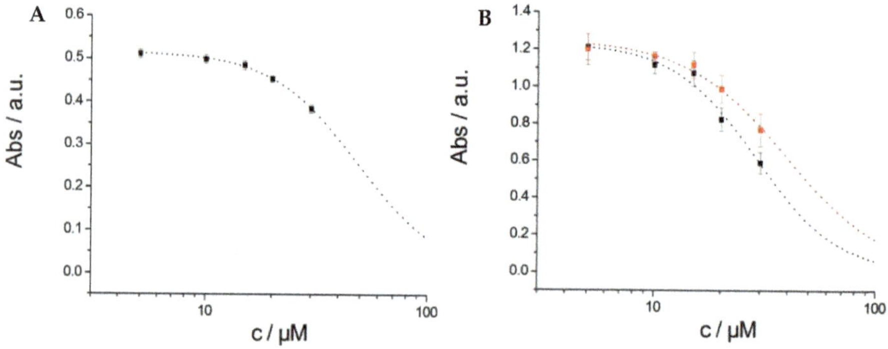

Figure 7. Cytotoxicity of 1 (**A**) and 2 (**B**) as determined in MCF-7 cells after a three days exposure. **B**, lower trace upon additional irradiation at 366 nm.

The effect of rotenone as inhibitor of complex I of the respiratory chain on the formation of the superoxide anion radical formation in HeLa cells (Hauck et al., 2009) was followed with 2 (Fig. 8). The initial fluorescence and after 100 or 90 sec, respectively, is shown in the absence and presence of rotenone. The time course of this decay as measured in 0.5 sec intervals is shown as well. Clearly, inhibition of complex I significantly increases ROS formation. The figure also reveals that the intracellular distribution of 2 is not as enriched in mitochondria as had been found with 1. Evidence that quench did not result, or at least not largely, from photobleaching came from experiments in which the shutter was closed during measurements. However, some non-specific redox reactions cannot be excluded.

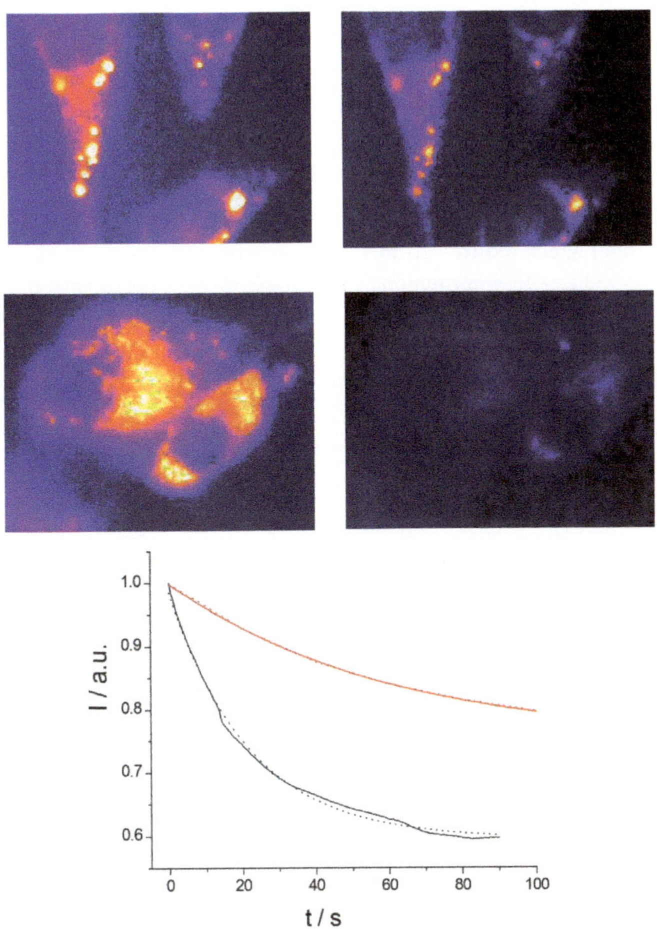

Figure 8. False color representation of the fluorescence decay of 2 upon trapping of ROS in HeLa cells in the absence (upper panels) and presence of 10 μM rotenone (lower panels), 0 and 100 sec after addition of the inhibitor as well as the time course of these changes (bottom panel).

2.3. Detailed studies with *p*-nitrostilbene-*tert*-butyl-nitrone, 1

Spin-trapping of mainly superoxide anion radicals formed in mitochondria under various conditions was followed in HeLa cells as previously (Hauck et al., 2009) and as well for better comparison with data employing 2. Fig. 9 shows the fluorescence decay after 100 sec in the absence of any inhibitor. All subsequent data are based on this control (Figs. 10 - 15). Results are summarized in Fig. 16.

The inhibitors of complexes I, rotenone, and complex III, antimycin A, have the strongest effect, particularly in combination. KCN is known to inhibit complex IV (Leavesley et al., 2010) and oligomycin blocks ATP synthesis by binding to the oligomycin sensitivity conferring protein, OSCP, of the F_0 moiety of F_1F_0-ATP synthase (Xu et al., 2000). The latter leads to an increased membrane potential and drives the respiratory chain backwards. In the absence of sufficient substrates for NADH this is known to produce superoxide anion radicals in complex I as in complex III via reaction of oxygen with the coenzyme Q semiquinone radical (Muller et al., 2004).

Rather interesting is the effect of the protonophore CCCP (carbonyl-cyanide *m*-chloro-phenyl hydrazone) which reduces the membrane potential (Nieminen et al., 1994) and thus, apparently also reduces ROS formation. The half-life of the fluorescence in its presence is almost doubled as compared to the control corroborating the opposite effect of oligomycin.

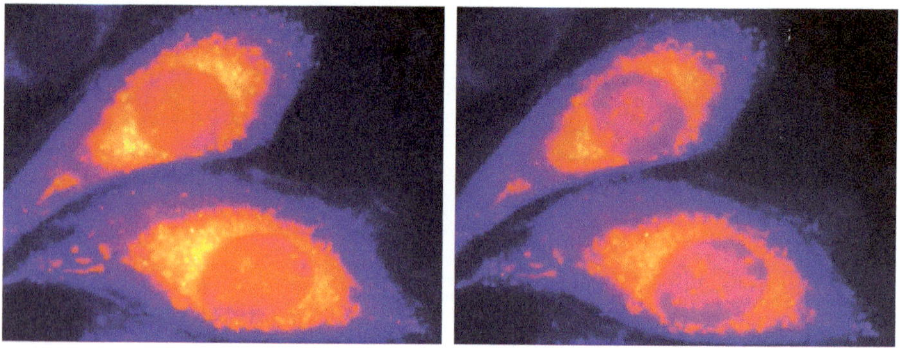

Figure 9. Fluorescence decay of 1 in HeLa cells within 100 sec in the absence of any inhibitor of respiration

2.4. 4-Pyrrolidine-1,8-naphthylimido-methylphenyl-*tert*-butyl-nitrone, 3

In order to allow for excitation at longer wavelengths a third compound was synthesized according to the scheme in Fig. 17, again a *tert*-butyl-nitrone, but with 4-pyrrolidine-1,8-naphthylimido-benzylidene as fluorescent moiety. 4-(1,3-Dioxacyclopent-2-yl)benzonitrile **B** was synthesized according to Ouari et al., 1998 in good yield and in the last step the formation of the nitrone 3 was carried out according to Kim et al., 2007.

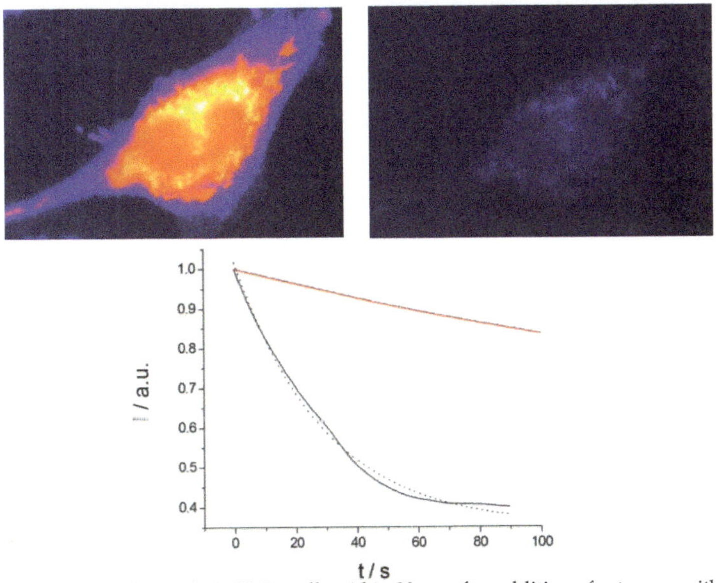

Figure 10. Fluorescence decay of **1** in HeLa cells within 90 sec after addition of rotenone with $t_{1/2}$ = 32 sec, control 176 sec assuming exponential decay (dotted lines)

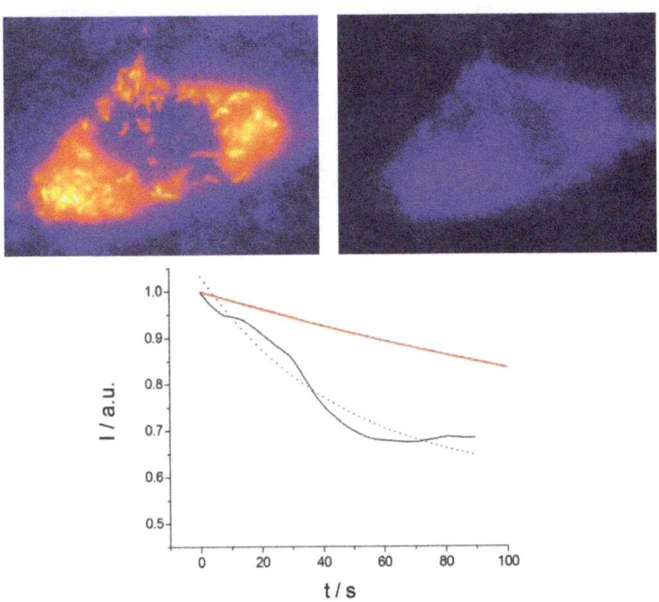

Figure 11. Fluorescence decay of **1** in HeLa cells within 90 sec after addition of antimycin A with $t_{1/2}$ = 20 sec, control 176 sec assuming exponential decay (dotted lines)

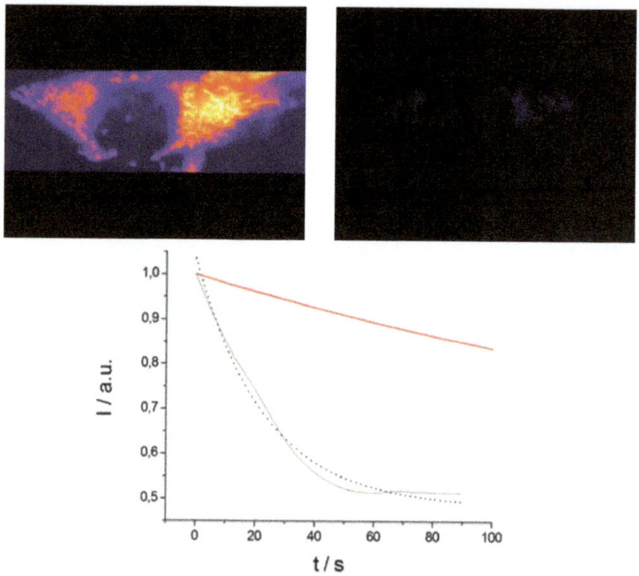

Figure 12. Fluorescence decay of 1 in HeLa cells within 90 sec after addition of rotenone and antimycin A with $t_{½}$ = 16 sec, control 176 sec assuming exponential decay (dotted lines)

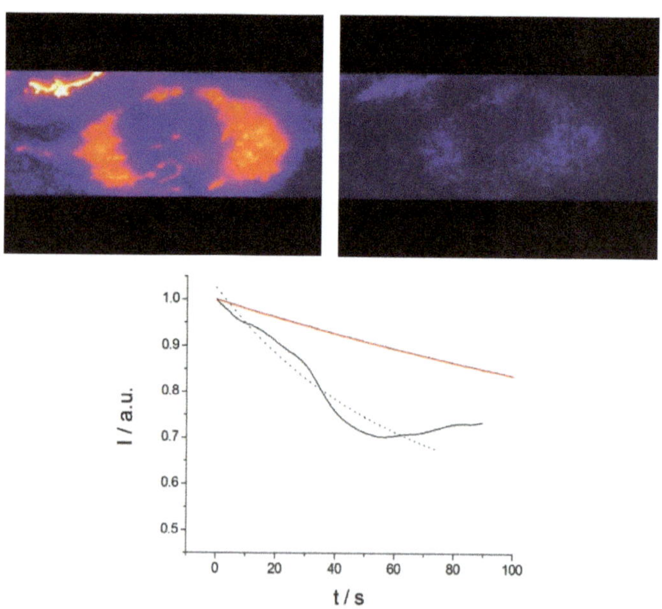

Figure 13. Fluorescence decay of 1 in HeLa cells within 90 sec after addition KCN with $t_{½}$ = 40 sec, control 176 sec assuming exponential decay (dotted lines)

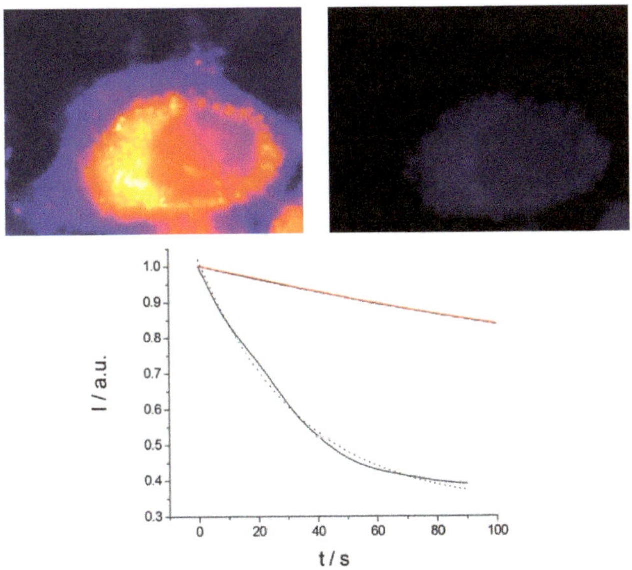

Figure 14. Fluorescence decay of 1 in HeLa cells within 90 sec after addition of oligomycin with $t_{½}$ = 23 sec, control 176 sec assuming exponential decay (dotted lines)

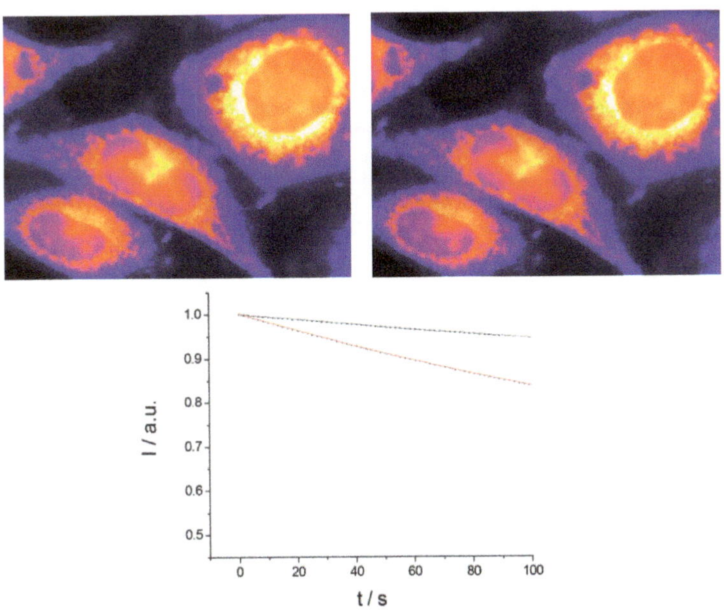

Figure 15. Fluorescence decay of 1 in HeLa cells within 90 sec in the presence of CCCP with $t_{½}$ = 336 sec, control 176 sec (red) assuming exponential decay (dotted lines)

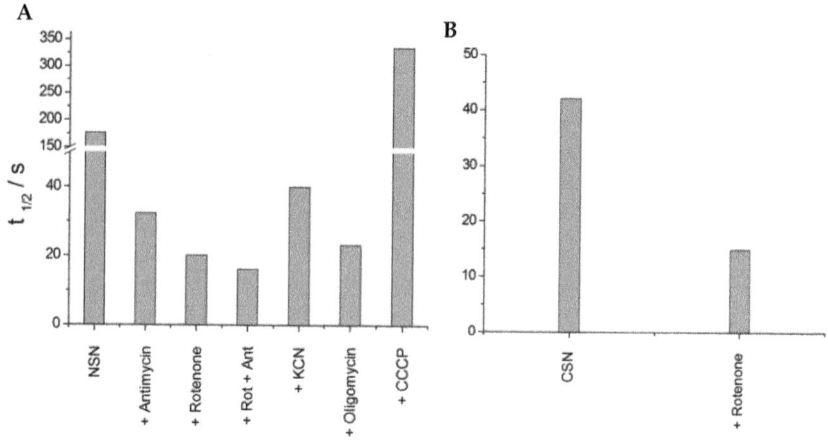

Figure 16. Effect of inhibitors of mitochondrial respiration on ROS formation as monitored by 1 (**A**) and 2 (**B**)

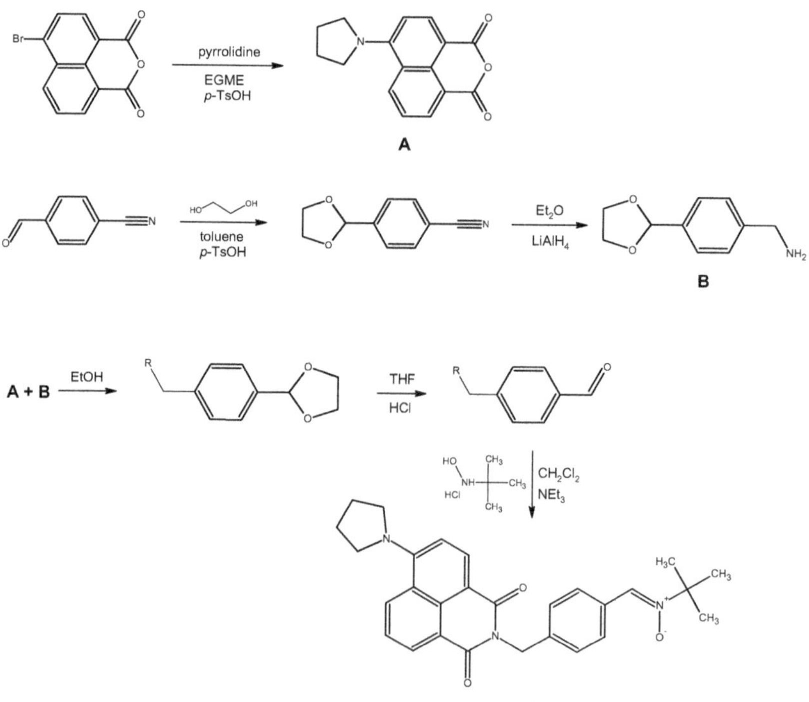

Figure 17. Synthetic scheme for the synthesis of 4-pyrrolidine-1,8-naphthylimido-methylphenyl-*tert*-butyl-nitrone, 3

Absorbance and fluorescence emission spectra are shown in Fig. 18. Fig. 18 A shows the fluorescence spectrum (red) with a maximum at 522 nm upon excitation at 488 nm and the UV/VIS spectrum with a maximum at 457 nm (black). Fig. 18 B illustrates the fluorescence spectrum (red) with a maximum at 514 nm upon excitation at 405 nm. Excitation at 488 nm is well in the range of standard confocal laser microscopes and certainly, not likely to lead to radiation damage in biological systems. Within the concentration range feasible due to limitations in solubility, 3 was non-toxic as studied so far in MCF-7 cells up to 5 μM (data not shown).

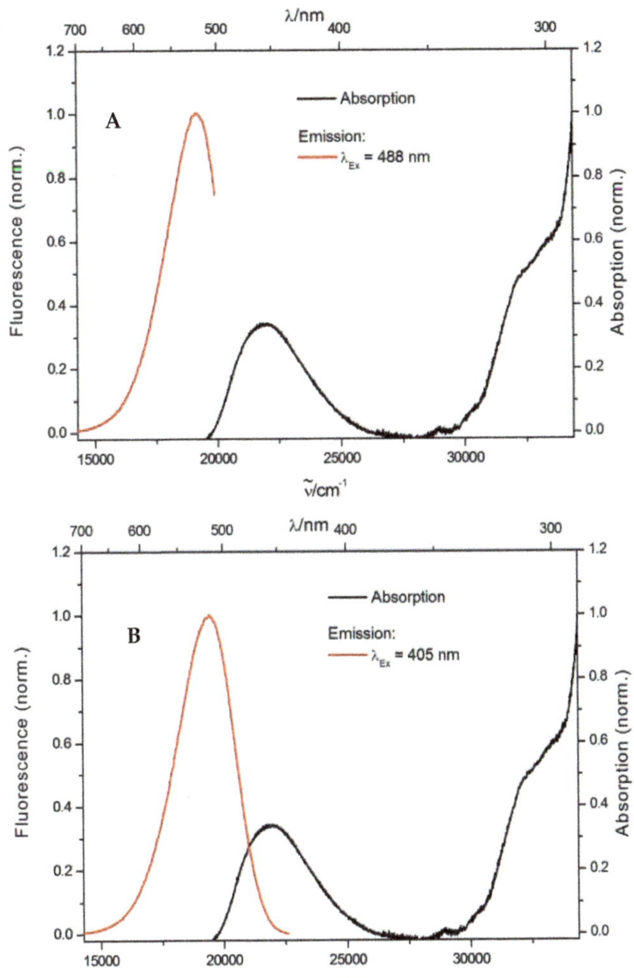

Figure 18. Absorption and fluorescence emission spectra of 3. Excitation at 488 nm (**A**) or 405 nm (**B**), respectively

Preliminary spin trapping experiments with **3** were carried out with a regular confocal microscope with excitation at 488 nm. Fig. 19 shows the control in the absence of inhibitors. As found before, complete quench occurs in about half an hour. Although the dye may accumulate again in mitochondria, we have not yet looked for co-localization with TMRE. Strikingly different are the effects of inhibitors as shown for antimycin A in Fig. 20. Corresponding experiments employing rotenone or both, rotenone and antimycin A as well as the Fenton reaction gave very similar results (data not shown). Quench is almost instantaneous,

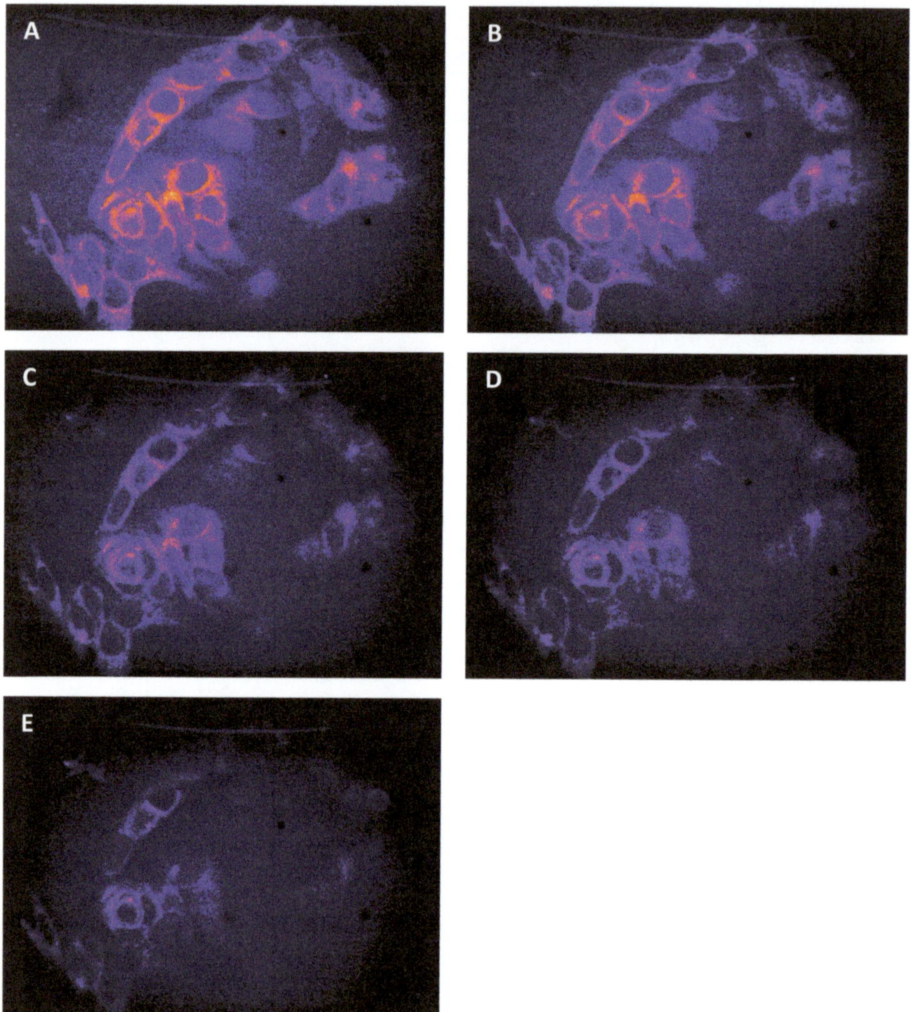

Figure 19. Fluorescence decay of <u>3</u> (1 μM) at selected times in MCF-7 cells: **A:** 0 min, **B:** 5 min, **C:** 10 min, **D:** 15 min, **E:** 20 min in the absence of inhibitors (control; laser intensity 80 % and exposure 800 ms)

but fluorescence recovered within four to five seconds and then decayed completely over the next 15 min. What could cause fluorescence recovery? The half-life of nitroxides as Tempone or Tempamine in viable systems is in the order of 30 min at most due to the reducing milieu in the cell (Berliner, 1991). Hence, the fluorescence being quenched by the radical could come back. There is however, also the possibility that differences are due to the experimental setup, *i.e.*, a continuous flow device for the medium at the ApoTome to which the inhibitor was added, whereas in these experiments a concentrated solution of antimycin A etc. was manually injected directly into the medium surrounding the adherent cells.

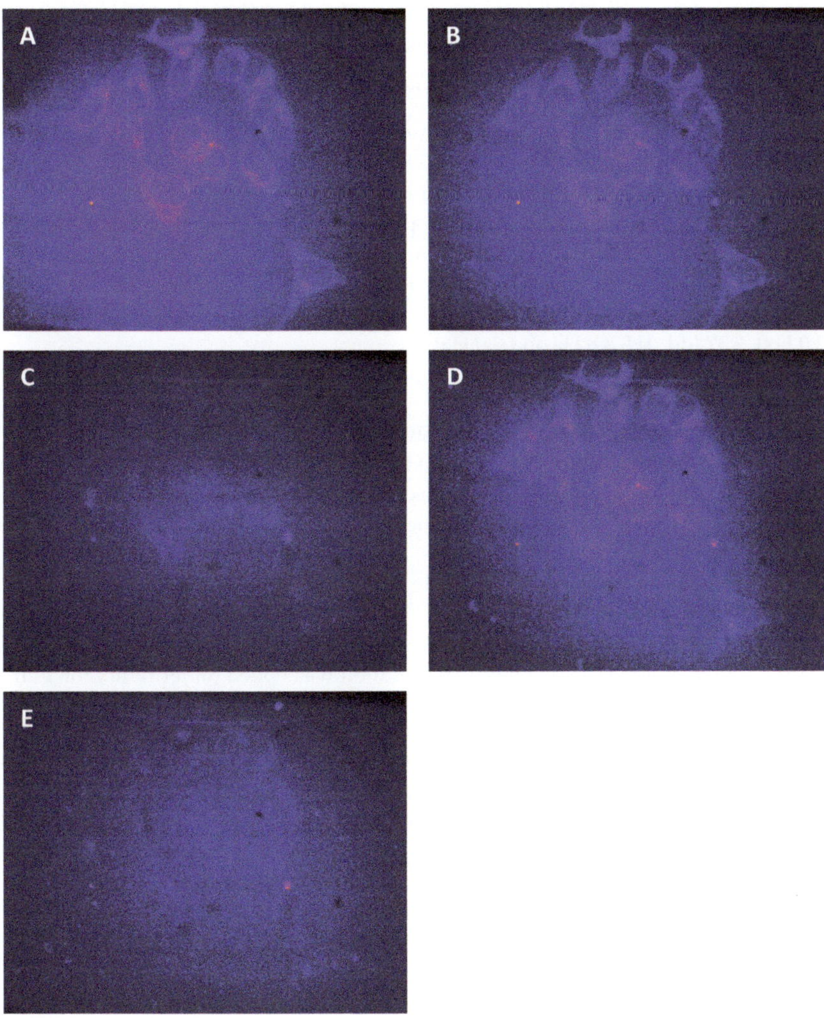

Figure 20. Fluorescence decay of 3 in MCF-7 cells at selected times after addition of 1 μl antimycin A (2 mg/ml) **A:** 0 sec, **B:** 1 sec, **C:** 2 sec, **D:** 4 sec, **E:** 13 min (laser intensity 80 % and exposure 800 ms)

3. Materials and methods

3.1. Synthesis of fluorescent nitrones

3.1.1. p-Nitrostilbene-tert-butyl-nitrone was synthesized as previously described (Hauck et al., 2009).

3.1.2. Coumaryl-styryl-tert-butyl-nitrone

2-(4-Trifluoromethyl-2-oxo-2H-chromen-7-yl)-*E*-vinyl-1-(N-(1,1-dimethyl))-phenyl-4-nitrone was synthesized according to the scheme in Fig. 4.

(Trifluoromethyl)-2-oxo-2H-chromene-7-diazoniumtetrafluoroborate

To a solution of 2.62 mmol of commercially available 7-amino-4-(trifluoromethyl)-coumarin (coumarin 151, Sigma-Aldrich, Taufkirchen, Germany) in 18 ml of hydrochloric acid (18%), a solution of sodium nitrite (2.72 mmol) was added dropwise and stirred for 1 h at 0 °C. Then, 41 mmol tetrafluoroboric acid was added dropwise over 1 h, with maintaining the reaction temperature lower than 0 °C. (Trifluoromethyl)-2-oxo-2H-chromene-7-diazonium-tetrafluoroborate (1.52 mmol, yield: 58 %) was separated and rinsed with cold water, methanol and ethyl acetate. ^{1}H-NMR: 6.45 (s, 1H), 6.51 (d, 1H, 2.4 Hz), 6.66 (dd, 1 H, 8.8 Hz and 2.0 Hz), 7.37 (dd, 1H, 8.8 Hz and 2.0 Hz).

4-Bromobenzaldehyde-dimethyl-acetal

4-Bromobenzaldehyde (94.9 mmol) was dissolved in 120 ml of dry methanol under an argon atmosphere, 96 µl titanium chloride was added and the solution stirred for 20 minutes. After addition of 240 µl trietylamine and 15 minutes of stirring, water was added and the product was extracted with diethyl ether. The organic phase was dried and the acetal isolated by evaporation of the solvent.

[2-(4-Dimethoxymethylphenyl)-vinyl]-trimethyl-silane

4-Bromobenzaldehyde-dimethyl-acetal (20.6 mmol) was dissolved in 20 ml of dry acetonitrile under an argon atmosphere. The solution was boiled under reflux for 24 h after addition of 51.6 mmol tetrabutylammonium acetate, 4 % (by mol) palladium(II) acetate and 41.3 mmol trimetyl-vinyl-silane. Then, 50 ml PBS buffer (pH = 7.4) was added and the product was extracted with diethyl ether. The organic phase was dried and the solvent removed by evaporation. The product was formed in 86 % yield (17.8 mmol). ^{1}H-NMR: 0.18 (s, 9H), 3.34 (s, 1H), 5.42 (s, 1H), 6.53 (d, 1 H, 19.07 Hz), 6.91 (d, 1H, 19.41 Hz), 7.45 (m, 2H).

(4-Trifluoromethyl-2-oxo-2H-chromen-7-yl)-E-2-vinyl]-benzaldehyde

2 mmol (Trifluoromethyl)-2-oxo-2H-chromene-7-diazoniumtetrafluoroborate and 2 mmol [2-(4-dimethoxymethylphenyl)-vinyl]-trimethyl-silane were dissolved in 40 ml of dry methanol and stirred for 5 h with 2 % (by mol) palladium(II) acetate (Hiyama reaction). Water was added and the product was extracted with diethyl ether. The solvent was removed under vacuum and the product was washed with ethyl acetate. ^{1}H-NMR: 7.28 (s, 2H), 7.33 (s, 1H), 7.72 (t, 2 H, 8.86 Hz), 7.94 (d, 2H, 8.51), 8.28 (d, 1H, 8.86), 8.45 (d, 1H, 8.51), 10.05 (s, 1H).

2-(4-Trifluoromethyl-2-oxo-2H-chromen-7-yl)-E-vinyl-1-(N-(1,1-dimethylethyl))-phenyl-4-nitrone (2, CSN)

1.25 mmol 4-[(4-Trifluoromethyl-2-oxo-2H-chromen-7-yl)-E-2-vinyl]-benzaldehyde and 1.85 mmol *N*-tert-butylhydroxylamine hydrochloride were dissolved in benzene and p-toluenesulfonic acid was added as catalyst. The solution was boiled in a Dean-Stark apparatus (reflux and water removal) for 12 h. Water was added and the product was extracted with ethyl acetate. ^{1}H-NMR: 1.66 (s, 1H), 6.78 (s, 1H), 7.22 (d,1H, 16.04 Hz), 7.28 (s, 1H), 7.30 (d, 1H, 16.04 Hz), 7.63 (d, 3H, 8.61 Hz), 7.73 (t, 2 H, 8.61 Hz), 8.20 (d, 1H, 8.61 Hz), 8.35 (d, 2H, 8.60 Hz).

3.1.3. *4-Pyrrolidine-1,8-naphthylimido-methylphenyl-tert-butyl-nitrone, 3*

4-Pyrrolidine-1,8-naphthylimido-methylphenyl-*tert*-butyl-nitrone was synthesized according to the scheme in Fig. 17.

6-(Pyrrolidin-1-yl)-benzo[de]isochromene-1,3-dione

To a solution of 37.4 mmol of commercially available 6-bromo-benzo[de]isochromene-1,3-dione (Sigma-Aldrich, Taufkirchen, Germany) in 50 ml of methyl glycol was added 56.7 mmol pyrrolidine and a catalytic amount of *p*-toluenesulfonic acid and was boiled under reflux for 6 h. After cooling to 0 °C the precipitation was filtered off and washed with cold ethanol. The crude product was recrystallized from n-hexane and dried. 6-(Pyrrolidin-1-yl)-benzo[de]isochromene-1,3-dione was formed in 90 % yield (33.7 mmol). $C_{16}H_{13}NO_3$ (267.3 g/mol) calc.: C, 71.90 %; H, 4.90 %; N, 5.24 %. found: C, 71.76 %; H, 4.88 %; N, 5.23 %. Mp.: 284 °C. ^{1}H-NMR (chloroform-d, 400 MHz): δ [ppm] = 2.07 – 2.19 (m, 4 H), 3.75 – 3.89 (m, 4 H), 6.77 (d, 1 H, 8.81 Hz), 7.52 (t, 1 H, 7.30 Hz), 8.31 (d, 1 H, 8.81 Hz), 8.50 (d, 1 H, 7.30 Hz), 8.64 (d, 1 H, 8.81 Hz). $^{13}C\{^{1}H\}$-NMR (chloroform-d, 100 MHz): δ [ppm] = 160.9, 153.2, 135.4, 134.0, 133.4, 133.0, 123.2, 122.3, 118.6, 108.8, 53.3, 26.1.

4-(1,3-Dioxacyclopent-2-yl)-benzonitrile

To a solution of 79.5 mmol of commercially available 4-cyanobenzaldehyde (Merck, Darmstadt, Germany) in 150 ml toluene, was added 0.32 mol of ethylene glycol and *p*-toluenesulfonic acid was added as catalyst. The solution was boiled in a Dean-Stark apparatus for 20 h. After complete removal of water, the solution was stirred for 20 h at room temperature. Then, 60 ml of a 5 % sodium bicarbonate aqueous solution was added. The organic layer was washed three times with 25 ml bidest. water each and dried over magnesium sulfate. The solvent was removed under vacuum and the remaining colorless liquid was crystallized from n-hexane and dried under reduced pressure. The hygroscopic product was formed in 95 % yield (75.9 mmol). $C_{10}H_9NO_2$ (175.2 g/mol) calc.: C, 68.56 %; H, 5.18 %; N, 8.00 %. found: C, 68.58 %; H, 5.12 %; N, 8.09 %. Mp.: -. ^{1}H-NMR (chloroform-d, 400 MHz): δ [ppm] = 4.02 – 4.11 (m, 4 H), 5.83 (s, 1 H), 7.58 (d, 2 H, 8.56 Hz), 7.66 (d, 2 H, 8.06 Hz). $^{13}C\{^{1}H\}$-NMR (chloroform-d, 100 MHz): δ [ppm] = 143.1, 132.2, 127.2, 118.6, 112.9, 102.4, 65.4.

4-(1,3-Dioxolan-2-yl)-phenylmethanamine

4-(1,3-Dioxolan-2-yl)-benzonitrile (42.2 mmol) was dissolved in 50 ml of dry diethyl ether under an argon atmosphere and was added dropwise at 0 °C to a well-stirred suspension of $LiAlH_4$ (84.5 mmol) in 100 ml of dry diethyl ether. The suspension was stirred at this temperature for 4 h, then the mixture was allowed to warm up to room temperature and was stirred for a further 18 h. Afterwards, the mixture was hydrolyzed by 35 ml of 95 % ethanol and then by 35 ml of 50 % ethanol. The ether layer was separated and the aqueous phase lyophilized. The resulting solid was extracted with 80 ml dichloromethane. The combined organic layers were dried over $MgSO_4$ and the yellowish-white solid product was obtained by evaporation of the solvent and crystallized in the fridge in 94 % yield (39.5 mmol). ^{1}H-NMR (chloroform-d, 400 MHz): δ [ppm] = 1.49 (s, 2 H), 3.87 (s, 2 H), 3.99 – 4.15 (m, 4 H), 5.80 (s, 1 H), 7.32 (d, 2 H, 8.19 Hz), 7.44 (d, 2 H, 8.19 Hz). $^{13}C\{^{1}H\}$-NMR (chloroform-d, 100 MHz): δ [ppm] = 144.4, 136.4, 127.1, 126.7, 103.6, 65.3, 46.3.

2-(4-(1,3-Dioxolan-2-yl)-benzyl)-6-(pyrrolidin-1-yl)-1H-benzo[de]isoquinoline-1,3(2H)-dione

7.11 mmol of 6-pyrrolidin-1-yl)-benzo[de]isochromene-1,3-dione and 10.7 mmol of 4-(1,3-dioxolan-2-yl)-phenylmethanamine were dissolved in 130 ml of ethanol. The mixture was boiled under reflux for 23 h and then cooled to 0 °C. The resulting orange solid was filtered off and washed four times with 20 ml of cold ethanol each. The crude product was recrystallized from toluene and dried over $MgSO_4$. The product was obtained as an orange solid in 88 % yield (6.26 mmol). $C_{26}H_{24}N_2O_4$ (428.5 g/mol) calc.: C, 72.88 %; H, 5.65 %; N, 6.54 %. found: C, 72.68 %; H, 5.70 %; N, 6.36 %. Mp.: 189 °C. ^{1}H-NMR (chloroform-d, 400 MHz): δ [ppm] = 2.05 – 2.12 (m, 4 H), 3.74 – 3.80 (m, 4 H), 3.95 – 4.09 (m, 4 H), 5.37 (s, 2 H), 5.78 (s, 1 H), 6.80 (d, 1 H, 8.61 Hz), 7.38 (d, 2 H, 7.83 Hz, 2 H), 7.51 (t, 1 H, 8.61 Hz), 7.56 (d, 2 H, 8.22 Hz), 8.41 (d, 1 H, 8.86 Hz), 8.56 (t, 2 H, 5.09 Hz). $^{13}C\{^{1}H\}$-NMR (chloroform-d, 100 MHz): δ [ppm] = 164.8, 164.0, 152.7, 138.9, 136.9, 133.6, 132.0, 131.2, 128.9, 126.4, 123.0, 122.6, 122.5, 110.5, 108.5, 103.6, 65.2, 53.2, 43.0, 26.1.

4-((1,3-Dioxo-6-(pyrrolidin-1-yl)-1H-benzo[de]isoquinolin-2(3H)-yl)-methyl)-benzaldehyde

To a solution of 2-(4-(1,3-dioxolan-2-yl)-benzyl)-6-(pyrrolidin-1-yl)-1H-benzo[de]isoquinoline-1,3(2H)-dione (8.40 mmol) in 100 ml THF, was added 25 ml of 3 % hydrochloric acid. The solution was boiled under reflux for 4 h and then cooled to 0 °C. Afterwards, the solution was brought to pH 8 with triethylamine upon which an orange precipitate was formed. The crude product was filtered off, washed three times with 20 ml of water each. The orange aldehyde was obtained in 98 % yield (8.22 mmol) and was stored under argon unless used immediately. $C_{24}H_{20}N_2O_3$ (384.4 g/mol) calc.: C, 74.98 %; H, 5.24 %; N, 7.29 %. found: C, 74.60 %; H, 5.27 %; N, 6.91 %. Mp.: 160 °C. ^{1}H-NMR (chloroform-d, 400 MHz): δ [ppm] = 2.05 – 2.14 (m, 4 H), 3.74 – 3.84 (m, 4 H), 5.43 (s, 2 H), 6.81 (d, 1 H, 8.61 Hz), 7.53 (t, 1 H, 7.43 Hz), 7.65 (d, 2 H, 8.22 Hz), 7.80 (d, 2 H, 8.22 Hz), 8.43 (d, 1 H, 8.61 Hz), 8.60 (t, 2 H, 7.43 Hz), 9.95 (s, 1 H). $^{13}C\{^{1}H\}$-NMR (chloroform-d, 100 MHz): δ [ppm] = 192.0, 163.9, 152.9, 144.9, 135.4, 133.8, 132.3, 131.4, 129.9, 129.1, 126.4, 123.1, 122.6, 122.2, 108.6, 53.2, 43.2, 26.1.

(Z)-N-(4-((1,3-dioxo-6-(pyrrolidin-1-yl)-1H-benzo[de]isoquinolin-2(3H)-yl-)methyl)-benzylidene)-2-methylpropan-2-amine oxide

A suspension of 4-((1,3-dioxo-6-(pyrrolidin-1-yl)-1H-benzo[de]isoquinolin-2(3H)-yl)-methyl)-benzaldehyde (3 mmol), N-(*tert*-butyl)hydroxylamine hydrochloride (6 mmol), triethylamine (6.6 mmol) and sodium sulfate (17.6 mmol) as desiccant in 50 ml CH_2Cl_2 was stirred for 9 days at room temperature under argon. Afterwards, the mixture was filtered off and the solid was washed three times with 20 ml CH_2Cl_2 each. The organic layer was dried and the crude product isolated by evaporation of the solvent. Then, the orange solid was recrystallized from ethanol and the nitrone was purified by column chromatography (200–400 mesh chromagel; solvent: methylene chloride/acetone/triethylamine, 12:1:0.1) in 38 % yield (1.14 mmol). ^1H-NMR (chloroform-d, 400 MHz): δ [ppm] = 1.57(s, 9 H), 2.07 – 2.10 (m, 4 H), 3.75 – 3.79 (m, 4 H), 5.37 (s, 2 H), 6.80 (d, 1 H, 8.80 Hz), 7.47 (s, 1 H), 7.51 (dd, 1 H, 8.56 Hz, 7.43 Hz), 7.56 (d, 2 H, 8.56 Hz, 2 H), 8.19 (d, 2 H, 8.56 Hz), 8.41 (d, 1 H, 8.80 Hz), 8.57 (m, 2 H). $^{13}C\{^1H\}$-NMR (chloroform-d, 100 MHz): δ [ppm] = 164.8, 164.0, 152.8, 140.0, 133.7, 132.1, 131.3, 130.0, 129.6, 128.8, 128.7, 123.0, 122.6, 122.5, 110.5, 108.5, 70.6, 53.1, 43.2, 28.3, 26.1.

3.2. UV/VIS and fluorescence spectroscopy

500 μM stock solutions of the spin traps in dimethyl sulfoxide containing 1 % $C_{12}E_9$ were employed. For fluorescence spectra the solutions were added to 25 mM phosphate buffer, pH = 7.2, with 1 % Triton-X 100 to a final concentration of 1 μM. To simulate photobleaching in the fluorescence microscope, the solutions were irradiated for three minutes with a blue LED light source (λ = 366 nm) and compared to samples without prior irradiation.

3.3. Cytotoxicity

Cytotoxicity of the spin traps was determined by the sulforhodamine assay. 170 μl of a cell suspension of 8000 cells/ml of Hela or MCF-7 cells were added to 20 ml of RPMI medium and 1 ml each was pipetted into a 24 wells plate and incubated for 48 h at 37 °C under 5 % CO_2. After this time, the cells were incubated for another 72 h with RPMI medium containing 0.5 % DMSO and the spin trap at six concentrations varying from 0 to 200 μM. The reaction was stopped by addition of 100 μl of 50 % trichloroacetic acid for 1 h at 4 °C and subsequently washed four times with cold water and then dried for 24 h at room temperature followed by addition of 250 μl of sulforhodamine B solution (0.4 % in 1 % acetic acid; Sigma, Taufkirchen). After 30 min wells were washed three times with cold water and another three times with 1 % acetic acid, dried for 24 h at room temperature and extinction measured at 570 nm after addition of 1 ml of 10 mM Tris base solution, pH 10.0.

In case of **3** concentrations were varied from 0 to 5 μM in RPMI medium containing 0.5 % DMSO and 0.5 % ethanol in addition.

3.4. Fluorescence microscopy measurements

These were carried out using a quasi-confocal microscope (Axiovert 440 equipped with an ApoTome, Carl Zeiss, Jena) as previously described (Hauck et al., 2009). For corresponding measurements with 3 a Nikon Eclipse E 600 confocal microscope equipped with a Hamamatsu ORCA-ER camera was employed. After 20 minutes of incubation with 1 μM spin trap and 0.5 % DMSO the cells were washed three times with RPMI medium and the coverslip was mounted on a chamber and put under the microscope. Imaging was achieved by laser excitation at 488 nm.

To determine the half-life of fluorescence, a representative cell was defined as region of interest (ROI) and the evolution of average intensity of the ROI was investigated in the presence and absence of various inhibitors of components of mitochondrial proteins.

4. Conclusions

At the moment, the *p*-nitrostilbene-*tert*-butyl-nitrone (1) seems to be best suited for investigations of ROS formation in mitochondria. Unfortunately, the cytotoxicity of the coumaryl derivative (2) limits it application potential. The third molecule, 4-pyrrolidine-1,8-naphthylimido-methylphenyl-*tert*-butyl-nitrone (3) still requires further detailed evaluation. If confirmed that it distributes fairly evenly throughout the cell it would nicely complement data from 1. With respect to the bioreduction leading to fluorescence recovery and time-dependent changes in fluorescence, although reproducible, we still have to make sure that this was not due to an experimental artifact employing different instrumentation.

Author details

Stefan Hauck, Yvonne Lorat, Fabian Leinisch, Christian Kopp, Jessica Abrossinow and Wolfgang E. Trommer
Department of Chemistry, TU Kaiserslautern, Germany

Acknowledgement

The authors thank Anneken Grün and André Karbach, Kaiserslautern, for their help in the syntheses as well as Dr. Lars Kästner, University of the Saarland, Homburg, for assistance with confocal laser microscopy and Yvonne Schmitt, Kaiserslautern, with fluorescence spectroscopy.

5. References

Berliner, L.J. (1991). Applications of *in vivo* EPR. In: *EPR Imaging and in vivo EPR*. Eaton, G.R., Eaton, S.S. & Ohno, Keiichi, eds. pp. 291-310. CRC Press, Boca Raton, FL, USA

Blough, N.V. & Simpson, D.J. (1988). Chemically mediated fluorescence yield switching in nitroxide-fluorophore adducts: optical sensors of radical/redox reactions. *J. Amer. Chem. Soc.* 110, 1915-1917

Bottle, S.E., Hanson, G.R. & Micallef, A.S. (2003). Application of the new EPR spin trap 1,1,3-trimethylisoindole N-oxide (TMINO) in trapping OH· and related biologically important radicals. *Org. Biomol. Chem.* 1, 2585-2589

Boveris, A. (1977). Mitochondrial production o superoxide radical and hydrogen peroxide. *Adv. Exp. Med. Biol.* 78, 67-82

Bystryak, I.M., Likhtenshtein, G.I., Kotel'nikov, A.I., Hankovsky, O. & Hideg, K. (1986). The influence of the molecular dynamics of the solvent on the photoreduction of nitroxy radicals. *Russian J. Phys. Chem.* 60, 1679-1683

Cadenas, E. & Davies, K.J.A. (2000). Mitochondrial free radical generation, oxidative stress and aging. *Free Radical Biol. Med.* 29, 222-230

Dlaskova, A., Hlavata, L. & Jezek, P. (2008). Oxidative stress caused by blocking of mitochondrial complex I $H^{(+)}$ pumping as link in aging/disease vicious cycle. *Int. J. Biochem. & Cell Biol.* 40, 1792-1805

Dröge, W. (2002). Free radicals in the physiological control of cell function. *Physiological Rev.* 82, 47-95

Farkas, D.L., Wei, M.D., Febbroriello, P., Carson, J.H. & Loew, L.M. (1989). Simultaneous imaging of cell and mitochondrial membrane potentials . *Biophys. J.* 56, 1053-1069

Han, D., Williams, E. & Cadenas, E. (2001). Mitochondrial respiratory chain-dependent generation of superoxide anion and its release into the intermembrane space. *Biochem. J.* 353, 411-416

Hauck, S., Lorat, Y., Leinisch, F. & Trommer, W.E. (2009). p-Nitrostilbene-*t*-butyl-nitrone: a novel fluorescent spin trap for the detection of ROS with subcellular resolution. *Appl. Magn. Reson.* 36, 133-147

Heyne, B., Beddie, C. & Scaiano, J.C. (2007). Synthesis and characterization of a new fluorescent probe for reactive oxygen species. *Org. Biomol. Chem.* 5, 1454-1458

Inoue, M., Sato, E.F., Nishikawa, M., Park, A.-M., Kira, Y, Imada, I. & Utsumi, K. (2003). Mitochondrial generation of reactive oxygen species and its role in aerobic life. *Current Med. Chem.* 10, 2495-2505

Janzen, E.G. (1971). Spin trapping. *Acc. Chem. Res.* 2, 279-288

Kalai, T., Hideg, E., Vass, I. & Hideg, K. (1998). Double (fluorescent and spin sensors for detection of reactive oxygen species in the thylakoid membrane). *Free Radical. Biol. Med.* 24, 649-652

Kim, S., de A Vilela, G. V. M., Bouajila, J., Dias, A. G., Cyrino, F. Z. G. A., Bouskela, E., Costa, P. R. R. & Nepveu, F. (2007). Alpha-phenyl-N-tert-butyl nitrone (PBN) derivatives: synthesis and protective action against microvascular damages induced by ischemia/reperfusion. *Bioorg. Med. Chem.* 15, 3572-3578

Kohen, R. & Nyska, A. (2002). Oxidation of biological systems: oxidative stress phenomena, antioxidants, redox reactions, and methods for their quantification. *Toxicol. Pathol.* 30, 620-650

Leavesley, H.B., Li, L., Mukhopadhyay, S., Borowitz, J.L. & Isom, G.E. (2010). Nitrite-mediated antagonism of cyanide inhibition of cytochrome c oxidase in dopamine neurons. *Toxicol. Sci.* 115, 569-576

Lozinsky, E., Shames, A.I. & Likhtenshtein, G. (2000). Dual fluorophore nitroxides: Models for investigation of intramolecular quenching and novel redox probes. *Recent Res. Devel. Photochem. Photobiol.* 2, 41-45

Muller, F.L., Liu, Y. & Van Remmen, H. (2004). Complex III releases superoxide to both sides of the inner mitochondrial membrane. *J. Biol. Chem.* 279, 49064-49073

Nieminen, A.L., Saylor, A.K., Herman, B. & Lemasters, J.J. (1994). ATP depletion rather than mitochondrial depolarization mediates hepatocyte killing after metabolic inhibition. *Am. J. Physiol. Cell Physiol.* 267, C67-C74

Oodyke, D.L. (1974). Monographs on fragrance raw materials. *Food Cosmet. Toxicol.* 12, 385-405

Ouari, O., Chalier, F., Bonaly, R., Pucci, B. & Tordo, P. (1998). Synthesis and spin-trapping behaviour of glycosylated nitrones. *J. Chem. Soc., Perkin Trans.* 2, 2299-2308

Pou, S., Bhan, A., Bhadti, V.S., Wu, S.Y., Hosmane, R.S. & Rosen, G.M. (1995). The use of fluorophore-containing spin traps as potential probes to localize free radicals in cells with fluorescence imaging methods. *FASEB J.* 9, 1085-1090

Severin, F.F., Severina, I.I., Antonenko, Y.N., Rokitskaya, T.I., Cherepanov, D.A., Mokhova, E.N., Vyssokikh, M.Y., Pustovidko, A.V., Markova, O.V., Yaguzhinsky, L.S., Korshunova, G.A., Sumbatyan, N.V., Skulachev, M.V. & Skulachev, V.P. (2010). Penetrating cation/fatty acid anion pair as a mitochondria-targeted protonophore. *Proc. Natl. Acad. Sci. U S A.* 107,163-168

Skehan, P., Storeng, R., Scudiero, D., Monks, A., McMahon, J., Vistica, D., Warren, J.T., Bokesch, H., Kenney, S. & Boyd, M.R. (1990). New colorimeric cytotoxicity assay for anticancer-drug screening. *J. Natl. Cancer Inst.* 82, 1107-1112

Turrens, J.F. (2003). Mitochondrial formation of reactive oxygen species. *J. Physiol.* 552, 335-344

Xu, T., Zanotti, F., Gaballo, A., Raho, G. & Papa, S. (2000). F_1 and F_0 connections in the bovine mitochondrial ATP synthase. *Eur. J. Biochem.* 267, 4445-4455

Walling, C. (1975). Fenton's reagent revisited. *Acc. Chem. Res.* 8, 125-131

Permissions

The contributors of this book come from diverse backgrounds, making this book a truly international effort. This book will bring forth new frontiers with its revolutionizing research information and detailed analysis of the nascent developments around the world.

We would like to thank Prof. Dr. Alexander I. Kokorin, for lending his expertise to make the book truly unique. He has played a crucial role in the development of this book. Without his invaluable contribution this book wouldn't have been possible. He has made vital efforts to compile up to date information on the varied aspects of this subject to make this book a valuable addition to the collection of many professionals and students.

This book was conceptualized with the vision of imparting up-to-date information and advanced data in this field. To ensure the same, a matchless editorial board was set up. Every individual on the board went through rigorous rounds of assessment to prove their worth. After which they invested a large part of their time researching and compiling the most relevant data for our readers. Conferences and sessions were held from time to time between the editorial board and the contributing authors to present the data in the most comprehensible form. The editorial team has worked tirelessly to provide valuable and valid information to help people across the globe.

Every chapter published in this book has been scrutinized by our experts. Their significance has been extensively debated. The topics covered herein carry significant findings which will fuel the growth of the discipline. They may even be implemented as practical applications or may be referred to as a beginning point for another development. Chapters in this book were first published by InTech; hereby published with permission under the Creative Commons Attribution License or equivalent.

The editorial board has been involved in producing this book since its inception. They have spent rigorous hours researching and exploring the diverse topics which have resulted in the successful publishing of this book. They have passed on their knowledge of decades through this book. To expedite this challenging task, the publisher supported the team at every step. A small team of assistant editors was also appointed to further simplify the editing procedure and attain best results for the readers.

Our editorial team has been hand-picked from every corner of the world. Their multi-ethnicity adds dynamic inputs to the discussions which result in innovative outcomes. These outcomes are then further discussed with the researchers and contributors who give their valuable feedback and opinion regarding the same. The feedback is then collaborated with the researches and they are edited in a comprehensive manner to aid the understanding of the subject.

Apart from the editorial board, the designing team has also invested a significant amount of their time in understanding the subject and creating the most relevant covers. They scrutinized every image to scout for the most suitable representation of the subject and create an appropriate cover for the book.

The publishing team has been involved in this book since its early stages. They were actively engaged in every process, be it collecting the data, connecting with the contributors or procuring relevant information. The team has been an ardent support to the editorial, designing and production team. Their endless efforts to recruit the best for this project, has resulted in the accomplishment of this book. They are a veteran in the field of academics and their pool of knowledge is as vast as their experience in printing. Their expertise and guidance has proved useful at every step. Their uncompromising quality standards have made this book an exceptional effort. Their encouragement from time to time has been an inspiration for everyone.

The publisher and the editorial board hope that this book will prove to be a valuable piece of knowledge for researchers, students, practitioners and scholars across the globe.

List of Contributors

Lawrence J. Berliner
Dept. of Chemistry and Biochemistry, University of Denver, USA

Günter Grampp and Kenneth Rasmussen
Institute of Physical and Theoretical Chemistry, Graz University of Technology, Graz, Austria

Alexander I. Kokorin
N.Semenov Institute of Chemical Physics RAS, Moscow, Russian Federation

Andrey Kh. Vorobiev and Natalia A. Chumakova
Department of Chemistry, M.V. Lomonosov Moscow State University, Leninskie Gory, Moscow, Russia

Boris Dzikovski and Jack Freed
National Biomedical Center for Advanced ESR Technology (ACERT), Department of Chemistry and Chemical Biology, Baker Laboratory, Cornell University, Ithaca, NY, USA

Elena Kovaleva
Ural Federal University, Russia

Leonid Molochnikov
Ural State Forest Engineering University, Russia

Rui Tamura, Yoshiaki Uchida and Katsuaki Suzuki
Graduate School of Human and Environmental Studies, Kyoto University, Japan

Toshihide Yamasaki, Fumiya Mito, Yuta Matsuoka, Mayumi Yamato and Ken-ichi Yamada
Faculty of Pharmaceutical Sciences, Kyushu University, Japan

Eugene M. Pliss and Alexander I. Rusakov
P.G. Demidov Yaroslavl State University, Russia

Ivan V. Tikhonov
Yaroslavl Branch of the Institute of Physics and Technology, Russian Academy of Sciences, Russia

Yaroslav Tkachev and Vladimir Timofeev
V.A. Engelhardt Institute of Molecular Biology/RAS, Moscow, Russia

Valery V. Khramtsov
Division of Pulmonary, Allergy, Critical Care & Sleep Medicine, Department of Internal Medicine, The Ohio State University, Columbus, OH, USA

Lev Weiner
Weizmann Institute of Science, Rehovot, Israel

Anatoly Filippov, Elena Chernikova, Vladimir Golubev, Ganna Gryn'ova, Ching Yeh Lin and Michelle L. Coote
Moscow State University named by M.V. Lomonosov, Russian Federation
ARC Centre of Excellence for Free-Radical Chemistry and Biotechnology, Research School of Chemistry, Australian National University, Australia

Vasily D. Sen', Alexei A. Terentiev and Nina P. Konovalova
Institute of Problems of Chemical Physics, Russian Academy of Sciences, Chernogolovka, Russian Federation

Stefan Hauck, Yvonne Lorat, Fabian Leinisch, Christian Kopp, Jessica Abrossinow and Wolfgang E. Trommer
Department of Chemistry, TU Kaiserslautern, Germany